AF403412

# An Introduction to Photonics and Laser Physics with Applications

# IOP Series in Advances in Optics, Photonics and Optoelectronics

## SERIES EDITOR

 **Professor Rajpal S Sirohi** Consultant Scientist

## About the Editor

Rajpal S Sirohi is currently working as a faculty member in the Department of Physics, Alabama A&M University, Huntsville, Alabama (USA). Prior to this, he was a consultant scientist at the Indian Institute of Science, Bangalore, and before that he held a chair and was professor in the Department of Physics, Tezpur University, Assam. During 2000–11, he was an academic administrator, being vice-chancellor to a couple of universities and the director of the Indian Institute of Technology, Delhi. He is the recipient of many international and national awards and the author of more than 400 papers. Dr Sirohi is involved with research concerning optical metrology, optical instrumentation, holography, and speckle phenomena.

## About the series

Optics, photonics, and optoelectronics are enabling technologies in many branches of science, engineering, medicine, and agriculture. These technologies have reshaped our outlook and our ways of interacting with each other and have brought people closer. They help us to understand many phenomena better and provide a deeper insight into the functioning of nature. Furthermore, these technologies themselves are evolving at a rapid rate. Their applications encompass very large spatial scales from nanometers to astronomical and a very large temporal range from picoseconds to billions of years. This series on advances on optics, photonics, and optoelectronics aims to cover topics that are of interest to both academia and industry. Some of the topics to be covered by the books in this series include biophotonics and medical imaging, devices, electromagnetics, fiber optics, information storage, instrumentation, light sources, CCD and CMOS imagers, metamaterials, optical metrology, optical networks, photovoltaics, free-form optics and its evaluation, singular optics, cryptography, and sensors.

## About IOP ebooks

The authors are encouraged to take advantage of the features made possible by electronic publication to enhance the reader experience through the use of color, animation, and video and by incorporating supplementary files in their work.

## Do you have an idea for a book you'd like to explore?

For further information and details of submitting book proposals, see iopscience.org/books or contact Ashley Gasque on Ashley.gasque@iop.org.

A full list of titles published in this series can be found here: https://iopscience.iop.org/bookListInfo/series-on-advances-in-optics-photonics-and-optoelectronics.

# An Introduction to Photonics and Laser Physics with Applications

**Prem B Bisht**
*IIT Madras*

**IOP** Publishing, Bristol, UK

© IOP Publishing Ltd 2022

All rights reserved. No part of this publication may be reproduced, stored in a retrieval system or transmitted in any form or by any means, electronic, mechanical, photocopying, recording or otherwise, without the prior permission of the publisher, or as expressly permitted by law or under terms agreed with the appropriate rights organization. Multiple copying is permitted in accordance with the terms of licences issued by the Copyright Licensing Agency, the Copyright Clearance Centre and other reproduction rights organizations.

Permission to make use of IOP Publishing content other than as set out above may be sought at permissions@ioppublishing.org.

Prem B Bisht has asserted their right to be identified as the author of this work in accordance with sections 77 and 78 of the Copyright, Designs and Patents Act 1988.

ISBN    978-0-7503-5226-0 (ebook)
ISBN    978-0-7503-5225-3 (print)
ISBN    978-0-7503-5227-7 (myPrint)
ISBN    978-0-7503-5235-2 (mobi)

DOI    10.1088/978-0-7503-5226-0

Version: 20220801

IOP ebooks

British Library Cataloguing-in-Publication Data: A catalogue record for this book is available from the British Library.

Published by IOP Publishing, wholly owned by The Institute of Physics, London

IOP Publishing, Temple Circus, Temple Way, Bristol, BS1 6HG, UK

US Office: IOP Publishing, Inc., 190 North Independence Mall West, Suite 601, Philadelphia, PA 19106, USA

...वसुधैव कुटुम्बकम् *...the whole earth is a family—( Mahopnishad)*

*Dedicated to people all over the world.*

# Contents

## Part II    Pulsed lasers and nonlinear optical applications

## 13    Laser spiking and $Q$-switching    13-1



## 14    Introduction to nonlinear optical phenomena    14-1

## 15   Second-order susceptibility, phase matching, and applications   15-1



## 16   Third-order nonlinear optical processes   16-1

## 17  Mode locking — 17-1



## 18  Characterization of ultrafast laser pulses — 18-1



## 19  Optical phase conjugation — 19-1

# Preface

Lasers are ubiquitous, from deep space communication to lab-on-the-chip to supermarket product scanning. Although they form an integral part of optics and photonics, the modern laser industry has contributed to interdisciplinary areas in scientific research and technology. Therefore, it is an appropriate time to make a self-sufficient, comprehensive text describing laser-related concepts available to beginners. This book aims to do just that. A list of books is provided at the end in an appendix for the reader who wishes to undertake comprehensive study in a particular area. A bibliography at the end of each chapter also connects the reader to original literature. The book is written with the intention of preparing a textbook for undergraduate and graduate students as well as reference material for any student working with lasers in the fields of optics, biosciences, and engineering. Sufficient mathematics, instead of details, have been given for the reader to be able to understand the topic.

The structure of the book is shown schematically in the following chart.

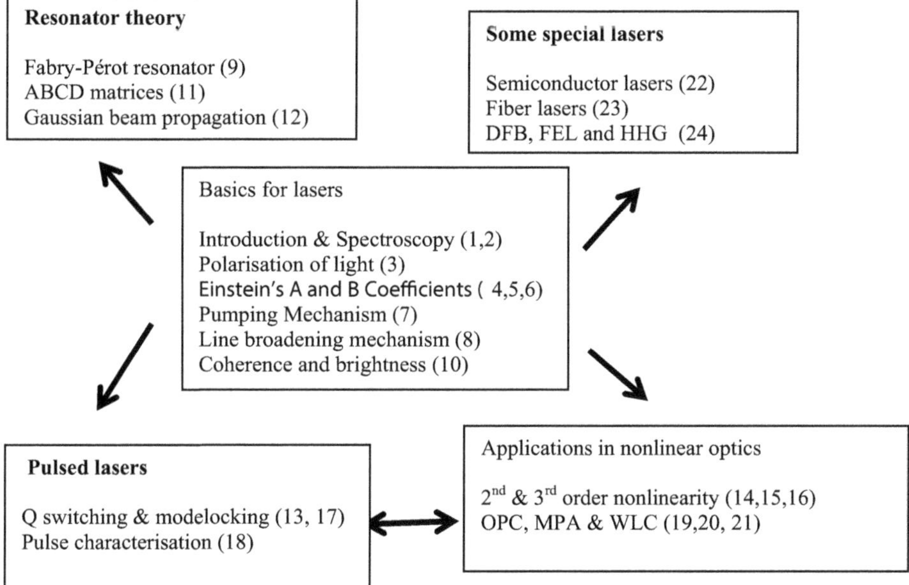



**Chart:** Schematic of the theme of the book. The chapter numbers are given in parentheses. Arrows indicate the interconnections between the chapters and the subsections.

As can be seen from the chart above, this book has been divided into two sections: **section I:** *Basics of photonics and lasers* and **section II:** *Pulsed lasers and nonlinear optics*. This book has five subsections, which can be grouped as follows:

- Basics of lasers (chapters 1–8 and 10);
- Resonator theory (chapters 9, 11, and 12);

- Pulsed lasers (chapters 13, 17, and 18);
- Nonlinear optical phenomena (chapters 14–16 and 18–21);
- Some special lasers (chapters 22–24).

The book starts with the basics of photons and their connection to light according to EM theory. The subject matter proceeds with the old and traditional subject of spectroscopy and subsequently develops toward the state of the art as the chapters progress. In part I we begin with a look at the basics of photons and the connection between optics and electrodynamics (chapter 1). This part helps the reader to review the study material required to understand the working principle of lasers and some of their applications. Chapter 2 in this part provides the required knowledge of spectroscopy and quantum mechanics. Chapters 4 to 12 are closely related and need to be studied in the given sequence. Chapters 13–21 of part II provide the background and applications of nonlinear optics, which was mainly developed following the invention of ultrafast pulsed lasers. A chapter on semiconductor lasers and one on fiber lasers are separately provided in this section. For the curious reader, a clear distinction between semiconductor lasers (diode lasers, vertical-cavity surface-emitting lasers (VCSELs)) and quantum cascade lasers (QCLs) is elucidated. Some other lasers with unique designs (mirrorless lasers) and lasers in the extreme ultraviolet and soft x-ray regions with pulse durations on the scale of attoseconds are described in chapter 24.

In each chapter, a few questions and end-of-chapter problems are included to aid in the understanding of the material. Most of the problems are not based on numerical answers. For a better understanding of the applications, extra informa-tion related to the topic is denoted by a ♣ symbol. Similarly, related information elsewhere in the book has been marked with a ♠ symbol.

Each chapter starts with a summary along with a diagram for those curious about the subject matter. Learning objectives are clearly outlined at the start of each chapter so that students can study on their own. I am conscious of the availability of additional reading material for the different topics spread over a large number of books/ journals. Therefore, full details of the references are given at the end of each chapter. Where appropriate, footnotes are given to help explain the concept further.

For the instructor, the material covered in this book can be used to make a course of about 45–50 classes. The book can also be split into two courses, as follows: (A) fundamentals of lasers and (B) ultrafast lasers and applications in photonics. For course (A), part I can be used with some selected material from part II. On the other hand, for course (B), chapters 13 to 24 must be preceded with introductory material from the previous section, depending on the level of the class. Additionally, some rescheduling may be necessary to interchange the sequence of chapter 17 (mode locking) with chapter 13 (Q-switching).

This book specifically targets the problems faced by research students and professionals in various fields, including biology. For example, the identification of materials based on *second-harmonic generation* or *two-photon absorption* is outlined towards the end of chapter 20. A word for researchers in biology: chapter 20 (multiphoton absorption) may be extremely useful, in addition to chapters 16–19.

Useful data from research papers have been provided as reference materials in a few tables for ready use.

I thank IIT Madras for permitting me to spend four months of sabbatical, during which I stayed in my home town of Champawat to initiate this humongous task. The climate of the Himalayan region and the company of the villagers was just excellent for this task. Besides the experiments with lasers at IIT Madras, my earlier stays at other institutes have created great interest in writing this book. Starting with DSB College (Kumuan University, Nainital), these include the Tata Institute of Fundamental Research, Mumbai, research visits to the Raja Ramanna Center of Advanced Technology (RRCAT) Indore, the Institute for Molecular Science (IMS, Okazaki), Kyoto Institute of Technology (Kyoto), Ludwig Maximilian University (Munich), the Optoelectronic Research Centre, University of Southampton (Southampton), and Dublin City University (Dublin). The academic training received from Professor H B Tripathi and Professor D D Pant as a graduate student followed by the interactions with Professors Keitaro Yoshihara, Satoshi Hirayama, Hrvoje Petek, Eberhard Riedle, and John Costello are gratefully acknowledged. At IIT Madras, the foundation laid down by Professor B M Sivaram and Professor J P Raina in developing ultrafast lasers also motivated the writing of this book. My colleagues Doctors G C Joshi, H C Joshi, K K Pandey, Sanjay Pant, Debi Pant, H Kandori, S Kumazaki, A Yartsev, S Kasiviswanathan, the staff at the instrument center of IMS and the scientists, Doctors S M Oak and K S Bindra of RRCAT, Indore also encouraged the idea of writing this book on several occasions. I thank all my former and present PhD students for the discussions on the topics of the book. A special mention is given to the students of various departments of IIT Madras over last two decades who took the courses I taught on lasers. I acknowledge the help in reading the first version of a few chapters given by Professor S N Thakur, Dr Srini Krishnamurthy (SRI international), Dr R Aravind and Dr Prabha Mandayam (IITM), and Dr Rama Chari (RRCAT). Thanks are due to Professor Anurag Sharma for readily agreeing to write the foreword of this book. Finally, I thank my wife Mamta and my sons Anupam and Sameer for their patience and for helping me in every way they could, during the course of this task.

8th March 2022

Prem B Bisht
IIT Madras, Chennai

# Author biography

## Prem B Bisht

**Prem B Bisht** is Professor of Physics at one of the Indian Institutes of Technology, IIT Madras at Chennai. His research interests include ultrafast laser spectroscopy and its application to nanomaterials with a special interest in noncollinear optical parametric amplifiers and fluorescence microscopy. Following his PhD in physics from Kumaun University, Nainital in 1991, Prem has been with IIT Madras since 1997 as a teacher and researcher. Prem has been a JSPS fellow and a member of the Indian Laser Association, the Optical Society of India, the Indian Association of Physics Teachers, the Indian Science Congress, and SPIE and is currently a senior member of Optica. Bisht's scientific career includes collaborations with several national and international laboratories on experimental physics using lasers. He has been associated with the organization of several national and international conferences in ultrafast optics at IIT Madras. He has four patents with two others filed. He has published 250 scientific papers, one edited book, and several book chapters. He has delivered about 100 talks at various institutes and conferences and has supervised 15 PhD students and over 35 UG/PG (Res) students to completion.

# Foreword

Photonics and optics have contributed greatly to the development of societies around the world and have provided solutions to many societal problems in recent times. Since the invention of the laser in 1960, progress is this field has been very rapid, leading to developments in nonlinear optics, laser spectroscopy, fiber optics and optical communications, ultrafast optics, optical sensing, and many other areas. These contributions to the welfare and development of mankind were recognized by the UN in their declaration of 2015 as the Year of Light and Light-based Technologies and their subsequent nomination of May 16 as the International Day of Light. Research and education in this field have also seen unprecedented growth over the last few decades. In particular, lasers are central to most developments that have taken place. Many developments and applications involving lasers are still confined to monographs and advanced texts and are therefore largely inaccessible to undergraduate and postgraduate students. There are many textbooks on laser and laser theory, but they do not include recent developments and applications. This new book fills this gap admirably. It describes the fundamentals of lasers, including the necessary basis of optics and photonics, and includes recent applications such as laser spectroscopy, nonlinear optics, ultrafast pulses, super-continuum generation, and fiber lasers.

This book has many notable features. First and foremost, the author, Professor Prem B Bisht, is a well-known experimentalist in the field with over 35 years of experience in laser spectroscopy and nonlinear optics. His hands-on expertise is amply reflected in the book by examples and descriptions of experiments drawn from his own laboratory. The mathematical details have been kept at the essential level and there is a greater focus on discussing physical understanding and practical details. Second, the material in the book has grown out of his experience of teaching courses at IIT Madras for several years, and the text is an outcome of the organic growth of the author's teaching and research experience. Thus, while the subject dealt with is very advanced, it is brought to an appropriate level for senior undergraduate and postgraduate students. Third, as mentioned above, there is an emphasis on applications and practical details. Some of the notable applications, in addition to the ones already mentioned, are the sensing technique of cavity ring-down spectroscopy and applications of optical parametric processes. Some of the applications are in interdisciplinary areas, such as imaging with nonlinear optics for biotechnology as well as fiber lasers and semiconductor lasers in the area of engineering. The applications of various spectroscopic methods will be useful to chemistry and materials science students. Finally, I personally like the format of this book, which has short sections and a large number of figures. These help the reader to directly focus on the desired topic while using this book as a reference book.

This book is a welcome and valuable addition to the educational literature on the subject and would benefit students in science and engineering who wish to learn about lasers and photonics and their applications.

March 15, 2022

Anurag Sharma
IIT Delhi

# Part I

Basics of photonics and lasers

**IOP** Publishing

An Introduction to Photonics and Laser Physics
with Applications

**Prem B Bisht**

# Chapter 1

# The photon and photonics

The full form of the acronym **'laser'** *is light amplification by stimulated emission of radiation.* The laser was invented in 1960 as a so-called 'tool looking for an application'. Within six decades, lasers have found applications in all walks of life, including industries based on them. Micromachining, waveguide fabrication, welding, cutting, nondestructive testing, and pulsed-laser deposition of thin films are some of their applications in materials science. In the field of electrical engineering, fiber optics has revolutionized the field of optical communication. In aerospace engineering applications involving jet and scramjet engines, studies of the mixing of fuel sprays require laser-induced fluorescence techniques. The defense, medical, and cosmetic fields, as well as scientific research, are interdisciplinary areas that make extensive use of lasers. Like the flow of electrons that completes *electronic circuits* in *electronics*, photons are related to *photonic circuits* in the field of *photonics.* Maxwell's equations suggest that light is an electromagnetic (EM) wave. Therefore, this chapter connects optics with EM theory. The figure shows an electric field ($E$, in the $y$ direction) and a magnetic field ($B$, in the $x$ direction), which are mutually perpendicular to each other. The EM wave is propagating in the $y$ direction; the details are given in this chapter.

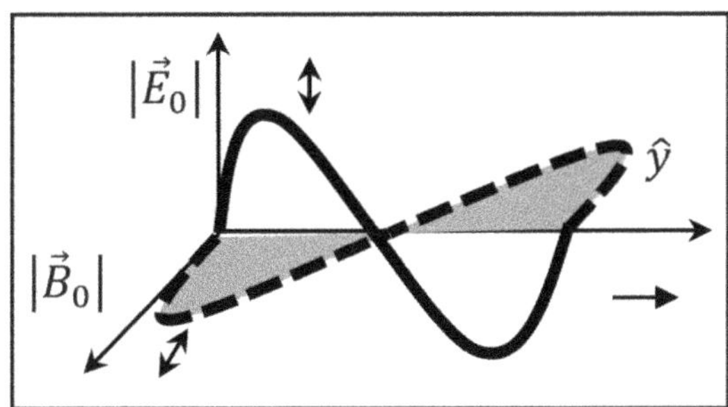

> **Learning objectives**
> **After reading this chapter, the learner will be able to:**
> Identify the various branches of photonics;
> Relate Maxwell's equation to optics;
> Define radiation pressure and the angular momentum of light;
> Explain radiation using accelerating charge;
> Describe the refractive index and dispersion of a material;
> Differentiate between electronic and photonic circuits.

## 1.1 The photon

Max Planck suggested the idea of energy packets known as 'quanta' in 1900. While Einstein called these packets 'energy particles' in 1905, it took until 1923 for this idea to be reinforced by the discovery of the Compton effect. These particles were named *photons* by Gilbert Lewis in 1926; this term denoted 'carriers of radiant energy'. Just as the electron is associated with electricity, light of wavelength $\lambda$ or frequency $\nu$ consists of photons of energy $h\nu$. Here, $h$ is Planck's constant.

The photon:

   (i) has no charge,

  (ii) is considered to have zero rest mass but

 (iii) carries momentum ($p = h/\lambda$),

 (iv) has a constant velocity $(c) \sim 10^8$ m s$^{-1}$ in vacuum,

  (v) carries a spin of 1 and thus follows Bose–Einstein statistics,

 (vi) is immune to EM noise (as it has no charge), and

(vii) does not undergo photon–photon interactions in linear optics.

The property of photons being *immune to EM noise* makes a light beam a special tool as compared to electrical circuits that are prone to picking up stray EM signals. Similarly, photon–photon interactions cannot take place under normal light levels. In the same way that *electronics* deals with the flow of electrons in electrical circuits, photon-related studies address *photonic circuits* that fall within the domain of *photonics*. This means that the two light rays can cross without interacting with each other. The photon–electron interaction, which falls within the domain of *light–matter interactions*, is an important area of interdisciplinary research. Chapter 2 introduces this aspect, along with spectroscopy.

## 1.2 Branches of photonics

This term photonics is used to describe the control of photons and the photon nature of light. This is one of the modern area of optics that deals with the technologies used to generate and harness light. 'Laser applications' is an interdisciplinary field that uses the principles of conventional optics, electromagnetism, spectroscopy, and quantum optics. Therefore, several subcategories (shown in figure 1.1) are encompassed by the term *photonics*. In addition, electrodynamics, which is a self-sufficient theory, is used in various areas of optical technologies.

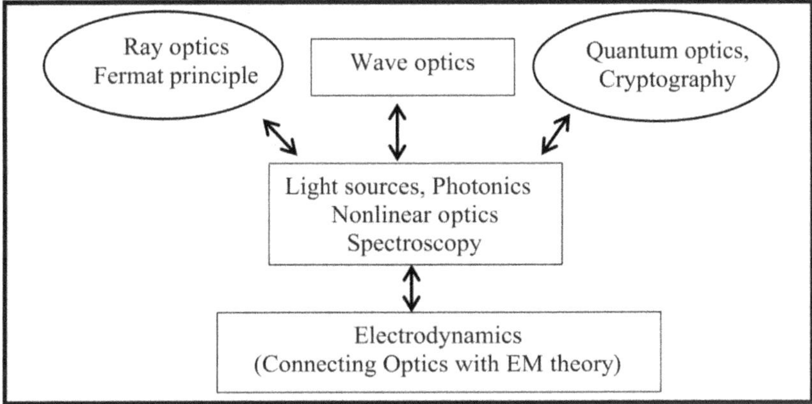


**Figure 1.1.** Interdisciplinary nature of photonics, illustrating some application areas.

### 1.2.1 Conventional optics

According to Fermat's principle, light rays travel along the path that can be traversed in the least time. When light propagates through large objects in which the wavelength of light is smaller than the dimensions of the object, its behavior can be explained by drawing light rays. We experience this in the form of reflection or refraction of light rays from a surface. Under such circumstances, the light rays follow the rules of geometrical optics. This method of understanding light falls under *ray optics*.

### 1.2.2 Electromagnetism and wave optics

Light is an electromagnetic wave and, as such, its electric and magnetic fields are represented in their vector forms (see section 1.3). However, in *wave optics*, the scalar representation of EM fields is sufficient. In Young's double-slit interference experiment, for example, Huygens' principle of secondary wavelets is used to explain wave interference. This can be achieved without taking the components of the EM field into account.

### 1.2.3 Quantum optics

Certain phenomena cannot be explained using EM theory. Optical phenomena that can only be explained by treating light as a stream of photons, such as coherent states and photon entanglement (♠ see chapter 15) fall within this category. The Mach–Zehnder interferometer (♠ see chapter 9) is used to experimentally test the basic theoretical proposals in quantum optics, such as the entanglement of photons and Bell's inequalities. Quantum cryptography also is the subject of *quantum optics*.

### 1.2.4 Light–matter interaction or quantum electronics

This topic addresses the interactions of light with matter. The phenomena of absorption and spontaneous and stimulated emission (i.e. the spectroscopy of atoms and molecules) are studied here. Nonlinear optics, bulk spectroscopy, and the spectroscopy of low-dimensional materials in the form of monolayers, quantum wires, or quantum dots fall into this category.

### 1.2.5 Optoelectronics

This is an area in which both electrical current and light are required for the operation of a device. The presence of electronics in the device controls the optical character of the device. Devices that fall into this category are electronic in nature but evolve light. Examples include the light-emitting diodes, solar cells, and display devices in modern equipment, including smartphones. Edison's bulb may fall into this category as well. Specialized centers are dedicated to this topic worldwide. This research area has immense applications in the photonics industry.

### 1.2.6 Electro-optics

An optical switch that operates the automated door of an elevator or an office falls into this category. In this field, the electronics in an item of equipment is used to control the device in combination with an optical effect. Electro-optic shutters fall into this category. In addition, the change in the optical response (absorption/transmission) of a material due to AC or DC electric/magnetic fields falls within this area. Examples of devices based on electro-optics include those based on the Kerr effect or Faraday rotation (♠ see chapter 3).

### 1.2.7 Light-wave technology

The whole of modern-day communication is based on data exchange with a large frequency bandwidth. This includes the communication used by television, the internet, and the telephone. Devices and systems that are used in optical communication and optical signal processing fall into this category. Fiber-optic communication is the key example in this field.

## 1.3 Maxwell's equations and their connection to optics

Optics is connected to EM theory through Maxwell's equations (MEs). The basic laws of reflection and refraction can be derived from the electric ($\vec{E}$) and magnetic ($\vec{B}$) field vectors via solution of the plane-wave equation. The set of Maxwell's four equations for EM fields in vacuum from classical electrodynamics are written as follows:

$$\text{(i). } \vec{\nabla}\cdot\vec{E} = \frac{\rho}{\varepsilon_0} \qquad\qquad \text{(ii). } \vec{\nabla}\cdot\vec{B} = 0$$
$$\text{(Gauss's law)} \qquad\qquad\qquad \text{(No name)}$$

$$\text{(iii). } \vec{\nabla}\times\vec{E} = -\frac{\partial\vec{B}}{\partial t} \qquad \text{(iv). } \vec{\nabla}\times\vec{B} = \mu_0\vec{J} + \mu_0\epsilon_0\frac{\partial\vec{E}}{\partial t}$$
$$\text{(Faraday's Law)} \qquad\qquad \text{(Ampère's law fixed by Maxwell)}$$

$$(1.1)$$

Here, $\epsilon_0$ and $\mu_0$, are known as the permittivity (in farad m$^{-1}$) and permeability (in Henry m$^{-1}$) of free space, respectively; $\rho$ is the charge density in Cm$^{-3}$, and $\vec{J}$ is the current density in A m$^{-2}$ in the region. It should be noted that these equations are of the *first order* in time and space. The equations do not have symmetry in either the $\vec{E}$ or $\vec{B}$ fields. By working on these equations a little, we can obtain symmetric equations for either of the fields, as follows. For instance, on taking the curl of the Faraday's law (equation (iii)),

$$\vec{\nabla} \times \vec{\nabla} \times \vec{E} = \vec{\nabla} \times (-)\frac{\partial \vec{B}}{\partial t} \equiv -\frac{\partial}{\partial t}(\vec{\nabla} \times \vec{B}).$$

Using the vector identity $\vec{\nabla} \times \vec{\nabla} \times \vec{E} = \vec{\nabla}(\vec{\nabla}.\,\vec{E}) - \vec{\nabla}^2 E$, the above equation can be rewritten as

$$\vec{\nabla}(\vec{\nabla}.\,\vec{E}) - \vec{\nabla}^2 E = -\frac{\partial}{\partial t}(\vec{\nabla} \times \vec{B}).$$

For charge-free ($\rho = 0$) and current-free ($\vec{J} = 0$) regions, we can use equations (i) and (ii) to write the following wave equation:

$$\frac{1}{c^2}\frac{\partial^2 \vec{E}}{\partial t^2} - \nabla^2 \vec{E} = 0. \tag{1.2}$$

Here, $c$ is the speed of light $\left(c = \frac{1}{\sqrt{\mu_0 \epsilon_0}}\right)$ in m s$^{-1}$. We have obtained equation (1.2) for one variable ($\vec{E}$) at the cost of using a *second-order* differential equation. Similarly, one can write the wave equation for the $\vec{B}$ field as well.

We can assume a general solution of equation (1.2) for $\vec{E}(\vec{r},t)$ in units of V m$^{-1}$ as

$$\vec{E}(\vec{r},\,t) = |\,E_0\,|\,\hat{n}\,\cos(\vec{k}.\,\vec{r} - \omega t + \phi). \tag{1.3}$$

Here, $E_0$ is the amplitude of the wave. For a simple case, we take a wave propagating in the $z$ direction (i.e. $\vec{k}.\,\vec{r} = kz$), as shown in the diagram below (figure 1.2). This is known as the plane-wave solution. The unit vector $\hat{n}$ indicates the direction of

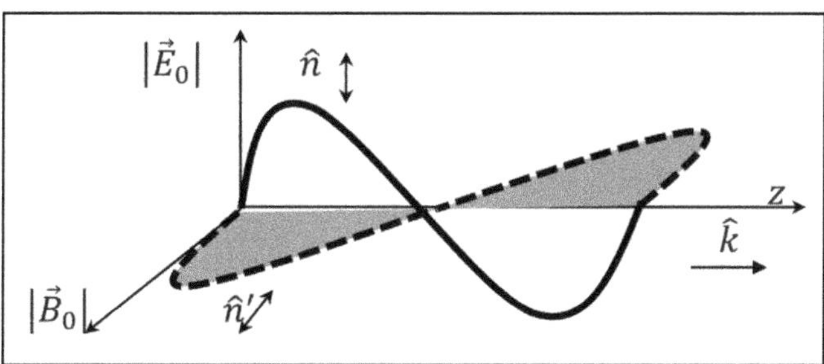

**Figure 1.2.** The electric field ($\vec{E}_0$), magnetic field ($\vec{B}_0$), and the direction of propagation of an EM wave ($\hat{k}$) make a triad. The polarizations of the field vectors are indicated by $\hat{n}$ and $\hat{n}'$, respectively.

oscillation of $\vec{E}$ and is known as the *polarization* of the field vector; $k$ is the propagation constant or the wavevector, i.e. the number of waves per unit length; $\omega$ is the frequency; and $\phi$ is the phase factor.

In Euler's form, equation (1.3) is the real part of equation (1.4) as follows:

$$\vec{E}(\vec{r},\, t) = E_0 \hat{n} e^{i(\vec{k}.\vec{r}-\omega t+\phi)}. \tag{1.4}$$

By using this solution for equation (1.2), we obtain

$$\frac{\partial}{\partial t}(e^{-i\omega t}) = -i\omega e^{-i\omega t}; \text{ and} \nabla(e^{i\vec{k}.\vec{r}}) = i\vec{k}(e^{i\vec{k}.\vec{r}}).$$

From these relations, we can effectively replace $\nabla$ with $i\vec{k}$ and $\frac{\partial}{\partial t}$ with $-i\omega$ for plane-wave equations. Now, using ME (iii), we can write the relation between the electric and magnetic field vectors with $\hat{k}$ as follows:

$$\frac{E(\hat{k} \times \hat{n})}{c} = B\,\hat{n}'.$$

This explains the *transverse nature of the EM wave* indicated in the diagram. The oscillation of the field vectors is perpendicular to the propagation direction of the wave. We recall that this is in contrast to sound waves, in which the rarefaction and densification of the medium's particles takes place in the propagation direction of the wave—for this reason, sound waves are generally called *longitudinal waves*.

The unit vector $(\hat{k})$ indicates the direction of propagation of the wave. From ME (ii), we can see that that the $\vec{B}$ field of an EM wave that has $\hat{n}'$ as its direction of oscillation is perpendicular to $\hat{k}$ (i.e., $\hat{n}' \cdot \hat{k} = 0$). Similarly, $\hat{n}$ and $\hat{k}$ are mutually perpendicular to each other, as can be seen in the figure.

The corresponding Maxwell's equations in matter are written by introducing the displacement vector ($\vec{D}$ in Cm$^{-2}$), the available charge and current densities, and the $\vec{H}$ field (in A m $-$ 1). The $\vec{D}$ field is defined as $\vec{D} = \epsilon\vec{E}$ and $\vec{H}$ is related to $\vec{B}$ according to $\vec{H} = \frac{\vec{B}}{\mu}$. Here, $\epsilon$ and $\mu$ are the permittivity and permeability of the *medium*. In metals, the current density $(\vec{J})$ is related to the conductivity $(\sigma)$ by $\vec{J} = \sigma\vec{E}$. In metals, the square root of the product of the quantities $\omega$, $\mu$, and $\sigma$ is defined as the inverse of the *skin depth* $(s)$ according to $s = \sqrt{\dfrac{2}{\omega\mu\sigma}}$ (♠ see question 7). Measured in units of nm, $s$ is the distance over which the amplitude of the EM waves decays to $1/e$ of its original value while propagating in the material with given parameters.

♣ The relation $\vec{J} = \sigma\vec{E}$ is known as 'Ohm's law'. The elementary form of this law is studied in high school. It states that the current $(I)$ across a resistor is proportional to the potential difference $(V)$ between the two ends of the resistor $(R)$, according to $V = IR$. As an exercise, one can obtain this relation by expressing $E$ in terms of V across a metal bar of length $L$ with resistance $R$ and conductivity $\sigma$. It is necessary to replace $|\vec{J}|$ with the current $(I)$ per unit area $(A)$ according to this definition.

**Exercise 1.1.** A linearly polarized plane EM wave is propagating in the $z$ direction, and its plane of polarization is the $x$ direction. The electric field of the wave has an amplitude given by $|E_0|$. The frequency of the wave is $\omega$, and its wave number is $k$. What are the electric and magnetic fields of the wave?

   **Solution:** The electric field vector of the wave $(\vec{E})$ is
$\vec{E} = |E_0| \, \hat{e}_x \cos(kz - \omega t)(Vm^{-1})$. The magnetic field vector is written as
$\vec{B} = \frac{(\vec{k} \times \vec{E})}{\omega} = \frac{1}{\omega} k |E_0| (\hat{e}_z \times \hat{e}_x) \cos(kz - \omega t)$.

   Therefore $\vec{B} = \frac{k}{\omega} |E_0|(\hat{e}_y) \cos(kz - \omega t)$ tesla (T). Note that we have ignored the phase part.

**Exercise 1.2. (a).** Show that the dimension of skin depth $(s)$ is that of length.

   **(b).** What is the typical value of $s$ for copper for the frequencies in the visible region $(\omega \sim 10^{15} Hz)$?

   **Solution:**

   **(a).** This is easy to verify if we take the dimension of charge to be $[Q]$. To get the dimension of skin depth $(s)$ as $[L]$, use the dimensions $[T^{-1}]$, $[M L Q^{-2}]$, and $[M^{-1}L^{-3}T Q^2]$ for $\omega$, $\mu$, and $\sigma$, respectively.

   **(b).** For copper, the value of $\sigma = 10^7 (\Omega cm^{-1})$. Using the value of $\mu = 10^{-6} N/A^2$ and an angular frequency for the visible region of $\omega = 10^{15} Hz$, we get the value of $s \sim$ 10 nm. This indicates that visible radiation can access the surface of the copper for distances up to the order of 10 nm!

**Exercise 1.3.** While ME (iii) is responsible for the generation of electricity, which equation is used in the mechanism of cooking by induction stove?

   **Solution:** We know that Faraday's law suggests that changing magnetic fields give rise to current; ME (iv) suggests that a changing electric field gives rise to a magnetic field. In an induction stove, when an AC current passes through the coil of a conducting wire, it induces a magnetic field within the skin depth of the ferromagnetic base of the cooking pan. The eddy currents induced by the magnetic field in the thick base of the cooking pan result in Joule heating due to its resistivity. This heat is responsible for cooking the food contents of the pan by heat conduction.

## 1.4 A few topics related to lasers and optics

### 1.4.1 Phase velocity and group velocity

The electric field of the wave (i.e. $\vec{E}(\vec{r}, t)$) is given in three dimensions by equation (1.3). For the one-dimensional case (i.e. along $z$), it can be rewritten as $\vec{E}(z, t) = |E_0| \hat{n}\cos(\vec{k} \cdot \vec{z} - \omega t + \phi)$. At a certain numerical value of the phase, the movement of the wave can be tracked so that the derivative $\frac{d\phi}{dt} = 0$. The plot of $\omega$ vs $k$ can be used to define two important concepts regarding the velocity of the wave: (i) the phase velocity $(\vec{v})$ of the waveform is defined as the ratio of $\omega$ and k according to $|\vec{v}| = \frac{\omega}{k}$; (ii) the instantaneous derivative $\left(\frac{d\omega}{dk}\right)$, on the other hand, comes into the picture when the waveform consists of slightly varying frequencies and wave vectors.

In this case, the waveform propagates as a superposition of two or more modulated waves known as a *wave packet*. The speed of the modulated signal is defined as the *group velocity*, $\left(v_g = \frac{d\omega}{dk}\right)$. For a nondispersive medium, $v_g$ remains equal to the magnitude of $\vec{v}$.

### 1.4.2 Power flow and Poynting vector

We know, for instance, that heat is also transported by sunlight to the Earth. An important aspect of electromagnetic (EM) wave transport is the propagation of energy. The energy densities in electric ($u_E$) and magnetic ($u_B$) fields are given by $u_E = \frac{\epsilon_0}{2}E^2$, and $u_B = \frac{1}{2\mu_0}B^2$, respectively. The energy density of the EM field is given by

$$u_{em} = \frac{1}{2}\left(\frac{1}{\mu_0}B^2 + \epsilon_0 E^2\right).$$

Maxwell's equations help in the derivation of an energy conservation equation. The energy flow per unit area per unit time is given by the Poynting vector ($\vec{S}$),

$$\vec{S} = \frac{1}{\mu_0}(\vec{E} \times \vec{B}). \tag{1.5}$$

### 1.4.3 Radiation pressure and angular momentum

The tail of a comet in orbit around the Sun always points away from the Sun. This happens due to the radiation pressure exerted by the momentum of the light. From a dimensional analysis, we can see that the magnitude of the linear momentum ($\vec{p} = \hbar\vec{k}$) imparted by a single photon of linearly polarized light is given by the ratio of the energy $\Delta U (=h\nu)$ absorbed to the speed of light, as follows:

$$\vec{p} = (h\nu/c)\hat{k}. \tag{1.6}$$

Alternatively, the momentum of a photon can be written in terms of its wavelength $\lambda$ as $p = h/\lambda$. Using equation (1.5), with $|\vec{E}| = c|\vec{B}|$, the density of the linear momentum in the EM field is given by

$$<\boldsymbol{P}_{em}> = \frac{\vec{S}}{c^2} = \epsilon_0(\vec{E} \times \vec{B}). \tag{1.7}$$

When a stream of $N$ photons per unit area per second falls perpendicularly on a perfectly black surface, it is assumed that all the photons are absorbed, thereby completely transferring their momenta to the surface. The irradiance ($I'$) in units of W m$^{-2}$ and the energy ($\Delta U$) absorbed by the area ($A$) in time ($\Delta t$) are related by $\Delta U = I' A \Delta t$. As the energy is completely absorbed, the gain in momentum based on equation (1.7) is

$$\Delta p = \Delta U/c \equiv I'A\Delta t/c.$$

The force ($F$) is defined as the rate of change of momentum, i.e., $F = \Delta p / \Delta t \equiv I'A/c$. Therefore, the radiation pressure ($P_r$) which is equivalent to the force per unit area is given by the following:

$$P_r = I'/c. \tag{1.8}$$

♣ The radiation pressure also has applications in the micromanipulation of particles using laser beams.

For a 100% reflecting surface, the photon undergoes a change of momentum equal to $2p$. Therefore, the pressure exerted on the surface is equal to $2I'/c$.

If a beam of circularly polarized light (♠ see chapter 3) is incident on an absorbing medium, the surface experiences a torque as predicted by classical EM theory. The magnitude of the torque $|\zeta|$ per unit area is given by

$$\zeta = \frac{I'}{\omega} \equiv \frac{Mh}{2\pi}.$$

Here, $M$ is the total number of photons. This indicates that the angular momentum ($\vec{l}_{em}$) of the photon is $h/2\pi$. This value of intrinsic momentum of photon is known as the *spin angular momentum* (SAM) and its value is taken as 1 unit. While for right circularly polarized light, the spin of the photon is parallel to the direction of propagation, it is antiparallel for left circularly polarized light.

In addition, the angular momentum carried by the *phase part of the wavefront* (also known as the *phase front*) of the light is called the orbital angular momentum (OAM). This is included in the expression for the cross-product of the radius vector $\vec{r}(r, 0, z)$ of the beam and $\vec{p}_{em}$ as follows:

$$\vec{l}_{em} = \vec{r} \times \vec{p}_{em} \equiv \epsilon_0[\vec{r} \times (\vec{E} \times \vec{B})]. \tag{1.9}$$

Equation (1.9) includes both the SAM and OAM contributions of light. Although the details are beyond the scope of this book, it is sufficient to mention that a light beam with an azimuthal phase dependence of $e^{il\phi'}$ in its cross section has an OAM value that is higher by several factors than the SAM. Here, $\phi'$ is the azimuthal coordinate and $l$ is an integer known as the *azimuthal mode index* or, the *order* of OAM.

♣ The transfer of SAM was first experimentally observed using a torque experienced by a suspended quarter-wave plate by Beth in 1936.

♣ For linearly polarized light, $<\vec{l}_{em}> = 0$. The same is true for unpolarized light, as this is considered to be a mixture of right circularly and left circularly polarized lights. The only difference is that for linearly polarized light, the mixture is coherent (♠ see chapter 3).

♣ It is also of contemporary interest that light beams with different orders of OAM can be produced; these are known as *vortex beams* and find applications in super-resolution light microscopy and photonics.

♣ The magnitude of the OAM of the vortex beam is said to be related to the *topological charge* of the beam by $\pm l|l|\hbar$.

### 1.4.4 Radiation emitted by the accelerated charge

For a single point charge $q$, the dipole moment can be given as $\vec{p}_{dip}(t) = q\vec{d}(t)$, where $\vec{d}$ is the position of the charge with respect to the origin. An oscillating dipole produces EM radiation in the perpendicular direction. The power (P) radiated by an accelerating charge with acceleration $a$ is given by the generalized Larmor formula as follows:

$$P \propto \frac{\mu_0 q^2 \gamma^6 a^2}{6\pi c}, \tag{1.10}$$

with $\gamma = 1/\sqrt{1 - v^2/c^2}$, where $v$ is the velocity of the relativistic particles. It can be seen that the power radiated by the charged particle is proportional to the square of the acceleration. The factor $\gamma^6$ indicates that the radiated power increases drastically for particles with speeds near to the speed of light.

**Exercise 1.4.** Write down the Poynting vector for the waves mentioned in exercise 1.1 by including the phase part, $\phi_1$.

**Solution:** The electric field in $V/m$ can be written as $\vec{E} = |E_0|\,\hat{e}_x \cos(kz - \omega t + \phi_1)$. By obtaining the corresponding $\vec{B} = \frac{(\vec{k} \times \vec{E})}{\omega}$ (T), the Poynting vector is given by

$$\vec{S} = \frac{1}{\mu_0}(\vec{E} \times \vec{B}) = \frac{k|E_0|^2}{\mu_0 \omega}(\hat{e}_z)\cos^2(kz - \omega t + \phi_1)(\text{Wm}^{-2}).$$

### 1.4.5 Refractive index

The refractive index ($n$) of a medium is defined as the ratio of the speed of the light in vacuum ($c$) to that in the medium ($v$). It should be noted that a change in the speed of the light wave results in a change in its wavelength. However, the oscillation frequency of a wave remains unchanged in any medium. We can also write the expression for the refractive index as follows:

$$n^2 = \frac{\epsilon\mu}{\epsilon_0\mu_0} \text{ or } \epsilon_r\mu_r. \tag{1.11}$$

For a vacuum, the values of the relative quantities $\mu_r$ and $\epsilon_r$ are taken to be unity. For most optical media (viz. transparent glass) other than ferromagnetic materials, the value of $\mu$ is taken to be the same as that of vacuum. Therefore, the relative permittivity $\epsilon_r$ is given by $n^2$. The quantity $\epsilon_r$ often denotes the dielectric constant of the material, which is related to the electrical susceptibility as follows:

$$\epsilon_r \equiv n^2 = (1 + \chi_e).$$

In this formula, $n$ is known as the high-frequency dielectric constant and is a function of the frequency. From here, an estimate of $\chi_e$ can be obtained, provided that the dielectric constant of a material is known.

### 1.4.6 Dispersion curve

In a non-conducting (or dielectric) isotropic medium, the electrons are bound to atoms. If a dielectric is placed in in external electric field ($\vec{E}$), the electronic charge ($q$) is displaced from its equilibrium position. From classical electrodynamics, we know that the induced polarization $\vec{P}_{\text{ind}}$ for a number of oscillators per unit volume, $N$, is given by

$$\vec{P}_{\text{ind}} = N\vec{p}_{\text{ind}}.$$

We can write the expression for a damped oscillation forced by an external time-varying field (in the $y$ direction) as

$$m\ddot{\vec{y}} + m\gamma\dot{\vec{y}} + K\vec{y} = q\vec{E}$$

or

$$\ddot{\vec{y}} + \gamma\dot{\vec{y}} + \omega_0^2\vec{y} = \frac{q\vec{E}}{m}, \tag{1.12}$$

where $K$ is the force constant, $m$ is the mass of the electron, $\omega_0$ is the effective resonance frequency ($=\sqrt{K/m}$) of the bound electrons, and $\gamma$ is the damping constant. Similar to the mass–spring system in mechanics, we assume a solution of equation (1.12) such as $\vec{y} = \vec{y}_0\, e^{i\omega t}$ to obtain the complex amplitude of oscillation as follows:

$$\vec{y}_0 = \frac{q\vec{E}}{m(\omega_0^2 - \omega^2 - i\gamma\omega)}.$$

As a result, an induced electric dipole moment $\vec{p}_{\text{ind}}$ is created in the y direction, as given by

$$\vec{p}_{\text{ind}} = \frac{q^2\vec{E}}{m(\omega_0^2 - \omega^2 - i\gamma\omega)}.$$

As a result of the incident $\vec{E}$, the $\vec{P}_{\text{ind}}$ in the dielectric is also given by $\vec{P}_{\text{ind}} = \epsilon_0\chi_e\vec{E}$, and we can obtain the value of $\chi_e$ as follows:

$$\chi_e = \frac{Nq^2}{\epsilon_0 m(\omega_0^2 - \omega^2 - i\gamma\omega)}.$$

The frequency dependence of the refractive index $n(\omega)$ leads to dispersion, which is represented as a complex quantity due to $n^2 = (1 + \chi_e)$ as follows:

$$n^2 = 1 + \frac{Nq^2}{\epsilon_0 m(\omega_0^2 - \omega^2 - i\gamma\omega)}. \tag{1.13}$$

For a large number of frequencies, this equation can be rewritten as

$$n^2 = 1 + \frac{Nq^2}{\epsilon_0 m} \Sigma_j \left( \frac{f_j}{\omega_0^2 - \omega_j^2 - i\gamma\omega} \right). \tag{1.14}$$

Here, $f_j$ describes the relative potential strengths of the oscillation frequencies. For small values of $\gamma$ (which are neglected), equation (1.14) can be written as follows:

$$n^2 = 1 + \frac{Nq^2}{\epsilon_0 m} \sum_j \left( \frac{f_j}{\omega_0^2 - \omega_j^2} \right). \tag{1.15}$$

This equation, when written in terms of the wavelength ($\lambda$), is known as Sellmeier's formula, an empirical relation proposed by Sellmeier in 1872 as follows:

$$n^2 = 1 + A_0 \left( \frac{\lambda^2}{\lambda^2 - \lambda_0^2} \right), \tag{1.16}$$

where $A_0 = \frac{Nq^2\lambda_0}{4\pi^2\epsilon_0 mc^2}$ is known as Sellmeier's coefficient. Equations (1.15) and (1.16) are extremely useful formulae used to estimate the refractive index data for various wavelength regions for applications in linear and nonlinear optics. The original dispersion relation, which did not take account of anomalous dispersion, was proposed by Cauchy as early as 1836 (♠ see question 11).

### 1.4.7 Normal dispersion and anomalous dispersion

For simple systems such as gaseous media, $n(\omega) \cong 1$, we can write ($n^2 - 1$) in equation (1.13) as $(n + 1)(n - 1) \approx 2n$. Near the resonant frequency, we can take $|\omega + \omega_0| \approx 2\omega_0$ and $|\omega - \omega_0| \ll \omega_0$. To separate $n(\omega)$ into refractive or real ($n'$) and absorptive or imaginary ($n''$) parts, we take $n = n' + in''$. So, from equation (1.13),

$$(n' + in'')^2 = 1 + \frac{Nq^2}{\epsilon_0 m(\omega_0^2 - \omega^2 - i\gamma\omega)}.$$

By rationalizing the second term and separating it into real and imaginary parts, we get

$$n'^2 - n''^2 = 1 + \frac{Nq^2}{\epsilon_0 m} \left( \frac{(\omega_0^2 - \omega^2)}{(\omega_0^2 - \omega^2)^2 + \gamma^2\omega^2} \right) \tag{1.17}$$

$$2n'n'' = \frac{Nq^2}{\epsilon_0 m} \left( \frac{\gamma\omega}{(\omega_0^2 - \omega^2)^2 + \gamma^2\omega^2} \right). \tag{1.18}$$

Here, $\gamma$ is related to the decay rate of the population. We can use these two relations to check the relation at the resonant frequency (i.e. $\omega = \omega_0$) as follows:

$$n'^2 - n''^2 = 1 \tag{1.19}$$

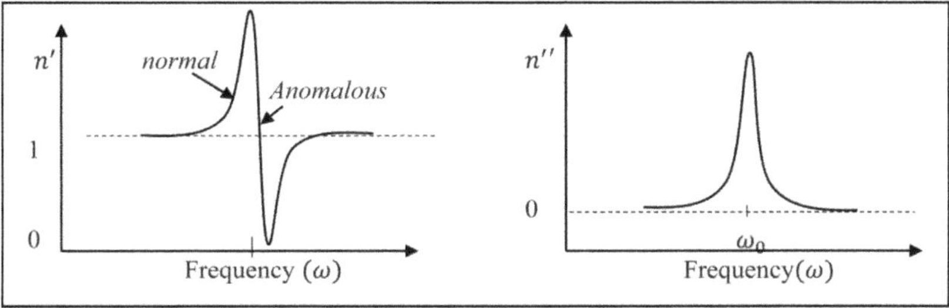



**Figure 1.3.** The real ($n'$) and imaginary ($n''$) parts of the refractive index as a function of the frequency ($\omega$). The *normal* and *anomalous* dispersion regimes are also indicated.

and

$$2n'n'' = \frac{Nq^2}{\epsilon_0 m\gamma\omega_0}. \tag{1.20}$$

Using equation (1.20), equation (1.19) can be rewritten as a quadratic equation, as follows:

$$n'^4 - n'^2 - \left(\frac{Nq^2}{2m\epsilon_0\gamma\omega_0}\right)^2 = 0 \text{ to give the roots of } n'^2 \text{ as } n'^2 = \frac{1}{2} \pm \frac{1}{2}\sqrt{1 + \left(\frac{Nq^2}{m\epsilon_0\gamma\omega_0}\right)^2}.$$

Only a positive root can represent real values. At very high frequencies, $n'$ will approach unity. The variables $n'$ and $n''$ are known as optical constants and are plotted in figure 1.3 based on equations (1.17) and (1.18). The plot of $n'$ vs $\omega$ is known as the dispersion curve. The value of $n'$ generally increases with frequency (known as normal dispersion), except for a narrow range where it falls sharply followed by a dip. This small region represents anomalous dispersion. The normal and anomalous dispersion regimes are indicated in the figure.

A plot of $n''$ against frequency shows a maximum at $\omega_0$, indicating the resonant frequency dependence of the absorption coefficient. The anomalous or, negative dispersion regions find applications in the dispersion compensation of ultrafast laser pulses and optical communication, respectively (♠ see chapters 22 and 23 for more details).

## 1.5 Comparison of an electronic circuit and a photonic circuit

A basic electronics circuit used to charge and discharge a capacitor is shown in figure 1.4. Here, a capacitor ($C$) is charged by the battery through the resistor ($R$) when the switch ($S_1$) is closed (and switch $S_2$ is open). It is discharged through the resistor when $S_1$ is open (and $S_2$ is closed).

A simple photonic circuit can be produced with the help of two polarizers (P1 and P2) placed in the path of a polarized light beam. Depending on the orientation of P2 (i.e. whether it is parallel or perpendicular to the optical axis of P1), a photodetector will record changes in the signal. This photonic circuit can represent the output (i.e. the light sensed by the detector) in terms of binary numbers using one when P1 is parallel to P2 and zero when P1 is perpendicular to P2. (♠ see chapter 3 for detailed description of the phenomenon of polarization).

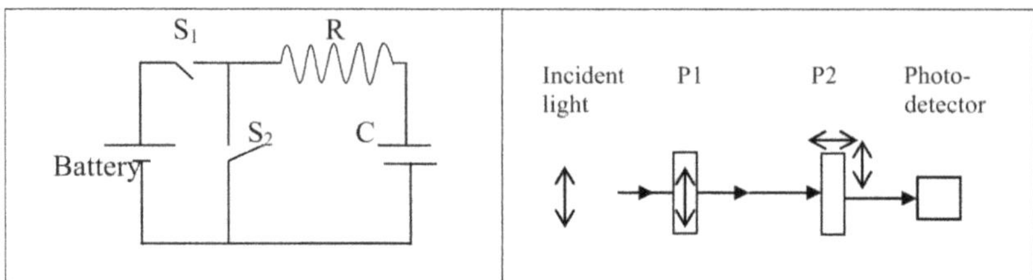

**Figure 1.4.** Typical electronic circuit used to *charge* and *discharge* a capacitor (*C*) through a resistor (*R*) and a battery (left panel). Representation of a photonic circuit (right panel) realized using polarized light (indicated by double arrows) with two polarizers (P1 and P2) and a photodetector.

**Table 1.1.** Notable Nobel prizes for laser-related work following the invention of the ruby laser in 1960 by Maiman.

| Year | Laser-related Nobel prizes |
|---|---|
| 1964 | Townes, Basov, and Prokhorov for the **maser–laser** principle. |
| 1971 | Gabor for the basic ideas of the **holographic** method. |
| 1981 | Nicolaas Bloembergen and Arthur Schawlow for **laser spectroscopy**. |
| 1997 | Claude Cohen-Tannoudji, Americans Steven Chu and William Phillips for the development of **methods to cool and trap atoms using laser light**. |
| 1999 | Ahmed Zewail for **ultrafast laser techniques** |
| 2000 | Zhores Alferov and Herbert Kroemer for **developing semiconductor heterostructures for continuous wave semiconductor diode lasers**. |
| 2001 | Eric Cornell, Wolfgang Ketterle, and Carl E. Wieman for **Bose–Einstein condensation**. |
| 2005 | Theodor Hansch and John Hall for the **development of laser-based precision spectroscopy**, known as the optical frequency comb technique. |
| 2009 | Charles K. Kao, Willard S. Boyle, and G. E. Smith for **fiber-optic communication and the CCD sensor** |
| 2012 | Serge Haroche and David J. Wineland for **experimental methods that enable the measurement and manipulation of individual quantum systems**. |
| 2014 | Isamu Akasaki, Hiroshi Amano, and Shuji Nakamura for the **invention of efficient blue light-emitting diodes** (Physics) |
| 2014 | Eric Betzig, Stefan W. Hell, and William E. Moerner for **stimulated-emission-depletion (STED)** microscopy using laser beams (Chemistry) |
| 2017 | Rainer Weiss, Kip S Thorne, and Barry C Barish for **the LIGO detector and the observation of gravitational waves** |
| 2018 | Arthur Ashkin, Gerald Mourou, and Donna Strickland for **optical trapping of particles and chirped pulse amplification of laser pulses** |

## 1.6 Nobel prizes related to lasers

When the laser was invented by Maiman in 1960, it was '*an invention searching for an application*'. Since then, lasers have contributed to the fundamental discoveries and applications of modern technologies that could not have been achieved otherwise. Table 1.1 gives some of the notable Nobel prizes for laser-related work in last six decades.

## Questions and problems

1. What is the difference between electro-optics and optoelectronics? Give one example of each.

2.    (i) With reference to the properties of the photon, why is light immune to EM noise?

     (ii) How do we eliminate external EM noise in experiments that use electronic devices?

3.    (i) Which of MEs is responsible for the generation of electricity, and which one describes the absence of magnetic monopoles?

     (ii) Estimate the speed of light using MEs in vacuum [Use $\epsilon_0 = 8.85 \times 10^{-12}\ C^2N^{-1}m^{-2}$ and $\mu_0 = 4\pi \times 10^{-7}NA^{-2}$].

4. Write down the *plane-wave solution* of a wave equation derived from MEs for charge-free, current-free regions. With reference to plane waves, what is the difference between the terms *wavefront* and *phase front*?

5. Using the solution of the wave equation in Euler's form, use ME (iii) to show that light is an EM wave and that $\vec{E}$ and $\vec{B}$ are perpendicular to each other.

6. If $\rho_f$ and $J_f$ are corresponding 'free' charge and current densities, respectively, $\overline{D}$ and $\overline{H}$ are the displacement vectors and $\overline{H}$ the field (often referred to as the magnetic field), write down the ME for a medium for which the permittivity and permeability are $\epsilon$ and $\mu$, respectively.

7. Assume that the displacement current term is too small to be neglected in ME (iv). For a charge-free and current-free region, let the wave equation be given by equation (1.2) and the speed of light in a medium be $=\sqrt{1/(\epsilon\mu)}$. Using the solution of the wave equation $\vec{E}(\vec{r}, t) = E_0\hat{n}e^{i(\vec{k}\cdot\vec{r}-\omega t)}$, obtain the value of the skin depth for a metal of conductivity $\sigma$.

8. Differentiate between the phase velocity and the group velocity for a wave. Show that in the absence of dispersion, both velocities are equal.

9. Describe how an accelerated charge radiates.

10. Half of the Nobel prize was awarded for work on the optical trapping of microparticles in 2018. What is the principle of optical trapping? ♣You may like to read the original paper by *Arthur Ashkin (PNAS*, 94 (10), (1997), pp 4853–4860). You will know that the particle size matters with respect to the wavelength of the light used for optical levitation. The scattering and gradient forces balance to trap the particle on the focussed laser beam.

11. The general form of Sellmeier's empirical relation equation (1.16) is given by $n^2 = 1 + \sum_j \left( \dfrac{A_j\lambda^2}{\lambda^2 - \lambda_j^2} \right)$. For $\lambda \gg \lambda_j$ and $A_j = A$, prove that it approaches Cauchy's formula $n = C_1 + \dfrac{C_2}{\lambda^2} + ..$ for $C_1 = \sqrt{(1 + A)}$ and $C_2 = \dfrac{A\lambda_0^2}{2\sqrt{(1 + A)}}$.

12. How do you define the normal and anomalous dispersion regimes?

13. In the dispersion curve, at what frequency is the absorption coefficient maximized?

14. For a photonic circuit for a logic gate that uses polarisers, write down the truth table, which is similar to that for electronic gates.

## Bibliography


[1] Beth R A 1936 Mechanical detection and measurement of the angular momentum of light *Phys. Rev.* **50** 115–25

[2] Fowles G R (ed) 1975 *Introduction to Modern Optics* 2nd edn (New York: Dover)

[3] Griffiths D J (ed) 1989 *Electrodynamics* 2nd edn (Englewood Cliffs, NJ: Prentice-Hall)

[4] Allen L, Beijersbergen M W, Spreeuw R J C and Woerdman J P 1992 Orbital angular momentum of light and the transformation of Laguerre–Gaussian laser modes *Phys. Rev.* A **45** 8185–9

[5] Sang X, Tienb E-K and Boyraz O 2009 Applications of two-photon absorption in silicon *J. Optoelectron. Adv. Mater.* **11** 15–25

**IOP** Publishing

# An Introduction to Photonics and Laser Physics with Applications

**Prem B Bisht**

# Chapter 2

# Light–matter interaction and the essentials of spectroscopy

The color of a substance, the intensity of a light source, and the decay time of an excited molecule are examples of features that are generally determined experimentally rather than calculated. When a photon of a certain energy interacts with an atomic or a molecular system, electrons are excited to the corresponding energy levels. This process is also known as the light–matter interaction. A rich variety of nomenclature has been developed to label energy states. An effort will be made here to explain a small portion in order to appreciate it. All this falls under the broad heading of spectroscopy, which is considered to be the mother of modern science. Experiments are also described here that led to the development of a basic understanding of the world at the atomic and molecular level. The figure shows a schematic of the fundamental process of the absorption of light and the subsequent photoluminescence (PL) spectrum of a material—an essential component of a laser. The details are given in this chapter. From the perspective of laser physics, the quantum mechanical detail will not be attempted here. Generally, atomic and molecular spectroscopy constitutes a full course. Therefore, the reader is advised to review the references given at the end of this chapter for a thorough study.

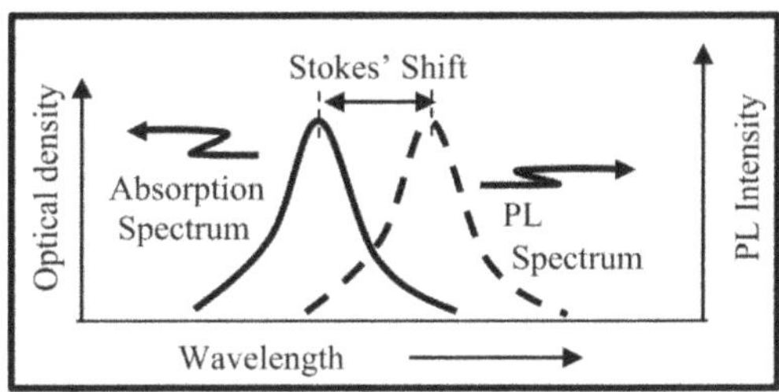


doi:10.1088/978-0-7503-5226-0ch2



© IOP Publishing Ltd 2022

**Learning objectives**
**After reading this chapter, the learner will be able to:**
Classify the EM radiation spectrum;
Relate Bohr's atomic model with Rutherford's experiments and the Franck–Hertz experiment;
Relate the Stern–Gerlach experiment and the photoelectric and Compton effects to the quantum mechanical picture;
Describe atomic and molecular transitions as well as energy states in solids;
Explain the Franck–Condon principle, the Stokes shift, and the Jablonskii diagram;
Define the Raman effect.

## 2.1 Light sources and types of spectra

Light sources have made important contributions to the development of modern science. We come across many types of light source every day. Before going into the details of spectroscopy, it is worth taking note of the types of spectrum emitted by light sources. There are three types of spectrum observed in spectroscopy as seen in table 2.1, viz. (i) line spectra, (ii) band spectra, and (iii) continuous spectra. The spectrum of a low-pressure mercury (Hg) arc contains the spectral lines of Hg (257.4 nm, 435.8 nm, etc). A sodium lamp gives a doublet or two close-lying spectral lines (589.0 nm and 589.6 nm). A household mercury tube light, on the other hand, consists of a mixture of the two types of spectra with lines (of Hg) and a continuous spectrum (due to phosphorescent material) of visible wavelengths. An incandescent bulb or a light-emitting diode (LED) produces a continuous spectrum. The spectra obtained from solutions or crystals are generally in the form of bands at room temperature. This is due to various broadening effects (♠ see chapter 8), including the smearing of the vibrational and rotational spectra. At very low temperatures, however, the vibrational structures can appear in the form of line spectra due to restriction of the thermal distribution of the population in the emitting state.

**Table 2.1.** Sources of line, band, and continuous spectra.

| Type of spectra | Source | Generally present in |
|---|---|---|
| Line | Mercury and sodium in gaseous form, molecules in supersonic free jets and at low temperatures | Atomic spectra or molecular spectra of cold, isolated molecules |
| Band | Crystals or solutions of molecular species | Absorption, transmission, and emission spectra in the condensed phase |
| Continuous | Halogen and mercury lamps, electric bulbs | Molecular gases at higher pressures |

## 2.2 Laser: a tool covering the EM spectrum

The acronym LASER stands for light amplification by stimulated emission of radiation. It is one of the most intense and monochromatic sources of light. As indicated in chart 2.1, several decades after its invention, the laser has become a tool of interdisciplinary fields in science and technology. Due to its range of wavelengths and versatility, hardly any field is untouched by lasers.

A summary of the EM spectrum in given in figure 2.1. Laser sources are available from the x-ray region to the far infrared (IR). Some popular lasers are indicated in the diagram along with an expanded portion of the visible region. The following subsections present some of the classic experiments and related information that have contributed to the development of the fields of laser optics and spectroscopy.

**Exercise 2.1.** What is the energy of (i) a wavelength of 600 nm and (ii) room temperature (300 K) in units of electron volt (eV)?

**Solution:** The energy ($E$)–wavelength ($\lambda$) relation is given by $= hc/\lambda$. Here, $h$ is Planck's constant and c is the speed of light. The energy corresponding to 600 nm in joules can be written as

$$\frac{6.6e - 34 \text{ Js} \times 3 \times 10^8 \text{ m s}^{-1}}{600 \times 10^{-9}\text{m}} = 3.32 \times 10^{-19}.$$

To convert this into eV, it has to be divided by the charge of the electron ($1.6 \times 10^{-19}$). Therefore, the corresponding energy is 2.08 eV. Similarly, for temperature (T), we use $E = kT$. Here, $k$ is the Boltzmann constant. When the room temperature of 300 K is converted into eV, it corresponds to 0.0258 eV or 25 meV.

## 2.3 Photoelectric effect

When light falls on a metal, electrons are emitted. Following this observation in late 19th century by Hertz, Einstein gave an explanation of the photoelectric effect in terms of a quantum of energy ($h\nu$), the frequency $\nu$, and the work function ($W$) of the metal, as follows: photoelectrons can only be emitted when the frequency of the

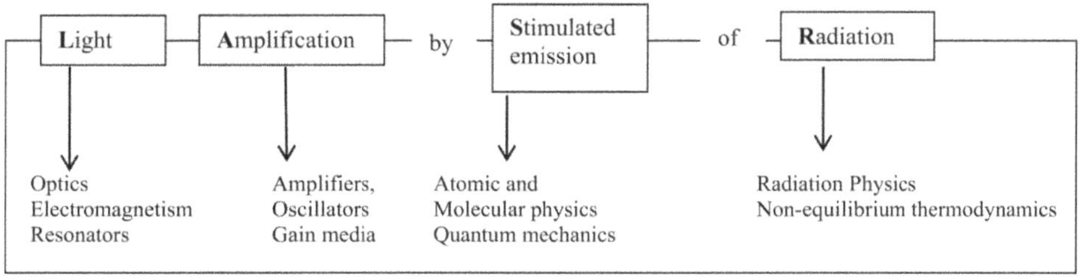


**Chart 2.1.** The full form of 'laser' and its interdisciplinary reach.

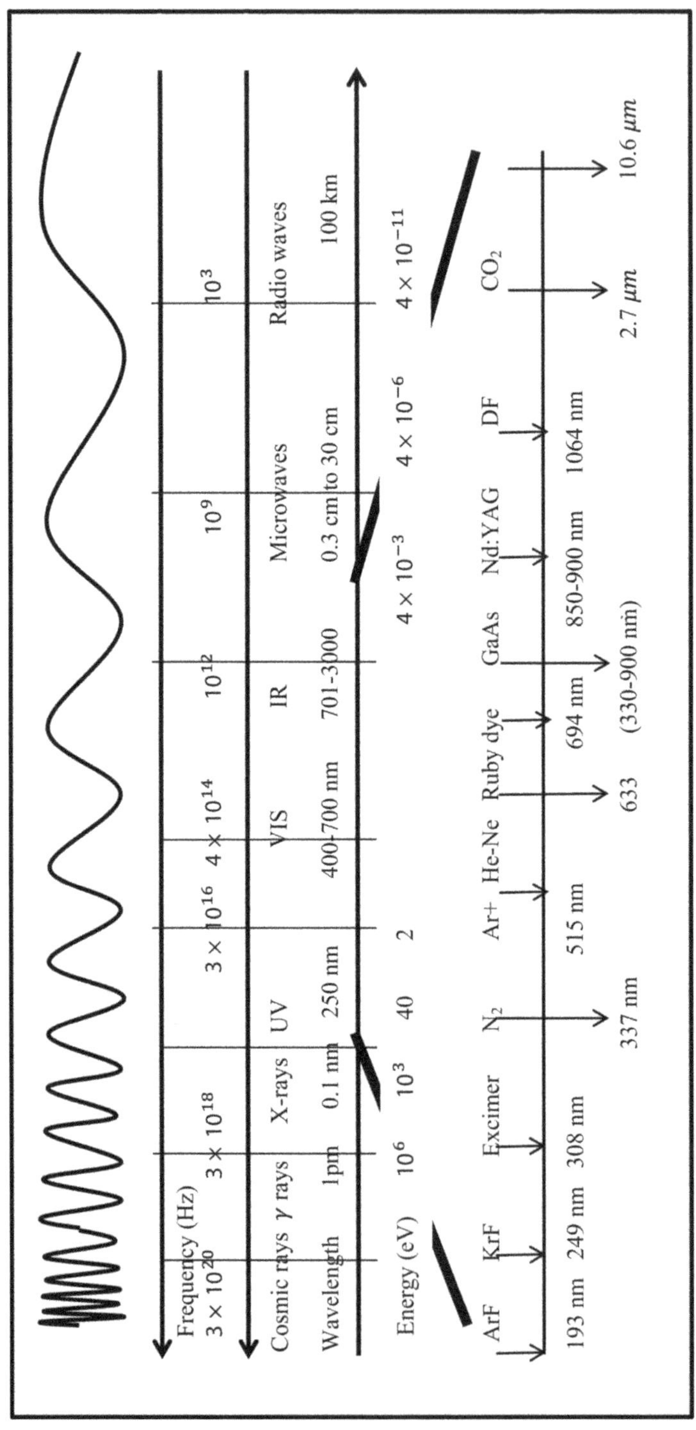


**Figure 2.1.** Schematic of the EM spectrum from cosmic rays to radio waves. The part of the spectrum corresponding to ultraviolet (UV), visible light, and IR (UV–vis–IR) is expanded to indicate some of the characteristic laser lines.

incident light is greater than the value of W. The kinetic energy ($K$) of the emitted photoelectrons follows the relation

$$K = h\nu - W.$$

The concept of light quanta or energy packets ($h\nu$) had already been proposed by Planck. Later, based on Maxwell's equations, the ratio of the energy ($E$) to the momentum ($p$) was found to be equal to the velocity of light (c). As a result, we have two famous relations for the energy of a photon, namely $E = h\nu$ and $E = pc$.

In 1923, de Broglie proposed a similar relation for matter waves, known as the wave–particle duality. The wavelength is now known as the de Broglie wavelength associated with a particle $as \lambda = \frac{h}{p}$. The lighter the particle, the larger the de Broglie wavelength. For the mass of a cricket ball or a human being, the de Broglie wavelength is negligible in contrast to that of the electron.

♣ Einstein wrote five seminal papers in 1905. One of them explained the photoelectric effect.

♣ In this chapter, $E$ (without a vector sign) denotes energy, following conventional use. This should not be confused with an electric field.


**Exercise 2.2.** For potassium, the photoelectric effect is observed when light with an intensity of ($I_l$) has a wavelength ($\lambda$) of 564 nm. What is the work function of potassium (use $h = 6.64 \times 10^{-34}$ J s)?

**Solution:** Using the relation $K = h\nu - W$ for $K = 0$ (photoelectrons are just emitted and have no kinetic energy), the work function of potassium can be obtained for the frequency corresponding to the threshold wavelength of 564 nm; the result is 2.2 eV.


**Exercise 2.3.** The de Broglie wavelength cannot be observed in everyday life. Can you demonstrate it for the case of a cricket ball weighing 163 g, thrown at a speed of 45 m s$^{-1}$? Compare it with the de Broglie wavelength of an electron moving at a speed of $10^6$ m s$^{-1}$ in a lattice.

**Solution:** The wavelength of the cricket ball ($\lambda_{\text{ball}}$) can be written as

$$\lambda_{\text{ball}} = \frac{h}{mv} \sim 9 \times 10^{-35} \text{ m}$$

Therefore, the de Broglie wavelength of the cricket ball is too small from the point of view of everyday life. As an exercise, check that for an electron, the de Broglie wavelength is of the order of the lattice spacing in crystals.


## 2.4 Rutherford's experiment

Rutherford's experiment in 1910 addressed the identification of the nucleus. The experiment was carried out by bombarding a gold foil with positive $\alpha$ particles (*He nuclei*). The fact that some of the $\alpha$ particles bounced back towards the source confirmed the presence of the positive charge in the tiny nucleus inside the atom.

## 2.5 Bohr's atomic model: atomic energy levels

The idea of discrete energy levels with stationary radii in atoms was proposed by Bohr in 1913 to explain the emission spectrum of hydrogen. The two important points of this model are:

(i) An electron of mass $m_e$ with a linear velocity of $v$ can only orbit the nucleus with certain discrete values of angular momentum $(\vec{L})$ in multiples of $n\hbar$ (i.e. $|\vec{L}| = m_e v r_n \equiv n\hbar$). Here, $r_n$ is the radius of the $n$th stationary orbit.

(ii) A change from one orbit to other is accompanied by either the emission or absorption of radiation; for example, between the first two orbits, this is $\Delta E = h\nu_{21}$, as shown in figure 2.2.

Using Newtonian mechanics and postulate **(i)** for an electron, the specific orbit of radius $r_n$ is given by

$$r_n = \left(\frac{4\pi\epsilon_0\hbar^2}{m_e}\frac{}{e^2}\right)n^2 \equiv a_0 n^2. \tag{2.1}$$

Here, $a_0$ is known as the Bohr radius ($= 0.5\mathring{A}$). This can now be used to derive the total allowed energy states ($E_n$) of the system, as follows:

$$E_n = -\left(\frac{e^2}{8\pi\epsilon_0 r_n}\right) \propto -\frac{E_1}{n^2}. \tag{2.2}$$

The negative sign indicates that the electron is bound. Here, $E_1$ (or the lowest energy—that of the ground state) is given by

$$E_n = -\left(\frac{e^2}{8\pi\epsilon_0 a_0}\right)\left(\frac{1}{n^2}\right) \approx -13.64\text{eV for } n = 1.$$

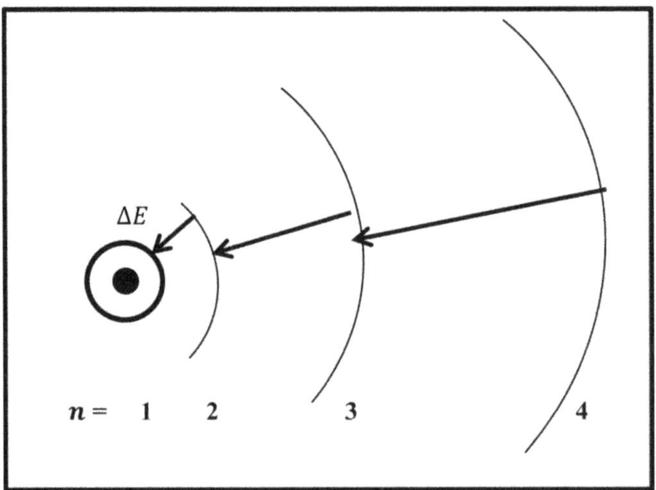


Figure 2.2. The atomic model proposed by Bohr for hydrogen-like atoms of stationary orbits. The dark spot at the center is the nucleus; it is surrounded by orbits denoted by the principal quantum number $n$. Arrows indicate the radiative transitions between energy levels.

For accurate energy values, we must use the reduced mass $m_r$ of the combined atom by including $m_e$ and the mass of the proton ($M_P$) as follows: $m_r = \frac{m_e M_P}{M_P + m_e}$. The lowest value (for $n = 1$) of the binding energy ($-13.64$ eV) when converted to a wavelength corresponds to about 91 nm. This suggests that any wavelength shorter than 91 nm can photo-ionize the hydrogen atom.

♣ For other hydrogen-like atoms, the ionization energy can be written by including the atomic number (Z) in equation 2.1, as follows: $E_n = -\frac{m_e Z^2 e^4}{8\epsilon^2 h^2} = -\frac{Z^2 E_0}{n^2}$.

For an electron to jump from a higher-energy orbit (2) to a lower energy (1), the frequency (in wavenumber units) for the change of energy ($\Delta E_{21}$) is given by

$$h\nu_{21} = \Delta E_{21}$$

$$\nu_{21} = -\frac{E_0}{h}\left(\frac{1}{n_2^2} - \frac{1}{n_1^2}\right). \tag{2.3}$$

The corresponding wavelength ($\lambda_{21}$) is given by

$$\frac{1}{\lambda_{21}} = R_H \left(\frac{1}{n_1^2} - \frac{1}{n_2^2}\right).$$

Here, $R_H = 1.06967758 \times 10^7$ m$^{-1}$ is known as the Rydberg constant for the hydrogen atom.

According to Bohr's theory, the energy levels of hydrogen atom were identified as a Lyman series (for transitions between higher states and the ground state $n = 1$), as shown in figure 2.3. The longest wavelength of the Lyman series ($n = 2$ to $n = 1$) in the hydrogen atom occurs in vacuum and under UV at 121.5 nm. Similarly, other

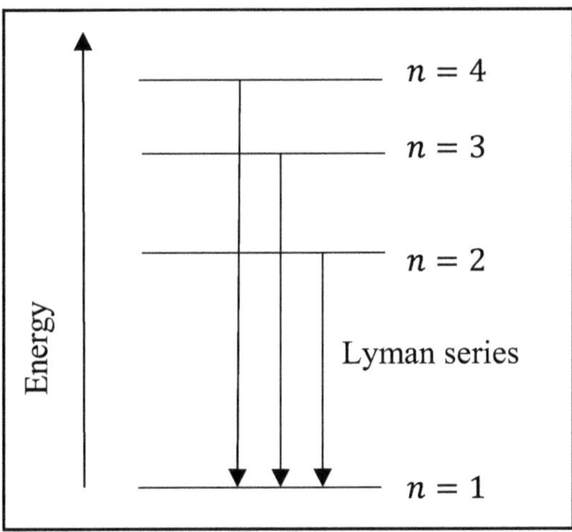


**Figure 2.3.** Atomic transitions corresponding to the Lyman series in the hydrogen atom in the UV–visible–IR region, as explained by Bohr's theory.

spectral lines known as the Balmer, Paschen, Brackett, and Pfund series, respectively, result from transitions from higher states to the second, third, fourth, and fifth levels, and these are observed at visible and IR wavelengths. These lines are observed in the spectra of stars, including the Sun, because their major constituent is hydrogen. Similarly, Bohr extended his calculations to other ionized elements in which only one electron is left (such as $He^+$, $Li^{++}$, etc.) and showed that many lines in the experimentally observed spectrum of the Sun could be assigned to these ionized species.

## 2.6 Franck–Hertz experiment

Bohr's theory was proved to be correct by a classic experiment performed by Franck and Hertz (1914), which showed that atoms could indeed only absorb and emit specific amounts of energy. Incidentally, this experimental design has a similarity with the basic design of gas lasers that work on the principle of electrical discharge. In chapter 6, you will learn more about this aspect in relation to He–Ne, nitrogen, and argon ion lasers, which came into existence only about half a century later.

In this experiment, (see figure 2.4, left panel) electrons are generated and accelerated towards the anode. The anode current (measured in mA) increases for an increase in the accelerating voltage, $V$. At some voltage $V$, the anode current decreases and shows a dip. It further starts to increase as a function of $V$ until the acceleration voltage is equal to 2 times $V$, at which point another dip is observed in the plot of anode current versus acceleration voltage (see figure 2.4(b)).

The dip in the anode current can be explained by the reduction in the number of electrons reaching the anode. The electrons collide with Hg atoms and lose their energy. However, the atoms absorb the energy in inelastic collisions, in which the energy gained is equal to the difference between the two atomic levels. There is a small retarding potential applied across the tube. Therefore, when electrons with energy corresponding to the potential $V$ (as expressed in eV) are absorbed by the atom, there is a dip in the recorded current. These dips are seen in the graph only at integral multiples of $V$.

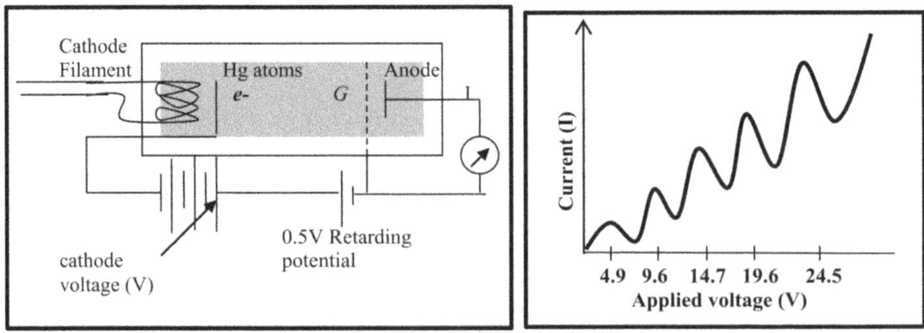


**Figure 2.4.** The experimental setup of Franck and Hertz (left panel). Here, $G$ is the grid and e– represents the electron. Graph of the observed current ($I$) vs applied anode voltage ($V$) (right panel).

This experiment confirmed the discrete nature of the energy levels proposed by Bohr. Franck and Hertz also measured the emission line at 253.7 nm, which matched the energy of the absorbing photons. The electrons that do not collide give rise to the current below the dip level. The energy of the absorbed and emitted photons (see section 1.1) in eV is given by

$$E = \frac{hc}{\lambda} = \frac{6.62 \times 10^{-34}\,\text{Js} \times 3 \times 10^{8}\,\text{m s}^{-1}}{253.7 \times 10^{-9} \times 1.6 \times 10^{-19}\,\text{C}} = 4.9\,\text{eV}.$$

## 2.7 Stern–Gerlach experiment: spin quantization

Experiments in the early 1920s discovered a new aspect of nature and found the simplest quantum property in existence. In the Stern–Gerlach (SG) experiment, a beam of silver atoms was passed through a nonuniform magnetic field. An image on a photographic plate (figure 2.5) was obtained as if the beam had split into two components in the direction of the field

This experiment discovered a unique feature of electrons, as follows: as seen in the periodic table of elements (chart 2.2, ♠ see section 2.9.6.1), silver has an atomic number of 47 and is a neutral atom. Forty-six out of its 47 electrons are paired up. The remaining unpaired electron contributes to the total spin angular momentum. The unpaired electron has a magnetic dipole moment ($\vec{\mu}$) given by

$$\vec{\mu} = g\frac{e}{2m_e}\vec{S} = g\mu_B\frac{\vec{S}}{h}.$$

Here, $\vec{S}$ is the spin angular momentum of an electron, $g$ is a constant known as the Landé $g$-factor, and $\mu_B = e\hbar/(2m_e)$ is the Bohr magnetron. Spin is intrinsic angular momentum and it is equal to ½ for electrons. Electrons are called spin-½ particles. Due to the inhomogeneity of the magnetic field along the $z$-axis, electrons experience the magnetic force and are deflected either up or down by a constant amount in approximately equal numbers. Apparently, the $z$-component of the electron's spin is quantized such that it can take only one of two discrete values, $+½$ or $-½$. We say

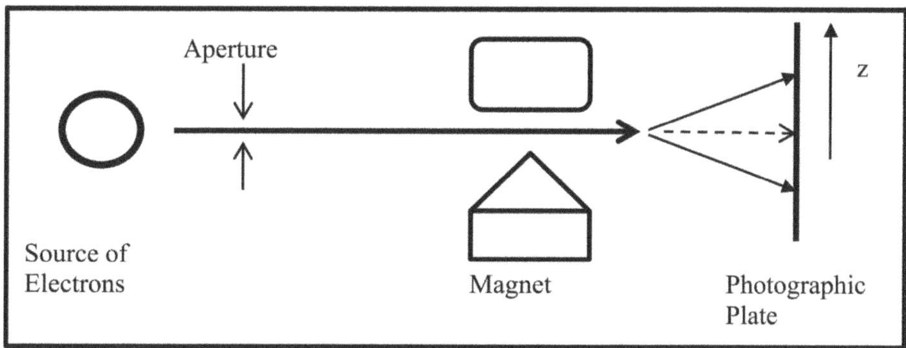


**Figure 2.5.** The Stern–Gerlach experiment. An oven filled with silver atoms with a single unpaired electron acts as a source of electrons. The magnetic field is inhomogeneous in the $z$-axis.

| 1A | 2A | 3B | 4B | 5B | 6B | 7B | 8B | 8B | 8B | 1B | 2B | 3A | 4A | 5A | 6A | 7A | 8A |
|---|---|---|---|---|---|---|---|---|---|---|---|---|---|---|---|---|---|
| 1<br>H<br>1.008 | | | | | | | | | | | | | | | | | 2<br>He<br>4.0026 |
| 3<br>Li<br>6.94 | 4<br>Be<br>9.0122 | | | | | | | | | | | 5<br>B<br>10.81 | 6<br>C<br>12.011 | 7<br>N<br>14.007 | 8<br>O<br>15.999 | 9<br>F<br>18.998 | 10<br>Ne<br>20.180 |
| 11<br>Na<br>22.990 | 12<br>Mg<br>24.305 | | | | | | | | | | | 13<br>Al<br>26.982 | 14<br>Si<br>28.085 | 15<br>P<br>30.974 | 16<br>S<br>32.06 | 17<br>Cl<br>35.45 | 18<br>Ar<br>39.948 |
| 19<br>K<br>39.098 | 20<br>Ca<br>40.078 | 21<br>Sc<br>44.956 | 22<br>Ti<br>47.867 | 23<br>V<br>50.942 | 24<br>Cr<br>51.996 | 25<br>Mn<br>54.938 | 26<br>Fe<br>55.846 | 27<br>Co<br>58.933 | 28<br>Ni<br>58.693 | 29<br>Cu<br>63.546 | 30<br>Zn<br>65.38 | 31<br>Ga<br>69.723 | 32<br>Ge<br>72.630 | 33<br>As<br>74.922 | 34<br>Se<br>78.971 | 35<br>Br<br>79.904 | 36<br>Kr<br>83.798 |
| 37<br>Rb<br>85.468 | 38<br>Sr<br>40.078 | 39<br>Y<br>88.906 | 40<br>Zr<br>91.224 | 41<br>Nb<br>92.906 | 42<br>Mo<br>95.95 | 43<br>Tc<br>98 | 44<br>Ru<br>101.07 | 45<br>Rh<br>102.91 | 46<br>Pd<br>106.42 | 47<br>Ag<br>107.87 | 48<br>Cd<br>112.41 | 49<br>In<br>114.82 | 50<br>Sn<br>118.71 | 51<br>Sb<br>121.76 | 52<br>Te<br>127.60 | 53<br>I<br>126.90 | 54<br>Xe<br>131.29 |
| 55<br>Cs<br>132.91 | 56<br>Ba<br>137.33 | 57-71 | 72<br>Hf<br>178.49 | 73<br>Ta<br>180.95 | 74<br>W<br>183.84 | 75<br>Re<br>186.21 | 76<br>Os<br>190.23 | 77<br>Ir<br>192.22 | 78<br>Pt<br>195.08 | 79<br>Au<br>196.97 | 80<br>Hg<br>200.59 | 81<br>Tl<br>204.38 | 82<br>Pb<br>207.2 | 83<br>Bi<br>208.98 | 84<br>Po<br>209 | 85<br>At<br>210 | 86<br>Rn<br>222 |
| 87<br>Fr<br>223 | 88<br>Rd<br>226 | 89-103 | 104<br>Rf<br>267 | 105<br>Db<br>268 | 106<br>Sg<br>269 | 107<br>Bh<br>270 | 108<br>Hs<br>277 | 109<br>Mt<br>278 | 110<br>Ds<br>281 | 111<br>Rg<br>282 | 112<br>Cn<br>285 | 113<br>Nh<br>286 | 114<br>Fl<br>289 | 115<br>Mc<br>290 | 116<br>Lv<br>293 | 117<br>Ts<br>294 | 118<br>Og<br>294 |

| 57<br>La<br>138.91 | 58<br>Ce<br>140.12 | 59<br>Pr<br>140.91 | 60<br>Nd<br>144.24 | 61<br>Pm<br>145 | 62<br>Sm<br>150.36 | 63<br>Eu<br>151.96 | 64<br>Gd<br>157.25 | 65<br>Tb<br>158.93 | 66<br>Dy<br>162.50 | 67<br>Ho<br>164.93 | 68<br>Er<br>167.26 | 69<br>Tm<br>168.93 | 70<br>Yb<br>173.05 | 71<br>Lu<br>174.97 |
|---|---|---|---|---|---|---|---|---|---|---|---|---|---|---|
| 89<br>Ac<br>227 | 90<br>Th<br>232.04 | 91<br>Pa<br>231.04 | 92<br>U<br>238.03 | 93<br>Np<br>237 | 94<br>Pu<br>244 | 95<br>Am<br>243 | 96<br>Cm<br>247 | 97<br>Bk<br>247 | 98<br>Cf<br>251 | 99<br>Es<br>252 | 100<br>Fm<br>257 | 101<br>Md<br>258 | 102<br>No<br>259 | 103<br>Lr<br>266 |

**Chart 2.2.** Periodic table of elements in the year 2018.

that the spin is either up or down in the $z$ direction. It should be noted that if electrons were like ordinary magnets, i.e. with random orientations of their magnetic moments, they would have shown a continuous spread on the photographic plate.

The SG experiment is considered to be an excellent learning tool in theoretical physics, especially in quantum optics and quantum computing, due to its ability to differentiate the quantized states. The polarization states of light (see chapter 3) are often also taken to be quantized states similar to the results of the SG experiment.

**Exercise 2.4.** In the original SG experiment, a beam of silver atoms from an oven was directed along the $y$-axis to the inhomogeneous magnetic field along the $z$-axis (i.e. the SGz setup), as shown in the figure below, to obtain two components (the Sz+ and Sz− components)

What happens when we now pass one of the beams though sequential SG experiments, either by using another SGz setup or a new SGx (field inhomogeneity along the $x$-axis) setup?

**Solution:** If the Sz− component is blocked and another SGz setup is introduced, we get the Sz+ component only. However, if after blocking the Sz− component, another SGx setup is introduced, we will get the Sx− and Sx+ components. The latter case is similar to introducing a polarizer at an angle to linearly polarized light.

## 2.8 Compton effect

In this experiment, a material was irradiated with a beam of x-ray photons of wavelength $\lambda$. It was found that besides the original photons, scattered radiation with a new wavelength $\lambda'(>\lambda)$ was also present at the detector. This new, longer wavelength was observed due to scattering from collisions with loosely bound (or free) electrons of the material. This experiment was first performed in 1923 by A H Compton.

The shift in the wavelength $(\Delta\lambda = \lambda' - \lambda)$ is independent of the material used, but is found to vary with the scattering angle $(\theta)$. $\Delta\lambda$ is given by $\frac{h}{mc}(1 - \cos\theta)$. Here, $m$ is the mass of the colliding particle (a free electron in this case). The factor $\frac{h}{mc}$ is known as the Compton wavelength $(\lambda_c)$) and has the dimension of length. The expression for the predicted $\Delta\lambda$ can be obtained by considering an elastic collision of photons with the free electrons at rest. In this way, the Compton effect confirms the particle nature of photons. The Compton effect is not observed in the visible region, as the energy of visible photons is small compared with the binding energy of even loosely bound electrons. Are you aware of a well-known similar effect (reported in 1928) in the visible region that changes the wavelength of incident light?

## 2.9 Quantum mechanical picture of matter

Some of the experiments discussed above revealed the structure of the atom and some other properties, such as the quantized spin of electrons. However, Bohr's classical theory was not sufficient to describe the electronic structure of larger atoms.

Therefore, a new theoretical approach was necessary. For example, a vibrating molecule can be considered to be a harmonic oscillator, and an electron restricted to a lattice structure can be viewed as a boundary-value potential problem. Due to the limitations of space, this section will present a brief account of the quantum mechanical picture of matter. For further details, the reader is encouraged to review the references presented at the end of this chapter.

### 2.9.1 Wave functions

To locate an electron, for example, we need its spatial coordinates, ($x$, $y$, and $z$) and the time, $t$. These can be represented in the form of a wave function $\psi$ as $\psi(x, y, z, t)$. The probability of locating an electron in a given volume element $dV'$ at time $t$ is given by $|\psi|^2 \, dV'$. The probability of finding an electron associated with an atom around the nucleus is one. Therefore, the probability distribution function is given by the volume integral of the scalar product of $\psi(x)$ and its complex conjugate $\psi^*(x)$, as follows:

$$\int \psi^*(x)\psi(x)\mathrm{d}V' = 1.$$

### 2.9.2 The Schrödinger equation

The wave function of a particle is subjected to restrictions, known as boundary conditions, based on a given problem. For example, a charge carrier trapped by the potential barrier of a semiconductor heterojunction ($\spadesuit$ see chapter 22) can be addressed here. This is related to the problem of an electron trapped in a metal. Both these physical situations can be approximated by a particle in an infinitely deep potential well, as shown in figure 2.6. The basic equation governing the behavior and dynamics of quantum states is the Schrödinger equation, which was introduced in 1926. The time-independent Schrödinger equation for an electron is written as

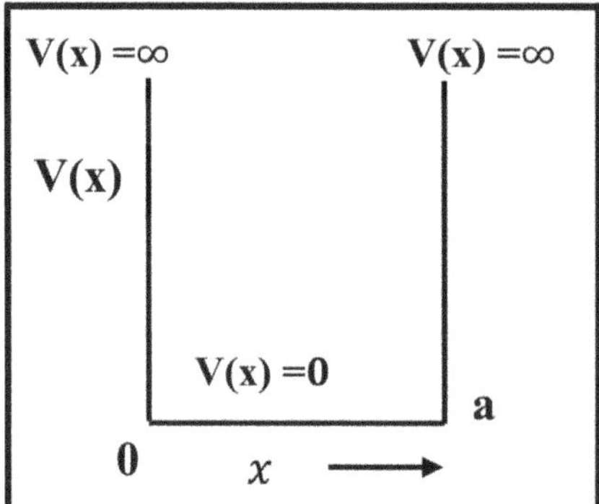


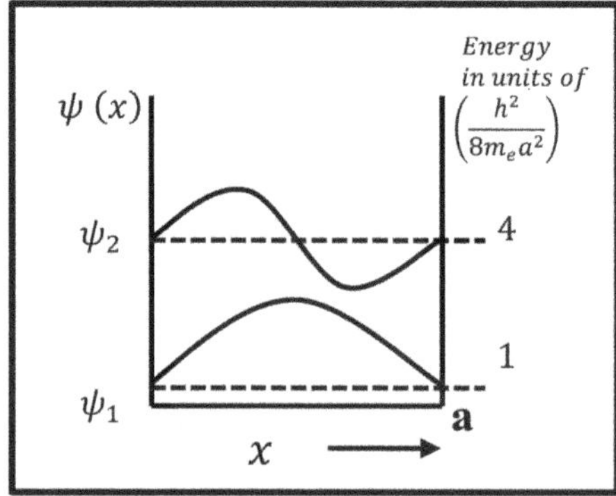


**Figure 2.6.** Infinitely deep one-dimensional potential well (left-hand panel): $V(x) = 0$ for $0 < x < a$ and $V(x) = \infty$ for $x = 0$, $a$. The right-hand panel gives the electron wave function of the ground and first excited states for the deep well, along with the energy.

$$\nabla^2 \psi(x) + \frac{2m_e}{\hbar^2}(E - V(x))\psi \quad (x) = 0. \tag{2.4a}$$

Here, $E$ is the total particle energy, $V(x)$ is the potential energy function that describes the environment of the particle, $m_e$ is the mass of any particle, and $\nabla^2$ is the second partial derivative (also known as the Laplace operator). The time-dependent Schrödinger equation is written as

$$i\hbar \frac{\partial \psi(x)}{\partial t} = -\frac{\hbar^2}{2m_e}\nabla^2 \psi(x) + V(x)\psi(x). \tag{2.4b}$$

Solutions of the Schrödinger equation occur only for certain discrete allowed values of E, known as eigenvalues ($E_n$). For each eigenvalue, there is an associated solution known as the eigenfunction.

### 2.9.3 Infinite quantum well

An electron in an atom is considered to be in an attractive Coulomb potential with the nucleus. This situation has been dealt with by solving the Schrödinger equation for a particle in a potential well (or box) of infinite height. Consider a one-dimensional infinite potential well of infinite depth and width $a$, as shown in figure 2.6. For the present demonstration, it is sufficient to consider the time-independent equation (2.4(a)).

For the one-dimensional potential well of figure 2.6, equation (2.4(a)) can be rewritten in the region of $V(x) = 0$ as the wave equation of the particle as:

$$\frac{d^2\psi \quad (x)}{dx^2} + \left(\frac{2m_e E}{\hbar^2}\right) \quad \psi \quad (x) = 0. \tag{2.4c}$$

We can assume a solution of partial differential equation (2.4(c)) such as $\psi(x) = A \sin kx + B \cos kx$, where $A$ and $B$ are constants and

$$k^2 = \left(\frac{2m_e E}{\hbar^2}\right). \tag{2.5}$$

According to the boundary conditions, the particle is confined within the deep well. Hence, the probability $|\psi(x)|^2$ of finding the particle outside the well is zero. Therefore $\psi(x)$ must also be zero at the boundaries as well, as the particle cannot penetrate the infinitely high potential barrier. At $x = 0$, $0 = A \sin 0 + B \cos 0$ implies that $B = 0$ at $x = 0$. It follows that at $x = a$, $0 = A \sin ka + 0 \cos ka$. This implies that $\psi(a) = A \sin ka$, indicating that either $A = 0$ or $\sin ka = 0$. In other words, $ka$ must be an integral multiple of $\pi$. Therefore, we can write

$\psi(x) = A \sin kx$, where $ka = n\pi$

or $k = n\pi/a$, $n = 1, 2, 3, \ldots$.

Therefore, substituting this for $k$ in equation (2.5), we get $\left(\frac{n\pi}{a}\right)^2 = \left(\frac{2m_e E_n}{\hbar^2}\right)$. The energies of the quantized energy states ($E_n$) are given by

$$E_n = \left(\frac{h^2}{8m_e a^2}\right)n^2. \tag{2.6}$$

These states for $n = 1, 2, \ldots$ are called stationary states, as shown in the right-hand panel of figure 2.6.

### 2.9.4 Stationary states and the coherent state

In quantum mechanics, the characteristic state of an electron is known as an eigenstate. These well-defined energy states are also expressed in Dirac notation as $|\psi_1\rangle$ (ket $\psi_1$) or $|\psi_2\rangle$ (ket $\psi_2$), as indicated in figure 2.7. As they are similar to the discrete energy states defined in Bohr's theory, these are also known as stationary states.

On the other hand, in a coherent state, the energy is not well defined. Rather, it is an intermediate state, or the state during a transition from one state to another state. Therefore, a coherent state is not a single state but is a superposition of $|\psi_1\rangle$ and $|\psi_2\rangle$. Hence, the wave function must include both states. The wave function is therefore a superposition of the two states with energies $E_1$ and $E_2$, respectively, as follows.

If $\Psi_1 = \psi_1 e^{-iE_1 t/\hbar}$ and $\Psi_2 = \psi_2 e^{-iE_2 t/\hbar}$, to write the wave function of a mixed state, we need the coefficients ($c_1$ and $c_2$).

$$\Psi = c_1\psi_1 e^{-iE_1 t/\hbar} + c_2\psi_2 e^{-iE_2 t/\hbar} \text{ and } \Psi^* = c_1^* \psi_1^* e^{iE_1 t/\hbar} + c_2^*\psi_2^* e^{iE_2 t/\hbar}.$$

We note that $E_2 - E_1 = h\nu_{21} \equiv \hbar\omega_{21}$ and the coefficients corresponding to particular states in the linear combination are complex quantities, thus allowing interference effects between the states, which can be written as

$$\Psi^*\Psi = c_1^* c_1\psi_1^*\psi_1 + c_2^* c_2\psi_2^*\psi_2 + c_1^* c_2 c_1\psi_1^*\psi_2 e^{-i\hbar\omega_{21}} + c_2^* c_1\psi_2^*\psi_2 e^{i\hbar\omega_{21}}.$$

We note here that a thorough understanding of quantum mechanics requires knowledge of linear algebra and linear vector spaces.

♣ P A M Dirac's bra-ket notation is a shorthand used to express the wave function and the state of a system. For example, $|+\rangle$ (read as 'ket plus') is used to represent a spin-up particle in its initial state; $\langle - |$ (read as 'bra minus') is used to represent the state of a particle with spin down. In general, $\langle f,|, i\rangle$ (read as 'bra $f$ ket

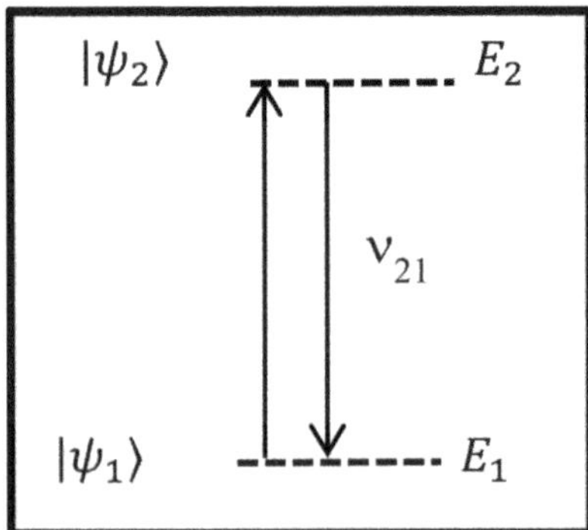


**Figure 2.7.** Stationary states with energies $E_1$ and $E_2$ with a transition frequency of $\nu_{21}$.

$i'$) is the probability amplitude for the transition to a final state $\langle f|$ from an initial state $|i\rangle$. This is similar to the expression $|\psi|^2$ given above.

### 2.9.5 Heisenberg's uncertainty principle

Unlike the situation in classical mechanics, this principle suggests that certain pairs of observables, such as distance $(x)$ and momentum $(p)$ or energy $E$ and time $t$, have a minimum uncertainty (represented by $\Delta$) of the order of $\hbar \left(=\frac{h}{2\pi}\right)$ in their product, according to $\Delta x \Delta p \geqslant \frac{\hbar}{2}$ or $\Delta E \Delta t \geqslant \frac{\hbar}{2}$. If the value of one of the variables is certain, its eigenfunction is specified. Pairs of observables which do not follow commutation relations obey the uncertainty principle because the two operators cannot have the same eigenfunctions. This principle also has to be seen in the context of the wave–particle duality principle for matter waves as well as for photons.

A state with a finite lifetime $\Delta t$ will posses an uncertainty in its energy (and hence its frequency). This is the reason for lifetime broadening ($\spadesuit$ see chapter 8). Another consequence of this principle is that at the lowest temperature, even a molecule with no vibrations still has a nonzero momentum. To precisely locate the nucleus with certain $\Delta x$, momentum uncertainty $(\Delta p)$ implies a nonzero energy at such temperatures.

### 2.9.6 Atomic quantum states

In atoms the spectrum of emitted radiation results from electron transitions from one discrete energy state to another. The pattern of emitted radiation can be predicted and understood by quantum mechanics. Even the simplest atom, such as hydrogen, has several lines, as shown in figure 2.3. The quantum mechanical treatment agrees well with Bohr's theory for the hydrogen atom. For larger atoms (even He and Ne), a large number of transitions are possible.

Spectroscopy helps us in two ways: first, it helps us to identify the various energy levels; second, it tells us the set of rules governing allowable transitions between two states. Given the complexity of identifying the large number of energy levels and their exact values, their characterization by experiments and their analysis via quantum mechanics are tasks for spectroscopists. We accept the labels assigned by expert spectroscopists as final data. However, we will summarize the salient feature of atomic spectroscopy. Based on this foundation, we will introduce the molecular levels so as to provide an understanding of the dynamics of laser physics, while also helping the reader to understand the basic nomenclature for energy states. For complete and comprehensive knowledge, the reader is referred to the specialized textbooks on atomic and molecular spectra [1, 2].

#### 2.9.6.1 Electronic quantum numbers and the periodic table of elements
We know the quantum number $n$ from Bohr's theory. It takes the values 1, 2, 3..., $\infty$, with the result that the orbital radius varies according to $n^2$ (see equation 2.1). The Heisenberg uncertainty principle suggests that discrete orbits with well-defined energies do not exist, but rather implies a probability of density distribution with varying degrees of electron presence. In quantum mechanics, the quantum number $n$

(known as the principal quantum number) represents the approximate radial distance of an electron from the nucleus. It is also denoted by the letters K, L, M, etc for $n = 1, 2, 3...$, respectively.

Additional quantum numbers describe an electron moving in a spherically symmetrical force field. The quantum number $n$ is accompanied by an azimuthal quantum number, $l$, due to spherical symmetry. $\vec{l}$ gives the angular momentum of an electron in the orbit and it can take any integral value from zero to $n - 1$. Accordingly, electrons are said to be in s ($l = 0$), p ($l = 1$), d ($l = 2$), f ($l = 3$), and so on. For instance, an electron with $n = 2$ and $l = 1$ is said to be in the state 2p subshell. The number of electrons in each shell and the sequence in which subshells are filled are given by Hund's rule. For example, the maximum number of electrons in a given $l$ shell is given by s: 2, p: 6, d: 10, f: 14, etc.

Chart 2.2 gives the periodic table of elements that was developed during the last century, following Mendeleev's version, created in 1869. Here, the elements are placed in rows according to increasing atomic number. Elements with similar properties are placed in the same column. For example, column 8A contains the inert gases (the subshells of these elements are filled). Such elements are stable and inactive in their ground states. However, in their excited states, some of these may be highly reactive, as, for example, in the excimer laser (♠ see section 2.9.7.5 in this chapter). The trivalent transition-metal ions, such as chromium ($Cr^{+3}$) and titanium ($Ti^{+3}$), and rare-earth ions, such as neodymium ($Nd^{+3}$), erbium ($Er^{+3}$), samarium ($Sm^{+2}$), and thulium ($Tm^{+2}$), have longer lifetimes that help them to achieve a metastable state (♠ see section 6.2.2) when embedded in crystals.

In order to find elements with properties that are suitable for lasing, it is useful to know their ground-state electronic configurations and their notations. For example, the electronic configurations of the ground states of He and Ne atoms, respectively, are given by He: $1s^2$; Ne: $1s^2, 2s^2 2p^6$.

Similarly, silver (Ag), which was mentioned in the context of the SG experiment in section 2.7 above, has the electronic distribution $1s^2, 2s^2 2p^6, 3s^2 3p^6 3d^{10}, 4s^2 4p^6 4d^{10}, 5s^1$. As verified by the SG experiment, the electron possesses an intrinsic angular momentum, or spin, denoted by the spin quantum number, ($s = \pm\frac{1}{2}$). In the presence of a magnetic field, for example in the Zeeman effect, the orbital and spin angular momenta become space quantized; they are described by the magnetic quantum numbers $m_l$ and $m_s$ and take certain values as follows: $m_l$ can take the values $l, l - 1, l - 2... -l$ and $m_s$ takes a value of either $+\frac{1}{2}$ or $-\frac{1}{2}$, as it can have either a parallel or an antiparallel component along the applied field. Several types of state can take similar energy values; these are known as degenerate states. Nevertheless, due to Pauli's exclusion principle, the set of all four quantum numbers ($n, l, m_l, m_s$) is never the same.

*2.9.6.2 Multiple electron systems: orbital and spin angular momentum*
Let us introduce the angular momentum of an atom, $\vec{J}$. It is divided into two components, orbital angular momentum ($\vec{L}$) and spin angular momentum ($\vec{S}$). $\vec{L}$ and $\vec{S}$ denote the resultant energy states of the individual $l$ and $s$ vectors, respectively. Their magnitudes are given by

$$\left|\overrightarrow{L}\right| = \sqrt{l(l+1)}\,\hbar \text{ and } \left|\overrightarrow{S}\right| = \sqrt{s(s+1)}\,\hbar\overrightarrow{J} = \sqrt{j(j+1)}\,\hbar.$$

The quantum number $j$ can take values from $l + s$ to $l - s$ in steps of one. The possible value of $j$ is equal to the smaller of the two numbers $2s + 1$ or $2l + 1$. As $l$ and $s$ interact magnetically, in the absence of a magnetic field, $\overrightarrow{j}$ is conserved in magnitude and direction.

### 2.9.6.2.1 L–S and j–j coupling

For lighter atoms in the periodic table and for most atoms, the Russell–Saunders coupling between $\overrightarrow{L}$ and $\overrightarrow{S}$ known as LS coupling ($\overrightarrow{J} = \overrightarrow{L} + \overrightarrow{S}$) is applicable. For this coupling, the values of $\overrightarrow{J}$ lie between $L + S$ and $L - S$. On the other hand, there are cases for which the individual values of $l$ and $s$ do not couple strongly and result in $\overrightarrow{L}$ and $\overrightarrow{S}$, respectively. Under such circumstances, for each electron, the individual values of $l$ couple with the individual values of $s$ and subsequently combine to give $\overrightarrow{j}$. This results in a coupling scheme known as j–j coupling (note the small letters in 'j–j coupling'). In atoms that possess many electrons, the inner shells are completely filled, hence the coupling between $\overrightarrow{L}$ and $\overrightarrow{S}$ does not exist. When the electrons in the outermost shell are considered, j–j coupling comes into effect. As LS coupling is applicable to large numbers of atoms, it forms the basis of the nomenclature of atomic spectroscopy, as summarized below.

With the above information, we can now try to understand the atomic states that are written as

$$^{2S+1}L_J \tag{2.7}$$

Here, the orbital quantum numbers $L = 0, 1, \ldots$ are labeled by the capital letters $S$, P, D, etc. as follows:

| L | 0 | 1 | 2 | 3 | 4 |
|---|---|---|---|---|---|
|   | S | P | D | F | G |

In equation (2.7) above, a superscript to the left of the letter $L$ indicates the multiplicity of the term due to possible orientations of the spin, i.e. $(2S + 1)$. The multiplicity values 1, 2, and 3, are known as singlet, doublet, and triplet, respectively. As mentioned above, $\overrightarrow{J}$ is the total angular momentum of the atom.

**Exercise 2.5.** Explain the transitions denoted by the state $^4$F.

**Solution:** From the aforementioned discussion, the letter $F$ tells us that the value of the angular momentum $L$ is 3. To explain all the possible transitions associated with the given state, $^4$F, we use the following procedure.

First, let us compare the given state $^4$F with equation (2.7) above, i.e. $^{2S+1}L_J$. Since $2S + 1 = 4$, the value of $S$ is 3/2.

From the values of $L$ and $S$, we obtain $J$. The values of $J$ range from $(L + S)$ to $(L - S)$. Therefore, as $L + S = \frac{9}{2}$ and $L - S = \frac{3}{2}$, the four possible multiplets are $J = 9/2, 7/2, 5/2$, and $3/2$. These substates are designated $^4$F$_{9/2}$, $^4$F$_{7/2}$, $^4$F$_{5/2}$, and $^4$F$_{3/2}$, respectively. Just for information, $^4F_{3/2}$ is one of the energy levels of the $Nd^{+3}$: yttrium aluminum garnet (YAG) laser ($\spadesuit$ see figure 6.6 in chapter 6).

### 2.9.6.2.2 Hyperfine splitting

Finally, the nucleus of the atom possesses an intrinsic angular momentum or nuclear spin, described by a quantum number $\vec{I}$. The magnitude of $\vec{I}$ is an integral or half-integral multiple of $\hbar$. The proton has a nuclear spin quantum number of $\vec{I} = \frac{1}{2}$. Like the combination of $\vec{L}$ and $\vec{S}$, $\vec{J}$ and $\vec{I}$ combine to give the total angular momentum $\vec{F}$ of the whole atom. $\vec{F}$ can take values from $J + I$ to $J - I$. Since the magnetic moment of the nucleus is smaller by three orders of magnitude than that of the electron, this split is much smaller than that of the spin–orbit coupling. It is known as hyperfine splitting, as it is observed only at very high spectral resolutions.

### 2.9.6.3 Atomic spectra: selection rules

Hund's rules of maximum multiplicity, which are important for the selection rules of atomic transitions, are as follows:
1. For a given electronic configuration, the lowest-energy term is the one which has the highest spin multiplicity.
2. For terms that have the same spin multiplicity, the term that has the highest orbital angular momentum has the lowest energy.
3. (i) If the unfilled subshell is exactly half-filled or more than half-filled, the level with the highest $J$ value has the lowest energy.
   (ii) If the unfilled subshell is less than half-filled, the level with the lowest $J$ value has the lowest energy.

Spectroscopic selection rules imply the 'allowed' transitions in any absorption or emission process. Spectroscopy qualitatively addresses the interaction between EM radiation and atoms or molecules. An atom or molecule can be considered to be an oscillating electric dipole that has static and dynamic dipole moments. When its mode of oscillation and an interacting electric field have the same frequency and phase, a transfer of energy to the chemical species (and thus a transition) can take place. It has been observed that changes in the dipole moment are responsible for the transitions. The transition of the dipole moment from an initial state $\psi_i$ to a final state $\psi_f$ is given by

$$\mu_x = \int \psi_f^*(x)\mu_x(x)\psi_i(x)dV' \tag{2.8}$$

where $\mu_x$ is the dipole moment in the direction of the electric field (in this case, along the $x$-axis). It can be shown that there is a nonzero probability of a transition from $\psi_i$ to $\psi_f$ when equation (2.8) is nonzero. Within the dipole approximations, the selection rules are determined using the transition dipole and the total energy eigenfunctions. The selection rules based on the dipole approximations are as follows:

$$\Delta l = \pm 1, \text{ and } \Delta L = 0, \pm 1, \Delta J = 0, \pm 1 \text{ and } \Delta S = 0$$

The transition rule for $\Delta l$ is the change in the orbital angular momentum of a single electron, while the other rules refer to the vector sum of all the electrons in an atom.

### 2.9.7 Molecular quantum states

It is obvious that we should expect much more complicated spectra for molecules. Let us look at the simple cases of diatomic molecules such as $H_2$, $N_2$, and CO. In the diatomic molecule, in addition to the angular momentum of the electron, two additional angular momenta are associated with vibration of the two nuclei along the molecular bond and the rotation of the molecule about an axis perpendicular to the molecular bond. The ability of a molecule to vibrate and rotate about an axis gives rise to additional states. Therefore, each ground or excited electronic state will possess vibrational as well as rotational levels.

For molecules with $N$ atoms, there are $3N$ degrees of freedom. There are three translational directions of motion and three rotations about the three axes. Therefore, the remaining number, $3N - 6$, involves vibrational motion. This number is $3N - 5$ for linear molecules, as they can have only two axes of rotation. Depending on the energy difference between the two states, EM radiation transitions can take place. The total energy of the system can be described as the sum of the electronic, vibrational, and rotational energies:

$$E_{\text{total}} = E_e + (E_v + E_r). \tag{2.9}$$

The separate energies corresponding to the motion of nuclei and that of electrons is given by Born–Oppenheimer approximation, which also includes the tiny amount of nuclear spin energy ($E_{\text{nuc}}$).

#### 2.9.7.1 Rotational energy

Following Herzberg's nomenclature [2], $I$ is introduced here as the moment of inertia of the molecule and $J$ as the rotational quantum number ($I$ is not to be confused with the nuclear spin, and $J$ is not to be confused with the total orbital electronic quantum number introduced in the previous section). The energy of a rigid rotator is quantized as

$E_r = hcB(J + 1)$, where $J = 0$, 1, 2, …. . $B$ is known as the rotational constant and is given by

$$B = \frac{h}{8\pi^2 cI}. \tag{2.10}$$

We know that molecules are not rigid rotors. In the classical picture, the distance between atoms expands with increasing rotational energy (figure 2.8) due to centrifugal force. Therefore, the energy is modified by introducing the centrifugal distortion constant, $D$, as follows:

$$E_r = hc[B(J + 1) - DJ^2(J + 1)^2]. \tag{2.11}$$

As a result of equation (2.11), for real molecules, as the rotational levels increase, their energy gaps continue to increase.

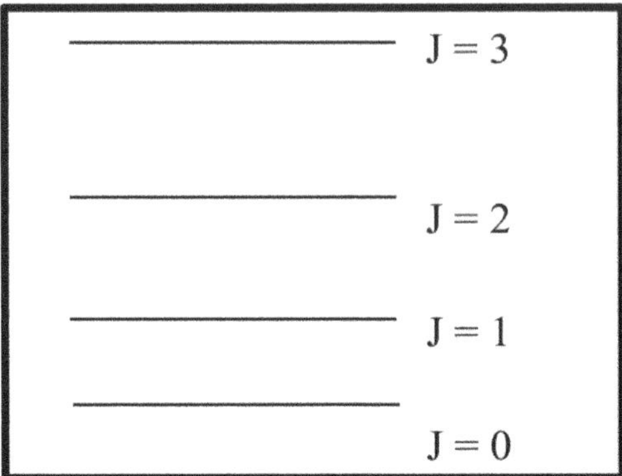



**Figure 2.8.** Schematic of the relative energies of the first four rotational levels of a molecule. The energy gap between sublevels increases due to the flexible nature of the molecule.

### 2.9.7.2 Vibrational energy

The vibration of a molecule is considered by treating it as a harmonic oscillator of frequency $\nu$. The allowed vibrational energies (figure 2.9) are given in terms of a vibrational quantum number $v$ (with $v = 0, 1, 2, 3...$) by

$$E_v = h\nu\left(v + \frac{1}{2}\right). \tag{2.12}$$

It is more accurate to describe a vibrating, real diatomic molecule as an anharmonic oscillator with higher terms in its energy, as follows:

$$E_v = hc\left[\omega_e\left(v + \frac{1}{2}\right) - \omega_e x_e\left(v + \frac{1}{2}\right)^2 + \omega_e y_e\left(v + \frac{1}{2}\right)^3 + ...\right], \tag{2.13}$$

where $\omega_e = \frac{\nu}{c}$ is expressed in $\mathrm{cm}^{-1}$, $\omega_e x_e$, $\omega_e y_e \ll \omega_e$, and $x_e$ and $y_e$ are anharmonicity constants. Equation (2.13) causes a reduction of the vibrational energy between the two subsequent vibrational levels, eventually reaching a continuum of states.

As for example, the vibrational modes of a linear molecule, $CO_2$, are shown in figure 2.10. It can be seen that there are three types of vibrational mode; these are known as the symmetric, antisymmetric (stretching), and bending modes. The bending mode is degenerate because the nuclei can move in any of the two perpendicular planes containing the molecular axis. Frequencies corresponding to these fundamental modes and their overtones or combined modes appear in the absorption or emission spectra.

### 2.9.7.3 Selection rules and the labeling of molecular energy states

A molecule can also vibrate while it is rotating. Therefore, appropriate corrections are used for the constants $B$ and $D$ in equation (2.11). The spacing between the rotational levels is smaller than that of the vibrational levels. The transitions occur from a rotational sublevel of one vibrational level to the rotational sublevel of another vibrational level. Such transitions are known as ro-vibrational transitions;

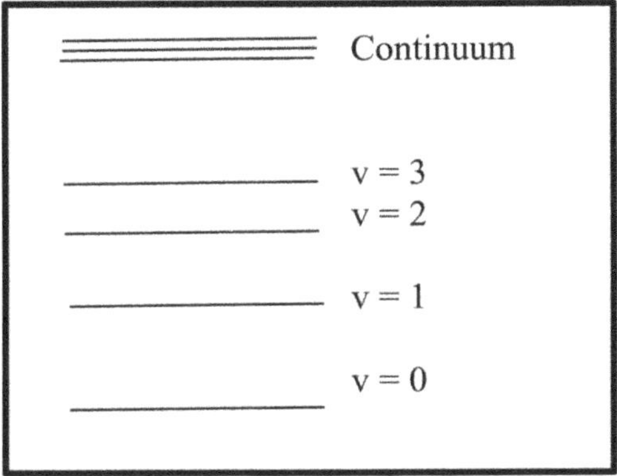



**Figure 2.9.** Relative energies of the first four vibrational levels. The energy gap between the vibrational levels is reduced with increasing energy due to anharmonicity.

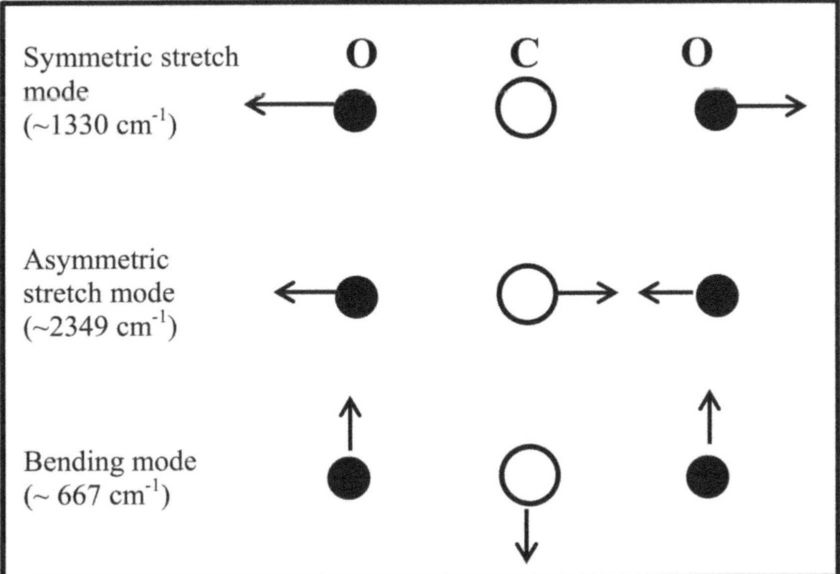



**Figure 2.10.** Vibrational modes of the $CO_2$ molecule. The arrows indicate the direction of movement of the atom (its return follows the same path in reverse). The corresponding vibrational frequencies are given in parentheses.

they follow the $J$-selection rules, and their branches are named as follows (apart from $\Delta v = \pm 1$):

the $\Delta J = 0$ $Q$ branch,

the $\Delta J = +1$ $R$ branch, and

the $\Delta J = -1$ $P$ branch.

In most diatomic molecules, the $Q$ branch is not present in the ro-vibrational spectra (the NO molecule being the exception). Similarly, in linear triatomic molecules, for example in $CO_2$, the two laser transitions (10.6 μm and 9.6 μm) correspond to the P

**Table 2.2.** Electronic energy-level nomenclature in molecules.

| Orbital angular momentum ($\Lambda$) | 0 | 1 | 2 | 3 |
|---|---|---|---|---|
| Electronic state | $\Sigma$ | $\Pi$ | $\Delta$ | $\Phi$ |

and R branches, respectively, while the Q branch is not observed. In homonuclear diatomic molecules (with no permanent dipole), such as $H_2$, $N_2$, $O_2$, and so on, none of the ro-vibrational transitions are observable (except in the Raman spectrum of the molecule). Transitions of $\Delta v = 2$ occur but are very rare.

In the multi-electron system of molecules, coupling of the rotational and electronic modes also takes place. The electronic energy levels were historically labeled using a similar approach to that used for atoms. Here, the letter $\Lambda$ is used for the component of orbital angular momentum along the molecular axis. The electronic states are denoted by Greek letters (such as $\Sigma$, $\Pi$ ....) in place of the English alphabet (S, P) as indicated in table 2.2.

For any value of $\Lambda$, the rotational quantum number $J$ can take the values $\Lambda$, $\Lambda + 1$, $\Lambda + 2$.... The spin determines the multiplicity of the electronic state, which is given by $2S+1$, as in the case of atoms. For $S = 0$ we have singlet states such as $^1\Sigma$, $^1\Pi$, ...., and for $S = 1/2$ we have doublet states such as $^2\Sigma$, $^2\Pi$, and so on. The lowest (ground-state) electronic manifold is given the name X, followed by the multiplicity as a superscript, and then the spectroscopic formula for the configuration, for example $^1\Sigma$. The excited states are labeled A, B, C, ...., usually in order of their identification. For example, if the ground state of the CO molecule is labeled $X^1\Sigma$, the first excited state can be designated $A^1\Pi$. The total angular momentum is labeled using either $g$ (*gerade*) for even or $u$ (*ungerade*) for odd. Thus, the states listed above might be expressed as $X^1\Sigma_g$ and $A^1\Pi_u$.

It should be mentioned that there are often exceptions to the selection rules or even the nomenclature of the levels. The identification of the transition levels, even in simple molecules, is a task that is continuously being carried out in both experimental and theoretical work. A molecular species can have transitions as a neutral molecule, an ion, or a dimer. Therefore, a large variety of rules for electronic transitions exist, which are outside the scope of this book. It is sufficient to say that there are transitions between different vibrational and rotational levels within an electronic state and the constants described above depend on the energy state.

**Exercise 2.6.** The energy difference between the two levels in a $CO_2$ laser is 0.117 eV. What is the wavelength of the laser output?

**Solution:** We again use the energy–wavelength relation, $E = hc/\lambda$. where

$$hc = 1.99 \times 10^{-25}\text{J m}.$$

The wavelength is given by $\lambda$ (nm) = hc/energy in joules

$$= \left(\frac{1.99 \times 10^{-25}}{0.117}\right)/1.6 \times 10^{-19}\text{nm} = 10.6\,\mu\text{m}.$$

### 2.9.7.4 Franck–Condon principle

Even though there are several electronic states for polyatomic molecules, a typical diagram representing two electronic states is given below. Figure 2.11 plots the energy as a function of the internuclear distance ($r$) of a molecule. In an electronic state, each vibrational level (say $v_0$ in figure 2.11) is associated with a set of rotational levels (drawn in between $v_0$ and $v_1$). Generally, in the condensed phase (solids, liquids), broader spectral lines (due to the overlap of broadened rotational levels) result in band spectra. However, at lower temperatures and in the isolated conditions of a supersonic free jet, well-resolved vibrational structure can be seen in the form of sharp line spectra (♠ see table 2.1). The Franck–Condon principle states that all electronic transitions must take place along a vertical line, so that the nuclear coordinates are unchanged (i.e. $r$ is a constant). The shape of the electronic state is an inverted bell-shaped curve with an energy minimum near the first vibrational level, $v_0$.

For example, in figure 2.11, the absorption of a photon of frequency $\nu_a$ takes the molecule from $v_0$ (in $S_0$) to the higher vibrational states (a quasi-continuum of states in this case) in $S_1$. At room temperature, due to thermal distribution, there is a process of vibrational relaxation to the lowest vibrational level ($v_0'$) of $S_1$. The emission transition shown from $v'_0$ vertically down to $v_1$ is the most probable path, according to the Franck–Condon principle. The difference between the absorption ($\nu_a$) and emission frequencies ($\nu_e$) is known as the Stokes shift (♠ see section 2.9.7.7). This shift arises because excitation energy is lost in the relaxation processes before emission takes place. It has immense consequences in laser physics and spectroscopy for all gas, liquid, and solid-state lasers.

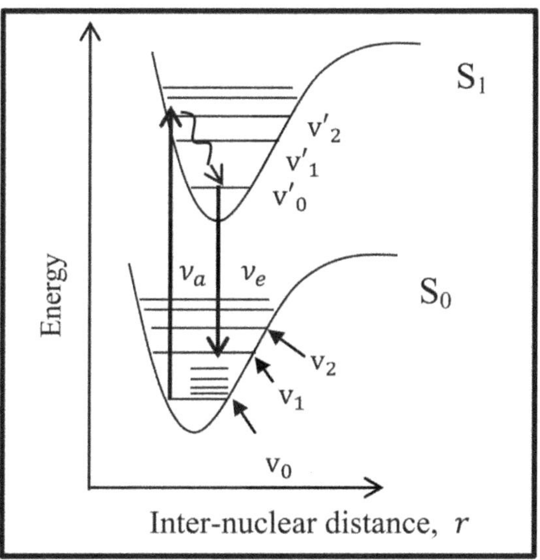


**Figure 2.11.** Electronic states ($S_0$ and $S_1$) with vibrational states ($v_0$, $v_1$, $v_2$, and their primes, respectively) and rotational states (horizontal lines between $v_0$ and $v_1$). The vertical upward- and downward-pointing arrows indicate the Franck–Condon transitions.

### 2.9.7.5 Excimer/exciplex state

Let M denote an atom which is a monomer. Due to the van der Waals force, two atoms can come together in the ground state to form a new molecule, known as a dimer (MM). Alternatively, an excited atom (M*) can interact with an atom in the ground state, and the newly formed species is known as an *excimer* (M*M). When an excited atom (M*) interacts with a different atom in the ground state (X), the resulting species is known as an *exciplex* (M*X). Alkali halide lasers such as Xe*Cl and Xe*F lasers are known as excimer lasers, which is a misnomer (as per the definition just given, they are 'exciplex' lasers). In these lasers, the laser transition takes place to the repulsive ground state, as the exciplex is unstable in the ground state.

### 2.9.7.6 Jablonskii diagram for dye molecules

Let us consider the energy-level diagram of an organic molecule (known as a Jablonskii diagram) shown in figure 2.12. In the case of polyatomic molecules with

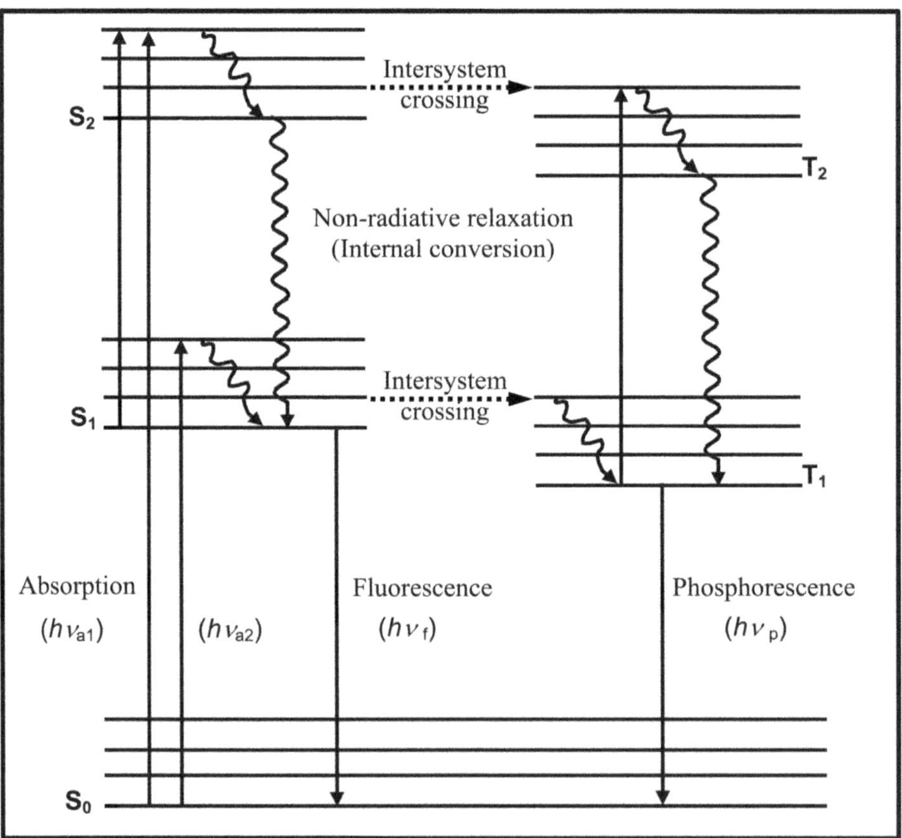


**Figure 2.12.** Jablonskii diagram for a polyatomic molecule indicating various processes; $h\nu_{a1}$ and $h\nu_{a2}$ represent the absorbed photon energies between $S_0$ and the higher singlet states $S_1$, $S_2$, ..., while $h\nu_f$ and $h\nu_p$ represent the fluorescence ($S_1 \to S_0$) and phosphorescence ($T_1 \to S_0$) photon energies, respectively. Internal conversion (IC) is a non-radiative transition within the same type of electronic state, while intersystem crossing (ISC) refers to non-radiative transition from a singlet to a triplet state, such as from $S_1 \to T_1$.

an even number of electrons, the electron spins (up and down) are all paired and the ground singlet state 'manifold' is designated $S_0$. When one of the electrons in the highest filled orbital is excited to the lowest vacant orbital, its spin orientation can remain unchanged and the resulting excited state is designated $S_1$. If the electron excited to the vacant orbital reverses its spin, such that it becomes parallel to that of its partner, the resulting excited state has a total spin of $S = 1$ and it is designated $T_1$. Thus, the excitation of an electron from the highest filled orbital to the lowest vacant orbital gives rise to a singlet state $S_1$ and a triplet state $T_1$, of which the triplet has the lower energy. If the electron from the highest filled orbital is excited to the next vacant orbital, the resulting singlet and triplet states are designated $S_2$ and $T_2$, respectively. Thus, increasing subscripts represent higher degrees of electronic excitation.

In figure 2.12, the singlet states are depicted by $S_0$, $S_1$, $S_2$, ..., and the triplet states are indicated by $T_1$, $T_2$, ... . Due to Pauli's exclusion principle, since two electrons of identical spin cannot occupy the same state, the first triplet state is $T_1$. The electronic levels are widely distributed in energy due to the vibrational and rotational states of the molecules. Only the lowest vibrational and rotational states of the molecule are populated at room temperature; they follow the Boltzmann distribution ($\spadesuit$ see chapter 4). The absorption of photons with energy $h\nu_{a1}$ and $h\nu_{a2}$ results in an increase in the population of the excited state. The transitions between states are depicted as vertical lines to illustrate the instantaneous nature of light absorption or emission. In the visible region, optical transitions occur instantaneously within a cycle of the incident EM field, i.e. in about $10^{-15}$ s. This time scale is too short for significant displacement of the nuclei. Depending on the energy of the incident photon, a molecule (also known as a chromophore or a fluorophore) is usually excited to some higher electronic level. With a few exceptions, condensed-phase molecules rapidly relax to the lowest level of the excited $S_1$ state by IC through vibrational relaxation. This process generally occurs in $\sim 10^{-12}$ s (1 ps) or less. The $S_1$ state decays with a fluorescence lifetime that is typically of the order of 1–10 ns.

Molecules in the $S_1$ state can also undergo a spin conversion to the first triplet state, $T_1$. Emission ($h\nu_p$) from $T_1$ is known as phosphorescence and is generally shifted to longer wavelengths relative to fluorescence ($h\nu_f$). The conversion of $S_1$ to $T_1$ is called intersystem crossing (ISC). Transition from $T_1$ to the singlet ground state ($S_0$) is forbidden. Therefore, rate constants for triplet emission are several orders of magnitude slower (of the order of *ms* to *s*) than those for fluorescence.

$\spadesuit$ An undergraduate experiment based on measuring the lifetime with a *wrist-watch* can be performed using a phosphorescent sample excited with ordinary light sources available in the visible region. A fluorescent dye, such as fluorescein in boric acid, forms a glass when heated to about 200 °C and becomes a good phosphorescent material with a lifetime of several seconds.

### 2.9.7.7 *Stokes shift*

For a material such as an organic dye dissolved in a solvent, the absorption and photoluminescence (PL) spectra are schematically shown in figure 2.13. The Stokes shift is defined as the difference between the peak frequencies of the absorption and

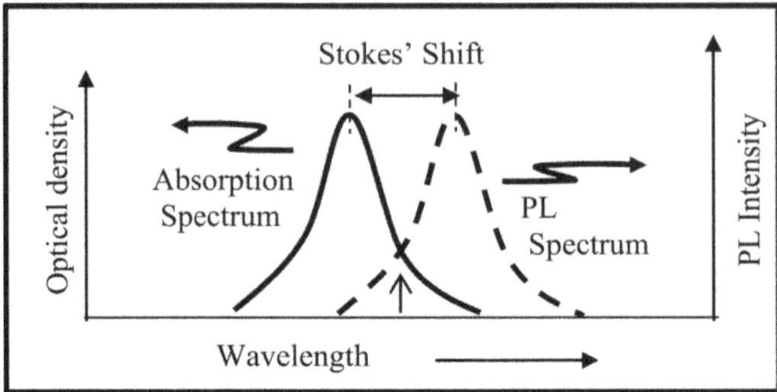


**Figure 2.13.** Schematic diagram of the Stokes shift in a typical dye solution. The vertical arrow at the intersection point of the two spectra corresponds to the (0,0) transition.

the PL spectra. This 'red shift' of the PL spectrum arises because of the loss of excitation energy due to various non-radiative processes shown in the Jablonskii diagram (figure 2.12). As a convention, the transition associated with the lowest vibrational level of the ground state to that of the excited state is known as the (0,0) transition. In the condensed phase, this corresponds to the intersection point between the emission spectrum and the absorption spectrum.

Organic dyes are complex molecules and are useful as lasing materials in the near-UV, visible, and near-IR regions. Dyes can be used as economical lasers that offer the particular property of easy tunability over a wide range of wavelengths (♠ see chapter 24) and are also used for mode-locking applications (♠ see chapter 17). The dyes are bleached very fast in continuous operation and need to be replaced quite often. However, dye lasers have made a significant contribution to the development of modern ultrafast laser spectroscopy since the 1980s. The first femtosecond laser pulse was generated by a dye laser.

**Exercise 2.7.** What do you understand by the Stokes shift in the condensed phase? Can we observe the Stokes shift in the emission spectra of cooled molecules (say, in 'supersonic jet expansion')?

**Solution:** As indicated in figure 2.13, the Stokes shift is the difference between the peak frequencies (or wavelengths) of the absorption spectrum and the emission spectrum of a molecule. At low temperatures (or in supersonic jet expansion), non-radiative processes are drastically reduced, as the molecules do not have sufficient energy for vibration or rotation. Therefore, there is no Stokes shift in the the resulting emission spectra. The (0,0) transition in this case corresponds to the highest-wavelength (lowest-energy) excitation. However, in the presence of any excited-state reaction, such as an excited-state energy transfer, this shift can still be observed.

*2.9.7.8 Quantum yield of photoluminescence*
An excited molecule can relax back to the ground state by emitting photons with a rate constant of $k_r$ or (*non-radiatively*) through collisions with a rate constant of $k_{nr}$.

**Table 2.3.** Quantum yield and lifetime of standard samples.

| System | QY | Lifetime |
| --- | --- | --- |
| R6G in ethanol | 0.94 | 3.5 ns |
| Rose Bengal in methanol | 0.11 | 534 ps |

The quantum yield ($QY$) of fluorescence (or photoluminescence) is defined as the ratio of the number of photons (or quanta) emitted to the number absorbed by a fluorophore[1]

$$QY = \frac{\text{number of quanta emitted}}{\text{number of quanta absorbed}}$$

Since both the rate constants $k_r$ and $k_{nr}$ depopulate the excited state, $QY$ can also be calculated as the fraction of the fluorophores that decay by emitting photons, as follows:

$$QY = \frac{k_r}{k_r + k_{nr}}.$$

If the non-radiative decay rate is much slower than the radiative decay rate, the value of $QY$ can be close to unity. In terms of decay rates, the observed fluorescence lifetime ($\tau_0$) can be written as

$$\tau_0 = \frac{1}{k_r + k_{nr}}. \tag{2.14}$$

In the absence of non-radiative processes ($k_{nr}$), the intrinsic lifetime ($\tau_r$) is given by $1/k_r$. Then the $QY$ is expressed as the ratio of the observed fluorescence lifetime to the radiative lifetime as $\tau_0/\tau_r$. Table 2.3 gives the $QY$ values and the lifetimes for two standard fluorophores.

### 2.9.8 Energy–momentum diagrams in semiconductors

The energy ($E$)–momentum ($\hbar k$) diagram for a semiconductor is shown in figure 2.14. This follows the wave vector ($k$)-selection rule (or vector ($\vec{k}$) = 0 rule). The consequence of this rule is that only *direct bandgap* semiconductors (such as GaAs) have radiative transitions and can therefore be used as laser media. On the other hand, Si and Ge, which have *indirect bandgaps* are not useful lasing materials[2]. In *indirect bandgap* semiconductors, a phonon-assisted transition results in the emission of a photon from the conduction band (CB) to the valence band (VB).

For a free electron, the relation between $E$ and $k$ can be approximated as $E_k = \frac{\hbar^2 k^2}{2m_e}$ (♠ which is similar to equation 2.5) and is a parabola, as shown in

---

[1] A fluorophore is a molecule that fluoresces when it absorbs photons.
[2] In recent years, porous silicon and boron-doped silicon have been found to be luminescent.

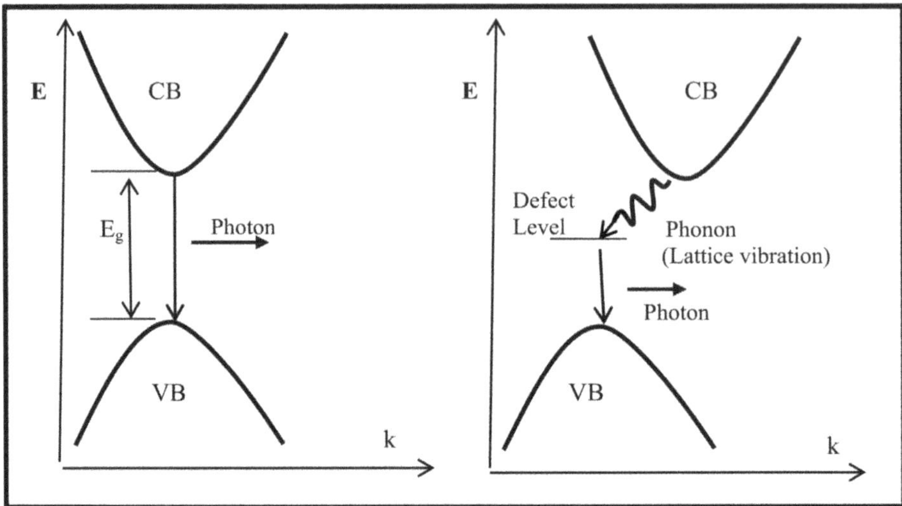



**Figure 2.14.** *E–k* diagram and transitions in direct (left) and indirect bandgap (right) semiconductors. VB and CB denote the valence and conduction bands, respectively.

figure 2.14. Here, $m_e$ denotes the mass of the electron in the CB. In the VB, electrons cannot move freely. Due to a majority of holes, the curve in the VB is a negative parabola. Holes behave as if they have positive charge but negative mass; see chapter 22 for more details.

### 2.9.9 Energy levels in solids: doped insulators

In solid-state lasers (such as ruby or $Nd^{+3}$: YAG), the active medium consists of impurity ions (guests) embedded within a solid (host) lattice. Here, the impurities replace the host ions in the crystal; therefore, the energy levels involved in laser transitions are those of the embedded ion. However, the host lattice is equally important, as its physical properties (such as thermal conductivity) are important in determining the power levels at which the laser can be operated. In addition, the host lattice modifies the impurity ion energy levels as follows: there are basically three main interactions to be considered, namely (i) the *Coulomb force* acting between electrons, (ii) the *crystal field* interaction, and (iii) *LS* coupling. Due to these interactions, the split of the energy levels required for lasing varies according to the selected ion.

As mentioned in section 2.9.6.1, the *transition-metal* ions (see chart 2.2) and the *rare-earth* ions are the two most important types of ion. In transition-metal ions (e.g. $Ti^{+3}$, $Cr^{+3}$) the low-lying energy levels of interest are due to the presence of electrons in the unfilled 3d subshells. For example, the unfilled subshell in the electronic configuration of the transition metal $Cr^{+3}$ contains $3d^3$. Here, we will have crystal field interaction, as the 3d electrons are not shielded. Remember that the ion is situated in the lattice and is surrounded by the *host*, with which it will *interact*. This is known as *crystal field* interaction. Rare-earth ions, on the other hand, are characterized by the presence of electrons in the unfilled 4f subshell. The electronic configuration of rare-earth ions (for example, $Nd^{+3}$) has the configuration

$$1s^2 2s^2 2p^6 3s^2 3p^6 3d^{10} 4s^2 4p^6 4d^{10} 4f^3 5s^2 \quad 5p^6.$$

In rare-earth ions, the 4f subshell is physically well shielded by other electrons in the outer orbits. The split of the energy levels is decided by these interactions in each guest–host system.

## 2.10 Raman spectroscopy

Raman scattering is inelastic scattering, first reported by Raman and Krishnan in 1928. When a photon is incident on the molecule, two simultaneous processes of absorption and emission define the resultant scattering. The incident photon takes the molecule to a virtual excited state with a lifetime of $10^{-17}$ s, which emits a scattered photon. This is also known as a two-photon event, as both the incident and scattered photons exist simultaneously. If the molecule returns to the original vibrational level at the end of the event, the scattering process is called elastic or Rayleigh scattering. However, the return of the molecule to a different vibrational level at the end of the two-photon event is called inelastic scattering. If the final vibrational level of the molecule has a higher energy than the initial level, the Raman scattering is called Stokes Raman, and in the reverse case it is called anti-Stokes Raman.

In other words, Raman scattering involves a shift in frequency from that of the incident photon, $\omega_\nu$, to that of the scattered photon. The observed shift in frequency is equal to that of the Raman active vibrations of the molecule. When the frequencies of the generated radiation are lower than the frequency of the incident photon, the corresponding new lines in the spectra are known as Stokes ($\omega_s$) lines, and when the generated frequencies are higher, the corresponding new lines are called anti-Stokes ($\omega_{as}$) lines (♠ see chapter 16 for details).

♣The first experiments on the Raman effect were performed with the help of a *prism* spectrograph and sunlight. Spectrographs with *gratings* provide linear dispersion. Can you differentiate the spectra obtained by the two (♠ see question 2)?

## Questions and problems

1. (i) What is the ratio of the energy of microwaves and x-ray photons? (ii) Microwave ovens are used to cook food, while x-ray machines are used for the diagnosis of disease in the human body. Why it this so? (iii) Why do we use x-rays to determine crystal structure? Could we also use visible light for this task?
2. What are the differences in the spectrum of an incandescent bulb and that of a mercury tube light as seen through (i) a prism spectrograph (ii) a grating spectrograph?
3. Why can a triplet not have the ground state?
4. What is the ground state of the hydrogen atom in Bohr's model?

5. What is the energy of the photons absorbed in the Franck–Hertz experiment? How does this experiment support Bohr's model of discrete stationary orbitals?

6. You are given a light source and two polarizers. Can you make an observation similar to that of the Stern–Gerlach experiment using light (♠ see chapter 3)?

7. What is a Compton wavelength? Can we observe the Compton effect at a wavelength of 1064 nm? why?

8. With reference to molecular orbitals, describe the roles of the highest occupied molecular orbital (HOMO) and lowest unoccupied molecular orbital (LUMO) in spectroscopy.

9. The transition denoted by the Lyman series, as shown in figure 2.3, also contain transitions from upper states to the $n = 1$ state. Similarly, transitions occur for other states as well (hence, these are known as 'series'). Draw a new diagram with the transitions for all series, including the Pfund series, for which the lower state is $n = 5$.

10. The Balmer series corresponds to transitions in which the lower state is $n = 2$. Find the energy corresponding to the lowest-energy and highest energy photons emitted.

11. Consider a 1D finite potential well of depth $V$ and width $2a$, as shown in figure 2.7. Solve the Schrödinger equation in one dimension to obtain the energy states.

12. Find the de Broglie wavelength corresponding to a person with a mass of 100 kg moving at a velocity of $1000\ \mathrm{m\ s^{-1}}$ and for an electron moving with a velocity of $c/1000$. Comment.

13. Explain the transitions denoted by $^1D_2$ and $^4P$ in terms of spin, orbital, and total angular momentum quantum numbers.

14. With reference to the excimer laser, define the terms monomer, dimer, and excimer.

15. Describe the Franck–Condon principle in molecular spectroscopy.

16. What are direct and indirect band semiconductors? Give an example of each.

17. If the radiative rate and non-radiative rates are equal and and have a value of $1 \times 10^9\ \mathrm{s^{-1}}$, what is the quantum yield of the system? What is the observed lifetime?

18. Will the *Coulomb force* be dominant in the case of transition-metal ions? How about the *crystal field interaction* in the case of rare-earth ions?

19. Comment on the following: $Nd^{+3}$ ions used as a dopant in glass will have a different laser wavelength than those used as a dopant in YAG.

## Bibliography

[1] Herzberg G and Spinks J W T 1944 *Atomic Spectra and Atomic Structure.* (New York: Dover)

[2] Herzberg G 1950 *Molecular Spectra and Molecular Structure: Spectra of Diatomic Molecules* (Princeton, NJ: Van Nostrand)

[3] Wilson J and Hawkes J F B 1989 *Optoelectronics: An Introduction* 2nd edn (Upper Saddle River, NJ: Prentice-Hall)

[4] Benwell 1994 *Fundamentals of Molecular and Spectroscopy* (New Delhi: McGraw-Hill Education (India) Pvt Limited)

[5] Lakowicz J R 2013 *Principles of Fluorescence Spectroscopy* (New York: Springer US)

[6] Raman C V and Krishnan K S 1928 A new type of secondary radiation *Nature* **121** 501

[7] Serway R A 1986 *Physics for Scientists and Engineers* 3rd edn (Philadelphia, PA: Saunders College Publishing)

[8] Bisht P B, Petek H, Yoshihara K and Nagashima U 1995 Excited state enol-keto tautomerization in salicylic acid: a supersonic free jet study *J. Chem. Phys.* **103** 5290–307

[9] Sailaja R and Bisht P B 2007 Tunable multiline distributed feedback dye laser based on the phenomenon of excitation energy transfer *Org. Electron.* **8** 175–83

**IOP** Publishing

# An Introduction to Photonics and Laser Physics with Applications

**Prem B Bisht**

# Chapter 3

## Polarization of light

Nature has several phenomena related to the polarization of light. This is due to the fact that the amount of light reflected at the boundary between two surfaces depends on the polarization of the incident beam. For example, it is possible to have no reflected light for a particular combination of incident angle and surface refractive index. Light scattered by matter is generally polarization sensitive. Different polarizations travel at different velocities and undergo different phase shifts. Some materials have the ability to rotate the plane of linearly polarized light. Other materials can rotate the polarization in the presence of external electric and magnetic fields. The polarization of light is the key to the photonics industry. Various aspects of polarization that are useful in other chapters are described in this chapter. We also discover that several materials have more than one refractive index ($n$) at a single wavelength. For example, the diagram shows two values, $n_o$ and $n_e$, away from the optical axis. One of them ($n_e$) has an angular ($\theta$) dependence. The details are given in this chapter.

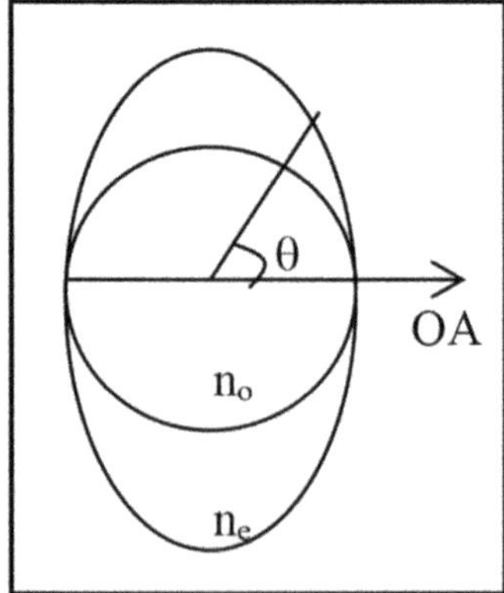

**Learning objectives**
**After reading this chapter, the learner will be able to:**
Identify the types of polarization;
Understand the Jones vector and matrix representation;
Describe methods of generating polarized light;
Explain dichroism, birefringence, and total internal reflection;
Describe Fresnel's equations and Brewster's angle;
Understand the DC Pockels effect, the Kerr effect, and the magneto-optic effect (Faraday effect).

# 3.1 EM waves and linearly polarized light

Figure 3.1 shows a string attached to a wall. If a transverse wave is generated in the string, it travels down the string as shown. A slit has been placed in its path in such a way that the wave passes through. If another slit oriented at 90 degrees to the first slit is added to the path of the wave, the wave does not pass through the second slit. This demonstration gives an idea of the transverse oscillation of a wave, which is similar to that of an EM wave. In the case of light, the transverse oscillations are known as its polarization states.

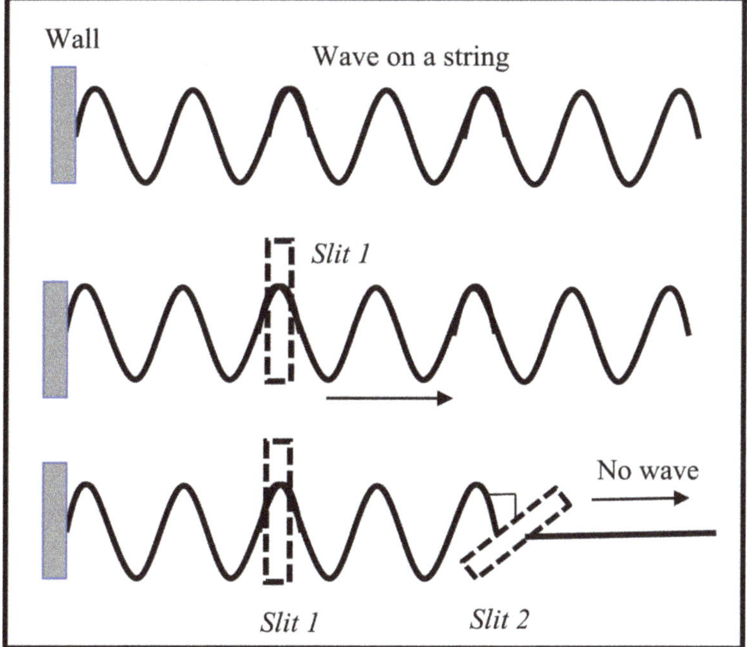


**Figure 3.1.** The transverse wave of a string attached to a wall (top curve) passes through *Slit* 1 with amplitude oscillations parallel to the slit (middle curve). If *Slit* 2 is introduced perpendicular to *Slit* 1, the wave cannot pass through (bottom).

As discussed in section 1.3, light is composed of mutually perpendicular electric ($E$) and magnetic ($B$) waves, as shown in figure 3.2. Let us consider the electric field given by equation (1.3), $\vec{E}(r, t) = |E_0|\, \hat{n}\cos(\vec{k}.\,\vec{r} - \omega t + \delta_1)$. Here, $\hat{n}$ is the direction of oscillation of the electric field vector, while the direction of the wave is given by $\vec{k}$. The corresponding magnetic field vector ($B(r,t)$) is given by $\vec{B}(r, t) = |B_0|\, \hat{n}'\cos(\vec{k}.\,\vec{r} - \omega t + \delta_1)$ with $\hat{n}' \times \vec{k} = \hat{n}$. Here, $\hat{n}'$ is the oscillation direction of the magnetic field.

The wave in figure 3.2 is said to be linearly polarized or plane polarized. For a plane polarized light wave, at any given point, the tip of the field vector oscillates along a line. As described in chapter 1, the magnitude of $E$ is stronger by a factor of $c$ (the speed of light) than $B$ ($|B| = |E|/c$). Therefore, in the interaction of light with matter, the strongest term is usually the electric dipole interaction. This leads to the fact that in most discussions, the polarization of $E$ is taken to be the polarization of the EM wave.

Light from incandescent bulbs, light-emitting diodes, arc lamps, the Sun, and many other sources is generally randomly polarized or unpolarized. This means that the $E$ vectors associated with the waves of such sources have no preferred direction of orientation, i.e. they change rapidly with time. On the other hand, most lasers produce light with highly oriented electric fields and fall into the category of polarized light sources.

Consider the wave in figure 3.2, given by $\vec{E}(y, t) = a\cos(ky - \omega t + \delta_1)\hat{z}$, directly reaching the eye of an observer. Figure 3.3 shows the behavior of the tip of $E$ as a function of time when polarized light propagates in a medium. Initially, the observer

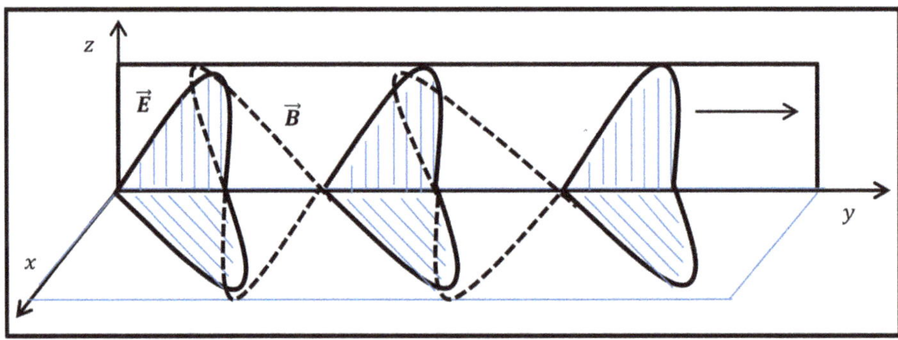

**Figure 3.2.** An EM wave traveling in the $y$ direction. The direction of oscillation of $E$ (solid line) is along $z$, while that of $B$ (dashed line) is along $x$.

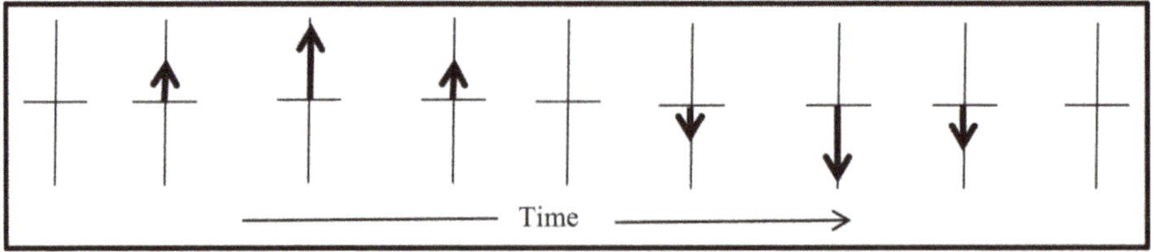



**Figure 3.3.** The tip of the electric field vector of figure 3.2 as seen by the observer.

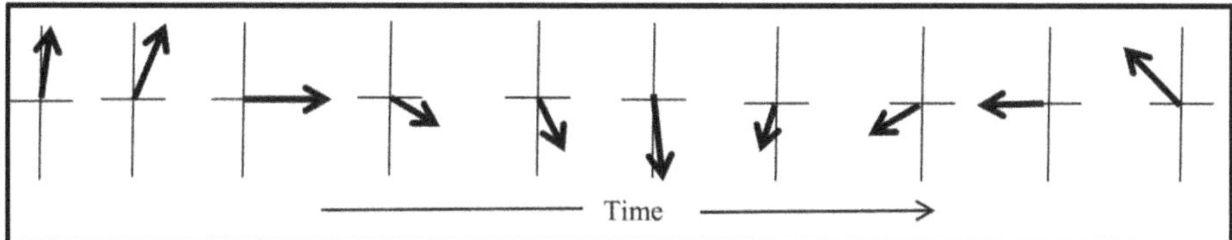


**Figure 3.4.** The tip of $E$ vectors for the unpolarized light as seen by the observer.

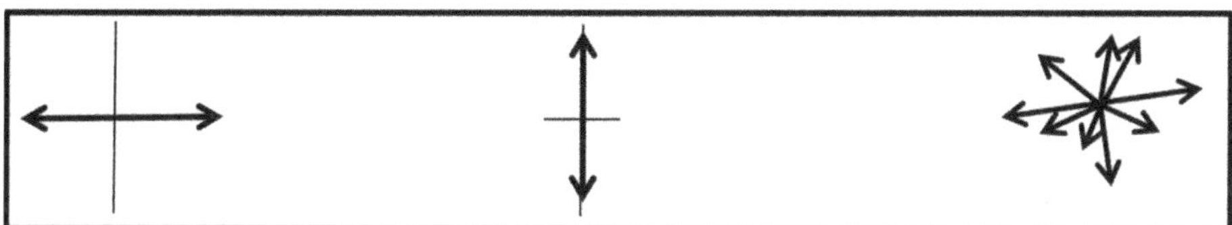

**Figure 3.5.** Representation of the orientation of $E$ for horizontally polarized (left), vertically polarized (middle), and unpolarized (right) light.

sees no electric field. However, as time passes, the observer will see the electric field vector grow towards the $+z$ direction and then diminish, grow in the $-z$ direction and then diminish, and so on.

The oscillations of unpolarized light are seen as shown in figure 3.4. Based on this understanding, the oscillations of the $E$ vectors for horizontally polarized, vertically polarized, and randomly polarized light are shown in figure 3.5.

## 3.2 Types of polarization

### 3.2.1 Ordinary and extraordinary light beams

In the interaction of polarized light with matter, we initially consider homogeneous media. These can be divided in two main categories *viz.* isotropic and anisotropic. For example, a crystalline material is anisotropic if the properties of the material (specifically, its refractive index) vary as a function of its crystallographic orientation. On the other hand, polycrystalline materials are isotropic because, although a single grain of the material may be anisotropic, the net properties of a large collection of grains average out the anisotropy. Isotropic media have the same refractive index in all directions. Therefore, there is no change in polarization as beams propagate through it.

In anisotropic crystals, the light can propagate at different speeds in different directions. Let us introduce the permittivity tensor $\epsilon_{ij}$, where $i, j$ represent Cartesian coordinates. We know that the displacement vector $(D)$ of a dielectric can be expressed in index form in terms of the electric field $(E)$ as $D_i = \sum_j \epsilon_{ij} E_j$. However, as a result of the symmetry properties of the crystalline material, we can choose an appropriate coordinate system. This permits conditions in which the off-diagonal elements of the $\epsilon_{ij}$ tensor can be made to vanish, so that only the three components $\epsilon_{11}$, $\epsilon_{22}$, and $\epsilon_{33}$ remain to be considered. Accordingly, we have three direction-dependent refractive indices, as follows:

$$n_1 = \sqrt{\frac{\epsilon_{11}}{\epsilon_0}}, \; n_2 = \sqrt{\frac{\epsilon_{22}}{\epsilon_0}}, \; \text{and } n_3 = \sqrt{\frac{\epsilon_{33}}{\epsilon_0}}.$$

Here, $\epsilon_0$ is the permittivity of free space. Crystals are known as *'uniaxial'* if the crystal symmetry permits $n_1 = n_2$. In such a case, we can introduce an *'ordinary refractive index'* $n_0$ ($= n_1 = n_2$) and an *'extraordinary* refractive index' $n_e(= n_3)$. Such crystals (i.e. those with $n_0$ and $n_e$) are known as birefringent materials. This fact is used to generate polarized light for various applications, as discussed later in this chapter (sections 3.6 and 3.7). In the case of birefringent crystals, horizontally polarized light traveling along the optic axis of the crystal is often known as ordinary light (an *o-ray*) and the vertically polarized light is known as the extraordinary light beam (or *e-ray*). Isotropic media can also be made anisotropic by applying an external field.

### 3.2.2 Classification of polarized light

In general, three types of light polarization are encountered in optics. These are: linearly polarized, elliptically polarized, and circularly polarized light. Linearly polarized light can have two further mutually perpendicular types of light known as plane polarized (*p*-polarized) and vertically polarized (*s*-polarized).

♣ In German, 'senkrecht' means perpendicular, hence the letter *s* was historically used for vertically polarized light.

In order to define these polarizations, we need to know the plane of incidence, i.e. the plane on which the incident, transmitted, and reflected lights as well as the normal are drawn (♠ see section 3.4.3). The field oscillations of *p*-polarized light are in the plane of incidence. In this case, the magnetic field vector oscillates in the perpendicular direction; hence, it is also known as transverse magnetic (TM) polarization. Similarly, *s*-polarized light is also known as a transverse electric wave. As with any vector, $\vec{E}$ can be split into its $x$ and $y$ components along orthogonal axes, as shown in figure 3.6.

*Linearly polarized* light can be described by the maximum amplitude of the field and the angle of the polarization vector relative to the axis. Therefore, any light of

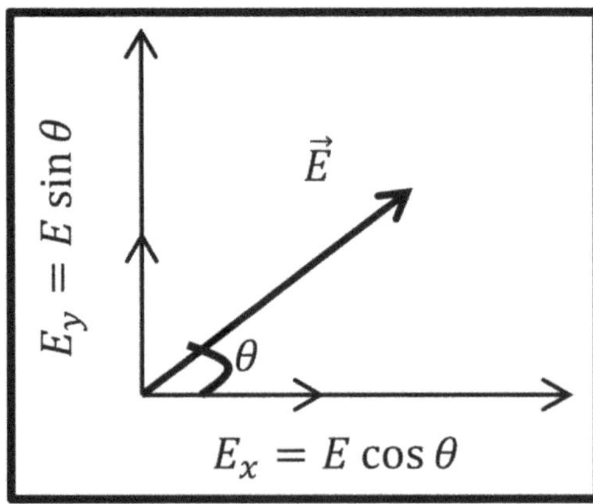


**Figure 3.6.** The $x$ and $y$ components of an electric field $\vec{E}$ that makes an angle $\theta$ with the $x$-axis.

frequency $\omega$ can be a combination of o-rays and e-rays. Let $a$ and $b$ respectively be the components along the $x$-axis ($E\cos\theta$) and the $y$-axis ($E\sin\theta$), and let $\phi$ be the phase difference between the two.

The e-ray can be expressed as

$$x = a \sin(\omega t + \phi), \tag{3.1}$$

while the o-ray is given by

$$y = b \sin \omega t. \tag{3.2}$$

Using trigonometric relations, we can rewrite equation (3.1) as

$$\frac{x}{a} = \sin \omega t \, \cos\phi + \cos \omega t \, \sin\phi.$$

By assigning values to $\sin\omega t$ and $\cos\omega t$ and rearranging, we can write this as

$$\left(\frac{x}{a}\right)^2 + \left(\frac{y}{b}\right)^2 - \frac{2xy}{ab}\cos\phi = \sin^2\phi. \tag{3.3}$$

This is the general equation for an ellipse, where $a$ and $b$ are the major and minor axes. From equation (3.3), three states of polarization (SoPs) and their variants can immediately be obtained.

**Exercise 3.1.** What are the various polarization states of light that can be obtained using equation (3.3) for the following values of the phase difference between its two components?

(i) $\phi = 0$, (ii) $\phi = \pi/2$ and $3\pi/2$, and (iii) $\phi = \pi$.

**Solution:** *Case (i).* For $\phi = 0$, equation (3.3) becomes $\left(\frac{x}{a}\right)^2 + \left(\frac{y}{b}\right)^2 - \frac{2xy}{ab} = 0$ or $\left(\frac{x}{a} - \frac{y}{b}\right)^2 = 0$. This is the equation for a straight line according to $y = \frac{b}{a}x$. This is shown in panel A of figure 3.7 and is known as *linearly polarized* light.

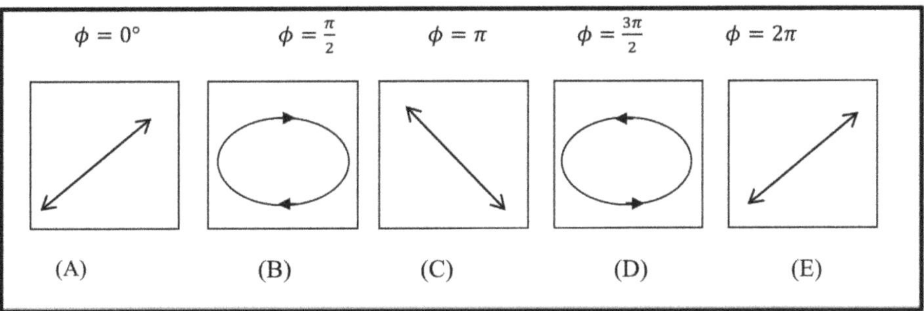


**Figure 3.7.** Various SoPs can be obtained using the value of the phase difference $\phi$ between the two components of light. (A), (C) and (E): plane polarized, (B) and (D): elliptically polarized.

***Case (ii).*** For $\phi = \frac{\pi}{2}$, equation (3.3) becomes $\left(\frac{x}{a}\right)^2 + \left(\frac{y}{b}\right)^2 = 1$. This is the equation for a symmetric ellipse, as shown in panel **B** of figure 3.7. As the wave advances in time, the projection of the electric field vector onto the $x$–$y$ plane traces out an ellipse over time. The vector rotates clockwise in time. This is known as *right elliptically* polarized light. For $\phi = \frac{3\pi}{2}$, this becomes *left elliptically* polarized light, as the projection of the electric vector rotates anticlockwise over time. Figure 3.7 shows the polarizations of the beam for various values of $\phi$.

***Case (iii).*** In case (ii) above, for $a = b\ (= a,\ say)$, we get $x^2 + y^2 = a^2$, the equation for a circle. For equal amplitudes, elliptically polarized light becomes *right circularly* polarized (RCP) and *left circularly* polarized (LCP) light, as shown in figure 3.8.

♣ As a general case, depending on the value of $\phi$, we can obtain various SoPs of light rotated by a fixed amount. This property is used in making wave retarders (half-wave plates and quarter-wave plates) for a given wavelength (♠ section 3.9). This is achieved using the birefringent property of crystalline materials (♠ see sections 3.6 and 3.7).

**Exercise 3.2.** Two linearly polarized plane EM waves (namely wave 1 and wave 2) are propagating in the $z$ direction, and their planes of polarization lie in the $x$ and $y$ directions, respectively. The electric fields of the two waves have equal amplitudes, given by $|\,E_0\,|$. The frequency of each wave is $\omega$, and its wave number is $k$. If the phases of these waves are given by $\phi_1$ and $\phi_2 (= \phi_1 + \pi/2)$, respectively, identify the SoPs of the individual electric fields and their superposition.

**Solution:** For wave 1, the electric field vector is $\vec{E_1} = |\,E_0\,|\,\hat{e}_x \cos(kz - \omega t + \phi_1)$. Similarly, for wave 2, the electric field vector is given by $\vec{E_2} = |\,E_0\,|\,\hat{e}_y \cos(kz - \omega t + \phi_2)$. If $\phi_2 = \phi_1 + \frac{\pi}{2}$, the superposition of the two waves is written as

$$\vec{E} = |\,E_0\,| \left( \hat{e}_x \cos\left(kz - \omega t + \phi_1\right) + \hat{e}_y \cos\left(kz - \omega t + \left(\frac{\pi}{2} + \phi_1\right)\right) \right)$$

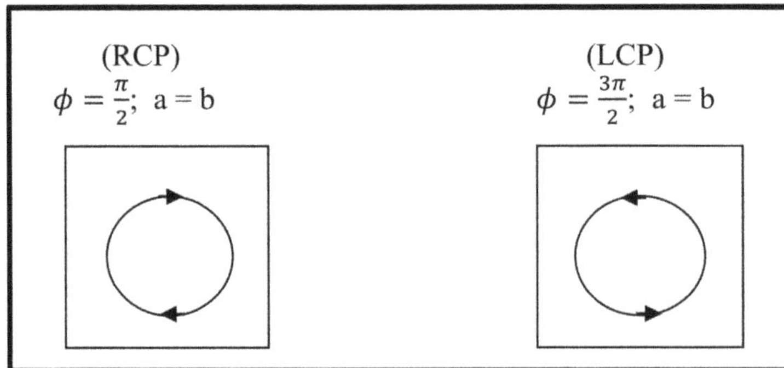


**Figure 3.8.** *Right circularly* polarized (RCP) and *left circularly* polarized (LCP) lights have equal amplitudes but different values of $\phi$ as indicated.

Or

$$\vec{E} = \mid E_0 \mid \left[ \hat{e}_x \cos \left(kz - \omega t + \phi_1\right) - \hat{e}_y \sin \left(kz - \omega t + \phi_1\right)\right].$$

The tip of the resultant amplitude $E$ that propagates over time in the forward direction from the source rotates in the clockwise direction. However, when viewed looking into the beam, it will appear that the opposite is the case, i.e. the beam will appear to have rotated anticlockwise. Hence, the superposition of the two waves results in LCP light.

## 3.3 Jones vector representation of polarization

### 3.3.1 Jones vector

The SoPs of light were formulated by Clark Jones in 1941 in the form of the *Jones vector* and the *Jones matrix*. This was followed by Hans Mueller in 1943 who presented the matrix representation of the mixed states of light using Stokes vectors (created by Sir George G Stokes in 1852). Let us look at the vector representation of light. Consider a case similar to that shown in figure 3.6: the electric field of a light ray with two components $E_x$ and $E_y$ is described by a function of the wave vector $(k)$ and time $(t)$, as follows:

$$\vec{E}(k, t) = \hat{e}_x E_x(k, t) + \hat{e}_y E_y(k, t). \tag{3.4a}$$

We can rewrite this using its components in Euler's form $E_x = E_{ox}e^{i(kz - wt + \phi_x)}$ and $E_y = E_{oy}e^{i(kz - wt + \phi_y)}$:

$$\vec{E}(k, t) = \hat{e}_x E_{ox}e^{i(kz - wt + \phi_x)} + \hat{e}_y E_{oy}e^{i(kz - wt + \phi_y)}$$

$$= e^{i(kz - wt)}\left[\hat{e}_x E_{ox}e^{i\phi_x} + \hat{e}_y E_{oy}e^{i\phi_y}\right] \tag{3.4b}$$

$$= \vec{E}_0 e^{i(kz - wt)}. \tag{3.4c}$$

The term in the square brackets of equation (3.4b) represents the complex amplitude of the plane wave with $E_0 = \sqrt{E_{0x}^2 + E_{0y}^2}$ in equation (3.4c). At an angle $\alpha$, the amplitude becomes

$E_0 = \sqrt{E_0^2(\cos^2 \alpha + \sin^2 \alpha)}$ . Using equations (3.4a)–(3.4c), the linearly polarized light can be defined in terms of a vector known as the Jones vector $(J)$, as

$$\vec{J} = \frac{1}{E_0}\begin{pmatrix} E_{ox}e^{i\phi_x} \\ E_{oy}e^{i\phi_y} \end{pmatrix} \equiv \frac{1}{E_0 e^{i\phi_x}}\begin{pmatrix} E_{ox} \\ E_{oy}e^{i(\phi_y - \phi_x)} \end{pmatrix}.$$

If we take the phase difference $\phi_y - \phi_x = \phi$, for an initial phase of $\phi_x = 0$, we can write

$$\vec{J} = \frac{1}{E_0}\begin{pmatrix} E_{ox} \\ E_{oy}e^{i\phi} \end{pmatrix}. \tag{3.5}$$

### 3.3.2 Jones vectors for some common types of light

Table 3.1 gives the Jones vectors for various types of commonly observed light. (i) Based on equation (3.5), for circularly polarized light the phase difference is $\pi/4$ and the amplitudes of both components are equal.

$$e^{\pm\, i\phi} = \cos\,(\pi/4) \pm\, i\,\sin\,(\pi/4) = \frac{1}{\sqrt{2}}(1 \pm\, i).$$

(ii) For elliptically polarized beams, the phase difference could still be $\pi/4$, but the amplitudes of the two components are not equal. Further, for equal amplitudes, a difference in the phases of the two components also affects the SoP. Based on equation (3.5) the Jones vectors for various types of light are given in table 3.1.

**Table 3.1.** Jones vectors for types of light with various SoPs.

| Polarization state | Direction of $\vec{E}$ | View of the tip of $\vec{E}$ | Jones vector $(\vec{J})$ |
|---|---|---|---|
| Linear | $x$ direction (also known as transverse magnetic (TM) or $p$-polarization) | | $\begin{pmatrix}1\\0\end{pmatrix}$ |
| | $y$-direction (Also known as transverse electric (TE) or $s$-polarization) | | $\begin{pmatrix}0\\1\end{pmatrix}$ |
| | At an angle $\zeta$ | | $(\cos\zeta\,\sin\zeta)$ |
| Circular (viewed into the beam) | Right | | $\frac{1}{\sqrt{2}}\begin{pmatrix}1\\+i\end{pmatrix}$ |
| | Left | | $\frac{1}{\sqrt{2}}\begin{pmatrix}1\\-i\end{pmatrix}$ |
| Elliptical (viewed into the beam) | Right | | $\frac{1}{E_0}\begin{pmatrix}E_{0,x}\\+iE_{0,y}\end{pmatrix}$ |
| | Left | | $\frac{1}{E_0}\begin{pmatrix}E_{0,x}\\-iE_{0,y}\end{pmatrix}$ |

All drawings are in the right-handed Cartesian coordinate system.

### 3.3.3 Jones matrix for a medium

To model the *effect of a medium* on the SoP of a light beam, we use *Jones matrices.* Suppose that A is a transfer function (i.e. a matrix) representing the medium; we can write the SoP in the form of a Jones vector for an input ($J_{in}$) and an output ($J_{out}$) as

$$J_{out} = A \, J_{in}.$$

Assuming a 2×2 matrix $A = \begin{bmatrix} A_{11} & A_{12} \\ A_{21} & A_{22} \end{bmatrix}$ for the medium, we can write the components of the output polarized field components as

$$E_{1x} = A_{11}E_{0x} + A_{12}E_{0y}$$

$$E_{1y} = A_{21}E_{0x} + A_{22}E_{0y}.$$

If we define the Jones matrix for an $x$ polarizer to be $A_x = \begin{bmatrix} 1 & 0 \\ 0 & 0 \end{bmatrix}$, the equation (that includes the Jones matrix) for the Jones vector is written as

$$J_{out} = A_x J_{in} \equiv \begin{bmatrix} 1 & 0 \\ 0 & 0 \end{bmatrix}\begin{bmatrix} E_{0x} \\ E_{0y} \end{bmatrix} \equiv \begin{bmatrix} E_{0x} \\ 0 \end{bmatrix}.$$

$J_{out}$ describes the output SoP after light passes through the $x$ polarizer. Similarly, the Jones matrix for a $y$ polarizer is defined as $A_y = \begin{bmatrix} 0 & 0 \\ 0 & 1 \end{bmatrix}$.

Let us consider a plane wave, as follows

$$\vec{E}(z, t) = \left(\hat{e}_x E_{ox} + \hat{e}_y E_{oy}\right)e^{i(kz - wt)}. \tag{3.6}$$

♣ **New basis vectors $\hat{e}_1$ and $\hat{e}_2$:** Let us define the new basis vectors $\hat{e}_1$ and $\hat{e}_2$. We define the transmission axis of the **polarizing medium** in the direction of the unit vector as $\hat{e}_1$ (making an angle $\theta$ to the $x$-axis) and we define its absorption axis to be along $\hat{e}_2$ (perpendicular to the transmission axis), so that their inner product is given by $\hat{e}_1 . \hat{e}_2 = 1$. The unit vectors along the $x$- and $y$-axes are expressed as follows, respectively:

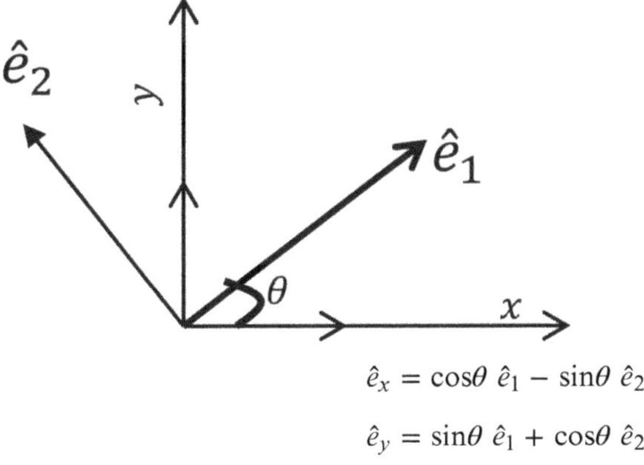


$$\hat{e}_x = \cos\theta \, \hat{e}_1 - \sin\theta \, \hat{e}_2$$

$$\hat{e}_y = \sin\theta \, \hat{e}_1 + \cos\theta \, \hat{e}_2$$

or

$$\hat{e}_1 = \cos\theta \; \hat{e}_x + \sin\theta \; \hat{e}_y$$

$$\hat{e}_2 = -\sin\theta \hat{e}_x + \cos\theta \; \hat{e}_y.$$

The plane wave equation (3.6) can be written in terms of the new basis vectors as defined in the box, $\hat{e}_1$ and $\hat{e}_2$, as

$$\vec{E}(z, t) = \left[(\cos\theta \; \hat{e}_1 - \sin\theta \; \hat{e}_2) \; E_{ox} + (\sin\theta \; \hat{e}_1 + \cos\theta \; \hat{e}_2) \; E_{oy}\right] e^{i(kz-wt)}$$

$$= \left[\cos\theta \; \hat{e}_1 \; E_{ox} + \sin\theta \; \hat{e}_1 \; E_{0y} - \sin\theta \; \hat{e}_2 E_{ox} + \cos\theta \; \hat{e}_2 \; E_{oy}\right] e^{i(kz-wt)}.$$

The plane wave now is nothing but the incident wave ($\vec{E}_I(z, t)$):

$$\vec{E}_I(z, t) = (E_1\hat{e}_1 + E_2\hat{e}_2) \; e^{i(kz-wt)} \tag{3.7}$$

with

$$E_1 = E_{ox}\cos\theta + E_{0y}\sin\theta \tag{3.8a}$$

and

$$E_2 = -E_{ox} \sin\theta + E_{0y}\cos\theta. \tag{3.8b}$$

We now introduce the effect of the medium, as follows. We assume that $E_1$ is transmitted and $E_2$ is absorbed. Let us multiply $E_2$ by a constant, $\gamma$, to account for the absorption. Therefore, after traversing the medium, the transmitted plane wave ($\vec{E}_T(z, t)$) now obeys the following equation:

$$\vec{E}_T(z, t) = (E_1\hat{e}_1 + \gamma E_2\hat{e}_2) \; e^{i(kz-wt)}. \tag{3.9}$$

Expressing $\hat{e}_1$ and $\hat{e}_2$ in terms of unit vectors along $x$ and $y$ (see the box above) gives

$$\hat{e}_1 = \cos\theta \; \hat{e}_x + \sin\theta \; \hat{e}_y \tag{3.10a}$$

and

$$\hat{e}_2 = -\sin\theta \hat{e}_x + \cos\theta \; \hat{e}_y. \tag{3.10b}$$

From equation (3.9), the transmitted wave can be written as

$$\vec{E}_T(z, t) = [E_1 \left(\cos\theta \; \hat{e}_x + \sin\theta \; \hat{e}_y\right) + \gamma E_2\left(-\sin\theta \hat{e}_x + \cos\theta \; \hat{e}_y\right)] \; e^{i(kz-wt)}.$$

Substituting the values from equations (3.8a) and (3.8b) yields

$$\vec{E}_T(z, t) = \left[\left(E_{0x}\cos\theta + E_{0y}\sin\theta \right)\left(\cos\theta \; \hat{e}_x + \sin\theta \; \hat{e}_y\right)\right.$$

$$\left. + \gamma\left(-E_{0x}\sin\theta + E_{0y}\cos\theta \right)\left(-\sin\theta \hat{e}_x + \cos\theta \; \hat{e}_y\right)\right] e^{i(kz-wt)}$$

$$= \Big[\big( E_{0x}\cos^2\theta\,\hat{e}_x + E_{0y}\sin\theta\,\cos\theta\,\hat{e}_x + E_{0x}\cos\theta\,\sin\theta\hat{e}_y + E_{0y}\sin^2\theta\hat{e}_y \big)$$

$$+ \gamma\big( E_{0x}\sin^2\theta\hat{e}_x - E_{0y}\cos\theta\,\sin\theta\,\hat{e}_x - E_{0x}\cos\theta\,\sin\theta\,\hat{e}_y + E_{0y}\cos^2\theta\,\hat{e}_y e^{i(kz-wt)}\big)\Big]$$

or

$$\vec{E}_T(z,\,t) = \hat{e}_x[E_{0x}(\cos^2\theta + \gamma\sin^2\theta) + E_{0y}(\sin\theta\cos\theta - \gamma\sin\theta\cos\theta)]$$

$$e^{i(kz-wt)} + \hat{e}_y[\{E_{0x}(\cos\theta\sin\theta - \gamma\sin\theta\cos\theta)\} \tag{3.11}$$

$$+ \{E_{0y}(\sin^2\theta + \gamma\cos^2\theta)\}]\,.$$

We can write equation (3.11) in matrix form as

$$\vec{E}_T(z,\,t) = P(\theta)\begin{bmatrix} E_{0x} \\ E_{0y} \end{bmatrix} e^{i(kz-wt)},$$

where

$$P(\theta) = \begin{bmatrix} \cos^2\theta + \gamma\sin^2\theta & (\sin\theta\cos\theta - \gamma\sin\theta\cos\theta) \\ (\sin\theta\cos\theta - \gamma\sin\theta\cos\theta) & (\sin^2\theta + \gamma\cos^2\theta) \end{bmatrix}.$$

For $\gamma = 0$ (no absorption),

$$P(\theta) = \begin{bmatrix} \cos^2\theta & \cos\theta\sin\theta \\ \cos\theta\sin\theta & \sin^2\theta \end{bmatrix}.$$

This is the general equation for a polarizer kept at an angle.

♣ For $\gamma = 1$, equation (3.11) behaves as if there is no polarizer, i.e. it becomes the same as equation (3.6).

**Exercise 3.3.** Show that $P(\theta + \pi) = P(\theta)$.

**Solution:** By replacing the value of $\theta$ with $(\theta + \pi)$ from the transformation of the sin and cos functions, it is easy to show that $P(\theta + \pi) = P(\theta)$. This relation indicates that a polarizer behaves in the same manner when it is rotated by 180°.

## 3.4 Methods of generating polarized light

### 3.4.1 Wire-grid polarizer in the microwave region

One of the common and traditional methods used to obtain polarized light in the microwave region is to use a material that transmits waves whose electric vectors are aligned in the desired direction. This is one of the easiest ways to understand the principle of the polarizer. We know that the wavelength of microwaves is of the order of cm. Let us consider a mesh made of copper wire with a thickness of about 100 μm and a grid spacing of a few cm, as shown in figure 3.9.

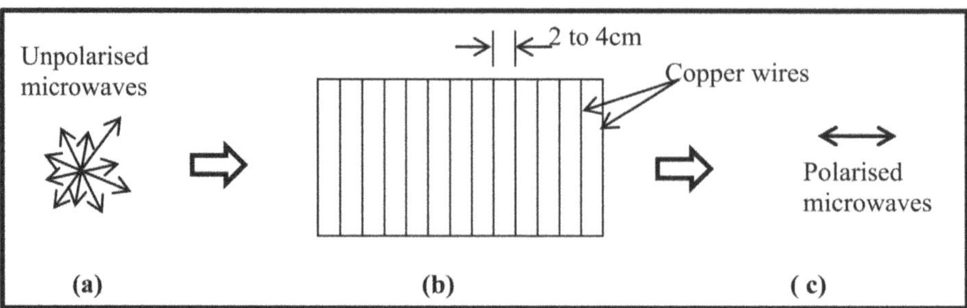


**Figure 3.9.** Incident unpolarized microwaves (A), falling on a grid made of copper wire (B) that transmits horizontally polarized microwaves (C).

When unpolarized microwaves fall on this mesh, the component of the incident field along the length of the wires (i.e. the parallel component) is absorbed. This process results in Joule heating in the metal. On the other hand, the component perpendicular to the wires passes through (although attenuated), as the wires are extremely thin.

As described in the previous section, if the light incident on any polarizer is linearly polarized, the amount transmitted by the polarizer depends on the angle $\theta$ between the electric field vector of the light wave and the polarizer axis. If $A$ is the amplitude of the incident electric field, the transmitted amplitude is $A\cos\theta$, and the transmitted intensity is given by Malus' law as $I = I_0\cos^2\theta$, where $I_0$ is the maximum irradiance transmitted at $\theta = 0$.

### 3.4.2 Long polymer chain

For visible light, a wire-grid polarizer is extremely difficult to fabricate due to the limitations imposed by the dimensions of the wire. However, plastic sheet polarizers contain chain polymers coated with *conductive molecules* such as iodine. The chain polymers are stretched in one direction so that long chains are distributed parallel to the direction of stretch. For instance, iodine is able to move the valence electrons along the polymer chain (but not perpendicular to the chain), thus providing conductivity for the absorbed component of the electric field for visible light (figure 3.10). In this case, the electric field parallel to the molecules is absorbed. When light with perpendicular polarization strikes the polymer, the electrons do not move and hence the component is not absorbed but rather transmitted.

### 3.4.3 Polarization based on reflection

Laser light is generally polarized when it emerges from the cavity. This is due to the fact that the oscillation of the electric field vector is constrained to a particular polarization and no component of the perpendicular component is permitted to oscillate in the cavity. This is usually achieved by placing glass cut at a certain angle so that only a particular polarization can pass through, while the other is completely reflected. To understand this mechanism, let us look at the reflection and refraction of the field vectors across an interface between two refractive indices.

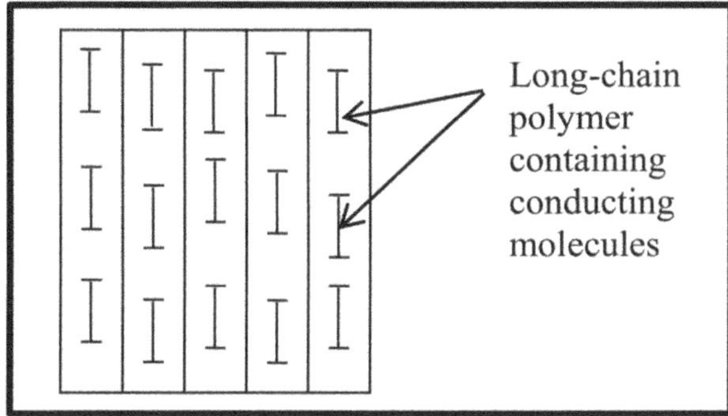



**Figure 3.10.** Representation of polarizers made of long-chain polymers containing conductive molecules aligned vertically.

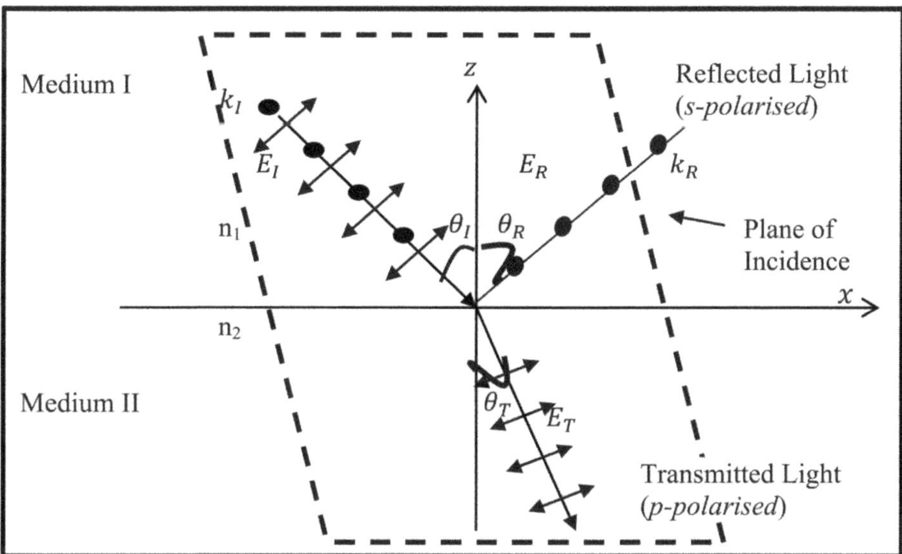



**Figure 3.11.** Unpolarized light incident on a surface from medium I. Part of the light is reflected (*s*-polarized) and the rest is transmitted (*p*-polarized) into medium II. The plane of incidence is shown by a dashed box.

We define the *plane of incidence* as a plane formed by the incident, reflected, and transmitted wave vectors along with the normal to the surface (e.g. the $x$–$z$ plane in figure 3.11). Here, an interface is shown between two media (I and II) with refractive indices of $n_1$ and $n_2$, permittivities of $\epsilon_1$ and $\epsilon_2$, and permeabilities of $\mu_1$ and $\mu_2$, respectively. An incoming wave impinges at an arbitrary angle $\theta_I$ on a reflecting/transmitting boundary consisting of the normal (along the $z$-axis) and the horizontal ($x$–$y$) plane. The wave has three components with wave vectors $k_I$, $k_R$, and $k_T$. The boundary conditions for the corresponding electric ($E_I$, $E_R$, $E_T$) and magnetic ($B_I$, $B_R$, $B_T$) field vectors that have the same angular frequency ($\omega$) in the tangential (in the $x$–$y$ plane) and the normal (along the $z$-axis) are given by the following:

$$\text{(a). } \epsilon_1\left(\overrightarrow{E_I} + \overrightarrow{E_R}\right)_z = \epsilon_2\left(\overrightarrow{E_T}\right)_z$$

$$\text{(b). } \left(\overrightarrow{B_I} + \overrightarrow{B_R}\right)_z = \left(\overrightarrow{B_T}\right)_z$$

$$\text{(c). } \left(\overrightarrow{E_I} + \overrightarrow{E_R}\right)_{x,\,y} = \left(\overrightarrow{E_T}\right)_{x,\,y}$$

$$\text{(d). } \frac{1}{\mu_1}\left(\overrightarrow{B_I} + \overrightarrow{B_R}\right)_{x,\,y} = \frac{1}{\mu_2}\left(\overrightarrow{B_T}\right)_{x,\,y} \qquad (3.12)$$

Equation (3.12) suggests that if there are no free charges or currents at the surface, the normal components of the fields are continuous. Similarly, the tangential components of fields for a charge-free, current-free interface are continuous across the interface. For example, for a *p-polarized wave*, equation (3.12a) can be written as follows for the normal component of the electric field:

$$\epsilon_1(E_I \sin \theta_I + E_R \sin \theta_R) = \epsilon_2(-E_T \sin \theta_T).$$

As the magnetic field has no $z$-component, equation (3.12b) does not play any role. For the tangential components of the electric field (TM polarization), equation (3.12c) becomes

$$E_I \cos \theta_I + E_R \cos \theta_R = E_T \cos \theta_T. \qquad (3.13)$$

Expressing $B$ in terms of $E$ in equation (3.12d) gives

$$\frac{1}{v_1\mu_1}(E_I - E_R) = \frac{1}{\mu_2 v_2}E_T. \qquad (3.14)$$

Here, $v_1$ and $v_2$ are the velocities in media I and II, respectively.

With the law of reflection ($\theta_I = \theta_R$) and Snell's law of refraction $\left(\frac{\sin\theta_I}{\sin\theta_T} = \frac{n_2}{n_1}\right)$, equation (3.14) gives

$$(E_I - E_R) = \frac{v_1\mu_1}{v_2\mu_2}E_T.$$

If

$$\frac{v_1\mu_1}{\mu_2 v_2} = \beta = \frac{n_2\mu_1}{n_1\mu_2}$$

then

$$E_I - E_R = \beta E_T. \qquad (3.15)$$

Using these equations, (3.13) can be rewritten as

$$E_I + E_R = E_T\frac{\cos \theta_T}{\cos \theta_I}.$$

If

$$\frac{\cos\theta_T}{\cos\theta_I} = \alpha'$$

then

$$E_I + E_R = \alpha' E_T. \tag{3.16}$$

Using equations (3.15) and (3.16), the reflected and transmitted amplitudes can be written in the form of *Fresnel's well-known equations* (for the case of polarization in the plane of incidence), as follows:

$$E_R = \left(\frac{\alpha' - \beta}{\alpha' + \beta}\right) E_I,$$

$$E_T = \left(\frac{2}{\alpha' + \beta}\right) E_I. \tag{3.17}$$

Since $\alpha'$ depends on $\theta_I$, it decides the amplitudes of the transmitted and reflected waves. There is a special angle of incidence, $\theta_B$ (known as *Brewster's angle*) for which the light is transmitted completely, i.e. $E_T = E_I$ in equation (3.17). The coefficient of reflection $(r_s)$ is defined as $r_s = \left(\frac{E_R}{E_I}\right)_{TE}$ and the coefficient of transmission is defined as $t_s = \left(\frac{E_T}{E_I}\right)_{TE}$. For $n = \frac{n_1}{n_2}$, it is easy to show that the reflectance $R_s(=\mid r_s \mid^2)$ is given by $R_s = \left[\frac{(n-1)}{n+1}\right]^2$. (♠ Also see problem 11.)

**Exercise 3.4.** In figure 3.11, the two media (I and II) have refractive index values of $n_1$ and $n_2$, respectively. If the *p*-polarized beam is incident at the surface at an intermediate angle known as the Brewster angle ($\theta_B$), there is no reflected beam. Considering $\mu_1 \cong \mu_2$, $\beta = \frac{n_2}{n_1}$, and $\alpha' = \beta$ in equation (3.17), obtain the following relation

$$\tan \theta_B \cong \frac{n_2}{n_1}. \tag{3.18}$$

**Solution:** Snell's law is given by $n_1 \sin\theta_I = n_2 \sin\theta_T$. With $\theta_I = \theta_B$ and $\theta_I + \theta_T = 90°$, for reflected and transmitted beams to be at right angles to each other, $\theta_R + \theta_T = 90°$. Snell's law becomes, $n_1 \sin\theta_B = n_2 \sin(90 - \theta_B)$ or

$$\tan \theta_B = \frac{n_2}{n_1}.$$

The observation made by D Brewster in 1815 was that the angle between the reflected and transmitted light is 90° when the light is incident at $\theta_B$. We will use this relation: $\alpha' = \frac{\cos\theta_T}{\cos\theta_I}$. By identifying $\theta_T = \frac{\pi}{2} - \theta_I$ and $\theta_I$ as $\theta_B$, we can write $\cos\theta_B = \frac{\sin\theta_B}{\beta}$.

We can now write

$$\alpha' = \frac{\sqrt{(1 - \sin^2\theta_T)}}{\cos \theta_I} = \frac{\sqrt{\left(1 - \left(\frac{n_1^2}{n_2^2}\right)\sin^2\theta_B\right)}}{\cos \theta_B}.$$

Rearranging this (and with $\alpha' = \beta$), we get $\sin\theta_B = \dfrac{\beta}{\sqrt{1+\beta^2}}$. For the given definition of $\beta$ (i.e. the ratio of the refractive indices of the two media and $\mu_1 \cong \mu_2$), $\tan\theta_B \cong \dfrac{n_2}{n_1}$. This property of the angular dependence of the light polarization is used in several devices.

**Exercise 3.5.** An uncoated glass block which has a refractive index of 1.45 has one of its faces cut at the Brewster angle ($\theta_B$); what is the value of $\theta_B$ for the glass block?

    **Solution:** As can be seen from equation (3.18), the value of $\theta_B$ is 55.4°. Such a glass block or plate cut at $\theta_B$ is known as a Brewster window. Brewster windows are used, for example, in laser cavities to achieve polarized laser beams, as outlined at the beginning of this section.

    ♣ There are two more Fresnel equations for polarization perpendicular to the plane of incidence.

### 3.4.4 Polarization based on dichroism

A *dichroic* crystal has different *absorption coefficient* values for the parallel ($\alpha_{\parallel}$) and the perpendicular ($\alpha_{\perp}$) components of light. Polarized light can be obtained with the help of *dichroic* and anisotropic crystals, as indicated in figure 3.12. Here, we deal with the bulk absorption property of natural anisotropic crystals, which is different from that of stretched polymer polarizers (section 3.4.3). Here, the light is split into two polarizations while propagating in anisotropic crystal. One of the polarizations is absorbed by the crystal itself. Crystals such as tourmaline have this property.

### 3.4.5 Polarization based on double refraction

#### 3.4.5.1 Birefringent crystals

The appearance of two refractive indices viz. *ordinary* ($n_0$) and *extraordinary* ($n_e$) in a *birefringent crystal* was introduced above in section 3.2.1. Polarization by double refraction depends on the fact that different polarizations of light (that have the same angular frequency) travel at different velocities inside a crystal. This effect is observed in birefringent crystals. Here, the incident light encounters different refractive indices as it travels across the crystal. In figure 3.13, schematic plots of $n_0$ and $n_e$ are given as a function of wavelength.

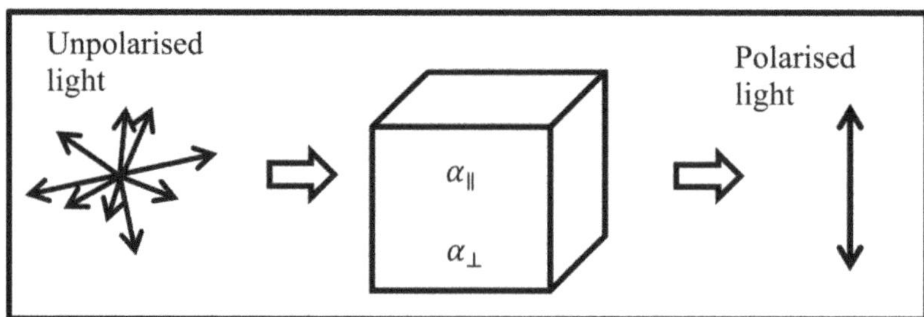


**Figure 3.12.** A crystal of tourmaline acts as a dichroic medium, presenting different absorption coefficients to incoming light.

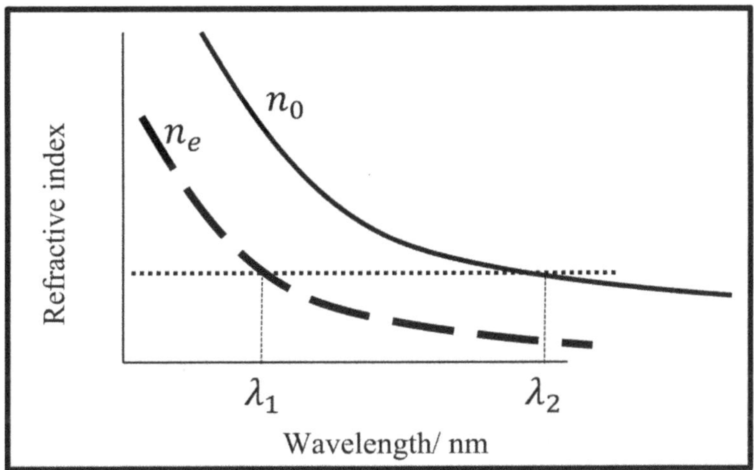

**Figure 3.13.** Indicative plots of $n_0$ and $n_e$ as a function of wavelength for a birefringent crystal. The intersection of the horizontal and vertical dotted lines indicates a common value of the refractive index ($n_0 = n_e$) at two wavelengths ($\lambda_1$ and $\lambda_2$, respectively).

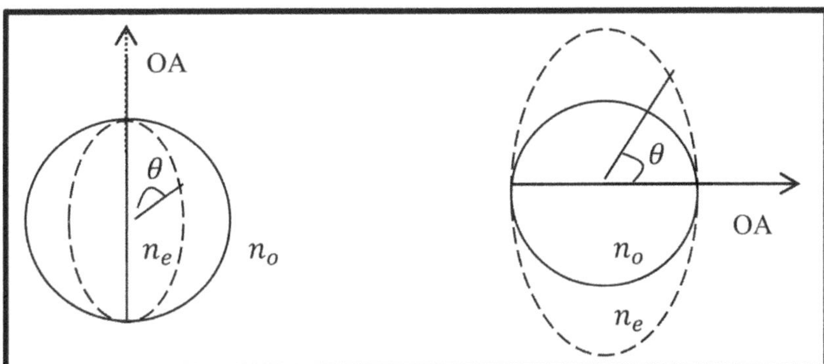


**Figure 3.14.** Representation of $n_0$ and $n_e$ in negative uniaxial (left) and positive uniaxial (right) birefringent crystals. $\theta$ is the angle relative to the optic axis (OA).

In a birefringent crystal there are two directions of light propagation that are perpendicular to each other; these are known as the 'fast' and 'slow' optic axes. Furthermore, there can be two kinds of birefringent crystals, known as *positive uniaxial* and *negative uniaxial* crystals, corresponding to the following two cases, respectively: (i) $n_0 > n_e$ and (ii) $n_0 < n_e$.

The value of $n_0$ remains constant across the propagation direction inside a birefringent crystal. On the other hand, the value of $n_e$ varies with the angle ($\theta$) from the crystal axis. In other words, while the value of $n_0$ is symmetrical around the optic axis and forms a circle in 2D, $n_e$ makes an ellipse, as shown in figure 3.14.

The properties of birefringent crystals can be summarized as follows. Birefringent crystals

    i. create polarization perpendicular to each other;

    ii. have an OA along which the value of $n_o$ is same as that of $n_e$;

    iii. have a value of $n_e$ that varies with the direction of propagation;

    iv. follow Snell's law for $n_0$ but not for $n_e$, i.e. $n_e$ is dependent on $\theta$.

    v. For negative uniaxial crystals $n_o > n_e$, i.e. $c/n_o < c/n_e$. The velocity of the ordinary (o-ray) wave is smaller than that of the extraordinary (e-ray) wave. The reverse is true for positive uniaxial crystals, as indicated in figure 3.14. However, the two velocities are equal along the OA in both types of crystal.

### 3.4.5.2 Nikol's prism

A popular example of polarization by double refraction is Nikol's prism. In addition to birefringence, here we also make use of the phenomenon of *total internal reflection* (TIR), which takes place in transparent materials.

Consider a ray of light traveling from a medium that has a higher refractive index, $n$ (say, water) to one that has a lower value (air, $n \sim 1$), as shown in figure 3.15. According to Snell's law, the refraction angle ($\theta_R$) is larger than the incident angle ($\theta_I$). When the incident angle is further increased, the refracted ray propagates along the surface of the medium. At this point, the incident angle is known as the critical angle ($\theta_c$). Any incident angle value beyond $\theta_c$ results in the reflection of the beam within the same medium, as shown in the figure. This is the phenomenon of TIR, which is also the principle behind light propagation in optical fibers (♠ see chapter 23).

Crystals such as calcite ($CaCO_3$) and quartz split and separate mutually perpendicular polarizations due to their birefringence. Nikol's prism consists of two prisms glued together as shown in figure 3.16. Here, the two prisms are joined with the help of a glue called Canada balsam, which has a refractive index slightly greater than that required to obtain the *o-ray* in calcite, but lower than that for the *e-ray*. Table 3.2 gives the values of $n_o$ and $n_e$ for crystals of calcite and quartz.

The crystal is cut at an angle such that the ordinary ray undergoes TIR at the glue layer joining the two prisms. Due to TIR, the o-ray bends from a linear path due to the lower refractive index of the glue. However, the e-ray is refracted through the prisms.

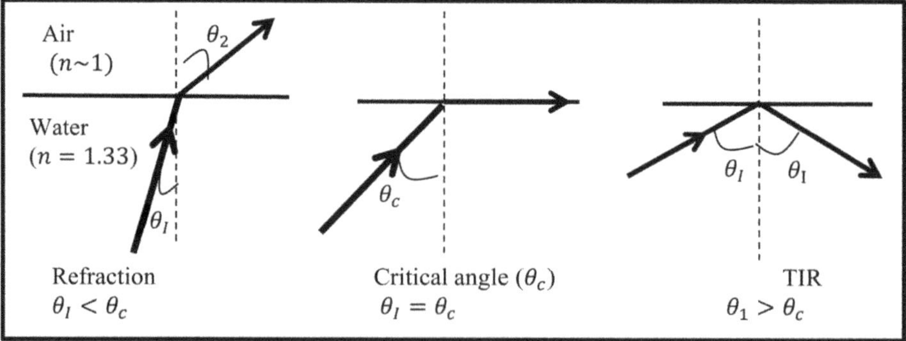


**Figure 3.15.** The left panel shows the phenomenon of the refraction of light at the water–air interface. TIR (right panel) is observed for angles of incidence greater than the critical angle, $\theta_c$ (middle panel).

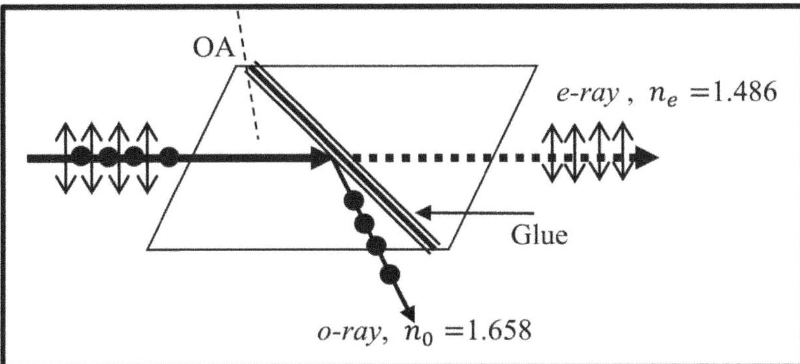


**Figure 3.16.** Nikol's prism splits incident light into mutually perpendicular polarized light beams. OA is the optic axis. Double arrows and filled circles indicate the two polarizations.

**Table 3.2.** Values of $n_o$ and $n_e$ for two types of crystal.

| Material | Type of crystal | $n_o$ | $n_e$ | $n_o/n_e$ |
| --- | --- | --- | --- | --- |
| CaCO$_3$ | Negative uniaxial | 1.658 | 1.486 | 1.188 |
| Quartz | Positive uniaxial | 1.544 | 1.553 | 0.994 |

## 3.5 Change of state of polarization

We learnt in section 3.2.2 that the phase difference between the two components of a linearly polarized wave can be used to rotate the SoP. The change in the refractive index of some materials due to the influence of external fields induces a phase change in a propagating beam, thereby affecting its polarization. In this section, we will encounter similar effects when a crystal is subjected to external electric or magnetic fields.

### 3.5.1 Electro-optic effects

Some optically isotropic substances (solids or liquids) exhibit *electro-optic (EO) birefringence* when placed in a strong DC electric field. Phase changes occur in light passing through such substances ($\phi$ in equation 3.3) under the stress of an electric field, due to the alignment of their molecules relative to the optic axis of uniaxial crystals or liquids. These changes are responsible for the Pockels effect (in 1893) or the Kerr effect (in 1875).

In the Pockels effect, a DC field ($E$) is applied parallel to the optic axis of the crystal (see figure 3.17 (left panel)). One type of device used to obtain phase changes in transmitted light is known as a Pockels cell. Here, the direction of the electric field is kept longitudinal i.e. parallel to the direction of the incident light. Devices based on the Pockels effect are also known as longitudinal EO modulators. In the absence of an electric field, as expected, there is no change in the SoP of a beam after it has

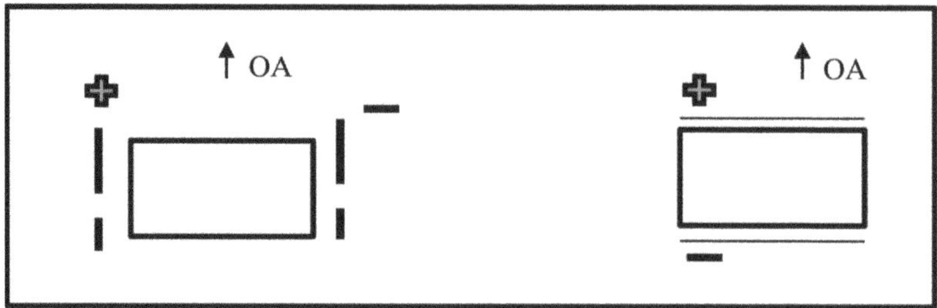


**Figure 3.17.** Schematic of the electro-optic effect: a DC Pockels cell (left panel) and a DC Kerr cell (right panel). Light is incident from the left in each case; OA denotes the optic axis of the medium.

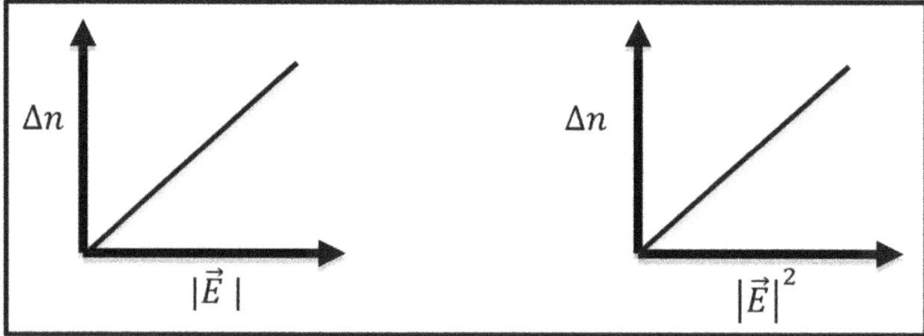


**Figure 3.18.** Schematic of the change of refractive index ($\Delta n$) due to an incident field: $\Delta n \propto E$ in the Pockels effect (left panel); $\Delta n \propto E^2$ in the Kerr effect (right panel).

passed through a Pockels cell. For low values of $E$, the birefringence is a linear function of the applied field (figure 3.18). Therefore, the Pockels effect is also known as the linear EO effect.

As shown in figure 3.17 (right panel), for higher DC fields (in order to observe the Kerr effect), a field is applied across a liquid such as nitrobenzene in the transverse direction relative to the incident light. Devices based on the Kerr effect are known as transverse EO modulators. Here, the magnitude of the birefringence at a wavelength $\lambda$ is proportional to the square of the electric field, as shown in figure 3.18. The Kerr coefficient ($K$) is defined in terms of the difference of the values of the refractive indices $\Delta n(= n_o - n_e)$, as follows:

$$\Delta n \propto \lambda KE^2. \tag{3.19}$$

The EO shutter used in chapter 13 also makes use of the Kerr effect. The values of the Kerr coefficient are of the order of $\sim 10^{-14}$–$10^{-18}$ m V$^{-2}$. Similarly, the values of the Pockels coefficient are of the order of $\sim 10^{-10}$m V$^{-1}$.

### 3.5.2 Magneto-optic effect

If a beam of light is passed through an isotropic dielectric medium in the presence of a magnetic field aligned in the same direction, the plane of the incident light is rotated by an angle. This occurs because the value of the relative permittivity

**Table 3.3.** Values of the Verdet constant for some materials.

| Material | $V$ (deg Oe$^{-1}$ cm$^{-1}$) |
| --- | --- |
| NaCl | 0.036 |
| Diamond | 0.012 |
| Flint glass | 0.030 |

becomes optically active in the direction of the applied magnetic field. This effect is also known as Faraday rotation. The amount of rotation of the SoP depends on the strength of the magnetic field ($B$) and the thickness ($t$) of the medium, as follows:

$$\theta = VBt. \tag{3.20}$$

Here, $V$ is known as the Verdet constant. Table 3.3 gives the values of $V$ for some materials.

## 3.6 Quarter-wave and half-wave plates

The property of birefringence is also used in making the quarter-wave and half-wave plates used to rotate the SoP of a beam. These wave plates are simply the optical flats of birefringent crystals of calcite or quartz of suitable thickness with faces cut parallel to the OA. As explained in section 2.2, the value of the phase difference between the two components can change the polarization of an incident beam. The thickness of the wave plate can be obtained for the required phase difference as described in the following exercise.

**Exercise 3.6.** For given values of $n_o$ and $n_e$, what thickness ($t$) of an optical flat is required to rotate the polarization of a wave of wavelength $\lambda$ by $\lambda/4$?

**Solution:** Optical flats that rotate polarization by an amount equivalent to one-quarter of a wavelength are known as *quarter-wave plates*. For a quarter-wave plate, the difference of the effective thickness $t_1$ is given by $n_o t_1 - n_e t_1 = \lambda/4$. The thickness of the plate can be found using $t_1 = \dfrac{\lambda}{4\,|n_o - n_e|}$. Similarly, when the phase difference between the *e-ray* and the *o-ray* is $\lambda/2$, the plate is known as a *half-wave plate*. For a half-wave plate, the thickness ($t_2$) is given by $t_2 = \dfrac{\lambda}{2\,|n_o - n_e|}$. Typical thickness values of the materials used for quarter-wave and the half-wave plates are given in table 3.4 for two wavelengths.

## 3.7 Polarized light in nature

Natural sources of light (e.g. sunlight, moonlight) are not strongly polarized. However, interaction with atmospheric layers or with air–water interfaces (such reflection by the sea's surface) can create strongly polarized light. Outdoor sports-men such as golfers, cyclists, and anglers use polarized sunglasses to reduce the glare

**Table 3.4.** Required thicknesses of wave plates for a typical value of $|\,n_o - n_e\,| = 10^{-5}$ at the fundamental and second-harmonic wavelengths of a Nd:yttrium aluminum garnet (YAG) laser.

| Wave plate | $\lambda$ (nm) | Thickness required (mm) |
|---|---|---|
| Quarter wave | 1064 | 26.6 |
| | 532 | 13.3 |
| Half wave | 1064 | 53.2 |
| | 532 | 26.6 |

reflected from water or road surfaces. On your next visit to a beach, you might like to carry a sheet polarizer to check the polarization of light reflected by the water and see whether you can explain it on the basis of the discussion in this chapter. If you can make out fine differences in reflection, you can also observe similar effects in the reflections from the leaves of a tree, or from the tiles of a floor. It is also now well proven that birds, insects, and bats use skylight polarization patterns to orient themselves.

## Questions and problems

1. Identify the SoPs of the electric fields whose $x$ and $y$ components are given by the following expressions: (i) $E_x = E_0\cos(kz + \omega t)$; $E_y = E_0\sin(kz + \omega t)$; (ii) $E_x = E_0\cos(kz - \omega t + \frac{\pi}{4})$; $E_y = (E_0/\sqrt{2})\,\sin(kz - \omega t)$.
2. What is the Brewster angle of a glass block that has a refractive index of 1.45 and is kept in air? How does it change when the block is placed in a liquid that has a refractive index of 5/3?
3. What is the Jones vector for plane polarized (linearly polarized) light?
4. Name four methods used to create polarized light. What is $s$-polarized light?
5. What are the properties of birefringent crystals?
6. The DC Kerr effect is observed when a material is kept in an electric field ($E$). Write down two differences between the DC Pockels effect and the DC Kerr effect.
7. An unpolarized beam of light that has a wavelength of 500 nm is incident on a linear polarizer ($P1$) whose SoP is vertical. The light beam then passes through a quarter-wave plate ($W$) with its optical axis rotated by $\pi/4$ from the vertical. $W$ is made of a material whose ordinary and extraordinary refractive indices are 1.658 and 1.558, respectively (a schematic is shown in figure 3.19). What is the SoP (i) after the wave plate? (ii) at the initial point after passing through $W$ followed by reflection by a mirror ($M$)?
8. Find the thickness of an optical flat used as a quarter-wave plate that yields the value $|\,n_o - n_e\,| = 10^{-5}$ for a wavelength of 632.8 nm.
9. With reference to the Jones vector formulation of light, to obtain the polarization of superimposed light beams, the amplitudes of the electric fields have to be multiplied after summing the vectors. Obtain the sum of right and left circularly polarized light of identical amplitude.

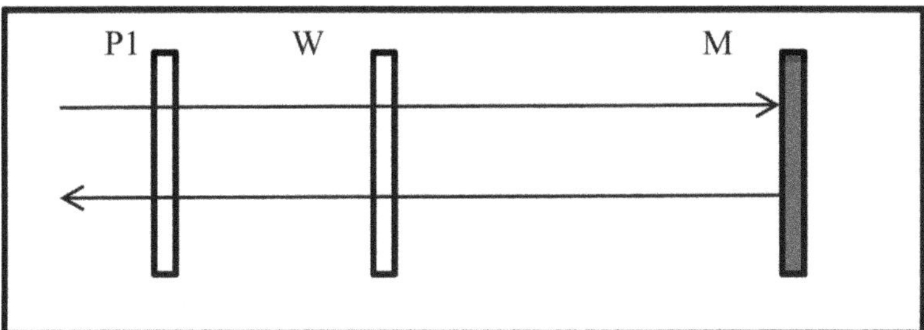


**Figure 3.19.** Figure for question 7.

10. From the general equation for the polarizer $P(\theta)$ at an angle $\theta$, show that $P^2(\theta) = P(\theta)$.

11. The field oscillations of $s$-polarized light are perpendicular to the plane of incidence; hence, it is known as a TE wave. The wave is incident on an interface between two media with refractive indices of $n_1$ and $n_2$. The coefficient of reflection $(r_s)$ is defined as $r_s = \left(\frac{E_R}{E_I}\right)_{TE}$ and the coefficient of transmission is given by $t_s = \left(\frac{E_T}{E_I}\right)_{TE}$. Show that for $n = \frac{n_1}{n_2}$, the reflectance $R_s(=\mid r_s \mid^2)$ is given by $R_s = \left[\frac{(n-1)}{n+1}\right]^2$.

12. How do you write the Jones matrices for the half-wave and quarter-wave plates?

13. From $E_R = \left(\frac{\alpha - \beta}{\alpha + \beta}\right)E_I$, when $\alpha = \beta$, then $E_R = 0$. This is the condition in which there is no reflected wave. Find the incident angle $(\theta_i)$ corresponding to this condition. What is the name of this angle?

## Bibliography


[1] Anderson J A 1908 The rotation of a crystal of tourmaline by plane polarised light *Nature* **78** 413

[2] Cronin T W and Marshall J 2011 Patterns and properties of polarized light in air and water *Philos. Trans. R. Soc. Lond. B Biol. Sci.* **366** 619–26

[3] Greif S, Borissov I, Yovel Y and Holland R A 2014 A functional role of the sky's polarization pattern for orientation in the greater mouse-eared bat *Nat. Commun.* **5** 4488

[4] Saleh B E A and Teich M C 1991 *Fundamentals of Photonics* (New York: Wiley)

[5] Menzel R P 2001 *Linear and Nonlinear Interactions of Laser Light and Matter* (Berlin: Springer)

[6] Fowles G R 1975 *Introduction to Modern Optics* (New York: Dover)

**IOP** Publishing

# An Introduction to Photonics and Laser Physics with Applications

**Prem B Bisht**

# Chapter 4

## Spontaneous and stimulated emission

The full form of the acronym **laser** is '*light amplification by stimulated emission of radiation*'. Historically, the term 'radiation' referred to the 'quantum' of energy emitted by a hot body. The radiation density when plotted against the wavelength had its maximum shifted to shorter wavelengths at higher temperatures. The theoretical model based on Wein's law was unable to explain this in the ultraviolet region. Planck explained this 'ultraviolet catastrophe' by suggesting an empirical formula known as 'Planck's radiation law', which stated that the number of radiation emitters in an energy state is given by the Boltzmann distribution at room temperature. The relationship between spontaneous and stimulated emission was given by Einstein in his famous paper of 1917. The diagram shows how an incident photon stimulates an atom in the excited state to emit a photon of the same frequency. You will learn in this chapter why and how the amplification of EM radiation was achieved in the microwave region (in the form of the *maser*) before the *laser* came into existence.

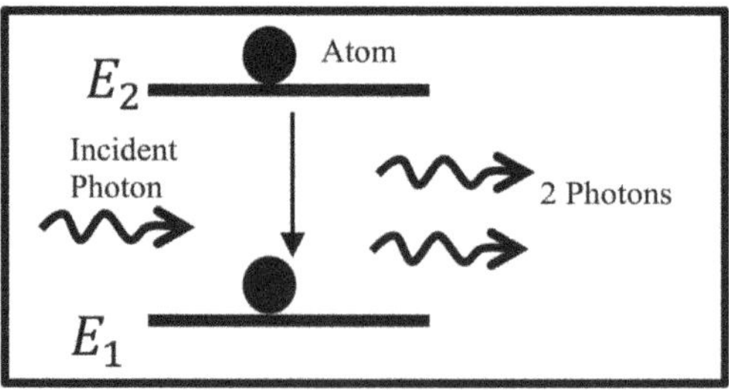


**Learning objectives**
Describe the temperature dependence of radiation density;
Identify the Boltzmann distribution;

> Write down the rate equations for a two-level system;
> Differentiate between spontaneous and stimulated emission rates;
> Describe Einstein's $A$ and $B$ coefficients;
> Relate Planck's law of radiation to the radiation density in a two-level system;
> Understand the concept of the lifetime of an excited state;
> Illustrate the frequency dependence of the ratio of stimulated and spontaneous emission.

## 4.1 Thermal radiation and Planck's law

### 4.1.1 Radiation density in a cavity

According to the first law of thermodynamics, a hot body tries to be in thermal equilibrium with its surroundings. Heat is radiated in the forms of temperature and EM radiation. Generally, black-body radiation is the EM radiation emitted by a body in thermodynamic equilibrium with its surroundings. For example, mosquitoes and other insects are able to sense the heat of the human body. IR-based temperature sensors are used in monitoring fever as well as in night-vision equipment. Stars with different temperatures appear in various colors. For example, Betelgeuse in the constellation of Orion is a red star with a temperature of about 3500 °C, while Rigel is blue (10 000 °C). Apart from the Sun, fire was also the source of heat and light that was directly accessible to humans on earth. Edison's incandescent filament bulb is a source of photons from the visible (light) region to the IR region (heat).

♣ If you have ever come across a bonfire or a furnace in a laboratory, you may be able to identify the different colors of fire, predominantly red (burning coals) to yellow (flames). Occasionally, part of the flame can be bluish as well—suggesting higher temperatures.

### 4.1.2 Density of modes in a closed container

We consider a cubic box with perfectly reflecting walls in 3D (figure 4.1). In such a cavity, a standing wave can be formed if the width ($L$) of the cube is an integral number of half wavelengths ($\lambda/2$) of the radiation present. For the wave vector $k$, this can be expressed as $\lambda/2 \equiv q\pi/k$, with $q = 1, 2, 3,\ldots$ . The corresponding frequency ($\nu_q$) is $\frac{qc}{2L}$. The components of wave vector along the three axes are given by $k_x = \frac{q_x\pi}{L}$, $k_y = \frac{q_y\pi}{L}$, and $k_x = \frac{q_z\pi}{L}$. Each mode characterized by the integers ($q_x$, $q_y$, and $q_z$) is represented by a point in $k$-space.

The frequency $\nu$ and the wave vector $k$ satisfy the resonance condition $k^2 = k_x^2 + k_y^2 + k_z^2 = \left(\frac{2\pi\nu}{c}\right)^2$. Here, the surface of the constant frequency $\nu$ is a sphere of radius $k$. The number of modes ($N$) within the frequency range is equal to the number of points within the radius $k$. $N$ can be large, but it can be approximated by the volume of one portion of the cube in the positive quadrant of a sphere of radius $q$ ($=k\pi/L$) as $\frac{1}{8}\left(\frac{4}{3}\pi q^3\right)$. For two states of polarization, it is multiplied by a factor of two. Therefore, the number of available modes in terms of $\nu$ can be given by

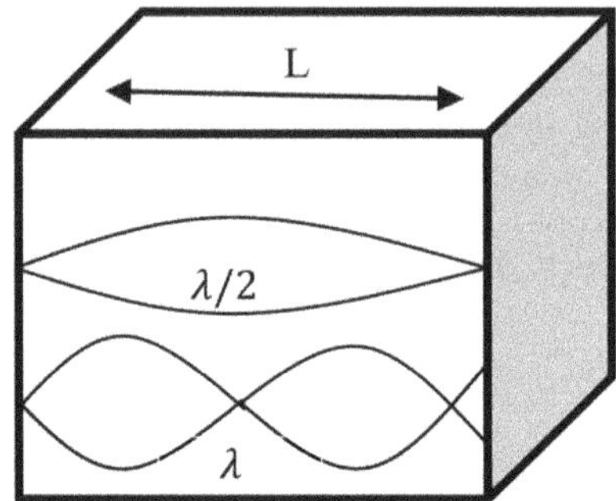


**Figure 4.1.** Cubic box of length $L$ with perfectly reflecting walls called a cavity, which can contain integral multiples of half wavelengths ($\lambda/2$).

$$N = 2 \times \frac{1}{8}\left(\frac{4}{3}\pi q^3\right) \equiv \frac{8\pi\nu^3}{3c^3}L^3.$$

Here, $L^3$ is the volume ($V$) of the cavity. Differentiating this expression gives the number of modes per unit frequency interval

$$\frac{\mathrm{d}N}{\mathrm{d}\nu} = \left(\frac{8\pi\nu^2}{c^3}\right)V. \tag{4.1}$$

The number of modes per unit frequency bandwidth per unit volume $\left(\frac{8\pi\nu^2}{c^3}\right)$ is known as the Jeans number. It is also known as the number of oscillators or the 'density of modes'.

**Exercise 4.1.** Find the mode density for a 1D cavity.
    **Solution:** Let the length of the 1D cavity be $L$ in the $x$ direction. We have $L = \frac{\lambda_x}{2}; \frac{2\lambda_x}{2}; \frac{3\lambda_x}{2}$; so that for any number $n$,
$$\frac{n\lambda_x}{2} = L; \lambda = \frac{2L}{n}$$
$$\nu_n = \frac{c}{2L}n, \left(\nu_{n+1} - \nu_n \equiv \partial\nu = \frac{c}{2L}\right).$$
The mode density in a length $L$ for two polarizations
$$g(\nu)\mathrm{d}\nu = \frac{2}{L}\left(\frac{(\nu + \mathrm{d}\nu - \nu)}{\partial\nu}\right) = \frac{4\mathrm{d}\nu}{c}$$

### 4.1.3 Wein's displacement law

The black-body radiation intensity $I(T)$ emitted from an opening in the cavity of a black body of temperature $T$ is given by an empirical relation obtained by Stefan (in 1879), and later theoretically derived by Boltzmann (in 1884). According to this law, the total radiation intensity (W m$^{-2}$) emitted by a body is proportional to the fourth power of $T$, as follows:

$$I(T) = \sigma T^4$$

where $\sigma = 5.68 \times 10^{-8}$ W m$^{-2}$ K$^{-4}$ is known as the Stefan–Boltzmann constant. This formula also includes another constant known as emissivity ($\epsilon_M(\nu)$), a quantity specific to the material. The value of $\epsilon_M(\nu)$ ranges from zero to one. Here, for a perfect black body, its value is taken to be one.

Classically, the average energy per mode is $k_B T$. Here, $k_B$ is the Boltzmann constant. The energy density of radiation per unit volume within a frequency range between $\nu$ and $\nu + d\nu$ is given by the Rayleigh–Jeans (RJ) law, as described below. From equation (4.1), the RJ law gives the energy density per unit frequency $\rho(\nu)$ as $k_B T$ multiplied by the number of modes per unit frequency interval per unit volume as

$$\rho(\nu) = \left(\frac{8\pi\nu^2}{c^3}\right)k_B T, \tag{4.2}$$

where the emitted radiation intensity is expressed as a function of frequency and temperature $I(\nu, T)$, as follows:

$$I(\nu, T) = \int \rho(\nu)d\nu.$$

As shown in figure 4.2, the experimentally obtained values of $I(\nu,T)$ show a peak wavelength that follows the classic Wein's displacement law given by

$$\lambda_{\max} T = 2.8978 \times 10^{-3}\,\text{m K}.$$

The $I(\nu, T)$ spectra obtained for a black body at three temperatures are shown in figure 4.2. The maximum of the curves ($\lambda_{\max}$) shifts to lower wavelengths (higher frequencies) as the temperature of the body is increased. This reveals the fact that

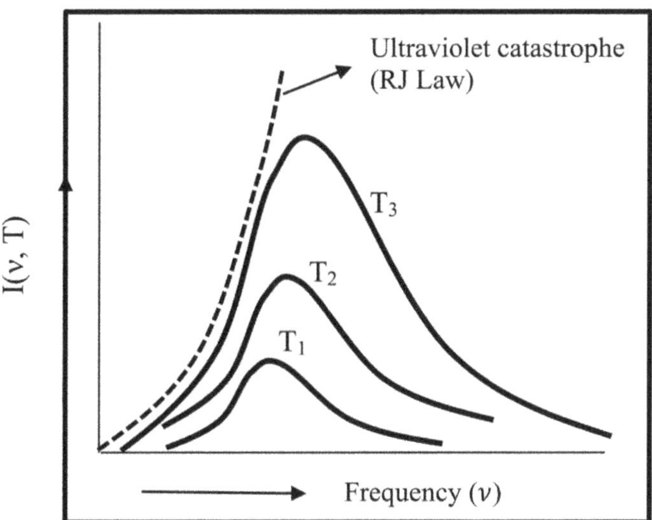


**Figure 4.2.** Plot of radiation intensity $I(\nu,T)$ versus frequency ($\nu$) for a body at three temperatures $T_1 < T_2 < T_3$. The dashed line shows the theoretical prediction obtained via the RJ law.

cold objects emit at longer wavelengths (or lower frequencies). Hot objects emit many more frequencies with a shift of the peak to shorter wavelengths.

The attempts to explain these results using classical theories were unsuccessful. Figure 4.2 also shows the curve of $I(\nu, T)$ predicted by the RJ law (equation (4.2)). One can see that it agrees with the experimental values at lower frequencies, but deviates drastically from the experimental values at higher frequencies or in the UV region. This deviation was popularly known as the 'ultraviolet catastrophe' and was later explained by Planck (♠ see section 4.3).

**Exercise 4.2.** An optoelectronics engineer in the department of climate protection has to design a device for *animal detection* to be used in the forest at night. What typical wavelength ($\lambda_{max}$) of radiation should be considered for the components of the device?

**Solution:** The body temperatures of animals range between 37 °C and 41 °C. For example, for humans, the typical body temperature is above 35 °C (or 308 K). Using Wein's law, we can obtain the wavelength, as follows:

$$\lambda_{max} = 2.8978 \times \frac{10^{-3} \text{ m K}}{308\text{K}} = 9.40 \times 10^{-6} \text{ m}.$$

Therefore, the device must have sensitivity in the IR region.

## 4.2 Boltzmann statistics

As described in chapter 2, in an atomic or molecular system, there are infinite set of discrete energy levels. The population ($N_j$) of an energy level with energy $E_j$ in thermal equilibrium is given by Boltzmann statistics, as follows:

$$N_j = \frac{g_j N_0 e^{-E_j/k_B T}}{\sum_i g_i e^{-E_j/k_B T}}. \tag{4.3}$$

Here, $\sum_i g_i e^{-E_j/k_B T}$ is known as the partition function ($z$). Equation (4.3) shows that when $E_j$ increases, $N_j$ decreases, as shown in figure 4.3. Here, $g_i$ and $g_j$ are the degeneracies, or the statistical weights of the two states. For high-density materials such as solids, the energy levels may be continuously distributed or separated—depending upon the state of the matter. The probability per unit energy $g(E)$ of finding a fraction of atoms excited to energy level $E$ is given by

$$g(E) = \frac{1}{KT} e^{-E/k_B T}.$$

The distribution of the population ($N$) in a state at temperature $T$ is given by the Boltzmann distribution:

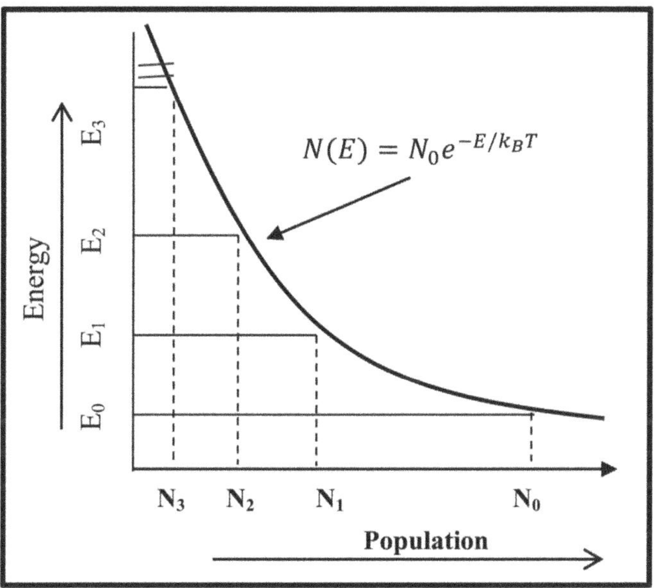



**Figure 4.3.** Boltzmann distribution of the population. The dashed lines correspond to the populations ($N_0$, $N_1$, $N_2$, and $N_3$) for indicated energy states ($E_0 < E_1 < E_2 < E_3$).

$$N(E) = N_0 e^{-E/k_B T} \tag{4.4}$$

Here, $N_0$ is the population in the ground state and $k_B$ is the Boltzmann constant ($= 1.3806488 \times 10^{-23}$ J K$^{-1}$). The higher particle states are also populated, depending on the temperature, as shown in figure 4.3.

We can write the ratio of the populations in two levels, 1 and 2, with energies $E_1$ and $E_2$ such that $\Delta E = E_2 - E_1$, as follows:

$$\frac{N_1}{N_2} = \frac{g_1}{g_2} e^{\Delta E/k_B T}. \tag{4.5}$$

**Exercise 4.3.** Find the population of first thermally excited level for a visible wavelength (600 nm) at room temperature.

**Solution:** We must convert the wavelength (in nm) to energy (eV) (♠ see exercise 2.1). For 600 nm, $E = 2.268$ eV, the population is given by equation (4.4). Let the populations in the ground and excited states be $N_0$ and $N_1$, respectively. Using equation (4.5),

$$\frac{N_0}{N_1} = \exp(\Delta E/k_B T)$$

$$\frac{N_0}{N_1} = \exp\left[\frac{2.268 \times (1.6 \times 10^{-19})}{\{(1.3\,806\,488 \times 10^{-23}) \times 300\}}\right]$$

$$= e^{87.6}$$

$$= 1.1 \times 10^{38}.$$

This shows that $N_0 \sim 10^{38} N_1$. At room temperature, the thermally populated value of the first excited state is smaller by a factor of $10^{-38}$ than that of the ground state. The ground-state population is dominant in this case.

**Exercise 4.4.** The wavelength of a ruby laser is 694.3 nm. What are the ratios of the populations in the ground ($N_0$) and excited states ($N_1$) corresponding to this wavelength at 300 K and 600 K?

**Solution:** We again use equation (4.5), $\frac{N_0}{N_1} = \exp(\Delta E/k_{\mathrm{B}}T)$.

For the temperature of 300 K, the value of the ratio

$$\frac{N_1}{N_0} = \exp\left[-\left(\frac{hc}{\lambda K_{\mathrm{B}}T}\right)\right]$$

$$= \exp\left[-,\frac{6.64 \times 10^{-34}\ \mathrm{Js} \times 3 \times 10^8 \mathrm{ms}^{-1}}{694.3 \times 10^{-9}\,\mathrm{m} \times 1.38 \times 10^{-23}\,\mathrm{JK}^{-1} \times 300\,\mathrm{K}}\right] = \exp[-69.27] = 8.26 \times 10^{-31}.$$ Similarly, at the higher temperature of 600 K, the ratio $\frac{N_1}{N_0}$ is $\exp[-34.63] = 9 \times 10^{-16}$. It can be estimated that at higher temperatures it will be easier to obtain population inversion (i.e. $N_1 > N_0$).

## 4.3 Planck's law of radiation

Apart from the ultraviolet catastrophe described in section 4.1.3, another problem with classical theory is that it predicts the total energy[1] density to be infinite, since all frequencies are possible. Physically, an infinite-energy EM field is not possible. In the year 1900, Planck described the spectral distribution of thermal radiation by introducing the idea of a quantum of energy $E$ at a given transition frequency $\nu$ between two levels $E_2$ and $E_1$:

$$E = h\nu.$$

Here, h is known as Planck's constant ($h = 6.62 \times 10^{-34}$ Js). The empirical function proposed by Planck for the radiation density within the frequency interval $d\nu$ is as follows:

$$\rho(\nu)d\nu = \frac{8\pi h\nu^3}{c^3}\left(\frac{1}{e^{h\nu/k_{\mathrm{B}}T} - 1}\right)d\nu. \tag{4.6}$$

The factor $1/(e^{h\nu/k_{\mathrm{B}}T} - 1)$ was reintroduced for the ideal Bose gas by Einstein in the form of Bose–Einstein statistics for identical and indistinguishable particles (called bosons, with integral spins) for a number density $\rho(E)$ at an energy $E$ as

---

[1] The total power per unit area $\mathrm{I} = \int_0^\infty I(\nu, T)d\nu$ diverges to $\infty$ when all frequencies are allowed.

$$\rho(E) = \frac{1}{(e^{(E-\mu)/k_BT} - 1)}. \tag{4.7}$$

Here, $\mu$ is known as the chemical potential for identical particles, and for photons, $\mu = 0$. Planck's law fits the experimental results shown in figure 4.2 quite well.

♣ Fermions, i.e. particles such as electrons (whose spin is multiples of 1/2) follow the Fermi–Dirac statistics given by $\rho(E) = \frac{1}{\left(e^{(E-\mu)/k_BT} + 1\right)}$.

**Exercise 4.5. (a)** Show that for long wavelengths, Planck's law reduces to the RJ law.
**(b)** What happens at shorter wavelengths?
**Solution: (a)** The exponential function in Planck's formula (equation (4.6)) can be expanded using a Taylor series for $f(x) = e^x$ i.e. $e^x = 1 + x + \frac{x^2}{2!} + \frac{x^3}{3!} + ....$

For longer wavelengths, or low frequencies, i.e. for small $x$ (ignoring the higher terms in the expansion), $e^x = 1 + x$.

Therefore, for $h\nu k_BT$, we have

$$\frac{1}{(e^{h\nu/k_BT} - 1)} \cong \frac{k_BT}{h\nu}.$$

Thus, at low frequencies, Planck's formula becomes $\rho(\nu)d\nu = \left(\frac{8\pi\nu^2}{c^3}\right)k_BT\,d\nu$, which is the RJ law (equation (4.2)).

**(b)** For shorter wavelengths or higher frequencies, Planck's law behaves as Wein's law. This is the case because, for $h\nu kT$, we can rewrite equation (4.7) as

$$\rho(\nu)d\nu \quad = \frac{8\pi h\nu^3}{c^3}e^{-\frac{h\nu}{k_BT}}d\nu.$$

The radiation intensity follows an exponential behavior, as shown in figure 4.2. This equation can be written in terms of wavelength as

$$E_\lambda \quad = \frac{8\pi hc}{\lambda^5}e^{-\frac{hc}{\lambda k_BT}}.$$

To find the maxima, we can differentiate this expression with respect to $\lambda$ and equate $\frac{dE_\lambda}{d\lambda}$ to zero. This is left as an exercise; it yields Wein's displacement law, $\lambda T = $ constant.

## 4.4 Einstein's $A$ and $B$ coefficients

### 4.4.1 Stimulated absorption coefficient

With reference to the Boltzmann distribution (figure 4.3), the exited states cannot have higher populations than the ground state. Nevertheless, there are ways in which the populations of the upper energy states can be made larger than that of the ground state. For example, on absorption of a photon by an atom, the thermal

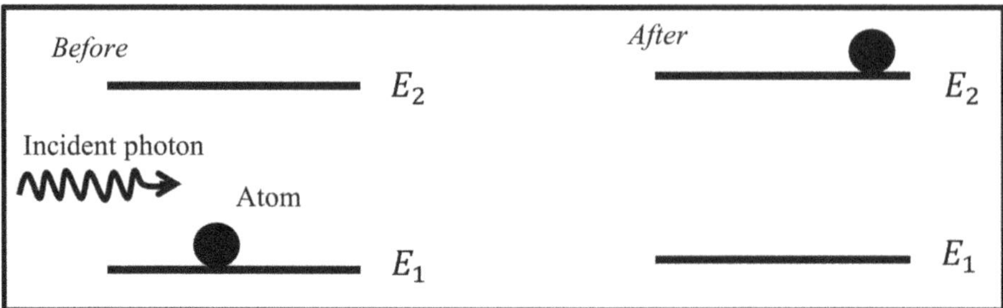


**Figure 4.4.** Process of absorption in an atomic system with two energy levels ($E_1$ and $E_2$): an atom before (*left*) and after (*right*) interacting with an incident photon.

equilibrium is disturbed. In 1917, Einstein combined Planck's law and the Boltzmann statistics and further developed the idea of light absorption and emission. To obtain the correct energy density of radiation, one must include a *stimulated emission probability* in dealing with the interaction of radiation with matter.

As shown in figures 4.4–4.6, consider the two energy levels of an atomic system $E_1$ and $E_2$ with populations of atoms denoted by $N_1$ and $N_2$, respectively. An atom can be placed in its higher level $E_2$ on absorption of a photon. The absorption of a photon is in addition to the thermal population, and hence it is also referred to as 'stimulated absorption'. This is due to the fact that EM radiation is required to stimulate the electrons to produce the short-lived entity—the excited state denoted by $E_2$ in figure 4.4.

The rate of change of the population in the lower state due to absorption is proportional to the population in the state $E_1$ and the incident photon density $\rho(\nu)$:

$$\frac{dN_1}{dt} \propto \rho(\nu)N_1.$$

The rate of change of $N_1$ can be rewritten by introducing a rate constant $B_{12}$, known as the *Einstein's B coefficient* of stimulated absorption, as follows:

$$\frac{dN_1}{dt} = B_{12}\rho(\nu)N_1. \tag{4.8}$$

### 4.4.2 Spontaneous emission

Consider figure 4.5, in which the atom is now in the higher state $E_2$. As $E_2 > E_1$, the atom decays to the lower state $E_1$ by emitting a photon of frequency $\nu$. This process is known as *spontaneous emission*.

In the condensed phase, the emission following optical excitation is also known as fluorescence or photoluminescence[2]. Overall, the population decay of the state $E_2$ in terms of another constant of proportionality known as the rate coefficient of spontaneous emission ($A_{21}$), is written as

---

[2] ♣To distinguish it from other types of emissions, see appendix A.

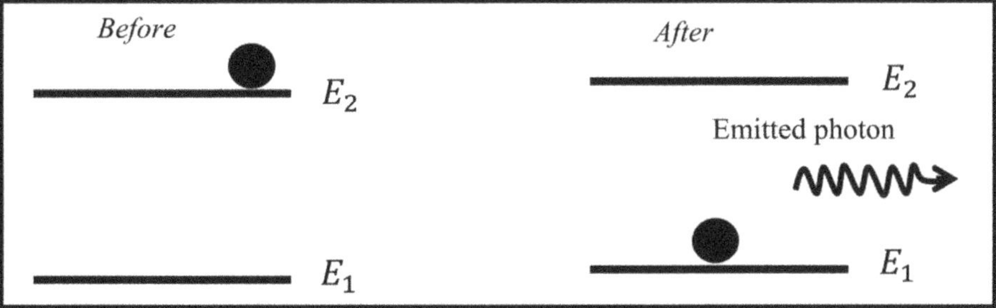


**Figure 4.5.** Schematic of the process of spontaneous emission in an atomic system with two energy levels ($E_1$ and $E_2$): an atom before (*left*) and after (*right*) spontaneously emitting a photon.

$$\frac{\mathrm{d}N_2}{\mathrm{d}t} = -A_{21}N_2. \tag{4.9}$$

The negative sign in this equation expresses the decrease of the population over time. Here, $A_{21}$ is also known as *Einstein's A coefficient* and has units of s$^{-1}$. The photons in the spontaneous emission process are emitted in random directions with arbitrary polarization and within a broad range of frequencies for allowed transitions.

The solution of differential equation (4.9) is a first-order decay that is well known in physics:

$$N_2 = N_2^0 e^{-A_{21}t}.$$

Here, $N_2^0$ is the population of the upper state at $t = 0$. The time in which the population falls to $1/e$ of its initial value is known as the *lifetime* of the spontaneous emission ($\tau_0$). It is the inverse of the decay rate $\left(= \frac{1}{A_{21}}\right)$ and is similar to that described in chapter 2 (♠ see section 2.9.7.8). For most molecules and semi-conductors, $\tau_0$ is of the order of nanoseconds ($10^{-9}$ s). For several other laser materials, such as rare-earth ions (e.g. $Cr^{+3}$ and $Nd^{+3}$), the lifetimes are found to be of the order of milliseconds ($10^{-3}$ s).

### 4.4.3 Coefficient of stimulated emission

The excited state can be populated by the absorption of a photon as well as by thermal excitation. The phenomenon of stimulated absorption (from a lower state) suggests that a process that is the reverse of stimulated emission should exist for the excited state.

Let us consider a situation in which an atom is in the higher state $E_2$ and a photon of same frequency $\nu$ is incident on the system, as shown in figure 4.6. There is a finite probability that the incident photon will stimulate the atom to resonantly undergo a transition to the lower state $E_1$ by emitting another photon. If this happens, then the two emitted photons will possess properties analogous to each other. This means that light consisting of these photons has one frequency (the property of

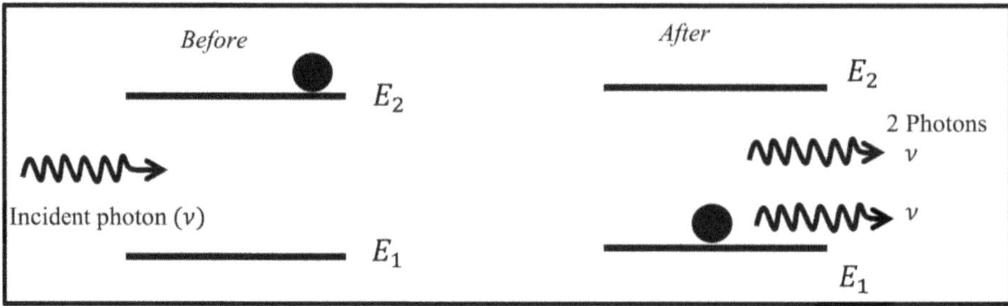


**Figure 4.6.** Schematic diagram of the process of stimulated emission in an atom when a photon of the same frequency ($\nu$) is incident on it, showing the situations before (*left*) and after (*right*) the photon strikes the system.

*monochromaticity*) and is in phase (the property of *coherence*). These properties are not associated with spontaneously emitted photons.

Let us also consider the coefficient of spontaneous emission ($A_{21}$), which is independent of the external field. It is a statistical function of space and time and there is no phase relationship in the light emitted by the atoms. The fluorescence lifetime of the excited state ($\tau_0$, ♠ see equation (2.13)) is an experimentally measurable quantity. On the other hand, the depletion of $E_2$ by $B_{21}$ is given by its dependence on external factors. In this case, the rate of change of population of the upper state due to absorption is proportional to the population in the state $E_1$ and the incident photon density $\rho(\nu)$. It is given by

$$\frac{\mathrm{d}N_2}{\mathrm{d}t} \propto \rho(\nu)N_2.$$

By introducing the rate coefficient of stimulated emission $B_{21}$, we can write this as

$$\frac{\mathrm{d}N_2}{\mathrm{d}t} = -B_{21}\rho(\nu)N_2. \tag{4.10}$$

The unit of energy density ($\rho_\nu$) per unit frequency interval is J s m$^{-3}$. Therefore, comparing the units of $A_{21}$ and $B_{21}\rho(\nu)$, the unit of Einstein's $B$ coefficient is m$^3$ J$^{-1}$ s$^{-2}$.

### 4.4.4 Rate equation analysis

The rates of stimulated absorption ($B_{12}\rho(\nu)N_1$), spontaneous emission ($A_{21}N_2$), and stimulated emission ($B_{21}\rho(\nu)N_2$) of a two-state system are indicated in figure 4.7. We now have sufficient information to work on the rate equations for a two-level system, as follows.

Using equations (4.8)–(4.10), we can write the rates of change of the populations of states $E_2$ and $E_1$ as

$$\frac{\mathrm{d}N_2}{\mathrm{d}t} = -B_{21}\rho(\nu)N_2 - A_{21}N_2$$

and

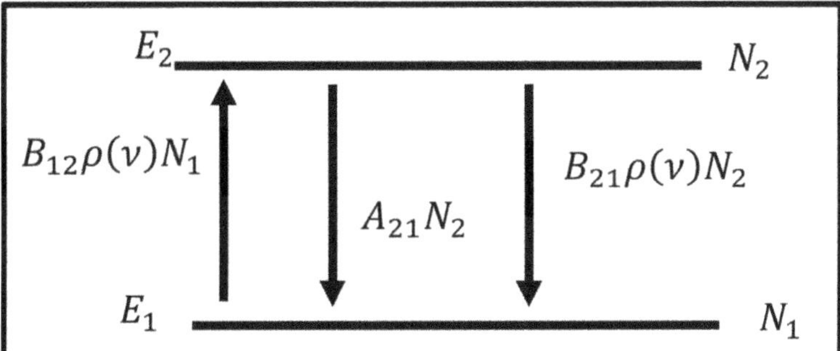



**Figure 4.7.** Two-level system used for analysis of the rate equation. $B_{12}\rho(v)N_1$, $A_{21}N_2$, and $B_{21}\rho(v)N_2$ are the rates of stimulated absorption, spontaneous emission, and stimulated emission, respectively.

$$\frac{\mathrm{d}N_1}{\mathrm{d}t} = B_{12}\rho(v)N_1,$$

respectively.

At equilibrium, the rate of increase of the upper state ($E_2$) via induced absorption is equal to its decay to the lower state ($E_1$) via the sum of the spontaneous and stimulated emission rates, as follows

$$B_{12}\rho(v)N_1 = B_{21}\rho(v)N_2 + A_{21}N_2.$$

Dividing this by $B_{21}\rho(v)N_2$ gives

$$\frac{A_{21}}{B_{21}\rho(v)} = \frac{B_{12}N_1}{B_{21}N_2} - 1.$$

From the Boltzmann distribution, substituting the value of $N_1/N_2$ from equation (4.5), we get

$$\frac{A_{21}}{B_{21}\rho(v)} = \frac{B_{12}}{B_{21}}\frac{g_1}{g_2}e^{\left(\frac{\Delta E}{k_{\mathrm{B}}T}\right)} - 1. \tag{4.11}$$

This can be rewritten as

$$\rho(v) = \left(\frac{A_{21}}{B_{21}}\right) \Big/ \left(\frac{B_{12}}{B_{21}}\frac{g_1}{g_2}e^{\left(\frac{\Delta E}{k_{\mathrm{B}}T}\right)} - 1\right). \tag{4.12}$$

By comparing this to the energy density of radiation at frequency $v$ in a cavity using Planck's law (equation (4.6)), we obtain

$$\left(\frac{A_{21}}{B_{21}}\right) = \frac{8\pi h v^3}{c^3} \tag{4.13}$$

and

$$\frac{B_{12}}{B_{21}}\frac{g_1}{g_2} = 1.$$

We assume here that either the split of the energy states is the same or that there is no split. Therefore, for the same degeneracies, i.e. $\frac{g_1}{g_2} = 1$, we obtain $B_{12} = B_{21}$. This indicates that the probabilities of absorption and stimulated emission for the transitions between the two states are equal. This is an important message, as it suggests that materials characterized by strong absorption coefficients can be expected to have strong emission coefficients, which is to say that they can be excellent laser materials. For example, organic dyes (such as rhodamine 6G) have strong absorption in the visible region and are excellent stimulated emitters. Two profound consequences of Einstein's coefficients are highlighted on rewriting the equations, as shown in the following subsections.

### 4.4.4.1 Relation between spontaneous and stimulated emission

From the relation given in equation (4.13), the lifetime of the upper level ($\tau_0 = 1/A_{21}$) can be related to the stimulated emission, as follows:

$$B_{21} = \frac{c^3}{8\pi h \nu^3 \tau_0}. \tag{4.14}$$

Therefore, if a material can be characterized by a spontaneous rate (by estimating the value of $\tau_0$), the stimulated rate can be found from equation (4.14). Notice that in terms of wavelength, equation (4.14) can be written as $B_{21} = \frac{\lambda^3}{8\pi h \tau_0}$.

**Exercise 4.6.** The dye rhodamine 6G in ethylene glycol has a lifetime of 3.5 ns. For an emission wavelength of 570 nm, what is the coefficient of stimulated emission? (Take the refractive index of the solvent to be 1.4.)

  **Solution:** Equation (4.14) gives the coefficient of stimulated emission as

$$B_{21} = \frac{c^3}{8\pi h \nu^3 \tau_0}.$$

The refractive index ($n$) of the solvent is 1.4. Therefore, the speed of light in this medium is $c/n$. The lifetime ($\tau_0$) is $3.5\times10^{-9}$ s. The frequency ($\nu$) corresponding to 570 nm is $5 \times 10^{14}$ Hz. Using the value of $h = 6.6 \times 10^{-34}$ J s, the value of $B_{21}$ obtained is $1.5\times10^4$ m$^3$ J$^{-1}$ s$^{-2}$.

### 4.4.4.2 The ratio and the magic of stimulated emission

Similarly, equation (4.11) can now be rewritten as follows:

$$\frac{A_{21}}{B_{21}\rho(\nu)} = e^{h\nu/k_{\mathrm{B}}T} - 1. \tag{4.15}$$

The quantity $\frac{A_{21}}{B_{21}\rho(\nu)}$ is purely a number and is known as the ratio ($R$) of spontaneous to stimulated emission coefficients for a system in thermal equilibrium with a radiation field. As can be seen from the above equation, its value varies with frequency and the temperature and is extremely important in the development of lasers. Table 4.1 gives typical values of $R$ at various temperatures for a wide range of the EM spectrum.

**Exercise 4.7.** Evaluate the ratio ($R$) of the rate of spontaneous radiation to that of stimulated radiation for a tungsten filament for the visible frequency range at 2000 K.

**Solution:** At $T = 2000$ K, for the visible frequency range ($5 \times 10^{14}$ Hz), the value of this ratio from equation (4.15) is approximately $e^{12}$ (or $1.5 \times 10^5$).

This is the reason why the majority of the radiation emitted by tungsten lamps takes the form of spontaneous emission and the probability of stimulated emission is minimal.

**Exercise 4.8.** Use equation (4.15) and assume a wavelength of 570 nm. If we wish to make the rate of spontaneous emission ($A_{21}$) equal to the factor $B_{21}\rho(\nu)$, what temperature could be used?

**Solution:** According to the question, to make the value of ratio $R$ in equation (4.15) equal one, we need

$$e^{h\nu/k_B T} = 2.$$

The frequency ($\nu$) corresponding to 570 nm is $5 \times 10^{14}$ Hz. Using the values $h = 6.6 \times 10^{-34}$ J s$^{-1}$ and $k_B = 1.38 \times 10^{-23}$ J K$^{-1}$, the value of

$$\frac{h\nu}{k_B T} = \frac{2.53 \times 10^4}{T}.$$

Therefore, $\frac{2.53 \times 10^4}{T} = \ln 2$,

or the required temperature $T = \frac{2.53 \times 10^4}{\ln 2} = 3.64 \times 10^4$ K

A glance at table 4.1 shows that in the microwave region, the level of stimulated emission dominates that of spontaneous emission, as the value of $R$ is small. This was the obvious reason for the invention of Microwave Amplification by Stimulated Emission of Radiation (MASER) before the invention of LASER. To obtain stimulated emission in the visible/IR region, the denominator of $R$ in equation (4.15) should be increased. This can be achieved by increasing the value of $\rho(\nu)$ between the two energy levels. This is the subject of the next chapter.


♣ The research groups of Townes (1955) and Basov and Prokhorov [4] independently demonstrated amplification in the microwave region in ammonia.

♣ Basov and Prokhorov [4] wrote a paper on their molecular generator for the journal '*Zhurnal Èksperimental'noi i Teoreticheskoi Fiziki*' (later known as '*Soviet Physics—JETP*'), which they submitted in December 1953, but this paper was delayed by a year due to a typographical error in $2\pi$ that needed to be corrected!

**Table 4.1.** The ratio ($R$) of spontaneous emission to stimulated emission at various temperatures for different frequencies.

| Wavelength[a] (nm) | Frequency[a] (Hz) | Value of $R$ at various temperatures | | | | |
|---|---|---|---|---|---|---|
| | | 1 K | 10 K | 100 K | 1000 K | 10000 K |
| 180000000 | $10^{10}$ | 0.008003 | 0.000797 | 0.000266 | $7.97 \times 10^{-05}$ | $7.97 \times 10^{-06}$ |
| 18000000 | $10^{11}$ | 0.082973 | 0.008003 | 0.002661 | 0.000797 | $7.97 \times 10^{-05}$ |
| 1800000 | $10^{12}$ | 1.219099 | 0.082973 | 0.026926 | 0.008003 | 0.000797 |
| 180000 | $10^{13}$ | 2894.794 | 1.219099 | 0.304344 | 0.082973 | 0.008003 |
| 18000 | $10^{14}$ | $4.15 \times 10^{34}$ | 2894.794 | 13.25353 | 1.219099 | 0.082973 |
| 1800 | $10^{15}$ | — | $4.15 \times 10^{34}$ | $3.46 \times 10^{11}$ | 2894.794 | 1.219099 |
| 180 | $10^{16}$ | — | — | $2.5 \times 10^{115}$ | $4.15 \times 10^{34}$ | 2894.794 |
| 18 | $10^{17}$ | — | — | — | — | $4.15 \times 10^{34}$ |

[a] See figure 2.2 for the ranges of wavelengths and frequencies in the UV, visible, and IR regions.

## Questions and problems

1. How do you represent the number of radiation modes per unit volume in a frequency range $\mathrm{d}\nu$ in free space, in terms of the frequency $\nu$ and the velocity of light, $c$?
2. The temperature of the Sun's surface is about 6000 K. What is the intensity of the radiation from the Sun?
3. What is the mechanism by which an incandescent bulb and a mercury light work?
4. The temperature of the tungsten filament of a light bulb is 3000 K. What is the peak of its light spectrum?
5. According to Planck's law of black-body radiation, what is the energy density of radiation between the frequencies $\nu$ and $\nu + \mathrm{d}\nu$?
6. The number of atoms of energy $E$ within a specific energy range $(\mathrm{d}E)$ can be given by $N(E)\mathrm{d}E = \frac{N}{KT}e^{-E/k_{\mathrm{B}}T}\mathrm{d}E$. Calculate the temperature at which you could see a considerable number of 'visible' photons.
7. Find the energy of a particle at a room temperature of ~300 K.
8. Calculated the mean number of photons excited in the radiation field mode of a thermal source that has a wavelength 400 nm at a temperature of 300 K.
9. For nondegenerate atomic levels of a system in thermal equilibrium, what is the ratio of Einstein's $A$ coefficient to Einstein's $B$ coefficient in terms of frequency $v$, the velocity of light $c$, and Planck's constant?
10. Write down the units of the rate constants of spontaneous emission $(A)$ and stimulated emission $(B)$. What is the radiative rate for a molecular system with radiative lifetime of 10 ns?
11. At a constant temperature, compare the ratio of the spontaneous and stimulated emission rates typically obtained for the microwave and UV regions of the EM spectrum.
12. An argon-ion laser works on a single line (514.5 nm) with an output power of 78.50 mW. Take the spectral purity to be 0.001 nm. What is the value of $A_{21}/B_{21}$ for the laser?

## Bibliography

[1] Einstein A 1917 Zur Quantentheorie der Strahlung *Phys Z.* **18** 121–8
[2] Gordon JP, Zeiger HJ and Townes CH 1955 The maser—new type of microwave amplifier, frequency standard, and spectrometer *Phys. Rev.* **99** 1264–74
[3] Maiman TH 1960 Stimulated optical radiation in ruby *Nature* **187** 493
[4] Basov N G and Prokhorov A M 1954 Application of molecular beams for the radio-spectroscopic study of rotational molecular spectra *Zh. Eksp. Teor. Fiz.* **27** 431–8
[5] Karlov NV, Krokhin ON and Lukishova SG 2010 History of quantum electronics at the Moscow Lebedev and General Physics Institutes: Nikolaj Basov and Alexander Prokhorov. *Appl. Opt.* **49** F32–46

**IOP** Publishing

An Introduction to Photonics and Laser Physics
with Applications

**Prem B Bisht**

# Chapter 5

# The Beer–Lambert law and the gain coefficient

The amount of light absorbed by a medium can be estimated by measuring the transmission of collimated light. Spontaneous emission spreads in all directions. Therefore, interrelating the quantities associated with the processes of absorption and emission is useful to obtain the decay rates of the excited energy states. This introduces the requirement for a relation between Einstein's $A$ and $B$ coefficients. Since we will deal with the energy density as well as the absorption and emission rates, we will work out the rate equations in this chapter. The measurement of the absorption spectrum of a material—a basic experiment for a student of spectroscopy—is introduced in this chapter. Knowledge of the absorption strength–known as the absorption cross section—is important in establishing whether a material is suitable for use as a gain medium in lasers.

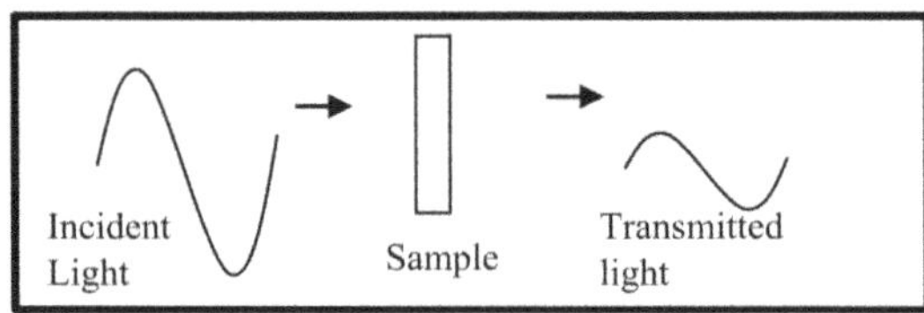


**Learning objectives**
**After reading this chapter, the learner will be able to:**
State the concept of absorption and the Beer–Lambert law;
Describe the absorption coefficient;
Define a gain medium;
Explain population inversion and the small-gain coefficient;
Relate the round-trip gain to gain saturation;
Understand the technique of cavity ring-down spectroscopy.

## 5.1 The Beer–Lambert law

As mentioned in chapter 2, spectroscopy has contributed immensely to the development of modern science, in particular, in ground-breaking experiments in the late 19th and early 20th centuries. One spectroscopic technique deals with the measurement of absorption spectra. For a two-level system, for example with $E_2 > E_1$, the Boltzmann law suggests that at a given temperature, the largest population of molecules lies in the lower energy state ($E_1$). Let us look at the experiment shown in figure 5.1. A broadband light source of various frequencies with a constant intensity $I_0$ is incident on the sample. The selection of a frequency is achieved using either a prism or a grating spectrograph (known as a monochromator). The sample is considered to be made of infinitesimally thin sheets of thickness $\Delta x$.

♣ The term *intensity* is defined as the power per unit solid angle per unit area of the source (♠ see chapter 10 for details of the solid angle). On the other hand, irradiance is the total power per unit area incident on the body.

The change of intensity ($\Delta I$) for subsequent sheets of the sample is proportional to their thickness ($\Delta x$) and the incident intensity at the particular sheet ($I(x)$):

$$\Delta I \propto \Delta x I(x)$$

or

$$\Delta I = -\alpha \quad \Delta x I(x).$$

Here, $\alpha$ is the proportionality constant known as the *absorption coefficient* (in $\text{cm}^{-1}$). It is related to the absorption strength of a material. For a sample in liquid form, it is further described as the product of the concentration ($C$) and the extinction coefficient ($\epsilon$) of the sample. The absorbing medium causes a loss of photons from an incident beam of initial intensity $I_0$. The detector is used to monitor the transmitted light intensity ($I$).

♣ In practice, with the help of a beam splitter, two beam paths are generated from the same source for the sample and the reference to measure $I$ and $I_0$, respectively.

We can rearrange the above equations as

$$\frac{\partial I}{\partial x} = -\alpha I(x). \tag{5.1}$$

On integrating we obtain the following form of the equation, known as the Beer–Lambert law.

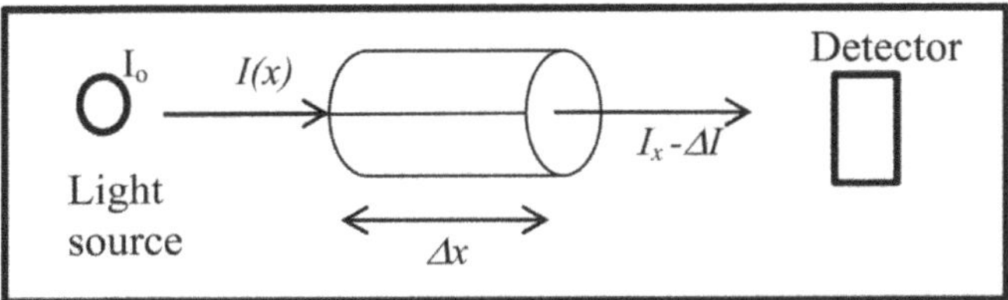


**Figure 5.1.** A collimated light beam of intensity $I(x)$ and frequency $\nu$ falls on a detector after passing through the sample with a thickness of $\Delta x$.

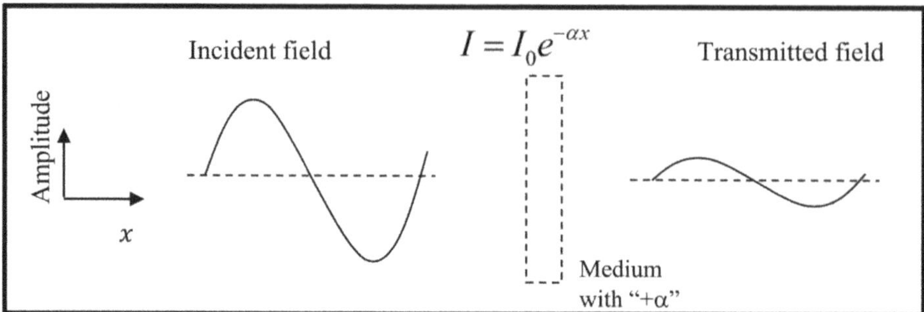



**Figure 5.2.** Effect of absorption by a medium on the amplitude of an incident EM wave.

$$\ln\left(\frac{I}{I_0}\right) = -\alpha x \text{ or } I = I_0 e^{-\alpha x}. \tag{5.2}$$

Here, $\alpha = \epsilon C$ for liquid solutions. The quantity $\ln\left(\frac{I_0}{I}\right)$ is also known as the optical density (OD) of the sample.

The basic idea is to measure the light transmitted by the sample to find its absorption strength across the frequencies of interest. The effect of the EM wave incident on the medium is schematically shown in figure 5.2. We can see that the amplitude of the incident light wave is reduced after passing through an absorbing medium.

**Exercise 5.1.** An optical density of unity is obtained for a sample that has a concentration of 0.1 M when the path length used is 1 mm. What are the other values of thickness and concentration that could be be selected to obtain samples with the same optical density?

**Solution:** The optical density (OD) is defined as $\ln(I_0/I)$. For the path length ($x$) and the concentration ($C$) of a sample, we know that its extinction coefficient ($\epsilon$) is a fixed quantity ($\alpha/C = \epsilon$). Equation (5.2) can be rewritten as $OD = \epsilon C x$, and we see that to maintain the same value of OD, the parameters $x$ and $C$ have to be adjusted suitably. Therefore, for the given situation, OD = 1 can be obtained, for example, using (i) $C = 0.5$ M with a sample path length of 0.2 mm, or (ii) $C = 0.01$ M with a path length of 10 mm, and so on.

## 5.2 Absorption coefficient

The irradiance $I(\nu)$ of a light source at a photon frequency is the power delivered by the source (or energy crossing in unit time) to the unit area of a body. Therefore, in terms of the energy density of the photons $\rho_\nu$, the speed of light $c$, and the refractive index $n$ of the medium, it is also written as $I(\nu) = \rho_\nu \frac{c}{n}$. A change in the irradiance, $\Delta I(\nu)$, is defined as the total number ($M$) of photons delivered by a source of an area ($\Delta A$) per unit time ($\Delta t$), as follows:

$$\Delta I(\nu) = \frac{Mh\nu}{\Delta A \Delta t}.$$

The energy density $(\rho_\nu)$ of $M$ photons for a volume $V$, is given by

$$\rho_\nu = \frac{Mh\nu}{V} = Nh\nu.$$

Here, $N$ is number of photons per unit volume. For sufficiently small $\Delta x$, we can rewrite the change in irradiance within the light beam as

$$\Delta I(\nu) = \frac{Mh\nu}{\Delta A \Delta t}\frac{\Delta x}{\Delta x} = \frac{Mh\nu}{\Delta V}\frac{\Delta x}{\Delta t} = \Delta Nh\nu\frac{\Delta x}{\Delta t}.$$

By replacing $\nu$ by $\nu_{21}$, which denotes the emission frequency between the two levels, we can express $\Delta N$ as

$$\Delta N = \frac{\Delta I(\nu)}{h\nu_{21}}\frac{\Delta t}{\Delta x}. \tag{5.3}$$

The beam travels through a volume element that has a thickness of $\Delta x$ and a unit sectional area of $\Delta A$. As described in chapter 4, after stimulated absorption rate $(B_{12}\,\rho\,(\nu)\,N_1)$, the molecules from the higher energy state decay to the lower state with a spontaneous emission rate of $(A_{01}N_2)$ and a stimulated emission rate of $(B_{21}\rho(\nu)N_2)$. As shown in figure 5.3, assuming that the spontaneous emission coefficient $A_{21}$ is very small compared to that due to stimulated emission $(A_{21} \ll B_{21})$, the net rate of photon loss from the beam per unit volume is given by the following:

$$-\frac{dN}{dt} = B_{12}\rho(\nu)N_1 - B_{21}\rho(\nu)N_2 \text{ so that}$$

$$-\frac{dN}{dt} = \left(N_1\frac{B_{12}}{B_{21}} - N_2\right)\rho(\nu)B_{21}. \tag{5.4}$$

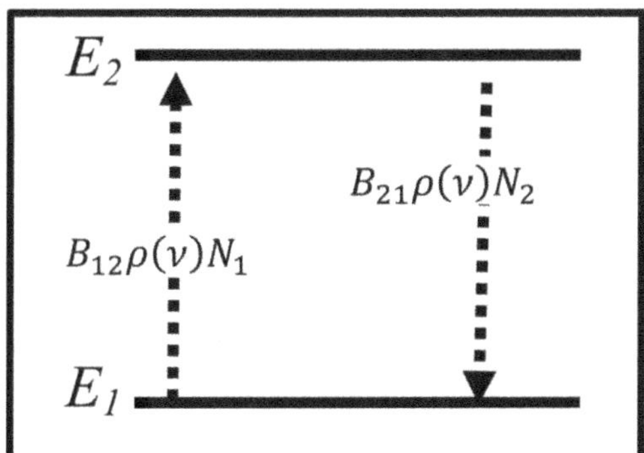


**Figure 5.3.** *Stimulated* absorption and absorption between two energy states $E_2$ and $E_1$; $N_2$ and $N_1$, respectively, are the corresponding populations. $B_{12}$ and $B_{21}$ are the coefficients of stimulated absorption and stimulated emission, respectively. The contribution of spontaneous emission has been neglected.

Equation (5.3) can be rewritten as

$$\frac{\partial N}{\partial t} = \frac{\partial I(\nu)}{\partial x}\frac{1}{h\nu_{21}}.$$

Using equation (5.1), we can rewrite this as

$$\frac{\mathrm{d}N}{\mathrm{d}t} = -\alpha I(x)\frac{1}{h\nu_{21}}.$$

Placing the value of irradiance $I(\nu)$, we get,

$$\frac{\mathrm{d}N}{\mathrm{d}t} = -\alpha\rho_\nu\frac{c}{n}\frac{1}{h\nu_{21}} \tag{5.5}$$

Comparing equations (5.4) and (5.5), and by using $\frac{B_{12}}{B_{21}} = \frac{g_2}{g_1}$, the absorption coefficient ($\alpha$) can be written as

$$\alpha = h\nu_{21}\frac{n}{c}\left(N_1\frac{g_2}{g_1} - N_2\right)B_{21}. \tag{5.6}$$

Equation 5.6 relates $\alpha$ to the coefficient of stimulated emission, $(B_{21})$. We already know from equation (4.13) that $B_{21}$ is related to the spontaneous emission rate coefficient, $(A_{21})$. Therefore, if we know the decay time of spontaneous emission $(\tau_0)$ of the system, the stimulated parameters can be estimated. From this relation, it is obvious that for a higher value of the absorption coefficient, the probability of stimulated emission is higher.

## 5.3 Gain media

We now know that atoms or molecules in the forms of gases, liquids, or solids give rise to spontaneous/stimulated emission when excited. Any suitable material which is capable of emitting with a high quantum yield, and can therefore be used to generate photons, is known as a *gain medium*. Examples of popular gain media are He–Ne (gas), dye solutions (liquid), and $Nd^{+3}$:yttrium aluminum garnet (YAG) (solid).

Equation (5.6) gives the absorption coefficient in terms of the populations of the two states involved in the transitions. This equation, when used in combination with the Beer–Lambert law, gives rise to an interesting insight about gain media. We see that the value of $\alpha$ is positive for $N_1 > N_2$ and becomes negative for $N_2 > N_1$. The latter is the situation for *population inversion* (see the next chapter). For instance, if we replace the negative value of $\alpha$ with k in equation (5.2), we obtain,

$$I = I_0 e^{kx}. \tag{5.7}$$

Equation (5.7) indicates that the irradiance increases exponentially with the thickness of the medium for a given value of $k$. Here $k$ is known as the *small-gain coefficient*, i.e. the gain at a given frequency of the emission spectrum.

## 5.4 Gain coefficient

Let us now consider figure 5.4 in view of equation (5.7) for negative values of the absorption coefficient ($\alpha = -k$). By comparing this figure with figure 5.2, we can schematically observe that the transmitted EM wave has an increased amplitude compared to that of the incident wave.

We know that when the absorption strength (also known as the oscillator strength) of the system is high, the amplitude of the interacting incident light is exponentially reduced by passing through the medium, as shown in figure 5.2. In contrast, for negative values of $\alpha$, the transmitted light gains photons from the gain medium, as shown in figure 5.4. The greater the oscillation strength, the larger the number of photons transmitted by the system. This point shows the importance of a prior estimate of the absorption parameters of a system in order to establish whether it is a suitable laser material for the given wavelength range.

♣ Richard Tolman proposed the idea of negative absorption in 1924.

♣ In terms of the transition dipole moment (♠ see equation (2.8)), the oscillator strength ($f$) is a dimensionless quantity given by $f = 4.319 \times 10^{-9} \int \epsilon(\nu)\mathrm{d}\nu$. Here, $\epsilon(\nu)$ is the extinction coefficient and the integration is performed over the absorption band.

To obtain a negative value of $\alpha$ (i.e. a positive value of $k$), we can rewrite equation (5.6) for a particular frequency $\nu_{21}$ as follows:

$$k = \frac{n}{c}\left(N_2 - N_1\frac{g_2}{g_1}\right)B_{21}h\nu_{21}. \tag{5.8}$$

Note that the difference between the populations (assuming an *inverted population* between $E_1$ and $E_2$) is now given by $\Delta N = N_2 - N_1\frac{g_2}{g_1}$. It should be noted that the transmitted amplitude has a net gain of photons and is now given by equation (5.7). This indicates that the amplitude is enhanced exponentially (i.e. amplified!). While $k$ is known as *small-gain coefficient*, the exponential factor $e^{kx}$ is termed the *overall gain coefficient* ($k_G$) or simply the *gain*.

From equation (5.8), the stimulated emission cross section ($\sigma_{es}$) is introduced, as follows:

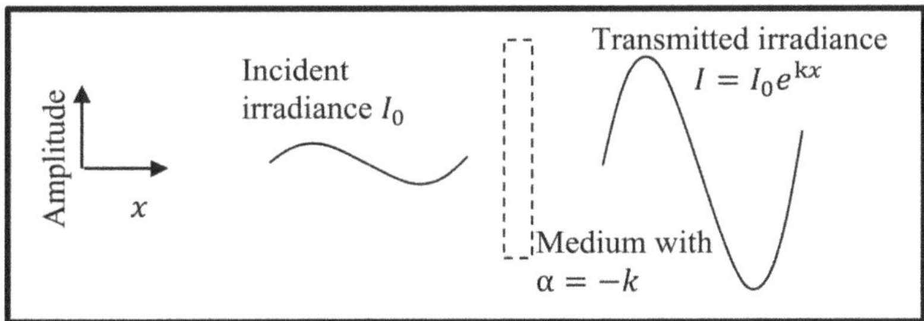


**Figure 5.4.** Schematic of the effect of a negative value of $\alpha$ on the amplitude of the incident EM wave. Compare this with figure 5.2.

$$k = \sigma_{es}\Delta N, \tag{5.9}$$

where $\sigma_{es} = \frac{B_{21}h\nu_{21}}{c}$. It can also be written in terms of $A_{21}$ from equation (4.13). Therefore, equation (5.7) can now be rewritten as

$$I = I_0 e^{\sigma_{es}\Delta N x}. \tag{5.10}$$

Lasers are also known as optical amplifiers[1]. We note here that an *ideal laser* requires a single optical transition between two energy levels. In practice, a correction by a spectral width factor is included in equation (5.8) (♠ see the next chapter).

**Exercise 5.2.** The stimulated emission coefficient of a gain medium is of the order of $10^{-19}$ cm$^2$. Assuming no losses, what is the value of the small-gain coefficient if the population inversion in the two states is equal to $10^{19}$ cm$^{-3}$ ?
   **Solution:** Using equation 5.9, we see that the value of $k$ is 1 cm$^{-1}$.

## 5.5 Round-trip gain

The value of the *small-gain coefficient $k$* is defined for a stimulated emission transition between two levels. As shown in figure 5.5, we now place the emitter in between two mirrors—a 2D cavity of length $L$. The reflectivities of the cavity mirrors (1 and 2) are $R_1$ and $R_2$, respectively. Let the initial photon density at mirror 1 be $\phi_1$. As shown below, after the first reflection, the photon density $\phi_2$ is given by the product of $\phi_1$ and the reflectivity of the first mirror:

$$\phi_2 = \phi_1 \times R_1. \tag{5.11}$$

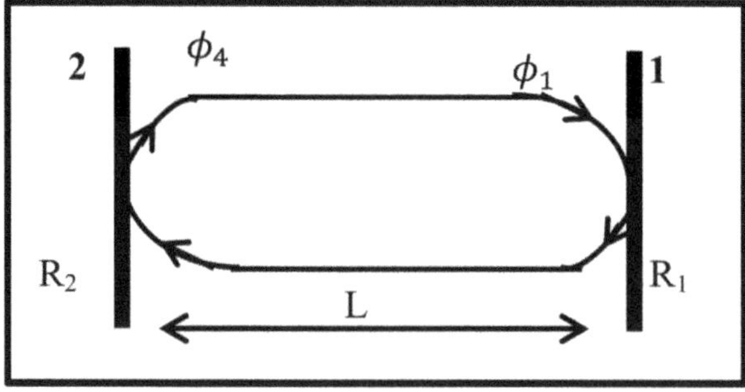


**Figure 5.5.** Schematic of a round trip in a cavity of length $L$ with mirrors (denoted by 1 and 2) of reflectivities $R_1$ and $R_2$, respectively. As explained in the text, $\phi_1$ and $\phi_4$ are the photon fluxes at the mirrors.

---

[1] In libraries, the books related to lasers can be found categorized as electrical amplifiers in the electrical section.

After traveling the length $L$, the photon flux ($\phi_3$) includes the gain of the medium and is given by

$$\phi_3 = (\phi_2 \times e^{kL}) \equiv \phi_1 \times R_1 \times e^{kL}.$$

Similarly, after the second reflection at mirror 2, on reaching mirror 1 (by completing a full round trip), the photon flux ($\phi_5$) is given by

$$\phi_5 = \phi_1 \times R_1 \times e^{kL} \times R_2 \times e^{kL}.$$

For sustained lasing, the initial and final photon densities must at least be identical. We can now define the *round-trip gain (G)* as the ratio of the initial and final photon densities, as follows:

$$G = \frac{\phi_5}{\phi_1} \equiv R_1 R_2 e^{2kL}. \tag{5.12}$$

**Exercise 5.3.** Find the length of the gain medium required to provide an overall gain of three for a small-gain coefficient value ($k$) of 0.4 cm$^{-1}$ at a given frequency. If the length of the gain medium is doubled, what is the overall gain?

**Solution:** In the absence of losses, the (overall) gain coefficient is given by $k_G = e^{kL}$.

Using the given values, $3 = e^{0.4L}$, we obtain $L = 2.75$ cm. Furthermore, if the gain length ($L$) is doubled, the value of $k_G$ is 9.02.

## 5.6 Gain saturation

Losses in the laser cavity arise from several factors. A few of them are: (i) reflection losses at the mirrors, (ii) scattering and absorption due to impurities in the active medium, and (iii) external parameters. The threshold value of the pump rate includes some of these losses. As a consequence, the value of $k$ is reduced. If we denote all such losses by $\gamma$, then the net value of the small-gain coefficient ($k'$) can be written as

$$k' = k - \gamma.$$

According to equation (5.12), if the value of $G$ is more than one, the oscillations will grow. However, if $G$ is less than one, the oscillations will die out. Therefore, for the steady state and for continuous-wave operation of the laser, the value of $G$ is taken to be unity. Therefore, equation (5.12) yields

$$R_1 R_2 e^{2k'L} = 1 \tag{5.13}$$

The situation in which the value of $G$ is unity is known as the condition of *gain saturation*. At the point of gain saturation, the number of photons extracted from the excited population is insignificant in comparison to the number of the excited molecules. This can be further understood as follows: as shown in figure 5.6, at the commencement of laser operation, the round-trip gain is greater the threshold value

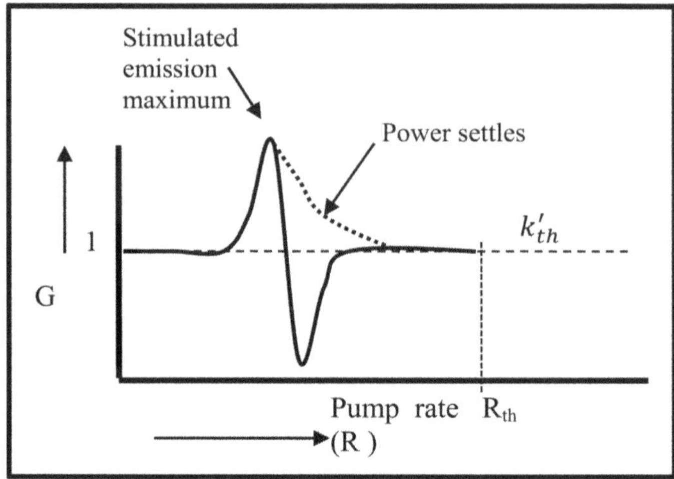


**Figure 5.6.** The output power stabilizes at a gain of unity due to a balance between the losses and the *small-gain coefficient* in the cavity.

(indicated by $G = 1$ in the figure). At the onset of the process of stimulated emission, the population inversion value starts to decrease. The existing losses in the cavity further decrease the gain. These processes affect the net value of the gain. As a result, the net round-trip gain may vary, as shown in figure 5.6. When the gain just balances the losses and the value of $G$ remains at unity over a period of time, the power stabilizes.

For $G = 1$, equation (5.13) can be rewritten as $k' = \frac{1}{2L}\ln\left(\frac{1}{R_1R_2}\right)$. In view of equation (5.9), in terms of population inversion, we have $\sigma_{es}\Delta N = k'$, so that $\Delta N = \frac{1}{2L\sigma_{es}}\ln\left(\frac{1}{R_1R_2}\right)$. Here, $\Delta N$ is known as the critical inversion.

**Exercise 5.4.** A He–Ne laser tube is 50 cm long and the reflectivities of the two mirrors are 100% and 99%, respectively. Assuming no losses, what is the value of the small-gain coefficient?

**Solution:** Assuming no losses, the small-gain coefficient is given by $k = \frac{1}{2L}\ln\left(\frac{1}{R_1R_2}\right)$. For the given values of $L = 50$ cm, $R_1 = 1$, and $R_2 = 0.99$, we obtain the value of $k = 1 \times 10^{-4}$ cm.

## 5.7 Applications of the Beer–Lambert law

The absorption coefficient (i.e. the parameter $\alpha$) contains vital information about the sample. The absorption strength of the sample (and hence the strength of the transition from the lower to the higher energy level) as well as the population in the upper level immediately after absorption can be estimated from $\alpha$. A number of modern sensing technologies for toxic and explosive materials are based on the measurement of $\alpha$. Therefore, the Beer–Lambert law has tremendous application in science and industry. Notice that the thickness of the sample plays an important role. In practical terms, the thickness should be sufficiently small that the intensity at the first layer ($\Delta x$) of the sample can be considered to be the same as that at the final layer of the sample. For this reason, the value of the optical density (OD) is generally kept at less than one. However, for extremely weakly absorbing systems, the thickness used can be as large as several meters. As mentioned above, $\alpha$ is a function of the concentration (C) and the extinction coefficient ($\epsilon$) of the sample. This means that the concentration or thickness of the sample can (and should) be controlled to achieve an OD value near unity. Therefore, by making use of capillaries several meters long or passing the light through the sample multiple times, it is possible to measure the absorption parameters of extremely dilute samples. The technique of cavity ring-down spectroscopy (CRDS), which is based on absorption coefficient, is used to detect explosives at the ppm level (♠ see question 7).

♣ Cavity ring-down spectroscopy (CRDS) measures the decay of light intensity using multiple passes through the sample.

## Questions and problems

1. (a). What is the significance of the thickness of the sample in Beer–Lambert's law?
   (b). The optical density of a sample is 0.5 for a path length of 1 cm and a concentration of 1 M. What is the molar extinction coefficient?
2. What is a gain medium? In a gain medium, the ratio of the inverted population of the two levels after optical pumping is $10^{25}$. At 300 K, what wavelength of light can be emitted by the system?
3. Can absorption by a system be related to amplification by the same system?
4. If $\gamma$ represents the total loss of the resonator, how is the threshold gain coefficient modified in terms of the length of the medium ($L$) and the reflectivity of the resonator mirrors $R_1$ and $R_2$?
5. For a dye laser at 540 nm, the stimulated emission cross section is $10^{-17}$ cm$^{-2}$. The population inversion is $10^{17}$ cm$^{-3}$. Calculate the small-gain coefficient, provided that there are no losses.
6. Define the term 'gain saturation'.
7. CRDS is a technique based on the phenomenon of absorption. It is used to quantitatively detect atomic or molecular species with high sensitivity [4]. The sample chamber made of a high finesse cavity (♠ see next chapter) is scanned repetitively thousands of times by a laser pulse. The intensity decay of the tunable transmitted laser pulse, i.e. the ring-down decay time ($\tau$) decreases exponentially. The decay time, as shown in the figure, is monitored through one of the mirrors. The rate of decrease is proportional to the reflectivity of the mirrors ($R$), the length of the sample chamber, and the absorption coefficient. The ring-down time $\tau$ in a round trip $t_r$ is given by

$\tau = \dfrac{t_r}{2(1 - R + \alpha L)}$. Here, $(1 - R)$ is the reflection loss. Furthermore, a plot of $1/\tau$ vs laser frequency shows that the sample's $\alpha$ is of the order of $10^{-8}$. Calculate the value of $\tau$ for a length of 0.5 m for a cavity with no absorption ($\alpha = 0$). What will happen if there are $10^3$ reflections?

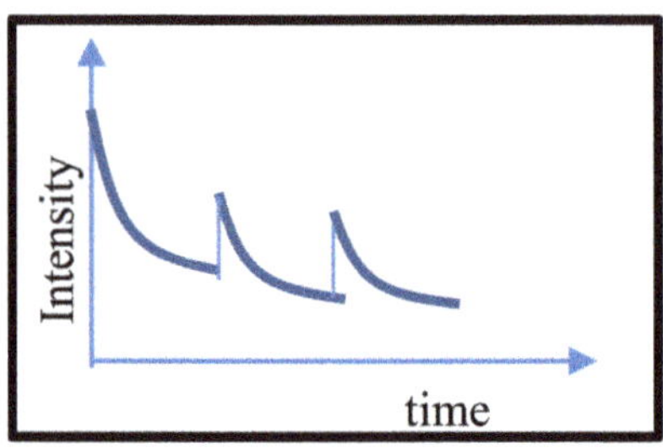


## Bibliography

[1] Rabek J F 1991 *Photochemistry and Photophysics* vol III (Boca Raton, FL: CRC Press)
[2] Tolman R C 1924 Duration of molecules in upper quantum states *Phys. Rev.* **23** 693–709
[3] Berden G, Peeters R and Meijer G 2000 Cavity ring-down spectroscopy: experimental schemes and applications *Int. Rev. Phys. Chem.* **19** 565–607
[4] O'Keefe A and Deacon D A G 1988 Cavity ring-down optical spectrometer for absorption measurements using pulsed laser sources *Rev. Sci. Instrum.* **59** 2544–51
[5] Stelmaszczyk K, Fechner M and Rohwetter P *et al* 2009 Towards supercontinuum cavity ring-down spectroscopy *Appl. Phys.* **B 94** 369–73

**IOP** Publishing

# An Introduction to Photonics and Laser Physics with Applications

**Prem B Bisht**

# Chapter 6

# Population inversion with moderate pumping

The decay rates between two energy levels of a multilevel system have a certain relation to each other. One of the upper energy states which has the slowest decay rate is known as the *metastable state*. This chapter introduces the concept of population inversion. The advantage of choosing a multilevel system (see the diagram) over two- or three-level laser systems is also described here. The population inversion schemes of a few lasers, viz. the $Nd^{+3}$:yttrium aluminum garnet (YAG), He–Ne, and argon-ion lasers are also described.

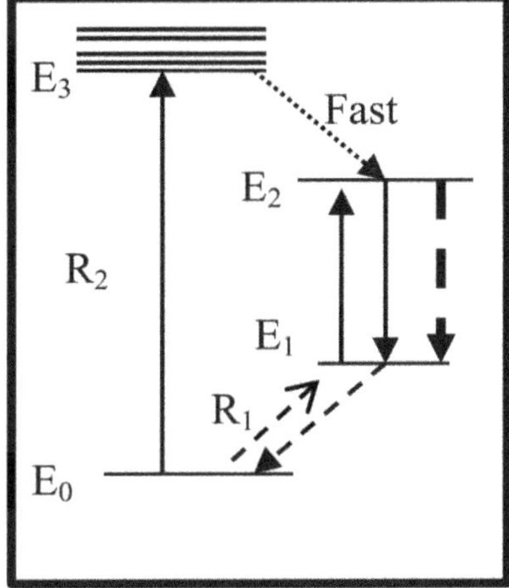

**Learning objectives**
**After reading this chapter, the learner will be able to:**
Write the rate equations for a system;
Compare two-level, three-level, and multilevel systems;
Illustrate the metastable state;
Explain the threshold pump rate and threshold population;
Describe lasers in the UV, visible, and near-IR regions: nitrogen, argon-ion, He–Ne, and $Nd^{+3}$:YAG lasers.

## 6.1 Population inversion schemes

The rate equation analysis and the Beer–Lambert law explained in the previous chapter led to the introduction of the *small-gain coefficient, k*. It turns out that for the case of a negative absorption coefficient ($\alpha$), the corresponding value of $\Delta N$ is known as a *population inversion* between the two energy states. In the following, we look at the efficiency of population inversion for the special case of systems with two or more levels.

### 6.1.1 Two-level system

In thermodynamic equilibrium, the population of the higher energy states is always smaller than that of lower energy states. Consider a system with two energy levels, $E_0$ and $E_1$. Figure 6.1 shows atoms[1] excited to $E_1$ by an external agent to a population represented by $N_1$.

The process of creating a larger population in the higher energy levels is known as 'pumping'. There are several pumping methods, such as optical, electrical, or chemical pumping (♠ see chapter 7). For the two-level system with populations

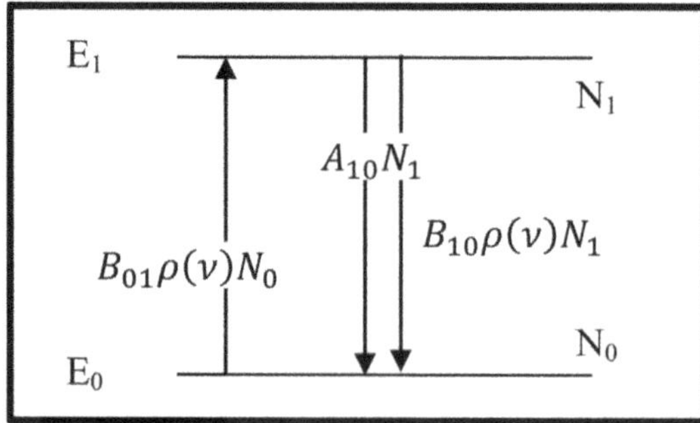


**Figure 6.1.** Two-level system with ground ($E_0$) and excited ($E_1$) states. $N_0$ and $N_1$ respectively are the corresponding populations. $B_{01}$, $B_{10}$, and $A_{10}$ are the coefficients of stimulated absorption, stimulated emission, and spontaneous emission, respectively.

---

[1] Atoms (such as He, Ne, Ar) or molecules (such as $N_2$, organic dyes).

$N_0$ and $N_1$ shown in figure 6.1, the total population ($N$) is given by $N = N_0 + N_1$, while the difference between their populations is $\Delta N = N_0 - N_1$. This implies that $N_0 = \frac{(N + \Delta N)}{2}$ and $N_1 = \frac{(N - \Delta N)}{2}$.

The rate of change of the upper level is

$$\frac{\mathrm{d}N_1}{\mathrm{d}t} = B_{01}\rho(\nu)N_0 - B_{10}\rho(\nu)N_1 - A_{10}N_1.$$

Here, $\rho_\nu$ is the photon energy density.

For the steady state $\left(\frac{\mathrm{d}N_1}{\mathrm{d}t} = 0\right)$, taking $B_{01} = B_{10}$ and $A_{10} = 1/\tau_0$, we can write

$$-B_{10}\rho(\nu)(\Delta N) - \left(\frac{N - \Delta N}{2}\right)\frac{1}{\tau_0} = 0$$

$$\Delta N = \frac{N}{1 + 2\tau_0 B_{10}\rho(\nu)}. \tag{6.1}$$

Upon pumping (i.e. the process of stimulated absorption), the atoms/molecules achieve the higher energy state. The atoms decay at the combined rate of spontaneous emission ($A_{01}$) and stimulated emission ($B_{10}\rho(\nu)$). In general, for $N_0 > N_1$, the rate of *stimulated absorption* will be higher, whereas for $N_1 > N_0$, the rate of *stimulated emission* will dominate. However, in the inevitable presence of spontaneous emission (which reduces the excited-state population such that $N_0 > N_1$), in any case, the maximum population density of the excited state can at best equal that of the ground state. We can see that irrespective of the value of the factor $2\tau_0 B_{10}\rho(\nu)$ in the denominator of equation (6.1), the value of $\Delta N$ is always positive. Therefore, achieving population inversion is not possible in a two-state system. Here, $B_{10}\rho(\nu)$ is the probability per unit time that atoms are excited to the upper level and is called the pumping rate $R$.

♣ In the absence of the phenomenon of population inversion, when a two-level system interacts with coherent electromagnetic (EM) radiation, Rabi oscillation can take place. Rabi oscillation is a process of the periodic absorption of a photon and its re-emission. The period of oscillation is called the Rabi period, and its reciprocal, known as the Rabi frequency, has various applications in quantum optics, quantum computing, and nuclear magnetic resonance.

This can also be illustrated by the following discussion. For a total population that is initially in the ground state ($N_0 \cong N$), equation (5.10) can be rewritten for absorption as

$$I = I_0 e^{-\sigma_{es}Nx}.$$

On pumping, $N_1$ becomes greater than $N_0$. Putting $N_1 = N - N_0$,

$$I = I_0 e^{\sigma_{es}\left(1 - 2\left(\frac{N_0}{N}\right)\right)Nx}. \tag{6.2}$$

Here, we see that as stimulated absorption begins to take place, the population increase in the upper level increases (and $N_0$ decreases). The value of $\frac{N_0}{N}$ starts decreasing until the population in the upper state reaches 50% of $N$. $\frac{N_0}{N} = 0.5$ implies that $N_0 = N_1$, for which the logic of the equation suggests that there cannot be any further absorption.

### 6.1.2 A low-efficiency scheme: the three-level system

Before we discuss multilevel systems, we will take a quick look at a three-level system (figure 6.2) such as the Ruby laser ($Al_2O_3$ in which 0.05% of the $Al^{+3}$ ions are replaced by $Cr^{+3}$ ions). Here, the initial excitation is to the $E_2$ level and stimulated emission takes place between the $E_1$ and $E_0$ levels.

In order to create population inversion between the $E_1$ and $E_0$ levels, it is necessary to obtain fast relaxation of atoms from $E_2$ to $E_1$. The final level after the laser transition is the ground state, which contains the highest population under conditions of thermal equilibrium. Therefore, a large population must reach the $E_2$ level before a sufficient population inversion can be established between $E_1$ and $E_2$. For instance, when a ruby crystal is excited, no laser output is obtained until the energy invested has pumped half of the population to the excited state. This stringent requirement makes the three-level system inefficient, as compared to four-level or multilevel systems.

### 6.1.3 Population inversion in a four-level system

Consider the four-level system shown in figure 6.3(a). The population distribution is given by the Boltzmann distribution (♠ see equation (4.3)) which follows an exponential decay. The upper level ($E_3$) schematically represents all the levels above and including itself. All these levels contribute to the population in $E_2$ through relaxation on a fast timescale. Therefore, as shown in figure 6.3(b), upon excitation to any higher level, a build-up of atoms takes place at this level. The lasing transition is shown between the $E_2$ and $E_1$ levels.

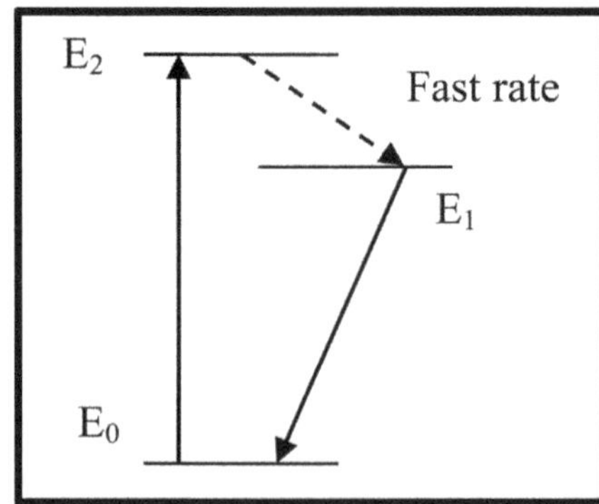


**Figure 6.2.** A three-level laser system with a ground level ($E_0$) and two excited levels ($E_1$ and $E_2$).

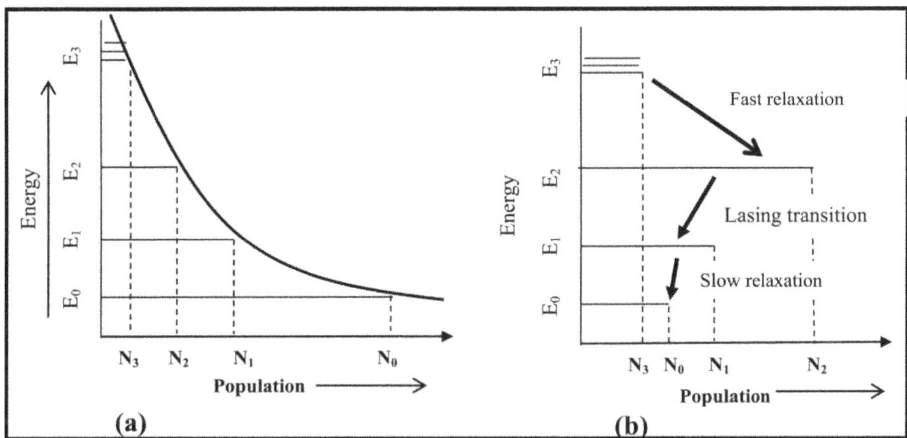


**Figure 6.3.** The populations in the various energy levels immediately before (a), and after excitation (b). Note the difference between the position of the dashed vertical lines indicating the populations of $E_3$ and $E_0$.

## 6.2 Rate equation analysis for a four-level system and a multilevel system

Let us now look at the rate equation of a system with four levels. Consider the rate of thermal equilibrium of the atoms to $E_1$, denoted by $R_1$. The direct external pumping rate of atoms/molecules to $E_3$ is $R_2$. The duration of the excitation is instantaneous, such that even the nuclear coordinates of the atom do not change. It should be noted that $E_3$ has been shown with a broad band consisting of several degenerate levels. Therefore, in the case of optical pumping, a broad-frequency light source can be used, such that absorption results in higher number of available states.

♣ A term frequently used by the laser community is the verb '*to lase*'. This refers to the conditions during laser alignment when the laser output appears, and one says 'It is lasing!'

♣ The materials used in lasers are could be in the liquid, gas, or solid phase. As mentioned in previous chapters, these are also known as gain media. Another name for such materials is '*lasing*' media.

As mentioned in the previous section, one of the essential conditions here is that the transition from $E_3$ to $E_2$ is fast. This condition assumes that the majority of the population of excited atoms immediately transfers to $E_2$. Therefore, the rate of feeding of level $E_2$ remains at $R_2$. As can be seen from figure 6.4, the additional rate of feeding this level is that of stimulated absorption $B_{12}\rho(\nu)N_1$. The level $E_2$ decays with a stimulated emission rate of $B_{21}\rho(\nu)N_2$ as well as a spontaneous emission rate of $A_{21}N_2$.

The rate of thermal excitation $R_1$ is too weak to be considered in the analysis for $E_1$ (see problem (6.1)). The decay rate of $E_1$ is $A_{10}N_1$. The energy gap between $E_2$ and $E_1$ decides the laser wavelength, while that between $E_1$ and $E_0$ decides the net population available in $E_1$. A larger gap between $E_1$ and $E_0$ helps to avoid the thermal population of the former.

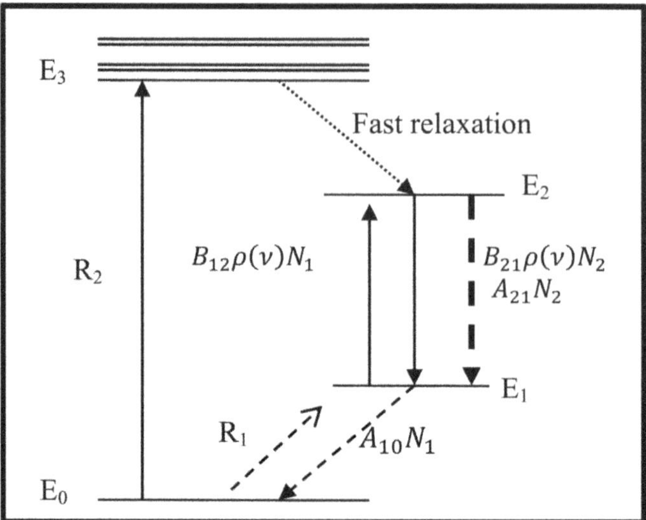


**Figure 6.4.** Energy-level diagram for a four-level system in a *lasing* material. The initial excitation is to the $E_3$ level, followed by fast relaxation to the $E_2$ state. The other symbols have their usual meanings.

Let us assume that the spontaneous decay times of levels $E_2$ and $E_1$ are $\tau_{21}$ and $\tau_{10}$, respectively. For the four-level system shown in figure 6.4, we can write the rate of change of population of $E_2$ as

$$\frac{\mathrm{d}N_2}{\mathrm{d}t} = R_2 - B_{21}\rho(\nu)N_2 - A_{21}N_2 + B_{12}\rho(\nu)N_1. \tag{6.3}$$

Similarly, the decay rate of level $E_1$ is given by

$$\frac{\mathrm{d}N_1}{\mathrm{d}t} = B_{21}\rho(\nu)N_2 + A_{21}N_2 - B_{12}\rho(\nu)N_1 - A_{10}N_1. \tag{6.4}$$

For steady-state operation of the lasing system, the change of population between the two levels is given by $\frac{\mathrm{d}N_2}{\mathrm{d}t} = \frac{\mathrm{d}N_1}{\mathrm{d}t} = 0$. By summing equations (6.3) and (6.4), we find

$$N_1 = \frac{R_2}{A_{10}}. \tag{6.5}$$

Equation (6.4) suggests that the population in the lower level is proportional to rate of pumping, $R_2$, and that it is also proportional to its spontaneous decay time (or lifetime $\tau_{10} = 1/A_{10}$).

Substituting $N_1$ from equation (6.5) into equation (6.3) yields

$$R_2 - B_{21}\rho(\nu)N_2 - A_{21}N_2 + B_{12}\rho(\nu)\frac{R_2}{A_{10}} = 0.$$

From this, the value of the population in $E_2$ can be obtained, as follows:

$$N_2 = \frac{R_2(1 + B_{12}\rho(\nu)/A_{10})}{(A_{21} + B_{21}\rho(\nu))}. \tag{6.6}$$

### 6.2.1 Population inversion between $E_2$ and $E_1$

Using equations (6.5) and (6.6) we can write the difference between the populations of the *upper level* $E_2$ and the *lower level* $E_1$ (or the population inversion, $N_2 - N_1 = \Delta N$) as

$$\Delta N = \frac{R_2(1 + B_{12}\rho(\nu)/A_{10})}{(A_{21} + B_{21}\rho(\nu))} - \frac{R_2}{A_{10}}$$

$$= \frac{R_2(A_{10} + B_{12}\rho(\nu)) - R_2(A_{21} + B_{21}\rho(\nu))}{A_{10}(A_{21} + B_{21}\rho(\nu))}$$

$$= \frac{R_2}{A_{10}(A_{21} + B_{21}\rho(\nu))}(A_{10} + B_{12}\rho(\nu) - (A_{21} + B_{21}\rho(\nu))).$$

Assuming that the split between the two levels is the same ($g_1 = g_2$), we can write $g_1 B_{12} = g_2 B_{21}$. The population inversion $\Delta N$ can now be rewritten as

$$\Delta N = \frac{R_2}{A_{10}(A_{21} + B_{12}\rho(\nu))}(A_{10} + B_{12}\rho(\nu) - A_{21} - B_{12}\rho(\nu))$$

$$= \frac{R_2}{(A_{21} + B_{12}\rho(\nu))}\left(1 - \frac{A_{21}}{A_{10}}\right). \tag{6.7}$$

Equation (6.7) gives the population inversion as a function of the net pumping rate ($R_2$) and the decay rates of levels $E_2$ and $E_1$.

**Exercise 6.1:** Calculate the *population inversion* value of a laser required to give a small-gain coefficient per unit frequency of 6 m$^{-1}$ between the levels $E_2$ and $E_1$, for which $\tau_{21} = 250$ ms, $\lambda = 1064$ nm, and the refractive index $n = 1.82$.

**Solution:** The small-gain coefficient $k$ per unit frequency is given by equation (5.8) in terms of $B_{21}$ as $k = \frac{n}{c}\left(N_2 - N_1\frac{g_2}{g_1}\right)B_{21}h\nu_{21}$. Assuming that the multiplicities of the two levels are the same ($g_1 = g_2$),

$$\Delta N = \frac{k}{n}\frac{c}{B_{21}} \times \frac{1}{h\nu_{21}}.$$

Using $\frac{A_{21}}{B_{21}} = \frac{8\pi h\nu_{21}^3}{c^3}$ along with $A_{21} = 1/\tau_{21}$ and $c = \nu\lambda$, we can express

$$B_{21} = \frac{\lambda^3}{8\pi h}\frac{1}{\tau_{21}}.$$

Hence for the given situation, $\Delta N = \frac{k}{n}\frac{8\pi}{\lambda^2}\tau_{21} = 1.8 \times 10^{13}$.

It should be noted that we need to take into account the broadening in the emission profile; this is addressed in chapter 8.

### 6.2.2 Metastable state

In equation (6.7), $A_{21} = \frac{1}{\tau_{21}}$ where $\tau_{21}$ is the lifetime of the state $E_2$. In order to obtain population inversion between $N_2$ and $N_1$, the quantity $\Delta N$ must be positive. This is possible only when $\frac{A_{21}}{A_{10}} < 1$ or $\tau_{10} < \tau_{21}$. This implies that for effective population inversion, $\tau_{21}$ must be large. As outlined at the beginning, the decay rate of $E_3$ is fast compared to the other rates. Therefore, in a lasing system, the state with the relatively longer lifetime ($E_3$ in this case) is also known as the *metastable state.*

### 6.2.3 Plot of output power after population inversion

From equation (6.6), we note that at the threshold value of $\Delta N(\Delta N_{\text{th}})$, the amount of the output power can be considered to be low. Therefore, at this point, by ignoring $\rho(\nu)$, we can rewrite $\Delta N_{\text{th}}$ in terms of the value of the threshold pumping rate ($R_{\text{th}}$) as follows:

$$\Delta N_{\text{th}} = \frac{R_{\text{th}}}{A_{21}}\left(1 - \frac{A_{21}}{A_{10}}\right). \tag{6.8}$$

For $\Delta N_{\text{th}}$ to be positive, $\left(1 - \frac{A_{21}}{A_{10}}\right) \sim 1$, so that

$$\Delta N_{\text{th}} = \frac{R_{\text{th}}}{A_{21}}. \tag{6.9}$$

Equation (6.9) indicates that $\Delta N_{\text{th}}$ is proportional to $R_{\text{th}}$. The left-hand panel of figure 6.5 shows a plot of $\Delta N$ versus the pump rate. It can be seen that only above the threshold value of $R_{\text{th}}$ (and hence $\Delta N_{\text{th}}$) can the power ($P$) output can be obtained from the laser. As shown in right-hand panel of figure 6.5, the output power then increases as the pump rate increases, and it remains proportional to the pump rate even beyond $\Delta N_{\text{th}}$.

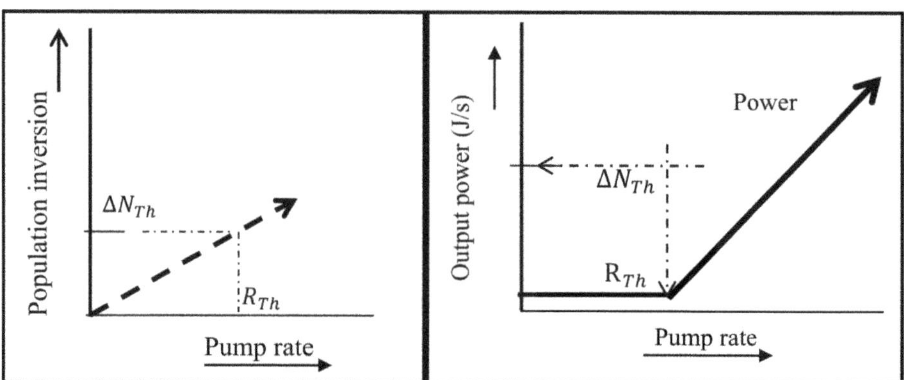


**Figure 6.5.** Plot of the pump rate versus population inversion (left-handpanel). The right-hand panel shows a plot of the pump rate versus the output power. $\Delta N_{\text{th}}$ is the threshold value of population inversion at a threshold pumping rate of $R_{\text{th}}$.

The power (in W) is also expressed in joules per second and has the dimensions of energy per unit of time. The total pumping power per unit of volume ($P_{th}$) at the threshold may be written as

$$P_{th} = \text{Energy} \times \text{rate of pumping}(R_{th})$$

or

$$P_{th} \propto R_{th}.$$

Using equation (6.9), $P_{th}$ can also be rewritten as

$$P_{th} = E_3 \times \Delta N_{th}\, A_{21} \equiv E_3 \times \frac{\Delta N_{th}}{\tau_{21}}.$$

Above the threshold, population inversion is followed by optical feedback. The process of population inversion is a necessary condition for laser operation but is not a sufficient condition. Optical feedback is necessary for laser oscillation in the laser cavity. Therefore, lasers are also often called 'oscillators'. The power of the laser beam obtained from an oscillator is subjected to further amplification, either in the same laser using an additional cavity or by feeding the beam into another laser's cavity for amplification. The system is then ready to give rise to various properties of a laser, such as directionality and mode structure. However, without the feedback, lasers can also work in superradiance mode (♠ see section 2.9.10), such as in nitrogen lasers.

Finally a word about the efficiency of lasers. Generally speaking, it is unnecessary for a gain medium with a high gain coefficient to be an efficient laser. The efficiency of the laser system is defined as the ratio of the output power to the input pump power. This ratio depends on how efficiently the pump power is converted into population inversion. For example, for optical pumping, we can see that the theoretical value of the maximum efficiency in figure 6.4 is $(E_2 - E_1)/(E_3 - E_0)$. This is equal to the ratio of the two frequencies. From a practical perspective, the efficiency is far less than that due to energy losses in converting the electrical energy into optical energy and also because of non-radiative transitions from $E_3$. The three-level system is considerably less efficient than the four-level system for the above reasons.

## 6.3 Typical laser systems

### 6.3.1 Nd:YAG laser

Neodymium ions ($Nd^{+3}$) doped in yttrium aluminum garnet (YAG) crystal are used in a popular laser—the $Nd^{+3}$:YAG laser. Figure 6.6 shows the energy-level diagrams and lifetimes associated with this gain medium. Here, the initial excitation takes the system ($Nd^{+3}$ atoms) to the $E_3$ state instantaneously. From $E_3$, the system relaxes to the $E_2$ state ($^4F_{3/2}$) within ~10 ns. This relaxation involves non-radiative processes such as thermal stabilization and lattice relaxation with reference to the external environment (a crystalline structure in this case). The lifetime of the $E_2$ state is about 0.5 ms and that of the $E_1$ state ($^4I_{11/2}$) is ~30 ns.

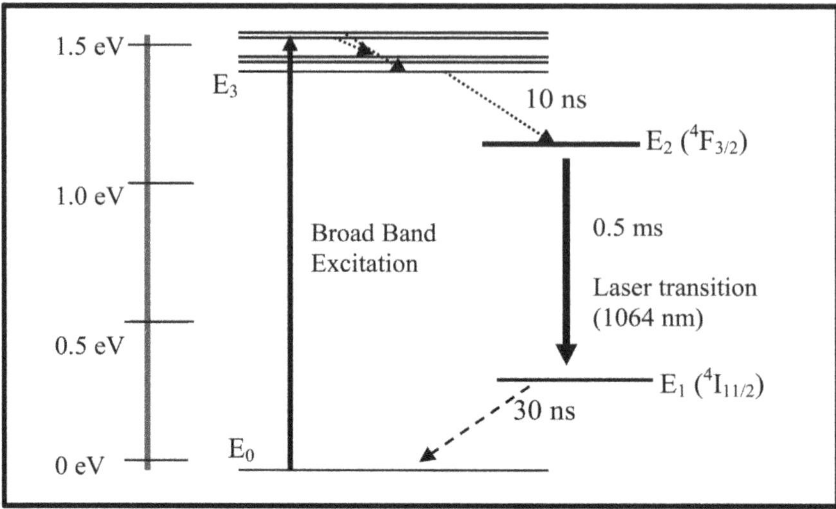


**Figure 6.6.** Energy-level diagram of a Nd:YAG laser. A broadband source (a flash lamp) is used as the excitation source to achieve a higher energy state ($E_3$). The lifetime values of the levels are also indicated along with the laser transition between $E_2$ and $E_1$.

Due to the slower decay of the $E_2$ state as compared to that of the $E_1$ state, a large number of excited atoms are collected in $E_2$. This results in population inversion between the $E_2$ and $E_1$ states, resulting in the transition corresponding to the 1064 nm laser line.

For the case of $Nd^{+3}$-doped glass (an amorphous structure), the difference between lasing energy levels is about 50 cm$^{-1}$, which is responsible for a 1054 nm line with a lower efficiency and a broader line width (♠ see chapter 8).

### 6.3.2 Helium–neon laser

In this laser, we have a special case of population inversion involving two different atoms. A unique method of achieving population inversion through the 'transfer of excitation energy' is used in the helium–neon (He–Ne) laser invented by Javan and co-workers. In the electrical discharge tube, the ratio of He and Ne gases is kept at approximately 10:1 with a partial pressure of about 3 Torr. The lasing transitions in this laser correspond to those of Ne atoms, as shown in figure 6.7. The electrical excitation takes place in both the He and Ne atoms. The electronic transitions in Ne are strongly coupled to the ground state but the He atoms are raised to the triplet state. These long-lived energy states of He serve as feeding levels for the energetically nearby Ne levels. As a result of the transfer of excited-state energy from the He levels, population inversion takes place in the energetically nearby levels of Ne, as shown in the diagram.

Initially, the He atoms are excited by collisions with electrons. These excited He atoms (with an energy of 20.61 eV) collide with Ne atoms and transfer their energy

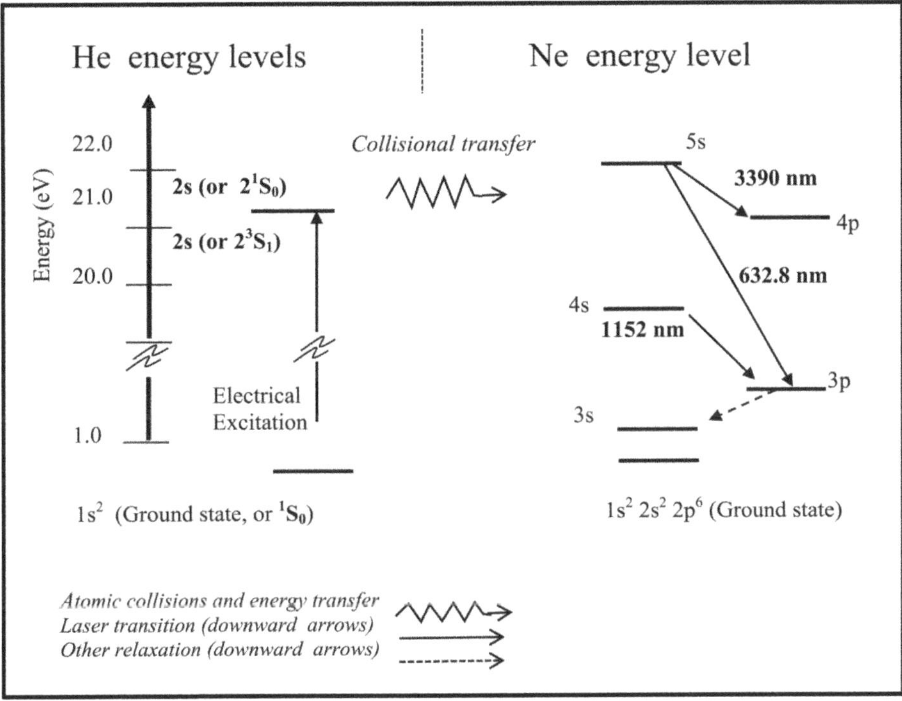


**Figure 6.7.** Schematic of population inversion in Ne atoms through collisional transfer from electrically excited He atoms in a He–Ne Laser. Some of the laser transitions are also indicated.

to a nearby degenerate energy levels of the latter (♠ see section 1.2: 'Collisions of the second kind').

$$He + e \rightarrow He^*$$

$$He^* + Ne \rightarrow He + Ne^*.$$

In the Ne energy levels, the upper-level transitions produce the visible He–Ne laser lines, viz. 632.8 nm. Some other laser lines with lower efficiency are also available, such as those at 1152 and 3390 nm. Several other lines, such as 543 nm (5s→3p), 1523, or 1118 nm (4s→3p) arise at the split level 3p with small variations in energy. Spontaneous emission from 3p and collisional relaxations bring the atoms back to the ground state. It should be noted that the lower levels of Ne do not receive any energy in the collision process. Therefore, there is an obvious population inversion in the Ne atoms. The notations for the energy levels are given in figure 6.7.

The electron distribution according to Hund's rule suggests that the outer subshell of the noble gas atoms is completely filled (e.g. $1s^2$ in He and $1s^2$, $2s^2$, and $2p^6$ in Ne). The total angular momentum (J) for such closed shells is zero. The ground state is designated $^1S_0$. The excited atomic states of Ne can be obtained by promoting one electron to either of the 3s, 3p, 3d, 4s, 4p, 4d, 4f, or 5s orbitals. However, the Ne levels cannot be described in terms of JJ coupling, which generally holds good for the heavier atoms in which spin–spin interactions are appreciable. The levels in simple notation as given in figure 6.7 are used for the observed laser lines.

In the Ne atoms, the laser transition in the IR (3390 nm) as well as the most prolific line in the visible region (632.8 nm) have a common upper level. The spectroscopic notation for this level is $^3S_2$. If the conditions (the reflective mirrors for feedback and cavity length matching) are not favorable for the IR line, the efficiency of the visible lines is increased. A common lower energy level ($^2P_4$) is shared by the lines at 1152 nm and 632.8 nm, which therefore compete with each other. The He–Ne system also provides example of several other laser lines (e.g. 543.500 nm in the green region) due to cascade transitions in laser media. He–Ne lasers are extremely stable, low-powered lasers that generally work for tens of thousands of hours. They are important tools for experimental laboratories and are used for optical alignment, in the calibration of spectrographs, and in interference experiments for sensing applications.

♣ Another gas laser that operates on a similar principle is the helium–cadmium (Cd) laser. Here, the lasing element is Cd, which is a metal at room temperature. While the He is ionized by an electric discharge, the Cd has to be evaporated by a heated filament. As the vaporized Cd has to be kept in the laser cavity at a certain pressure, the construction of the He–Cd laser is a bit more complex than that of the He–Ne laser. Continuous-wave outputs are obtained from He–Cd lasers at 322, 354, and 442 nm, which are more useful for UV–blue wavelength excitation than the outputs of He–Ne lasers which only have available wavelengths in the visible region.

**Exercise 6.2.** What is the power efficiency of a He–Ne laser that delivers a power of 1 mW at a DC operating voltage of 2500 V and draws a current of 1 mA ?

**Solution:** The power efficiency ($\eta$) is given by the ratio of the output power to the input power.

Therefore, $\eta = \frac{1\text{mW}}{VI} = 1 \times 10^{-3}/(2000 \times 1 \times 10^{-3}) = 0.05\%$

### 6.3.3 Argon-ion laser

The argon-ion ($Ar^+$) laser (figure 6.8) uses the transitions of $Ar^+$ ions and not those of the neutral Ar atom. To produce ionic argon (to remove an electron from the argon atom), and energy of ~16 eV is required. A high-current electrical discharge is used for electronic excitation in this case. Here, the intermediate level corresponding to the ground state of the $Ar^+$ ion has a longer lifetime. Further transfer with electron collisions takes place to the excited state (4p energy levels) of the $Ar^+$ ion. This is in contrast to the case of the He–Ne laser, in which the transfer of energy takes place between the equivalent energy levels of the two atoms.

Population inversion takes place between the 4p and 4s energy levels of ionic argon. A series of about 30 laser lines are observed in the UV and visible regions, among which the most intense transitions give rise to laser wavelengths of 488 and 514.5 nm. A radiative decay in timescales of ~50 ns with wavelengths in the 72 nm

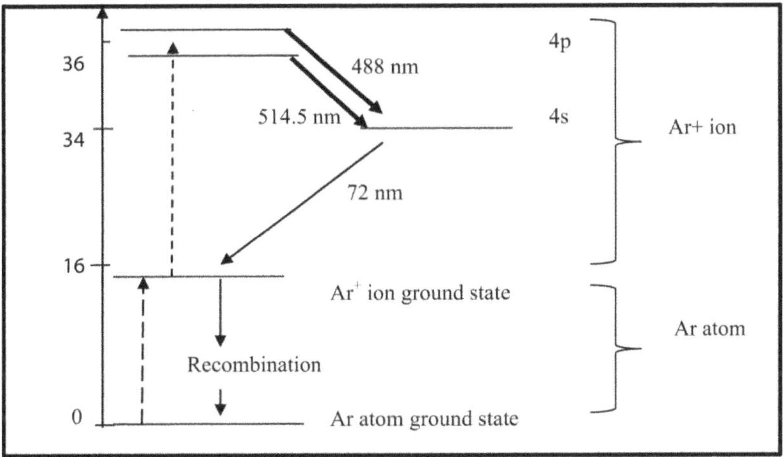

Figure 6.8. Schematic of some of the laser transitions (from the 4p to 4s levels) in the argon-ion laser.

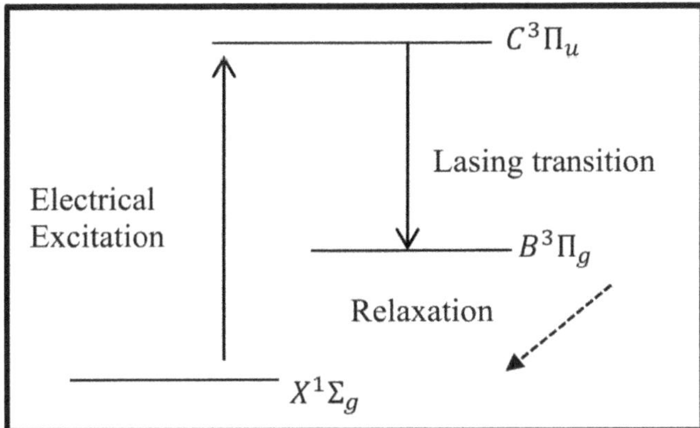

Figure 6.9. Energy levels involved in lasing in the $N_2$ molecule.

range brings the 4s levels to the ground state of the argon ion, which subsequently decays to the ground state of neutral argon through non-radiative processes. Apart from its uses in optics and spectroscopy, the $Ar^+$ ion laser is used in the UV range for the manufacture of fiber Bragg gratings. The UV lines are obtained by intra-cavity frequency doubling (♠ see chapter 15). The argon-ion laser can easily be actively mode locked (♠ see chapter 17 for mode locking).

### 6.3.4 Nitrogen laser and superradiance

One of the simplest lasers that can be fabricated in the laboratory is the molecular nitrogen ($N_2$) gas laser, which produces an output at 337 nm. Here, excitation is performed with the help of an electrical discharge, as shown in figure 6.9. The upper-state lifetime of $N_2$ is about 25 ns, while that of the lower state is 10 μs. Therefore, the population inversion required for lasing cannot be maintained for longer periods

**Table 6.1.** Lasing lines of an air laser based on optical pumping in deep UV (210–230 nm).

| Air component | Lasing line |
|---|---|
| Oxygen | 845 nm |
| Nitrogen | 870 nm |

in the upper level. As a consequence, this laser cannot be operated in continuous-wave mode. Therefore, the $N_2$ laser is also known as self-terminating laser.

♣ Nitrogen and oxygen are atmospheric gases. In order to achieve lasing in air, multiphoton pumping of air in deep UV (♠ see chapter 20 for multiphoton pumping) was recently carried out using the second harmonic of an optical parametric oscillator pumped by the third harmonic of a Nd:YAG laser (♠ see chapters 14–16 for details of the optical parametric oscillator and harmonic generation). The observed lasing lines for nitrogen and oxygen in the near-IR region achieved by optical pumping in deep UV are given in table 6.1.

The cavity of the $N_2$ laser does not use any mirrors for feedback and the laser operates on the principle of the superradiance amplifier. The phenomenon of superradiance is briefly described as follows. Let us take $V$ as the critical volume (area × critical length) of the emitter; the emission wavelength is $\lambda$. Spontaneous emission is a fundamental concept; it is generally believed that a number of excited molecules, $N$, will inevitably radiate randomly at a rate $\propto NV$. For a rod-shaped active material of diameter $D$, spontaneous emission will be take place into a solid angle (♠ see chapter 10) corresponding to the diffraction angle $\lambda/D$. The phenomenon of superradiance can only take place at the threshold value of the population inversion ($N_{th}$). Here, the emission may take place by a cooperative process in which the emitting atom influences others to emit at a rate faster than the spontaneous emission rate. As a result, superradiance is proportional to $(N_{th}V)^2$. Another distinction between spontaneous emission and superradiance is lies in the observed decay times. Spontaneous emission, as we know, follows an exponential decay over a lifetime $\tau_0$. Superradiance, on the other hand, decays faster, with a time–frequency bandwidth relation corresponding to $\tau_f/NV$. Superradiance should be differentiated from amplified spontaneous emission, in which fluorescence photons (typically in dye lasers) are amplified by the gain medium itself while traveling in the region of the remaining pumped region. Another differentiating aspect of superradiance is that the emitting dipoles are aligned in a single direction.

## Questions and problems

1. Find the population of the first thermally excited level for a visible wavelength (~500 nm) at room temperature.
2. Estimate the ratio of the populations ($N_2/N_1$) of two energy levels $E_1$ and $E_2$ such that a transition from $E_2$ to $E_1$ will give visible radiation (550 nm) at

room temperature. Assume that the degeneracies of the two levels are the same.

3. What differences in lasing properties are caused by changing the host from YAG to glass for an $Nd^{+3}$ ion dopant? (Hint: For a complete answer, you need to glance at chapter 8).

4. What are various transitions of the He–Ne laser? The He–Ne laser requires low current and a high anode voltage, but the argon-ion laser requires high current and a relatively lower voltage. Comment on this statement.

5. Why is the He–Ne laser unable to provide higher powers, like the argon-ion laser?

6. Draw plots for the population inversion versus pump rate for (i) a four-level laser system and (ii) a two-level laser system. Differentiate between the two, and indicate the output laser power in the plots.

7. Does the argon-ion laser provide any laser wavelengths in the UV region? Intra-cavity frequency doubling in the argon-ion laser can be used to generate a wavelength of 229 nm. [Hint: ♠ see chapter 14 for the principle of frequency doubling.]

# Bibliography


[1] Javan A, Bennett W R and Herriott D R 1961 Population inversion and continuous optical maser oscillation in a gas discharge containing a He–Ne mixture *Phys. Rev. Lett.* **6** 106–10

[2] Huestis D L 1982 Introduction and overview *Gas Lasers* ed E W McDaniel and W L Nighan (New York: Academic) **1** 1–34

[3] Phillips D T and West J 1970 The poor man's nitrogen laser *Am. J. Phys.* **38** 655–57

[4] Lofthus A and Krupenie P H 1977 The spectrum of molecular nitrogen *J. Phys. Chem. Ref. Data* **6** 113–307

[5] Diels J-C and Rudolph W 2006 Diagnostic techniques *Ultrashort Laser Pulse Phenomena* 2nd edn (Burlington: Academic) **9** 457–89

[6] Laurain A, Scheller M and Polynkin P 2014 Low-threshold bidirectional air lasing *Phys. Rev. Lett.* **113** 253901

**IOP** Publishing

An Introduction to Photonics and Laser Physics
with Applications

**Prem B Bisht**

# Chapter 7

# Pumping mechanisms and types of optical cavity

The process of excitation used to achieve high population densities in the excited state is known as 'pumping'. For example, optical pumping requires an intense light source that provides a large number of photons. In addition, the light source must have a wide spectral range covering the most of the absorption of the gain medium. The figure shows the spectrum of a xenon lamp used in Nd:yttrium aluminum garnet (YAG) lasers. This chapter gives a brief account of various pumping techniques used to achieve population inversion in laser gain media. The designs of some laser-head cavities are also described here.

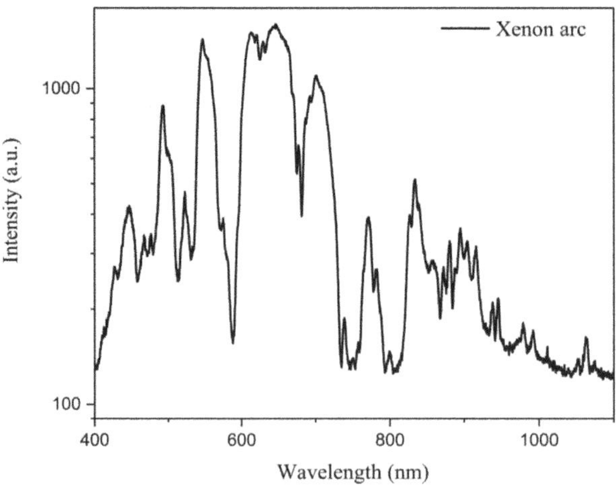

> **Learning objectives**
> **After reading this chapter, the learner will be able to:**
> Understand the methods of achieving high populations in excited states;
> List the types of pumping;
> Describe electrical and optical excitation;
> Identify the types of pump source;
> Identity the types of optical cavity and the various pumping geometries.

# 7.1 Pumping via electrical excitation

## 7.1.1 Collisions of the first kind

To achieve lasing, the gain medium needs to undergo a population inversion, as described in chapter 6. We know that population inversion can only be achieved by transferring a large amount of the ground-state population of the gain medium into its excited state. The process of excitation through which a source can supply sufficient electrons or photons to the gain medium is known as 'pumping'. The word *pumping* denotes the excitation mechanism used to achieve population inversion.

For instance, a strong continuous electric discharge between two electrodes is used for electric pumping. The electrical discharge known as glow discharge produces electrons at high speeds of $10^6$ m s$^{-1}$. The electrons transfer their energy to the atoms or molecules (denoted here by M) in inelastic collisions, resulting in the excited state of the monomer ($M^*$), as follows:

$$M + e \rightarrow M^*.$$

$M^*$ either emits a photon, as in the case of the Hg atom ($\spadesuit$ see section 2.6), or can be further ionized to a cation $M^+$. This type of collision is known as a collision of the first kind. This kind of discharge is preferred in gas lasers with single species, such as argon gas. A collection of ions and electrons with high energy produced using such an electrical discharge is known as a *plasma*; this is the *fourth state of matter*, which is different from the gaseous state. Plasma can also be generated with the help of lasers—this is known as laser-induced plasma and has applications in laser-induced breakdown spectroscopy (LIBS), which is used to characterize materials.

## 7.1.2 Collisions of the second kind

In a helium–neon (He–Ne) laser, a continuous electrical discharge is maintained in a mixed gas consisting of atoms of He and Ne in a chamber known as a *cavity*. The electrical discharge first excites the He atoms. As described in section 6.3.2, as well as emitting photons, these excited He atoms collide with Ne atoms in their ground state. These collisions raise Ne atoms to an excited electronic level as follows:

$$He + e \rightarrow He^*$$

$$He^* + Ne \rightarrow Ne^*.$$

The asterisk (*) denotes the excited state. We note here an important point about *automated* population inversion in Ne, as follows. Initially, the Ne atoms are in the ground state. Some of the energy levels of Ne have an energy equal to that of excited He levels. This creates a situation similar to two oscillators of the same frequency that are connected to each other. Under such circumstances, it is possible for energy to be transferred from the excited He level to a Ne level of equal energy. This type of resonant energy transfer is a *collision of the second kind*. We note here that there is no population in the other lower-energy levels of the Ne atom apart from those resonating with the excited He levels. Thus, on colliding with the excited He atoms, population inversion takes place between some of the energy levels of the Ne atoms. Such a mechanism is also used in the vibrational excitation of molecules in $CO_2$ lasers, in which the $CO_2$ is mixed with $N_2$. Pulsed electrical discharges are also used to produce electrical excitation in gas lasers such as the nitrogen laser (♠ see section 6.3.4).

### 7.1.3 Electrical pumping

A typical setup used for electrical pumping is shown in figure 7.1. Somewhat similar in design to the Franck–Hertz experiment (♠ see chapter 2) without the grid, the setup consists of two electrodes in the cavity that are connected to an electrical power supply. The gas-discharge tube located in the middle of the cavity is filled with a gain medium such as He–Ne gas. The two ends of the cavity have mirrors that make it an optical resonator (♠ see chapter 9). If required, the whole cavity can be cooled by an additional arrangement, for example, by circulating either air or deionized water to remove the Joule heating created by the electrical discharge.

## 7.2 Optical pumping

Since the beginning of civilization, humans have been aware of light from fire, lightning, oil lamps, resin torches, and candles as well as that from the Sun. Technological developments in last two centuries, especially after the advent of electricity, have made significant advances in the area of light sources.

We now know that attaining the excited state (with a population of $N_2$) involves optical excitation from a high-intensity light source or by electric discharge. The

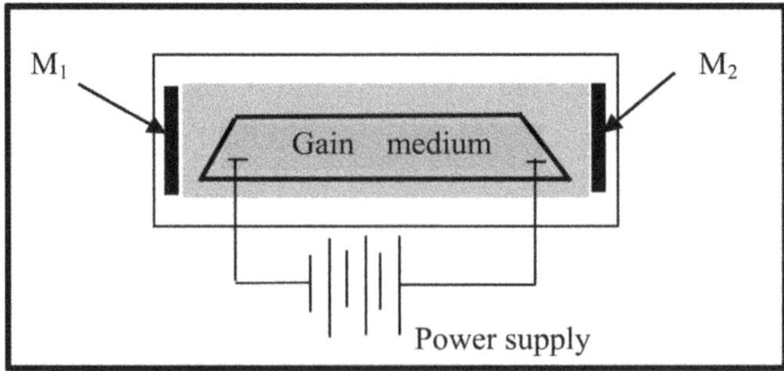


**Figure 7.1.** Schematic of a gas-discharge tube used for electrical pumping. The shaded region, known as the laser cavity, is enclosed by the mirrors $M_1$ and $M_2$.

pump power ($P_{\text{req}}$) required to produce an excited state with energy $h\nu_{12}$ over a volume $V$ of the gain medium is as follows:

$$P_{\text{req}} = h\nu_{12}\frac{dN_2}{dt}V.$$

Here, $\left(\frac{dN_2}{dt}\right)$ is the number of atoms per unit volume per unit time excited to the lasing level of the gain medium. If the optical pump power or the electrical input power is $P$, then the average pumping efficiency ($\eta_P$) of the system over the volume can be obtained as follows:

$$\eta_P = \frac{h\nu_{12}\int V\left(\frac{dN_2}{dt}\right)dV}{P}.$$

Here, $\frac{dN_2}{dt}$ depends on the initial conditions of the system. In other words, the total efficiency depends on the absorption coefficient of the gain medium at the frequency $\nu_{12}$ as well as on the pump source efficiency. The efficiency of the pump source in the case of optical pumping indicates the ratio of the availability of a number of photons at the absorption frequency of the gain medium to the consumed electrical power.

**Exercise 7.1.** A 20 W flash lamp provides light centered at a wavelength of 800 nm. If 20% of the photons are absorbed by the laser gain material, then how many photons are absorbed per second?

**Solution:** Only 20% of the photons are absorbed by the gain medium at a wavelength ($\lambda$) of 800 nm. The power absorbed by the gain medium
= 20% of 20 W = 4 W.

The number of photons per second ($n$) can be obtained using the relation
$nh\nu_{12} = 4$ Js$^{-1}$. Here, $h$ is Planck's constant. The frequency ($\nu_{12}$) corresponding to 800 nm is obtained as follows:
$\nu_{12} = c/\lambda = 3 \times 10^8/800 \times 10^{-9} = 3.75 \times 10^{14}$ s$^{-1}$. Therefore,

$$n = 4 \text{ Js}^{-1}/ \quad (6.6 \times 10^{-34} \text{ Js} \times 3.75 \times 10^{14} \text{ s}^{-1})$$

$$= 1.61 \times 10^{19} \text{ per second}$$

### 7.2.1 Lamps

Even before the laser was invented in 1960, considerable development had taken place in the field of light sources. The tungsten filament light bulb was one of the most popular incandescent light sources for more than a century, following its invention by Edison. Other gas-discharge lamps, such as xenon arc, xenon–mercury, low- and high-pressure mercury, sodium, and neon lamps provided the necessary instrumentation for scientific research and industry.

Due to pressure broadening mechanisms (♠ see chapter 8), high-pressure lamps exhibit broadband spectra along with the presence of characteristic lines (figure 7.2).

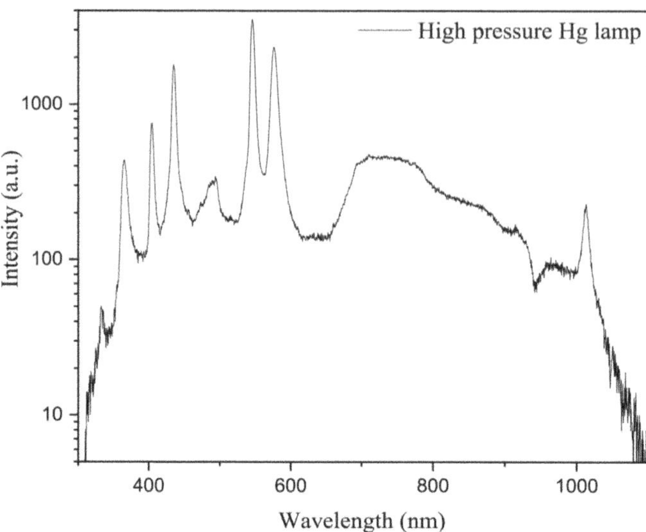


**Figure 7.2.** The spectrum of the high-pressure Hg lamp used as an excitation source in spectroscopy (recorded at IIT Madras).

Light produced by xenon and tungsten filament lamps contains a broad background. With the help of spectrographs, such light sources can provide the monochromatic light required for spectroscopy. Suitable lamps and arc lights are selected for optical pumping in lasers according to the absorption spectra of the gain material. It should be noted that pulsed excitations of the order of microseconds can be achieved with these lamps using simple electronic circuits.

In the first laser (a ruby laser), Maiman used a helical xenon flash lamp as the excitation source for optical pumping. Flash lamps used for optical pumping must have broadband spectra along with sharp lines that fall in the absorption region of the gain medium. This can be seen from a comparison of figure 7.3 (the spectrum of a xenon lamp) with figure 7.4 (the absorption spectrum of Nd:YAG). However, quite a lot of energy is wasted as heat by these intense light sources. Even though flash lamps are economical sources for optical pumping, they need to be changed after a few thousands of hours of operation due to the deterioration of their filaments or electrodes.

## 7.2.2 Light-emitting diodes and lasers

Several lasers are optically pumped by light-emitting diodes (LEDs) or other diode lasers (♠ see chapter 22). LEDs or solid-state diode lasers are efficient pumping sources, because their wavelengths are suitably matched with the absorption bands of various gain media. Diode-pumped solid-state (DPSS) lasers are popular in industry nowadays due to their robustness. Laser diodes are expensive, but they have compact designs and longer lifetimes than those of arc lamps.

In addition to DPSS lasers, other lasers often offer pumping advantages. Mode-locked $Nd^{+3}$:YAG lasers are used for the synchronous pumping of dye lasers to

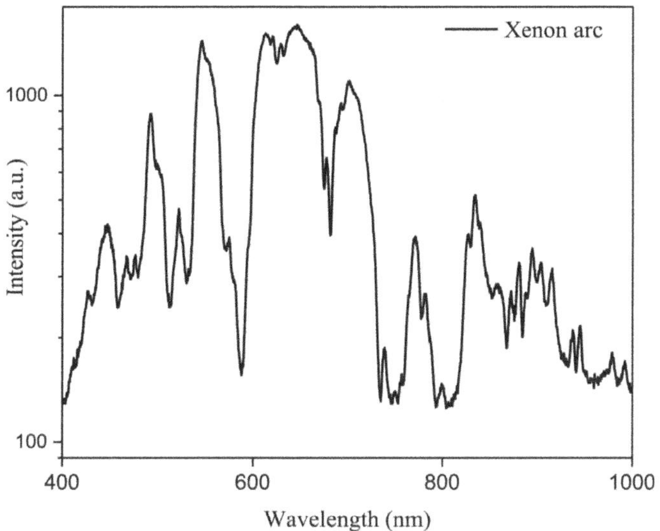

Figure 7.3. Spectrum of a xenon flash lamp used for pumping in a Nd:YAG laser (Recorded at IIT Madras).

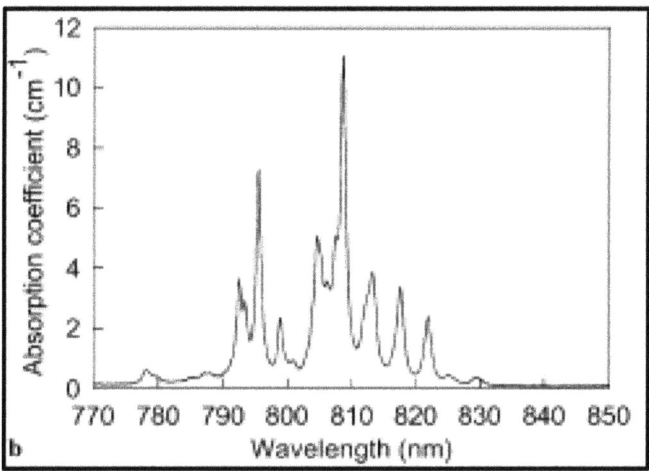

Figure 7.4. Absorption spectrum of $Nd^{+3}$:YAG in 1% ceramic. Reproduced from [13], copyright 2000 with permission of Springer.

obtain mode-locked pulsed outputs from them (♠ see chapter 17). Depending on their absorption spectra, dye lasers can be pumped by $Ar^+$ ion lasers, the second or third harmonics of $Nd^{+3}$:YAG lasers, nitrogen ($N_2$) lasers, or excimer lasers.

## 7.3 Thermal and gas-dynamic pumping

Thermal pumping is used in gas-dynamic lasers. Examples of these are lasers that use a gain medium of $CO_2+N_2$ along with catalysts such as He or $H_2O$. The adiabatic expansion of hot gas through a nozzle is used in gas-dynamic pumping.

Following expansion into a very low pressure chamber, due to sudden cooling, a non-equilibrium state is attained by the system. In this process, the upper vibrational level population is frozen but the lower vibrational level population is depleted through relaxation. This creates the population inversion between the two levels necessary for lasing.

## 7.4 Chemical pumping

In chemical pumping, population inversion is produced by a chemical reaction. Chemical lasers were also invented in early 1960s. When chemical pumping is adopted, there is no requirement for a continuous and large supply of electrical power. Chemical reactions create population inversion after releasing energy in an exothermal process and a substitution reaction. For instance, a reaction between hydrogen (H) and fluorine (F) begins in presence of an electric spark:

$$H_2 + F \rightarrow HF^* + H. \tag{7.1}$$

The excited state of the HF molecule ($HF^*$) will continue to be created along with free hydrogen until there are molecules of fluorine and hydrogen:

$$H + F_2 \rightarrow HF^* + H. \tag{7.2}$$

Some of the chemical reactions of molecular active media and their wavelengths are given in table 7.1.

As can be seen in the table, the all-gas-phase iodine laser is an example of an atomic chemical laser. Here, lasing takes place in excited iodine ($I^*$), which is produced via the interaction of singlet oxygen generated by a chemical reaction between chlorine ($Cl_2$), hydrogen peroxide ($H_2O_2$), and sodium hydroxide (NaOH) in the two steps shown below. Singlet oxygen is unstable and highly reactive, which helps to produce excited iodine.

Step 1: Production of singlet oxygen $O_2(^1\Delta)$:
$Cl_2 + H_2O_2 + 2NaOH \rightarrow O_2(^1\Delta) + 2H_2O + 2NaCl$.
Step 2: Production of the excited state of iodine ($I^*$):
$O_2(^1\Delta) + I \rightarrow O_2(^3\Sigma) + I^*$.

As shown in figure 7.5, the singlet oxygen ($O_2(^1\Delta)$) and molecular iodine ($I_2$) are admitted into the cavity in a carrier flow of Ar gas from the left. Lasing takes place in the transverse direction. Argon gas is also used to purge the mirrors.

**Table 7.1.** Active media, reactions, and laser wavelengths of some chemical lasers.

| Active medium | Chemical reaction | Laser wavelength(s) |
|---|---|---|
| HF (molecular) | $H_2 + F \rightarrow HF^* + H$ | 3.5 to 4.1 μm |
| DF (molecular) | $D_2 + F \rightarrow DF^* + D$ | 3.5 to 4.1 μm |
| HCl (molecular) | $HI + Cl \rightarrow HCl^* + I$ | 3.5 to 4.1 μm |
| HBr (molecular) | $H + Br_2 \rightarrow HBr^* + Br$ | 4 to 4.7 μm |
| I (atomic) | $I + O_2^* \rightarrow I^* + O_2$ | 1.31 μm |

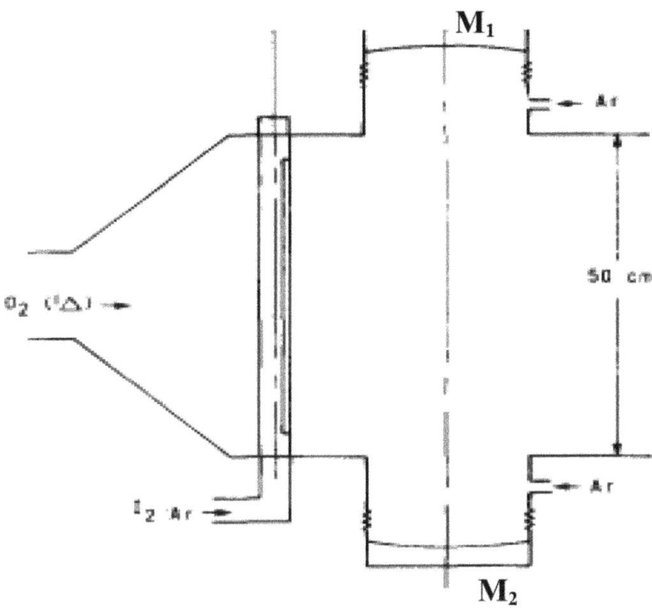


**Figure 7.5.** The laser cavity of a chemical oxygen–iodine laser (reprinted from [10], with the permission of AIP Publishing). The mirrors are indicated by $M_1$ and $M_2$.

♣ Chemical pumping has also been used in airborne lasers. During the proposed so-called 'Star Wars' program during the 1990s, laser cavities built inside the aeroplanes were used for delivering high-power laser beams from a chemical oxygen–iodine laser (COIL, see table 7.1).

## 7.5 Nuclear pumping

Nuclear reactions release gamma rays or particles as a result of radioactivity. In nuclear pumping, gamma rays or particles help to create an excited plasma of the gain medium. Some of the gain media that can be pumped by this method are gas combinations of He with Ar, Kr, or Xe.

## 7.6 Pump-cavity geometries

As outlined in the description of the gas-discharge tube (figure 7.1), the space between two highly reflective mirrors is known as the *laser cavity*. We introduce another term here, the *laser head*. In solid-state lasers such as the $Nd^{+3}$:YAG laser, this is situated at the center of the laser cavity. The laser head consists of a pumping system and a gain medium, as shown in figure 7.6. The laser cavity has two highly reflective mirrors known as the *'end mirror'* $M_1$ and the *'output coupler'* $M_2$.

♠The details of the parameters of the cavity, such as the radii of curvature of mirrors and the distance between them, are important for stable laser operation. The selection of stable as well as unstable laser cavity parameters is addressed in chapter 11.

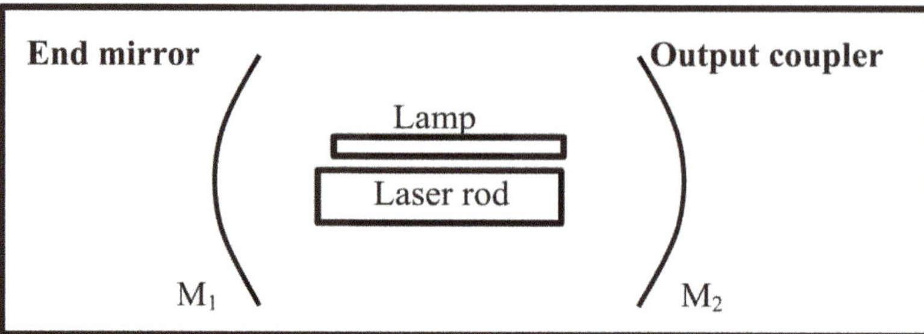


**Figure 7.6.** Positioning of a lamp and a laser rod for a typical optically pumped gain medium in a laser cavity consisting of two mirrors, $M_1$ and $M_2$.

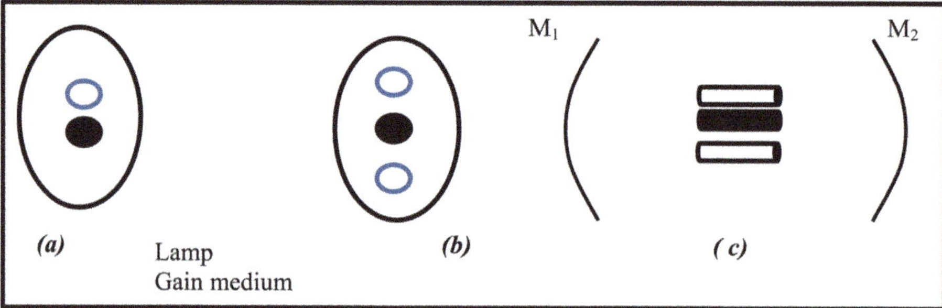


**Figure 7.7.** Side view of an elliptical laser head. (a) The lamp and the gain medium (*a laser rod*) are situated in an oscillator cavity. (b) A lamp with two rods: one each for the oscillator and the amplifier. (c) Top view of the laser head of panel (b) in a cavity with mirrors $M_1$ and $M_2$.

### 7.6.1 Optical side pumping

Most common solid-state lasers have side pumping in the sense that the gain medium (for example, the laser rod shown in figure 7.6) and the pump source (lamp) are kept side by side. Both of them are situated at the foci of the elliptical cavity, as shown in figure 7.7. The light rays originating from one focus of the ellipse are reflected to the other focus. The geometries can be such that there is a single lamp for an oscillator and an amplifier in a cavity, or a separate cavity can be provided for the amplifier.

### 7.6.2 Optical transverse pumping

Transverse optical pumping is used in distributed feedback (DFB) lasers. As shown in figure 7.8, with this type of pumping, the laser output can emanate from both sides of the gain medium. In this approach, the optical cavity is defined by an interference structure engraved on the gain medium itself (♠ see details in chapter 25), and the pump source is also a laser suitable for the absorption frequency of the gain medium.

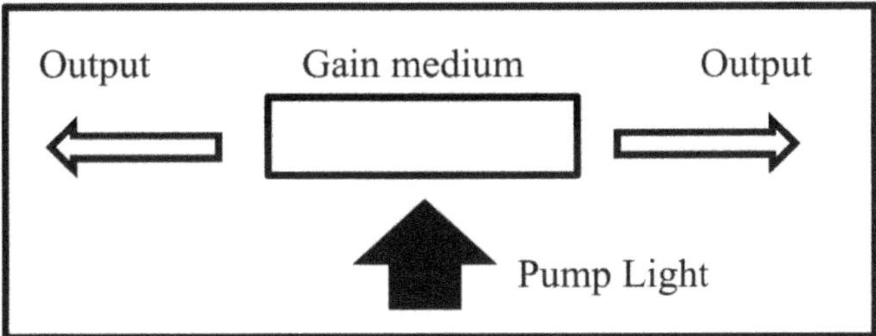



**Figure 7.8.** A schematic of the geometry used in DFB-type lasers, i.e. transverse optical pumping.

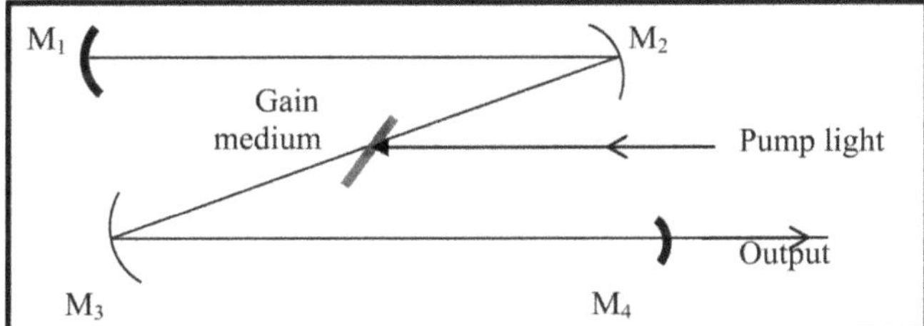



**Figure 7.9.** Bow tie-type geometry used in dye lasers and the Ti: Sapphire laser. $M_1$ is the end mirror and $M_4$ is the output coupler. $M_3$ and $M_4$ are mirrors used to fold the beam in the cavity.

### 7.6.3 Optical face pumping

Most of the laser cavities mentioned in previous sections have two mirrors. However, the cavity can be modified for applications that require a special mode or pulsed operation. In such cases, it is necessary to introduce extra mirrors and work with an extended cavity. The number of extra mirrors does not matter as long as the light beam continues to bounces back and forth between the *end mirror* and the *output coupler*.

In dye lasers, a thin dye jet is pumped by another laser in a *face-pumped* geometry. Feedback is provided using the *bow tie* type of laser cavity. The laser output is obtained as shown in figure 7.9. A similar geometry is used in the Ti:sapphire laser (♠ see chapter 17). In dye lasers, the dye jet is positioned at the Brewster's angles for the pump laser (an air-to-liquid jet) as well as for the dye laser (a liquid-to-air interface) to avoid unwanted reflections. In the figure, while $M_1$ and $M_4$ form the laser cavity, the additional mirrors $M_2$ and $M_3$ serve as folding mirrors. The latter are used in applications in order to obtain a pulsed output and further shorten the laser pulses (♠ see chapter 17 for details).

**Exercise 7.2.** Draw the configuration of a laser system consisting of an oscillator as well as an amplifier.

**Solution:** The schematic of *an oscillator–amplifier* system is shown in the chart below. For pulsed amplifiers, temporal synchronization between the oscillator and the amplifier is important. Such configurations are useful when laser pulses are to be amplified from nJ energy levels to mJ levels. The technique of chirped pulsed amplification (CPA) is used to obtain even higher energies (♠ see section 18.9).

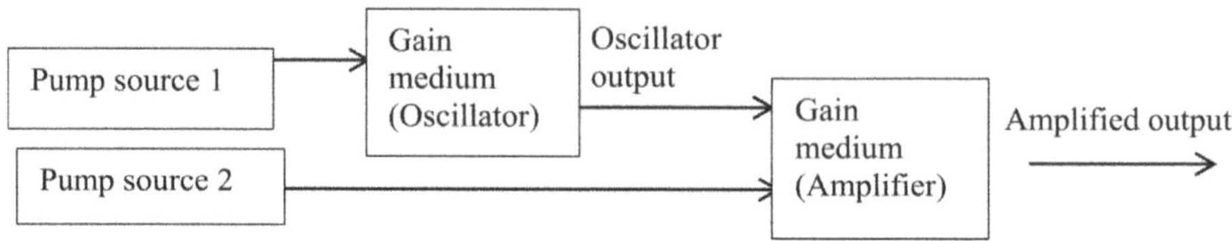


There are compact systems that use the same pumping source; in such systems a single flash lamp pumps the oscillator and the amplifier media, as indicated in figures 7.7(b) and (c).

### 7.6.4 Other optical pumping geometries

There are some other geometries in which the coupling of the pump power is done in more efficient way, such as *end* pumping in fiber lasers (♠ see the details in chapter 23), and *edge* pumping in slab lasers.

Although an attempt has been made to give a summary of pump sources and cavity geometries, their details are beyond the scope of this book. The reader is referred to the material listed in the bibliography for further information.

## Questions and problems

1. In a He–Ne laser, which element undergoes lasing transitions? How does the energy transfer process take place?
2. A rhodamine 6G dye laser is pumped by the second harmonic of a Nd:YAG laser with an output power of 10 W. Assume that the quantum yield of the luminescence of the dye is unity and that every pump photon produces one excited dye molecule. How many molecules are accumulated within the dye lifetime of 3.5 ns?
3. Give an example of each of: (i). an optically and (ii). an electrically pumped laser.
4. What laser gain material abundantly available in nature can be used in a laser?
5. Complete the columns of the following table corresponding to the pumping mechanisms and pumping geometries.

| Laser | Pumping mechanism | Pumping Geometry |
| --- | --- | --- |
| Nd:YAG laser | | |
| Argon-ion laser | | |
| He-Cd laser | | |
| Free-electron laser | | |
| Chemical oxygen–iodine laser | | |
| Semiconductor lasers | | |
| Fiber lasers | | |
| Dye lasers | | |
| Distributed feedback lasers | | |

# Bibliography


[1] Huestis D L 1982 Introduction and overview *Gas Lasers* ed E W McDaniel and W L Nighan (New York: Academic) pp 1–34

[2] Javan A, Bennett W R and Herriott D R 1961 Population inversion and continuous optical maser oscillation in a gas discharge containing a He–Ne mixture *Phys. Rev. Lett.* **6** 106–10

[3] Patel C K N 1965 Cw high power $N_2$–$CO_2$ laser *Appl. Phys. Lett.* **7** 15–7

[4] Phillips D T and West J 1970 The poor man's nitrogen laser *Am. J. Phys.* **38** 655–57

[5] Heard H G 1963 Ultra-violet gas laser at room temperature *Nature* **200** 667

[6] Lofthus A and Krupenie P H 1977 The spectrum of molecular nitrogen *J. Phys. Chem. Ref. Data* **6** 113–307

[7] Anderson J D 1976 *Thermodynamics and Vibrational Kinetics of the $CO_2$–$N_2$–$H_2O$ or He Gas Dynamic Laser* (New York: Academic) pp 15–33

[8] Kasper J V V and Pimentel G C 1965 HCl chemical laser *Phys. Rev. Lett.* **14** 352–54

[9] Yuryshev N P V *et al* 1984 Chemical oxygen-iodine laser utilizing low-strength hydrogen peroxide *Sov. J. Quantum Electron.* **14** 1138

[10] Benard D J, McDermott W C, Pchelkin N R and Bousek R R 1979 Efficient operation of a 100-W transverse-flow oxygen-iodine chemical laser *Appl. Phys. Lett.* **34** 40–1

[11] Manke G, Cooper C, Dass S, Madden T and Hager G 2003 A multi-watt all gas-phase iodine laser (AGIL) *34th AIAA Plasma Dynamics and Lasers Conf. Fluid Dynamics and Co-located Conferences* (Reston, VA: American Institute of Aeronautics and Astronautics)

[12] Svelto O and Hanna D C 1988 *Principles of Lasers* (Berlin: Springer)

[13] Lu J *et al* 2000 Optical properties and highly efficient laser oscillation of Nd:YAG ceramics *Appl. Phys. B* **71** 469–73

**IOP** Publishing

# An Introduction to Photonics and Laser Physics with Applications

**Prem B Bisht**

# Chapter 8

# Line-broadening mechanisms

In spectroscopy, we encounter absorption or emission profiles that represent transitions between two levels. Quite often, these transitions exhibit broad profiles. We know that a lack of resolution in spectrometers is responsible for the observed breadth. However, a limit is imposed on the spectral width by atomic or molecular transitions. This width arises due to several factors during emission from an excited energy level. The frequency ($\nu$) of the energy gap between two levels ($\Delta E$) is given by $\nu = \frac{\Delta E}{h}$. The emitted wave train experiences a change in its frequency. How does this happen? Do these changes take place uniformly or non-uniformly? For instance, the diagram shows the observed laser lines of $Nd^{+3}$ ions doped in a yttrium aluminum garnet (YAG) crystal and in glass. You will find answers to these questions in this chapter. Studies of line-broadening mechanisms are not only useful in laser physics but are also an interdisciplinary area of science, technology, and industry, including fiber-optic communication.

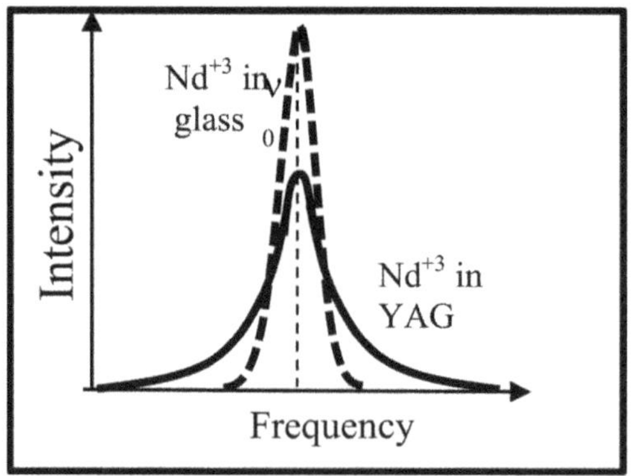

> **Learning objectives**
> **After reading this chapter, the learner will be able to:**
> Describe the resolving power of an instrument;
> Identify line broadening in a laser;
> Describe the Fourier transform of a function;
> Relate the electrons attached to atoms to a harmonic oscillator;
> Differentiate between homogeneous and inhomogeneous broadening;
> Distinguish the Lorentzian and Gaussian profiles;
> Recognize the Voigt profile.

## 8.1 The small-gain coefficient in practice

The small-gain coefficient $k$ per unit frequency is written in terms of Einstein's B coefficient in equation (5.8) as

$$k = \frac{n}{c}\left(N_2 - N_1\frac{g_2}{g_1}\right)B_{21}h\nu_{21}. \tag{8.1}$$

Theoretically, the frequency $\nu_{21}$ denotes a single frequency assigned to stimulated emission or an ideal laser transition. In practice, there are some variations that affect this assumption. Generally, the mechanism of transition is affected by the character of the host material, i.e. by its crystalline or amorphous nature. In addition, parameters such as variations in the temperature and pressure of the surroundings also contribute to the observed frequency width. These additional changes in frequency are very close to the central frequency $(\nu_0)$. Upon taking these factors into account, the experimentally obtained spectra are found to match with those predicted theoretically. In view of the above, we need to add a correction factor to equation (8.1). The information required to understand the mechanism of broadening phenomena associated with laser transitions or spectroscopic transitions in general is described in the following subsections.

## 8.2 Spectral resolving power

The resolving power $(R)$ of a spectrograph at a wavelength $\lambda$ is given by

$$R = |\,\lambda/\Delta\lambda\,|.$$

Here, $\Delta\lambda$ is the minimum separation betwwen the central wavelengths of two adjacent, just-resolved spectral lines with wavelengths $\lambda_1$ and $\lambda_2$, as shown in figure 8.1. Diffraction-limited line profiles with intensity $I(\lambda)$ are considered to be resolved if the individual curves cross each other as shown in the figure and show a dip known as the saddle point. Furthermore, the maximum intensity $(I)$ of either of the lines coincides with the first minimum of the diffraction of the other line. The saddle point lies between the two peak wavelengths at about 80% of the line intensity. This approach to resolving spectral lines is known as the Rayleigh

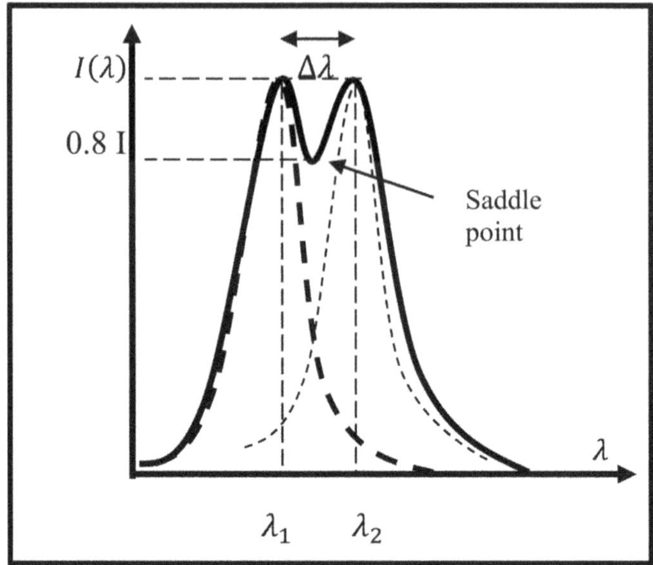


**Figure 8.1.** Schematic of two spectral lines with peaks at $\lambda_1$ and $\lambda_2$ separated by $\Delta\lambda$. Only the maximum and the first minimum of diffraction are shown for both lines.

criterion. It is useful in spectroscopy, and assumes that the intensity profile is the sum of several overlapping profiles.

**Exercise 8.1.** Sodium lamps emit a doublet and are used in various experiments in optics and spectroscopy. A diffraction grating is to be used with full illumination to resolve this doublet. How many grooves should the grating have?

   **Solution:** The sodium doublet has wavelengths of 588.996 nm and 589.593 nm. The average wavelength can be taken as 589.295 nm. For $\Delta\lambda = 0.597$ nm, the number of grooves $N = \frac{\lambda}{\Delta\lambda} = 987$.

**Exercise 8.2.** What is the experimentally observed line width of the He–Ne laser transition?

   **Solution:** Let us take the He–Ne laser transition at $632.8 \pm 0.002$ nm. We are assuming that the experimental resolution of the spectrograph is sufficient to resolve this. From $c = \nu\lambda$, we obtain

$$\Delta\nu = \left| -\frac{c}{\lambda^2}\Delta\lambda \right|.$$

For $\Delta\lambda = 0.002$ nm, the frequency difference ($\Delta\nu$) is $\Delta\nu = 1.5$ GHz.

$$\left[ \Delta\nu = \frac{3 \times 10^8 \times (0.002 \times 10^{-9})}{(632.8 \times 10^{-9})^2} = 1.5 \times 10^9\,\text{Hz} \right]$$

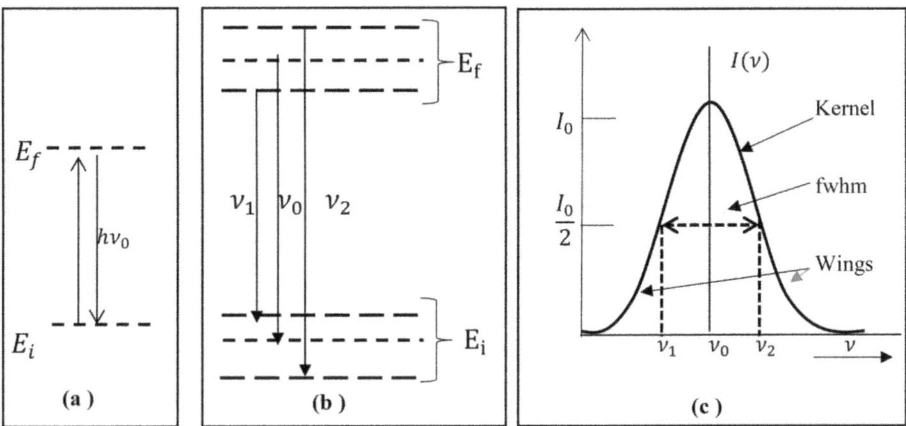


**Figure 8.2.** (a). Transition between two energy levels indicated by $E_f$ and $E_i$. Three typical transitions shown in (b) arise due to nondegenerate upper and lower states. As a result, the bell-shaped profile in (c) is indicated with its FWHM, kernel and wings.

## 8.3 Line broadening in He–Ne lasers

In deriving the expression for the small-gain coefficient, we assumed that all the atoms in either the upper or the lower level would be able to give rise to a (perfectly) monochromatic beam. From the above example, we observe that the line width of one of the transitions of the He–Ne laser, namely that at 632.8 nm, is 1.5 GHz. We therefore need information about the shape of the profile to be used for the correction by a factor $f(\nu)$ such that the expression for the small-gain coefficient (equation (8.1)) for the given frequency profile becomes

$$k = \frac{n}{c}\left(N_2 - N_1\frac{g_2}{g_1}\right)B_{21}h\nu_{21}f(\nu). \tag{8.2}$$

The broadened transitions generally have a typical bell-shaped profile, as indicated in figure 8.2. The energy corresponding to the central frequency of the transition is given by $E_f - E_i = h\nu_0$. If $I_0$ is the intensity of a given profile, the difference between the frequencies at half of the peak intensity $(\frac{I_0}{2})$ is known as the full width at half maximum (FWHM) and is denoted by $\delta\nu = \nu_2 - \nu_1$. The FWHM is often expressed in terms of the angular frequency: $\delta\omega\,(= 2\pi\delta\nu)$. The portion below the FWHM is known as the wings of the profile and the spectral portion within the half width is known as the kernel of the line.

The bell-shaped curves shown in figure 8.2 appear due to various transitions occurring between states with frequencies slightly higher and lower in energy, as shown in figure 8.2. The mechanisms responsible for the appearance of these adjacent frequencies are described in the following.

## 8.4 Harmonic oscillator

To understand the emission profiles, let us consider the photon-emitting atom to be a harmonic oscillator. In the Drude-Lorentz model of the atom, the electrons are considered to be tiny masses attached to springs—a model similar to that of a simple harmonic oscillator (SHO). The idea is to see whether there is any dephasing of the wave train or any change in the frequency of the photon during emission or absorption by the atom.

When an atom emits a photon of frequency $\nu_0$, the amplitude of the emitted wave train undergoes exponential dephasing. The restoring force for a spring with a spring constant $k$ for a displacement $x$ is $F_{\text{spring}} = -kx$. The equation of motion for this system is

$$\ddot{x} + \frac{k}{m}\, x = 0. \tag{8.3}$$

A solution of differential equation (8.3) is $x(t) = x_0 e^{-i\omega_0 t}$. Putting this in equation (8.3), we get $-m\omega_0^2 x + kx = 0$ with $\omega_0 = \sqrt{k/m}$ as the central circular frequency of the SHO. A damped oscillator (similarly to equation (1.9)) can be considered to be the superposition of several monochromatic oscillations:

$$\ddot{x} + \omega_0^2 x + \gamma_0 \dot{x} = 0. \tag{8.4}$$

The real part of the solution of equation (8.4) is $\quad x(t) = x_0 e^{\left(-\frac{\gamma_0}{2}\right)t}\cos\omega_0 t$ . Using the form of Euler's relations, this can also be expressed as $x(t) = Re\left(x_0 e^{-i\omega_0 t}e^{\left(-\frac{\gamma_0}{2}\right)t}\right)$. Here, $\frac{\gamma_0}{2}$ is a damping rate corresponding to uniform drag. A similar relation can be written for the radiation emitted by an atom in the form of the real part of the electric field $E(t)$ as

$$E(t) = E_0 e^{-i\omega_0 t}e^{(-\gamma_0/2)t} \text{ for } \quad t > 0. \tag{8.5}$$

This expression can be described as a superposition of monochromatic oscillators $e^{i\omega_0 t}$ with slightly different frequencies and amplitudes. The intensity is given by $I(t) = |E(t)|^2 = I_0 e^{(-\gamma_0)t}$ for $t > 0$. The exponential decay of the electric field can be seen in the figure below. The electric field decays at a rate of $\frac{\gamma_0}{2}$ while oscillating at a much higher frequency $\omega_0$ within the envelope.

We will use this understanding of a damped harmonic oscillator for the line-broadening mechanism in the following sections.

## 8.5 Broadening mechanisms

Generally, two major mechanisms are responsible for the broadening of spectral line profiles. One of these is known as an inhomogeneously broadened profile, which is the sum of several adjacent sharply peaked spectral lines. The second type, which contains a single broadened profile, falls under the category of homogeneous broadening.

### 8.5.1 Fourier transform

In order to obtain the frequency spread from the decay time of the atoms, we need to use the Fourier transform (FT) method. This is a mathematical expression that converts a function of frequency ($F(\omega)$) into that of time ($f(t)$). The FT has profound applications in science and technology. It is defined in the following two expressions.

The complex amplitude $F(\omega)$ is known as the FT of $f(t)$ if

$$F(\omega) = \int_{-\infty}^{\infty} f(t)e^{-i\omega t}dt. \tag{8.6}$$

The inverse FT, i.e. $f(t)$ is written as a superposition integral of harmonic functions of frequencies, as follows:

$$f(t) = \frac{1}{2\pi} \int_{-\infty}^{\infty} F(\omega)e^{i\omega t}d\omega. \tag{8.7}$$

$F(\omega)$ and $f(t)$ form a FT pair so that if one is known, the other can be determined. Some authors write the exponents with *opposite signs*.

In optics, let us consider a function with two variables in space (such as a 2D image), in which $f(x, y)$ represents a spatial pattern. The harmonic function $F(\nu) = Fe^{-i2\pi(\nu_x x + \nu_y y)}$ is a building block from which other functions can be composed. Here, the Fourier transform is between $f(x, y)$, and $F(\nu)$ can be obtained, where $(x, y)$ is the spatial coordinate of the 2D photograph; $\nu_x$ and $\nu_y$ are the lines per mm (the resolution of the image) in the $x$ and $y$ directions of the image plane, respectively. In a similar way, various frequencies of the damped harmonic oscillator can be expressed as a function.

### 8.5.2 Homogeneous broadening and the Lorentzian expression

According to equation (8.5), the electron begins to radiate at time $t = 0$ and a subsequent decrease in the electric field occurs at a rate of $\frac{\gamma_0}{2}$. The frequency components of the wave can be obtained by taking the FT of equation (8.5), as follows:

$$E(\omega) = \frac{1}{\sqrt{2\pi}} \int_0^{\infty} E(t)e^{i\omega t}dt$$

$$= \frac{1}{\sqrt{2\pi}} \int_0^{\infty} E_0 e^{-i\omega_0 t} e^{(-\gamma_0/2)t} e^{i\omega t}dt$$

$$= \frac{E_0}{\sqrt{2\pi}} \int_0^{\infty} E_0 e^{i[(\omega-\omega_0)+i(\gamma_0/2)]t}dt$$

$$E(\omega) = -\frac{E_0}{\sqrt{2\pi}} \left( \frac{1}{i[(\omega - \omega_0) + i(\gamma_0/2)]} \right).$$

As the intensity is the square of the amplitude $|E(\omega)|^2$, it can be written as the *Lorentzian* line shape

$$I(\omega) = I_0\left[\frac{\gamma_0/2\pi}{(\omega - \omega_0)^2 + (\gamma_0/2)^2}\right].\tag{8.8}$$

From this expression, we can take $f(\nu) = f(\omega)/2\pi$ with $f(\omega) = \dfrac{\gamma_0/2\pi}{(\omega - \omega_0) + (\gamma_0/2)^2}$ for equation (8.2). The homogeneous linewidth is expressed using a Lorentzian distribution (equation (8.8)) that has been normalized such that $\int_0^\infty I(\omega)d\omega = I_0$. Here, $I_0$ is the integrated intensity over the full profile.

The FWHM of the profile is the frequency width at which the intensity of the profile drops to half of the peak intensity on either side of the peak. Taking the maximum intensity at $=\omega_0\left(=\frac{2I_0}{\pi\gamma_0}\right)$, from equation (8.8), one can find the value of $(\omega - \omega_0)$ on either side, namely $\frac{\gamma_0}{2}$, and the FWHM of the homogenous profile is $\gamma_0$.

### 8.5.2.1 Natural line broadening

As described above for a damped harmonic oscillator, the very act of emission of a photon by an atom leads to an exponential dephasing of the amplitude of the wave train (see figure 8.3). This dephasing results in homogenous line broadening and is given by Lorentzian line shapes (equation (8.8)). As usual, the frequency between the energy gap of $\Delta E$ is given by $\Delta\nu = \frac{\Delta E}{h}$. Heisenberg's uncertainty principle states that if a system exists in an energy state for a time duration $\Delta t$, the energy of the state has an uncertainty of $\Delta E$, where $\Delta E \times \Delta t = \frac{h}{2\pi} = 10^{-34}$ Js. Here, $h$ is Planck's constant. For a limiting case, when $\Delta E = 0$, the system remains in the excited state for an infinite time. Generally, the lifetimes of the excited states are of the order of $10^{-8}$ s, which gives a value of $\Delta E = \frac{10^{-34}}{10^{-8}} = 10^{-26}$ J, resulting in a frequency uncertainty of $\frac{\Delta E}{h} \cong 10^8$ Hz. As this is inherent broadening related to the lifetime of the state, it is known as *natural line broadening* or *lifetime broadening*. This relation is also known as the *time–bandwidth product* ($\Delta\nu \times \Delta t = $ a constant), which manifests in ultrafast lasers in which the laser frequency bandwidths can be very large.

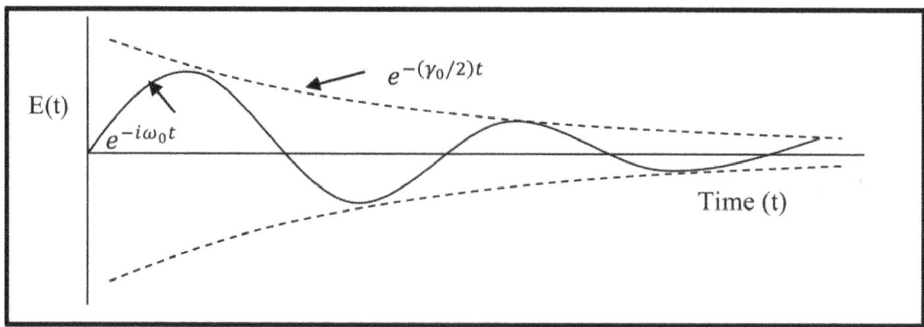


**Figure 8.3.** Exponential decay (dashed line) of the electric field amplitude with a decay rate of $\left(\frac{\gamma_0}{2}\right)$ of a damped harmonic oscillator (solid line) at frequency $\omega_0$.

**Exercise 8.3.** In which region of the EM spectrum is the natural line broadening contribution higher and why?

**Solution**: From the energy–time uncertainty discussion above, $\frac{\Delta E}{h} \cong 10^8$ Hz. For the visible region, in which the frequencies are of the order of $10^{14} - 10^{16}$ Hz, the value of broadening is very small. The natural line width contribution is the smallest among the other broadening mechanisms, such as Doppler or collisional broadening (see below). However, in the GHz frequency regions, the contribution of lifetime broadening is greater than that in the visible regions.

### 8.5.2.2 Collisional broadening

To understand collisional broadening, let us look at the well-known experimental observation of the spectral line of a low-pressure mercury lamp, which shows fine structure. Figure 8.4 shows the schematic behavior of the spectral lines of mercury at low and high pressures. It can be seen schematically that the spectral lines become broad at high pressure (♠ see figure 7.2 in the previous chapter) due to a kind of 'frictional force'. This broadening of the spectral lines is due to collisional broadening. It should be noted that the collisional frequency is much smaller than the optical frequencies.

At a given temperature $(T)$, if a slow-moving atom/heavy molecule suffers a collision during the emission of a photon, the phase of the wave train associated with the photon is suddenly altered, as shown in figure 8.5. At a fixed $T$, this effect leads to resonant or homogeneous broadening for all atoms. At high pressures, there are more collisions, hence the contribution of collisional broadening is higher. An atom suffers collisions from all possible directions and hence the displacement and momenta, on average, are equally probable in all directions (resulting in a null vector). This contributes to the homogenous line broadening. Such an assumption was made by Lorentz about the large number of collisions between atoms. This type

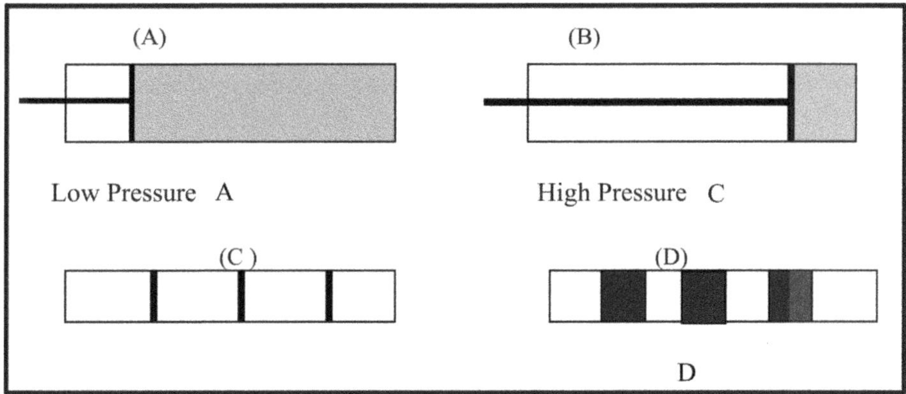


**Figure 8.4.** Schematic of mercury gas in a cylinder at low pressure (A) and its atomic spectrum (C). At high pressures (B), the spectral lines become broad due to collisions (D).

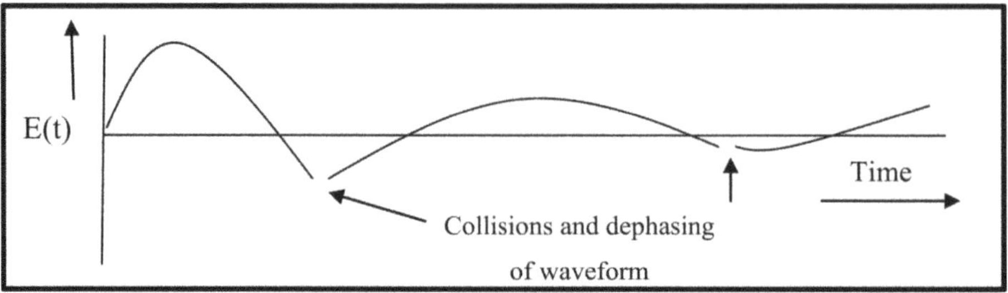


**Figure 8.5.** Collisions at higher pressure result in dephasing of the wave train.

of broadening effect is useful in high-pressure mercury lamps, in which it generates a continuum of light for applications requiring broadband light sources.

Collisional decay is a nonradiative pathway of deactivation. It not only reduces the efficiency of the emitting transitions (by reducing the population of the upper state) but also broadens the transition (increasing the emission linewidth), both of which are negative effects. Collisional broadening is proportional to the pressure and is also known as pressure broadening. Phonon collisions in solids and electron collisions in gases are examples of cases in which collisional broadening contributes to line profiles.

### 8.5.3 Inhomogeneous broadening and the Gaussian expression

From a classical perspective, the kinetic energy of a particle of mass $m$ in a gaseous system with an average velocity $v_x$ at a temperature $T$ is given by $v_x^2 = k_B T/m$. Here, $k_B$ is the Boltzmann constant. The Doppler shift in the frequency is given by $\nu = \nu_0\left(1 \pm \frac{v_x}{c}\right)$. Depending upon the velocity component, the widths of the individual line profiles add up to create a resultant broadened profile due to Doppler broadening (figure 8.6). The Brownian motion of the system contributes significantly to the resultant profile. Therefore, the higher the temperature of the surroundings, the greater the contribution due to the Doppler shift. Similarly, for molecules with lower mass, such as He and Ne, the Doppler shift of the emitted frequencies is higher. Such frequency shifts are responsible for the Doppler broadening of the observed profile.

The Maxwell–Boltzmann formula for a gas in thermal equilibrium at $T$ with a fraction of atoms $F(v)$ and a velocity distribution between $\nu$ and $\nu + d\nu$ along an axis (say the $x$-axis) is given by

$$F(v) = \sqrt{\left(\frac{m}{2\pi k_B T}\right)}\, e^{-\frac{m v_x^2}{2 k_B T}}\, dv_x. \qquad (8.9)$$

For the observer approaching the source, $v_x = \frac{c}{\nu_0}(\nu - \nu_0)$ and $dv_x = \left(\frac{c}{\nu_0}\right)d\nu$. Following equation (8.9), the intensity of the emitting state can be rewritten as

$$I(\nu) = \sqrt{\left(\frac{mc^2}{2\pi k_B T \nu_0^2}\right)}\, e^{-\frac{c^2 m(\nu - \nu_0)^2}{2 k_B T \nu_0^2}}\, d\nu.$$

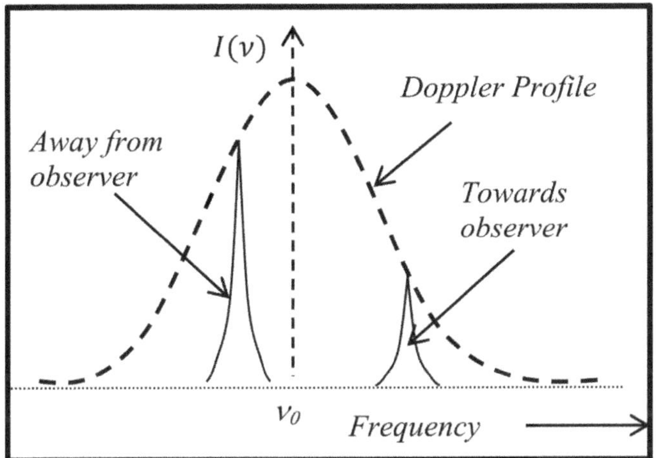


**Figure 8.6.** An inhomogeneously broadened profile resulting from an overlap of several peaks due to the individual line profiles of atoms moving towards or away from the detector.

If we take the pre-exponential factor $I_0$ as the constant of proportionality, the expression becomes

$$I(\nu) = I_0 \; e^{-\frac{c^2 m(\nu-\nu_0)^2}{2k_B T \nu_0^2}} \; d\nu. \tag{8.10}$$

The intensity is normalized again as $\int_0^\infty I(\nu)d\nu = I_0$. A plot of $I(\nu)$ vs $\nu$ is shown in figure 8.6.

The expression given in equation (8.10) shows a maximum at $\nu = \nu_0$. Assuming that there are no collisions, the Maximillian distribution results in an inhomogenously broadened Doppler profile—i.e. a Gaussian function. The effective width of the emission line can be determined by establishing the frequency shift $(\nu - \nu_0)$ with respect to the central maximum frequency $(\nu_0)$. The half intensity[1] for equation (8.10) is obtained as follows: where

$$\frac{I_0}{2} = I_0 \; e^{-A(\nu-\nu_0)^2} \text{ where } \frac{c^2 m}{2k_B T \nu_0^2} = A. \tag{8.11}$$

The FWHM of the line $\Delta\nu_{\text{fwhm}}$ is twice the value of $(\nu - \nu_0)$ (i.e. $(\nu - \nu_0) = \frac{\Delta\nu_{\text{fwhm}}}{2}$):

$$\frac{I_0}{2} = I_0 \; e^{-A\left(\frac{\Delta\nu_{\text{fwhm}}}{2}\right)^2}$$

or $A\dfrac{\Delta\nu_{\text{fwhm}}^2}{4} = \ln 2$ with $A = \dfrac{4\ln 2}{\Delta\nu_{\text{fwhm}}^2}$.

---

[1] At half the value of $x$ (i.e. half the width), the function $I(x)$ can be calculated using $\frac{I_0}{2} = I_0 e^{-Ax^2}$, so that $x = \sqrt{\frac{\ln 2}{A}}$; the FWHM is twice this value, so that $A = \frac{4\ln 2}{(\Delta\nu)^2}$.

Using the value of $A$ from equation (8.11), $\Delta\nu_{\text{fwhm}}$ can also be expressed as $\Delta\nu_{\text{fwhm}} = \frac{2\nu_0}{c}\sqrt{\left(\frac{2k_BT}{m}\right)\ln 2}$. Equation (8.10) can be rewritten as

$$I(\nu) = I_0 \sqrt{\frac{4\ln 2}{\pi \Delta\nu_{\text{fwhm}}^2}} \; e^{-\frac{4\ln 2}{\Delta\nu_{\text{fwhm}}^2}(\nu-\nu_0)^2}.$$

For the case of inhomogeneous broadening, in parallel to equation (8.8), we can now obtain the value of the function $f(\nu)$ for equation (8.2), as follows:

$$f(\nu) = \sqrt{\frac{4\ln 2}{\pi \Delta\nu_{\text{fwhm}}^2}} \; e^{-\frac{4\ln 2}{\Delta\nu_{\text{fwhm}}^2}(\nu-\nu_0)^2} \tag{8.12}$$

From the expression for $\Delta\nu_{\text{fwhm}}$, we can see that the Doppler broadening is proportional to the frequency $(\nu_0)$ of the transition. This confirms its greater contribution in the UV and visible regions as compared to that in IR frequencies. In addition, its inverse proportionality to the value of mass $m$ can be understood in the light of the fact that lasers with gain media composed of lighter molecules (such as the He–Ne laser) have a larger inhomogenous broadening contribution than those with heavier atomic masses (e.g. $CO_2$ lasers).

## 8.6 Correction of the small-gain coefficient

Using the expression for the Lorenztian profile (equation (8.8)) of $\nu = \nu_0$, the value of the function $f(\nu)$ can be obtained for equation (8.2), as follows:

$$f(\nu) = \frac{(\gamma_0/2\pi)}{(\gamma_0/2)^2} \equiv \frac{2}{\pi\Delta\nu} \sim \frac{1}{\Delta\nu}.$$

Similarly, using the expression for the Gaussian profile (equation (8.12)) of $\nu = \nu_0$, $f(\nu)$ can be written as follows:

$$f(\nu) = \sqrt{\frac{A}{\pi}}\, e^{-0} \equiv \frac{\sqrt{\frac{4\ln 2}{(\Delta\nu)^2}}}{\pi} \sim \frac{1}{\Delta\nu}.$$

In this way, the numerical value of $f(\nu)$ can be approximated by $1\backslash\Delta\nu$ in equation (8.2) of the gain medium for both types of broadened profile. We can now rewrite the corrected expression for the small-gain coefficient (equation (8.2)) in units of $\text{cm}^{-1}$ as

$$k = h\nu_{21}\frac{n}{c}\left(N_2 - N_1\frac{g_2}{g_1}\right)B_{21}h\nu_{21}\frac{1}{\Delta\nu}. \tag{8.13}$$

**Exercise 8.4.** For the value given in exercise 6.1, if the broadening at the transition frequency $(\Delta\nu)$ is 5 THz, calculate the value of the population inversion.

**Solution:** The value of the population inversion obtained by considering the broadening is given by

$$\Delta N = \frac{k}{n}\frac{8\pi}{\lambda^2}\tau_{21} \times \Delta\nu = 9 \times 10^{25}.$$

## 8.7 Voigt profile

We now know that collisional broadening results in homogeneous line widths and Lorentzian line profiles. On the other hand, inhomogeneous line widths are the consequence of Doppler broadening and follow a Gaussian expression. It is interesting to note that both profiles give rise to similar bell-shaped curves, but there are subtle differences. We know that the local variations of temperature, pressure, magnetic field, etc. and imperfections or impurities embedded in the crystals, for example $Ti^{+3}$ or $Cr^{+3}$ impurities in a sapphire ($Al_2O_3$, corundum) crystal, contribute to inhomogeneous broadening. At the same time, the individual Doppler-shifted lines contribute to slightly shifted resonant frequencies, resulting in inhomogeneous broadening. In some instances, the line widths of an inhomogeneous line profile (a Doppler contribution) and a homogeneous line profile (a collisional contribution, say) overlap, as they have nearly comparable spectral widths. Under such circumstances, both types of broadening have to be expressed by a single mathematical expression.

In such a case, each subgroup of atoms with a velocity component in a certain direction contributing to a Gaussian line profile may have the contribution due to homogenous broadening as well. The resulting line profile, which cannot be expressed in an analytic form, is a convolution of the Gaussian and the Lorentizian shapes. Such profiles are known as Voigt line shape profiles (figure 8.7).

Voigt profiles from the tail measurements of broadened spectral lines obtained from interstellar atmospheres are useful in the estimation of temperature, pressure, and other parameters. Their mathematical description is as follows: if $V(a, x)$ is the Voigt profile, we introduce two parameters $a$ and $x$ that are defined in terms of the FWHM of a Gaussian profile ($\Delta\nu_{\text{fwhm}}$) and the FWHM of a Lorentian profile ($\gamma_0$), respectively, as

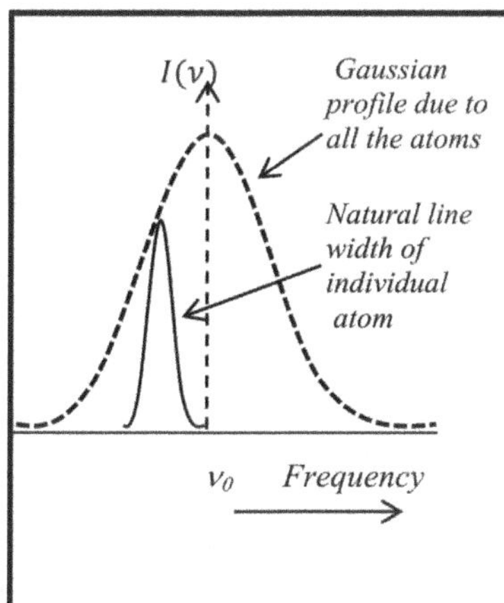


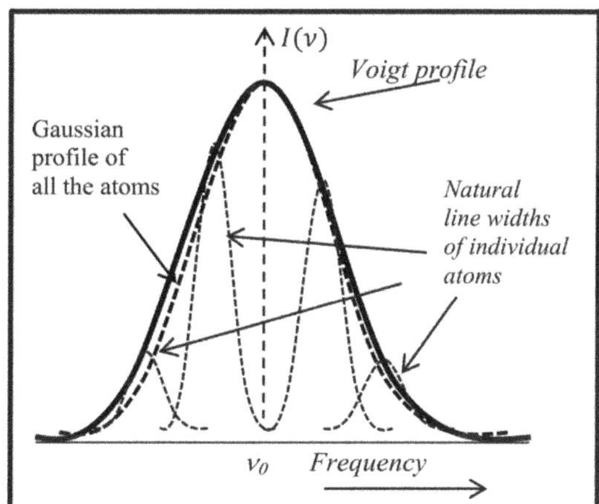


**Figure 8.7.** Example of a Voigt profile. The left-hand panel shows a schematic of the Gaussian profile (dashed curve) as well as the natural broadening due to individual atoms. The right-hand panel shows the Voigt profile, which is the combination of natural broadening due to several such atoms and the net Gaussian profile.

$a=\sqrt{\ln 2}\left(\dfrac{\gamma_0}{\Delta\nu_{\text{fwhm}}}\right)$ and $x=\sqrt{\ln 2}\,\dfrac{(\nu-\nu_0)}{\Delta\nu_{\text{fwhm}}}$. The line shape function is then defined by

$$I(\nu) = I_o\left(\frac{V(a,\,x)}{\Delta\nu_{\text{fwhm}}}\right)2\sqrt{\frac{\ln 2}{\pi}} \tag{8.14}$$

where

$$V(a,\,x) = \left(\frac{a}{\pi}\right)\int_{-\infty}^{\infty}\left(\frac{e^{-\nu^2}}{a^2 + (x-\nu)^2}\,\mathrm{d}\nu\right).$$

The Voigt profile reduces to a Lorentizian profile (with broader wings) when $a \cong 2$ and follows the Gaussian expression (dominating at the peak) for $0 < a < 1$. Therefore, when neither the Doppler nor the collisional-type broadening dominates, the two effects combine to give the Voigt profile described by equation (8.14).

♣ Second-order Doppler broadening also takes place (it is smaller by a factor of $10^{-4}$). It is observed in the special theory of relativity in terms of time dilation when objects move between two frames of reference. It is not particularly important here.

## 8.8 The effect of amorphous or crystalline hosts

A crystalline material has an ordered structure. The spectral lines caused by any dopant embedded in such a material (for example $Nd^{+3}$ in YAG) have a contribution due to homogenous broadening. On the other hand, if the host of a gain material is an amorphous solid (such as $Nd^{+3}$ in glass), then the broadening mechanism is dominated by inhomogeneous broadening. As shown schematically in figure 8.8, in practice, the ratio of the FWHM of $Nd^{+3}$:YAG to that of $Nd^{+3}$:glass is about 50:

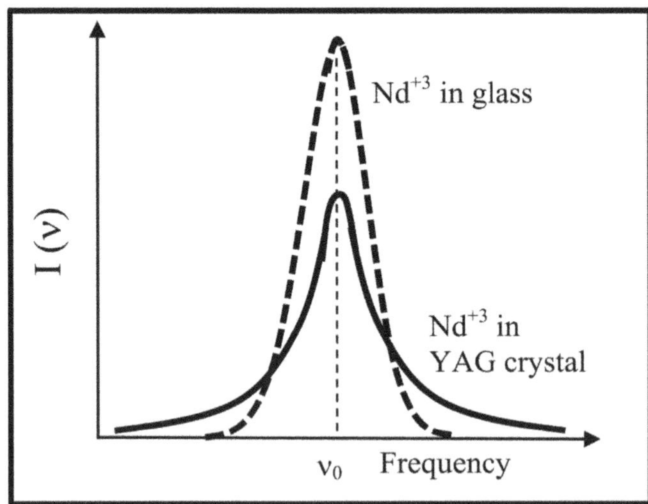


**Figure 8.8.** Schematic diagram of the spectral broadening of the gain spectrum of $Nd^{+3}$ doped in YAG (solid line) and in glass (dashed line).

$$\frac{\Delta\nu_{\text{fwhm}}(\text{Nd: Glass})}{\gamma_0(\text{Nd: YAG})} \cong 50.$$

It should be noted that for profiles with the same FWHM, the homogeneously broadened profile is more intense at the wings (see question 8.2).

## 8.9 Hole burning and the Lamb dip

On the basis of the aforementioned discussion, we now know that the dominant mechanism of inhomogeneous broadening of the absorption spectrum in, say, a gas, is Doppler broadening with a central frequency of $\nu_0$. If such a sample is simultaneously irradiated at a particular frequency $\nu_1$, a decrease in absorption (or a dip) is observed at this frequency. However, the rest of the absorption profile remains unaffected, as shown in figure 8.9. This dip is known as Bennett's hole. The phenomenon is known as hole burning.

In single-mode lasers (that have Doppler broadening), when the frequency of the mode is tuned to the peak of the gain curve (i.e. they are in resonance), there is a drop in the output power, compared to that obtained when the cavity is tuned away from resonance. This drop in the power is based on the principle of the phenomenon called the Lamb dip, as the phenomenon was explained by Lamb.

This phenomenon can easily be seen in a pump-probe experiment (i.e. using saturation spectroscopy). Consider an experimental setup (figure 8.10) in which an absorbent molecular vapor is enclosed in a cell. A tunable laser light with, say, wavelengths in the visible region (400–700 nm) enters the vapor cell from the right (this is known as the pump beam). Parts of the beam—known as the probe beams— pass through the cell such that one of them interacts with the portion of the cell crossing the pump beam. The other probe, which does not interact with the pump beam's path, is used as a reference. The difference between the two probe beams is detected by a detector.

Due to the Doppler effect, the changed frequencies of the absorbent atoms traveling with velocities in the $v_x$ and $-v_x$ direction are $\nu' = \nu_0\left(1 \pm \frac{v_x}{c}\right)$. The two frequencies matching the tunable probe light could be $\nu_1$ and $\nu_2$, corresponding to the frequencies shown in figure 8.11. Therefore, if we now measure the frequency

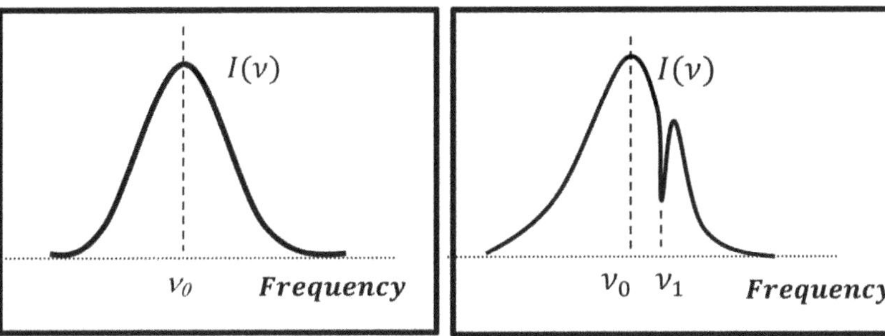


**Figure 8.9.** Left-hand panel: absorption spectrum of a gas with a peak at $\nu_0$. Right-hand panel: hole burning in a spectrum caused by passing a probe beam at a frequency of $\nu_1$ though a gas (see also figure 8.10).

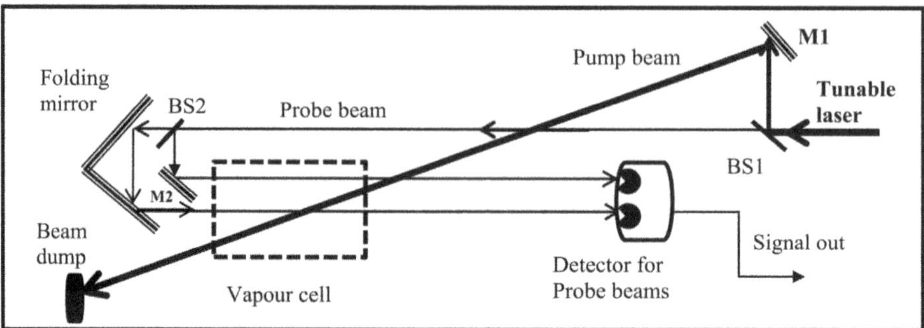



**Figure 8.10.** Schematic of saturation spectroscopy. M1, M2: mirrors, BS1, BS2: beam splitters. The pump beam passes through the vapor cell indicated by a dashed rectangle. The probe beam is split to obtain a reference signal before it is detected.

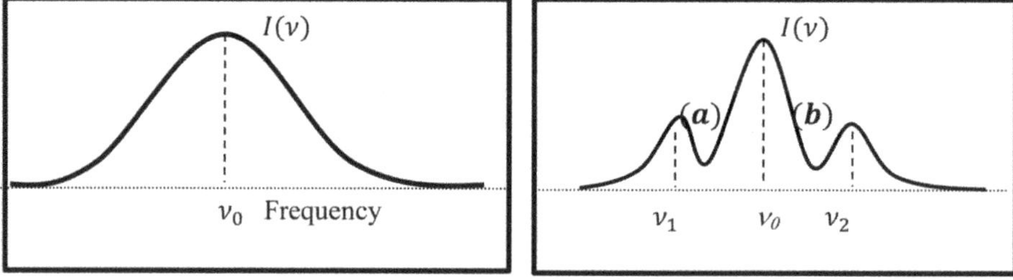



**Figure 8.11.** Absorption spectrum of a vapor cell in the absence (left-hand panel) and in the presence of probe light (right-hand panel) for the schematic setup shown in figure 8.10. The dips indicated by (a) and (b) are due to atoms traveling with velocities of $v_x$ and $-v_x$, respectively.

spectrum of the probe light using a spectrograph, there are two dips, ($a$) and ($b$), in its spectrum, as shown. These two dips correspond to the molecules traveling along the $x$-axis which can absorb light from the probe beam. In the right-hand panel of figure 8.11, the dip at $v_1$ is due to the velocities $-v_x$, while $v_2$ is due to $+v_x$.

**Exercise 8.5.** How can the laser output frequency be stabilized using the principle of hole burning?

**Solution:** Groups of atoms with equal and opposite velocities contribute to the Lamb dips ($v_1$, $v_2$) in figure 8.11. The Lamb dip has been used in the *stabilization of laser output.* Instead of the dips (($a$) and ($b$)) described above, the molecules traveling perpendicular to the $x$-axis (or those with no movement) contribute to a dip at the center frequency, as shown in figure 8.12.

In the case of laser cavities, the frequency and the FWHM of the oscillation can be changed by varying the external parameters of the cavity. For example, if the cavity length is changed by varying, say, the temperature of the cavity, the frequency dips can be shifted to the center of the profile. Under such circumstances, a single dip

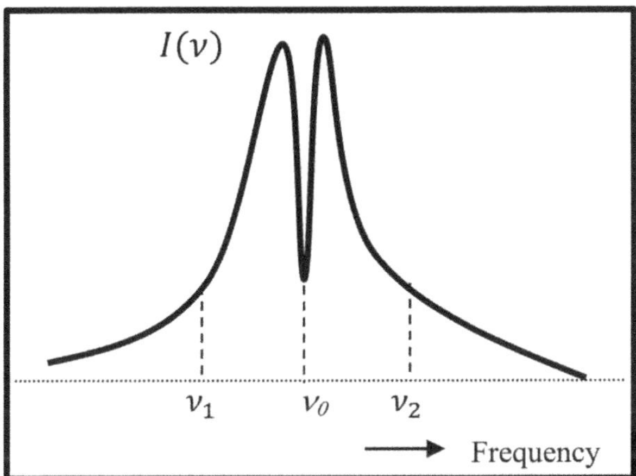


**Figure 8.12.** Line profile with a dip at the center in the absorption spectrum of the vapor cell in the presence of probe light with controlled cavity parameters for the schematic shown in figure 8.10. The movement of the atoms perpendicular to the axis of the cell has also been considered here.

(figure 8.12) with a higher strength can be observed, compared to the two dips mentioned above.

♣ Since any single deviation in either direction from the center of the laser line results in an increase in power, the effect can be used for frequency stabilization at the cost of a reduction of the power at the central minimum. This is done by providing feedback on adjusting the temperature of the system and to control the input parameters of the laser to maintain the frequency of the laser at the center of the spectrum.

## Questions and problems

1. Find the FWHM (in Hz) of a Nd:YAG laser by assuming that the line width of the laser is 0.002 nm. Compare this value with the FWHM of a nitrogen laser of the same line width.
2. Obtain the frequency ($\omega$) at which the intensity drops to one half of its maximum value for a Lorentzian profile. What is the FWHM at this frequency?
3. For a homogeneously broadened profile, show that $\int_0^\infty I(\omega)d\omega = I_0$.
4. In which region of the EM spectrum (UV–VIS or IR/microwave) is the contribution of Doppler broadening greater and why?
5. Out of two spectra obtained from two different samples, one is homogeneously broadened, while the other has Doppler broadening alone. However, both spectra have the same central frequency ($\nu_0$) and the same emission linewidth $\Delta\nu_{\text{FWHM}}$. Find the ratio of the intensities of the two cases at a frequency $\nu = 10 \times \Delta\nu_{\text{FWHM}}$.
6. In a pump-probe experiment, a continuous-wave broadband (400–700 nm) light beam (a probe) is collinearly propagating along a weak monochromatic (632.8 nm) beam (a pump) in an absorbent medium. What changes

do we expect in the transmitted spectrum of the probe beam (i) in the absence and (ii) in the presence of the pump beam?

7. Define the term 'hole burning' with reference to the pump-probe experiment in an inhomogeneously broadening system.

8. Two sources of spectrally broadened light have Lorentzian line shapes, widths $L_1$ and $L_2$, and central frequencies of $\omega_0$. Prove that the composite line is also Lorentzian, and obtain its width.

9. What is the origin of the correction factor of $1/\Delta\nu$ in the expression for the small-gain coefficient?

10. Experiments such as those involving laser cooling require highly frequency-stabilized lasers. What are the various ways of stabilizing the laser frequency? Is it possible to obtain a frequency-stabilized laser output by controlling the temperature of the cavity?

11. Identify the mechanism that is the major contributor to broadening in the following lasers: (i) the $CO_2$ laser (ii) the He–Ne laser, (iii) the Nd:YAG laser, and (iv) the Nd:glass laser.

12. Write down the expressions and sketch the emission line shapes of two species radiating at the same center frequency $\nu_0$ and the same emission-spectrum FWHM for the case in which one of them is homogenously broadened while the other is Doppler broadened.

13. Comment on the Voigt profile and its applications.

## Bibliography

[1] Demtröder W L 2014 *Spectroscopy* (Berlin: Springer)

[2] Preston D W 1996 Doppler-free saturated absorption: laser spectroscopy *Am. J. Phys.* **64** 1432–6

[3] Lamb W E 1964 Theory of an optical maser *Phys. Rev.* **134** A1429–450

[4] Bennett W R 1962 Hole burning effects in a He–Ne optical maser *Phys. Rev.* **126** 580–93

[5] Mitchell A and Zemansky M 1971 *Resonance Radiations and Excited Atoms* (Cambridge: Cambridge University Press)

**IOP** Publishing

# An Introduction to Photonics and Laser Physics with Applications

**Prem B Bisht**

# Chapter 9

## The Fabry–Pérot resonator

The interference phenomenon has been one of the cornerstones of optics. Young's classic double-slit experiment has been revisited for electrons in recent years. Today, several interferometers are available for various applications. For instance, while the Sagnac interferometer is used in laser gyroscopes, the Mach–Zehnder interferometer finds applications in optical communication systems and quantum-entanglement- related experiments. The Fabry–Pérot (FP) interferometer, invented in 1899, is used in high-resolution spectroscopy and, of course, as a laser cavity. The beauty of the FP resonator lies in is its ability to select spectral lines with high accuracy. Even without introducing a gain medium, the mode distribution and properties of a *passive* FP resonator can be obtained theoretically. The figure shows the sharp, well-separated resonances of a highly reflective cavity ($R = 0.9$). The properties of the FP resonator have provided a firm platform for optics research at the microscopic scale as well. The properties of its modes are described in this chapter.

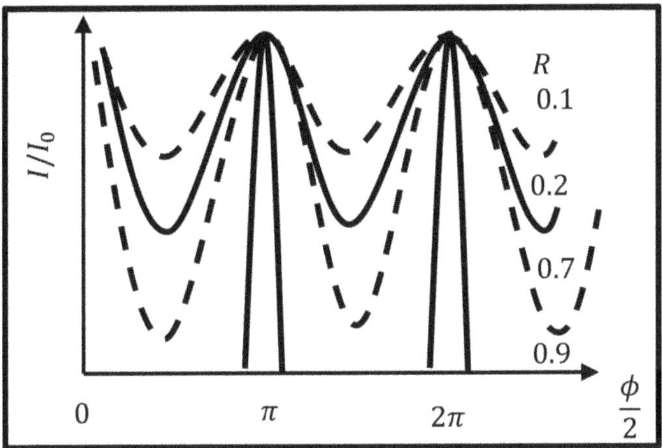






© IOP Publishing Ltd 2022

**Learning objectives**
**After reading this chapter, the learner will be able to:**

Define an active medium;
Estimate the number of modes of a cavity;
Characterize the longitudinal modes of a laser;
Describe the Airy function;
Identify the free spectral range, the full width at half maximum, the quality factor, and finesse;
Discover the FP etalon;
Define the Fresnel number and mode pulling.

# 9.1 Modes in a two-dimensional cavity

As described in chapter 4, the number of modes per unit frequency interval per unit volume in a three-dimensional (3D) black body is given by $\left(\frac{8\pi\nu^2}{c^3}\right)$. Qualitatively speaking, if we now truncate the 3D cavity in such a way that we are left with only two concave sides with a limited surface area in two dimensions (2D), the number of modes is reduced accordingly. Such a 2D cavity is similar to a laser cavity enclosed by two mirrors, as shown schematically in figure 9.1.

### 9.1.1 Active medium and gain bandwidth

A mixture of gases such as helium–neon (He–Ne) or a crystal such as ruby used in a laser is known as an *active medium* or a *gain medium*. The spontaneous emission spectrum profile of the active medium is known as the gain profile, as shown in figure 9.2 in an irradiance $((I(\nu))$ versus frequency plot. The full width at half maximum (FWHM) of the spectrum is referred to as the gain bandwidth.

At finite temperatures, as the spontaneous emission is spread over several frequencies due to the Boltzmann distribution, the gain profile is generally broad, although the line profile of ideal laser light is supposed to be sharply peaked at a

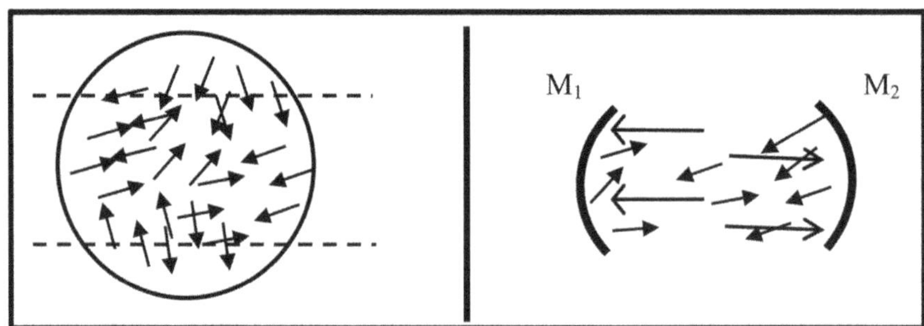


**Figure 9.1.** Schematic of the modes (shown by arrows) of a black body (left). After removing the portion of the cavity marked by the dashed lines as well as the front and back sides, a 2D cavity (right) represented by the mirrors $M_1$ and $M_2$ encloses fewer modes.

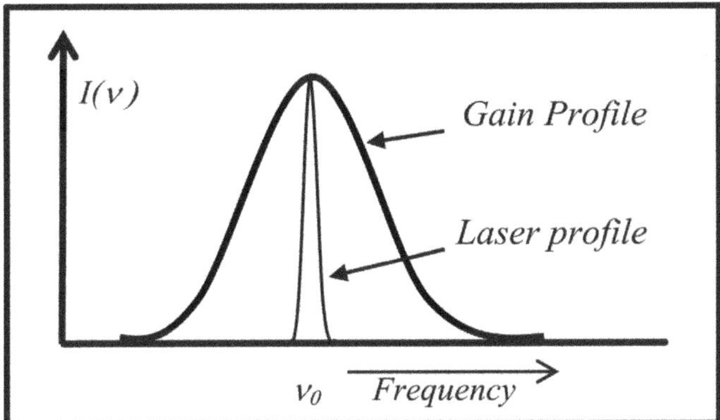


**Figure 9.2.** Schematic of the spectral profiles of the active medium and the laser. The laser profile is much sharper than indicated here.

frequency ($\nu_0$). The broadening mechanisms discussed in the previous chapter result in a certain FWHM.

### 9.1.2 Modes of a cavity

In general, the number of frequencies that can fit into a cavity is known as the number of *modes* of the cavity. For a cavity length of half a meter ($d$), the total number ($m$) of half wavelengths that can fit into the cavity can be obtained from $m = 2d/\lambda$. For a wavelength of 500 nm, the value of $m$ is approximately $2 \times 10^6$. The number of modes of a laser, however, is limited by the gain bandwidth of the active medium, as discussed in the following sections.

### 9.1.3 Free spectral range

A further selection of the number of modes is achieved by introducing an active medium into the laser cavity. In figure 9.3, the number of sharply peaked spikes represents the number $m$. The FWHM of each mode is represented by $\delta\nu$. The spacing between modes is known as the free spectral range (FSR), which is denoted by $\Delta\nu$ and given by $\Delta\nu = c/2d$, where $c$ is the speed of light. From exercise 8.2, we know that the spectral FWHM (also known as the *line width*) of a He–Ne laser output at $632.800 \pm 0.002$ nm is $1.5 \times 10^9$ Hz. The number of *longitudinal modes* of a laser is the number of modes that falls within the gain profile above the lasing threshold (or the loss line). It is easy to obtain the number of modes in the spectral line width of output of a typical He–Ne laser; it is equal to the ratio of the line width to the FSR.

**Exercise 9.1.** How many modes can fit into a 31.64 cm long cavity of a He–Ne laser for a wavelength of 632.8 nm?

    **Solution:** The number of modes in a cavity of length $L$ is given by $L = m\,\lambda/2$. Therefore, the answer is $m = \dfrac{2L}{\lambda} = \dfrac{2 \times 31.64 \times 10^{-2}}{632.8 \times 10^{-9}} = 10^6$ modes.

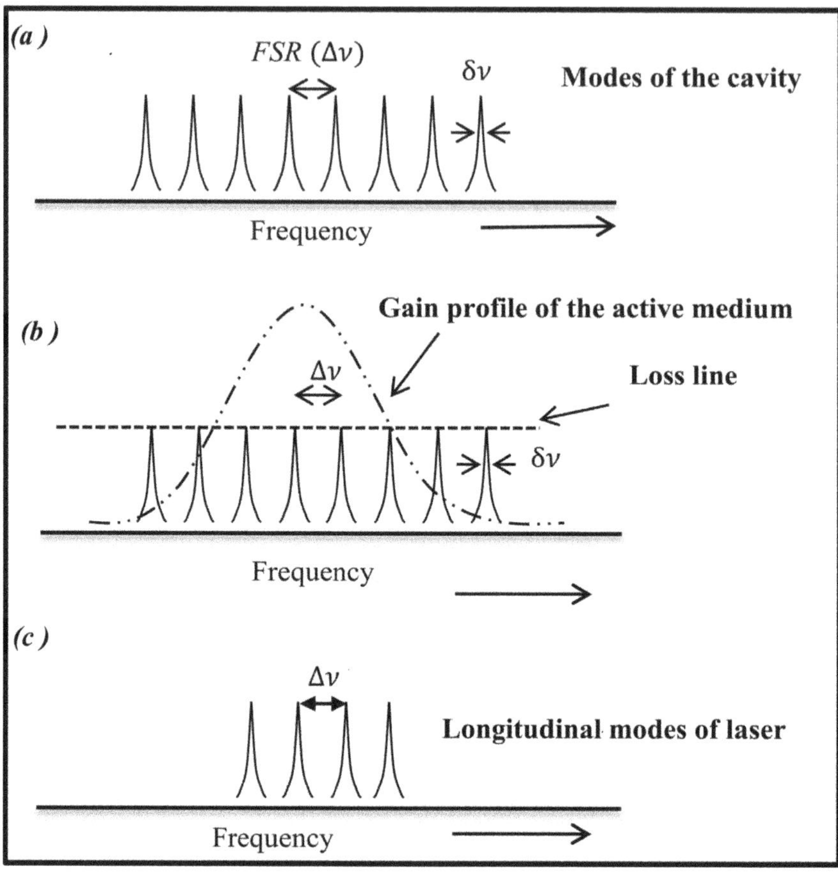


**Figure 9.3.** Selecting longitudinal modes by introducing an active medium: the top panel (a) shows the modes of a passive cavity. The middle panel (b) indicates the gain profile of the active medium and the threshold (dashed line) above which the gain overcomes the losses. The bottom panel (c) gives the number of longitudinal modes falling within the gain profile of the laser medium. Modes with a FWHM of $\delta\nu$ are separated by an FSR of $\Delta\nu$.

**Exercise 9.2. (i)** What is the FSR value of the cavity modes for a cavity length of half a meter?

**(ii)** Estimate the number of modes oscillating in a He–Ne laser output of 632.800 ± 0.002 nm.

**Solution: (i)** The FSR of the modes is given by the relation $\Delta\nu = \frac{c}{2d}$. The FSR for the given cavity is $3 \times 10^8$ Hz.

**(ii)** The number of modes can be estimated by realizing that within the laser line width, the modes are equally spaced by the FSR.

Therefore, the number of modes = Linewidth of the laser/FSR = $1.5 \times 10^9/3 \times 10^8 = 5$ modes.

## 9.2 Resonant cavity: the Fabry–Pérot resonator

To introduce the Fabry–Pérot (FP) resonator, we start with an example of a partially reflecting mirror, as follows.

### 9.2.1 A beam splitter in the path of the light

As shown in figure 9.4, consider a beam of light with a power of 10 W incident on a beam splitter BS1 with a transmission coefficient of 5%. The light transmitted by this setup is 0.5 W.

As shown in figure 9.5, if another beam splitter (BS2) with 5% transmission is added, the net transmitted power has to be equal to 0.025 W. It is interesting to note that the power transmitted can be made equal to that incident on BS1, i.e. 10 W, provided that we place BS2 'correctly'. Placing BS2 'correctly' is important to the basic understanding of a resonant FP cavity or an FP resonator, as described below. For the required output of 10 W, the intra-cavity power must be 200 W. In this case, the 95% power that is reflected back from BS2 hits BS1, and subsequently 5% of the reflected power is again transmitted back through BS1 towards the source! It appears that there are too many powers reflecting back and forth, and the law of conservation of energy is violated. However, we need to analyze the situation more carefully. If the two values of the powers returning back to BS1 are out of phase with each other, they cancel, with the result that no power returns to the source.

In fact, this is what happens in a resonator made of beam splitters, if we place BS2 'correctly' in the sense that the powers propagating forward have matching phases

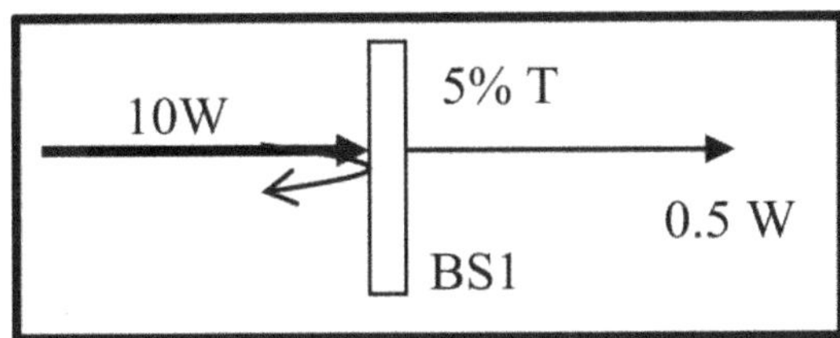


**Figure 9.4.** Illustration of a beam splitter (BS1) with a 5% transmission coefficient placed in the path of a light beam.

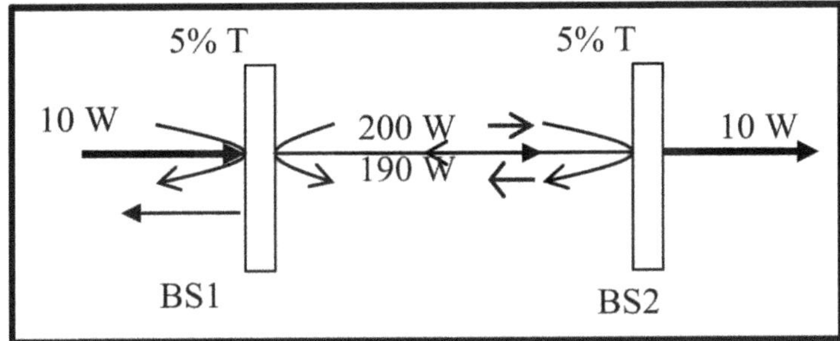


**Figure 9.5.** Schematic of light transmitted and reflected by two beam splitters.

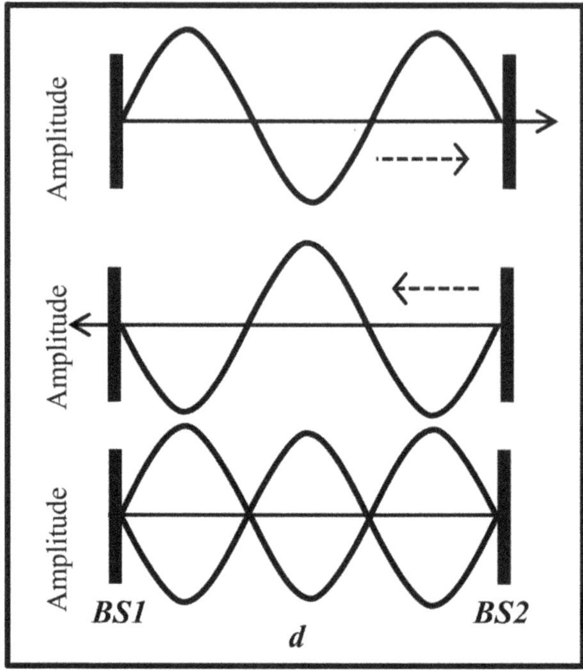


**Figure 9.6.** Forward-propagating wave (top) and backward-propagating wave (middle) in between two beam splitters (BS1 and BS2) separated by a distance $d$. Standing waves build up in the cavity when the forward and backward waves are in phase (bottom).

and those returning are out of phase, the cavity works according to the scheme shown in figure 9.6. This is the working principle of the FP resonator; we take a look at the theoretical aspects in the next section.

### 9.2.2 Phase difference between reflected and refracted light

From basic optics experiments such as Newton's rings, we know that reflection by a denser surface causes a phase change of $\pi$. The transmitted rays undergo no phase change. The front-coated beam splitters and mirrors have metal–dielectric coatings designed to reduce losses depending on the wavelengths. However, for a reflection from a metal–dielectric coated surface, the phase change can vary depending on the thickness and the incident wavelength. The transmission and reflection of an optical field ($E_0$) incident on a beam splitter are shown in figure 9.7.

For a single pass through the beam splitter, the transmitted component is $E_{0t}$, while the reflected component becomes $E_{0r}$. The coefficient of reflection is defined as $r = \frac{E_{0r}}{E_0}$ and that of transmission as $t = \frac{E_{0t}}{E_0}$ ($\spadesuit$ see section 3.4.3).

### 9.2.3 Field transmitted by a passive cavity: the Airy function

Now consider the optical field ($E_0$) incident on a pair of identical beam splitters whose coefficients of reflection and transmission are $r$ and $t$, respectively.

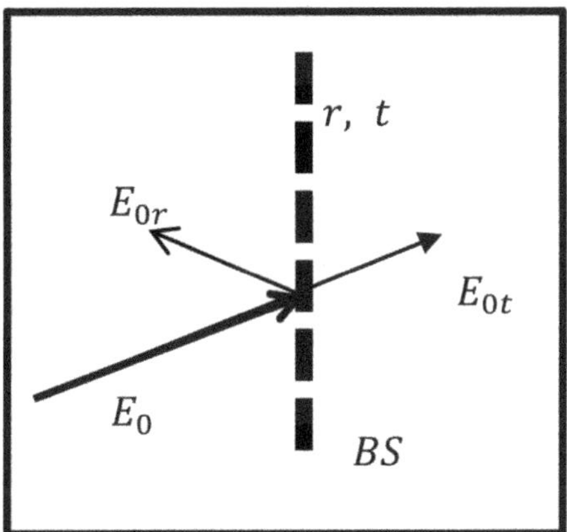


**Figure 9.7.** Reflection and transmission of an incident electric field ($E_0$) through a metal–dielectric coated beam splitter (BS). The coefficients of transmission ($t$) and reflection ($r$) influence the transmitted ($E_{0t}$) and reflected ($E_{0r}$) field amplitudes, respectively.

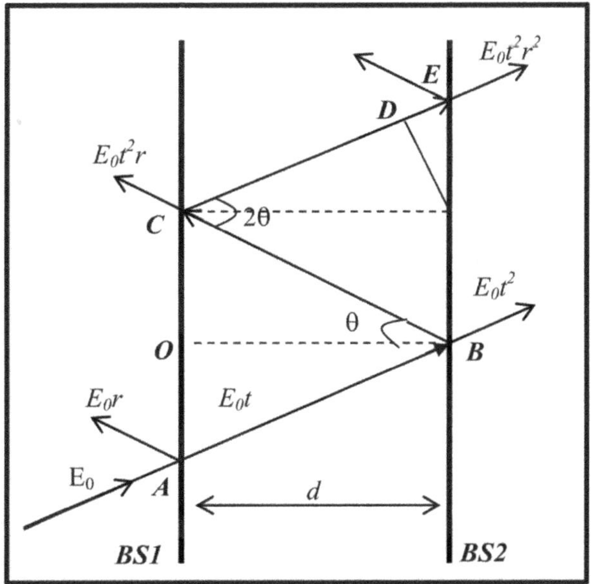


**Figure 9.8.** Field transmitted and reflected by a configuration of two identical beam splitters (BS1 and BS2) separated by a distance $d$. An incident field of amplitude $E_0$ strikes BS1 at point A. The reflection and transmission coefficients of the beam splitters are $r$ and $t$, respectively. See the text for details.

The distance between the beam splitters is '$d$', as shown in figure 9.8. For a beam incident at a small angle $\theta$, the path difference between two successively transmitted beams ($\Delta L$) is given by

$$\Delta L = \frac{d}{\cos\theta} + \frac{d}{\cos\theta}\cos 2\theta \equiv 2d\cos\theta.$$

Therefore, for the wave vector $k$, the phase difference $(k\ \Delta L)$ is given by $\phi = \frac{4\pi}{\lambda}d\cos\theta$.

The net transmitted electric field $(E_t)$ for the configuration given in figure 9.8 is

$$E_t = E_0 t^2 + E_0 t^2 r^2 e^{i\phi} + E_0 t^2 r^4 e^{i2\phi} + \dots \tag{9.1}$$

Here, $e^{i\phi}$ is the phase factor after two subsequent reflections. Equation (9.1) is a mathematical series, as follows:

$$E_t = E_0 t^2[1 + r^2 e^{i\phi} + r^4 e^{i2\phi} + \dots]$$

$$= E_0 t^2 \sum_{n=0}^{\infty} r^{2n} e^{in\phi}. \tag{9.2}$$

Equation (9.2) can be rewritten as

$$E_t = E_0 t^2 \frac{1}{1 - r^2 e^{i\phi}}.$$

The intensity $I(t)$ can be expressed in terms of $E_t$ as

$$I_t = E_0^2 t^4 \left[\frac{1}{1 - r^2 e^{i\phi}}\right]^2$$

$$= \frac{I_0 T^2}{(1 - Re^{i\phi})^2}.$$

Here, $R$ is the reflectivity $(r^2)$ and $T$ is the transmittivity $(t^2)$. In the absence of absorption by the optical elements, i.e. $R + T = 1$, we can express the denominator of the expression as

$$(1 - Re^{i\phi})^2 = (1 - R)^2\left[1 + \frac{4R}{(1 - R)^2}\sin^2\frac{\phi}{2}\right]$$

so that

$$I_t = \frac{I_0}{\left[1 + \frac{4R}{(1 - R)^2}\sin^2\frac{\phi}{2}\right]}$$

and

$$\frac{I_t}{I_0} = \frac{1}{\left[1 + F'\sin^2\frac{\phi}{2}\right]}. \tag{9.3}$$

Here, $F' = \frac{4R}{(1 - R)^2}$ is known as the *coefficient of finesse*. Equation (9.3) is known as the *Airy function*.

### 9.2.4 Effect of the reflectivity of beam splitters

The value of $(I_t/I_0)$ in equation (9.3) is maximal (i.e. one) for $\phi/2 = 0$, $\pi$, $2\pi$, and so on. A plot of the equation as a function of $\phi/2$ is given in figure 9.9. For large values of $R$, the ratio $I_t/I_0$ shows well-separated sharp peaks with a maximum value of one. This indicates that if the beam splitters are good reflectors ($R > 0.9$), the transmission intensity of the FP resonator will be equal to the incident intensity at discrete values of $\phi$. For lower values of $R$, the transmitted intensity shows overlapping peaks. This is shown by a few plots in figure 9.9 for specific $R$ values of 0.2 and 0.1.

### 9.2.5 Longitudinal modes

For the simplest case of $\phi/2 \to 0$, $\sin\phi/2 \to \phi/2$ from equation (9.3), we have

$$\frac{I_t}{I_0} = \frac{1}{\left[1 + F'\left(\frac{\phi}{2}\right)^2\right]}.$$

The transmitted intensity ratio $(I_t/I_0)$ can also be plotted against frequency $(\nu)$ as follows. We know that $\phi = \frac{4\pi}{\lambda}d\cos\theta$; so that for normal incidence $(\theta = 0)$ and for the $n$th mode, $\phi_n = \frac{4\pi d}{\lambda_n}$. Taking the difference $(\Delta\phi)$ between the two subsequent modes $\phi_{n+1}$ and $\phi_n$ (with $\lambda_n = \frac{\nu_n}{c}$ and $\lambda_{n+1} = \frac{\nu_{n+1}}{c}$), we get

$$\Delta\phi = \frac{4\pi d}{c}\Delta\nu. \tag{9.4}$$

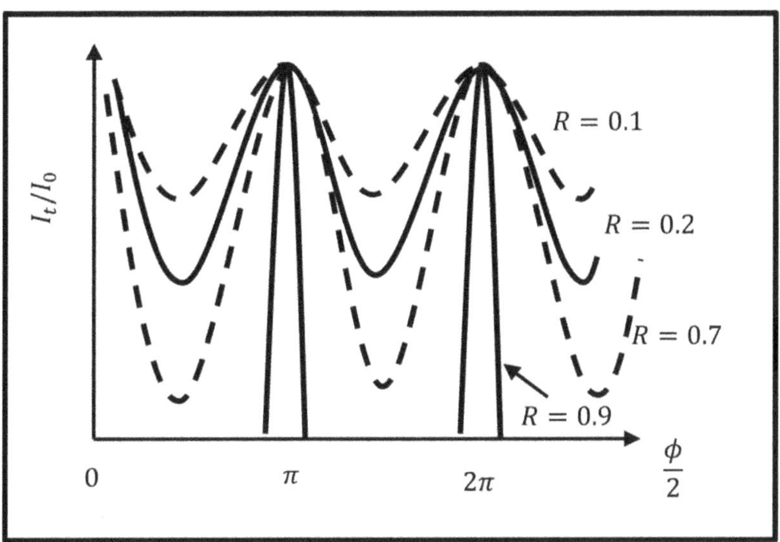


**Figure 9.9.** Typical plots of $I_t/I_0$ (Airy function) as a function of the phase change in a reflection ($\phi/2$) for various values of reflectivity ($R$) as indicated.

Equation (9.4) gives the value of the phase change ($\Delta\phi$) for a complete round trip (or two reflections in figure 9.8). From figure 9.9, we have $\Delta\phi = 2\pi$, so that

$$\frac{4\pi d}{c}\Delta\nu = 2\pi, \text{ or } \Delta\nu = \frac{c}{2d}.$$

Therefore, the frequency spacing ($\Delta\nu$) between the two peaks (or modes) is $c/2d$. This is defined in section 9.1.3 as the FSR of the cavity modes. Recall that the value of FSR in terms of the phase difference is $\Delta\phi$. You can obtain a plot of $I_t/I_0$ similar to figure 9.9 as a function of $\Delta\nu$ and compare it with panel (c) of figure 9.3. These well-separated peaks are known as the *longitudinal modes* of the FP resonator.

### 9.2.6 Properties of FP resonators

It should be noted that the modes of a cavity can be calculated even in the absence of the gain medium, i.e. they can be calculated for a *passive cavity*. We can identify the properties of the modes theoretically; later we will examine how these properties change on introducing the gain medium for an *active cavity* (♠ see section 9.5). To look at the properties of the FP resonator, we consider a cavity of identical mirrors with reflectivity $R$ as follows.

#### 9.2.6.1 The FWHM of a mode

It is desirable for laser cavities to have sharp, well-separated modes. Parameters such as *finesse* and the *quality factor* are used to characterize a cavity. A single-mode laser, for example, should have high values of both finesse and quality factor.

We now turn to the situation in which the transmitted intensity falls to half of its initial value on either side of the peak (i.e., $\frac{I_t}{I_0} = \frac{1}{2}$) in equation (9.3) in terms of the coefficient of finesse, $F'$:

$$\frac{1}{2} = \frac{1}{\left[1 + F'\left(\frac{\phi}{2}\right)^2\right]}, \text{ giving } \phi = \frac{2}{\sqrt{F'}}. \text{ To obtain the } FWHM \text{ of a mode, this value must}$$

be multiplied by a factor of two, i.e.

$$\text{FWHM}(\delta\nu) = \frac{4}{\sqrt{F'}} \tag{9.5}$$

As the FSR is one of the important parameters for FP cavities, this expression needs to be further modified to include the distance between the two mirrors, as described in following subsections.

#### 9.2.6.2 Finesse of a cavity

Recall that in equation (9.3), we defined a term called the *coefficient of finesse* ($F'$). Notice the prime on $F'$. In contrast, the quantity *'finesse'* of a cavity ($F$) is defined as the ratio of the FSR of the cavity to the FWHM of a mode, as follows:

$$F = \frac{\Delta\phi}{\text{FWHM}} = \frac{2\pi}{\left(4/\sqrt{F'}\right)} = \frac{\pi\sqrt{R}}{1 - R}. \tag{9.6}$$

Here, the value of reflectivity $R$, is taken to be the same for identical mirrors. If we have different values of $R$ for two mirrors (i.e. $R_1$ and $R_2$, respectively) the expression for finesse is written as $F = \dfrac{\pi(R_1R_2)^{1/4}}{1 - \sqrt{(R_1R_2)}}$. As can be seen from this expression, the value of $F$ is large and the denominator is small for highly reflective cavity mirrors.

### 9.2.6.3 Quality factor of a mode

The *quality factor* ($Q$) of a cavity mode is defined as the ratio of the energy stored to the energy lost per cycle.

$$Q = \frac{\text{Energy stored}}{\text{Energy lost per cylce}}.$$

In terms of the frequency, it is given by the ratio of the peak frequency ($\nu_0$) to the FWHM, i.e.

$$Q = \frac{\nu_0}{\text{FWHM}}.$$

As described in section 9.2.5, the FSR is given in terms of frequency by $\Delta\nu = \frac{c}{2d}$. Therefore, finesse can also be defined as $= \frac{\Delta\nu}{\text{FWHM}}$, so that $\text{FWHM} = \frac{\Delta\nu}{F}$.

Using equation (9.6), for identical mirrors we can write

$$\text{FWHM} = \frac{c(1 - R)}{2d\pi\sqrt{R}}. \tag{9.7}$$

Now, $Q$ can also be written as

$$Q = \frac{\nu_0}{\left(\frac{c(1 - R)}{2d\pi\sqrt{R}}\right)} = \frac{2\pi\nu_0 d\sqrt{R}}{c(1 - R)}. \tag{9.8}$$

It can be seen that the $Q$ values are high for highly reflective cavities. Therefore, we can say that a laser cavity is also a light storage device (albeit for brief spells of time).

♣ There is a story about ancient Egypt which tells that people tried to store sunlight in an earthen pot coated with reflective material on its inner wall.

**Exercise 9.3:** An FP resonator has identical mirrors separated by a distance of 10 mm. Take the medium between the mirrors to be air. Given mirror reflectivities of 99.69%, calculate the FWHM of the output modes.

**Solution:** In terms of FSR, the expression for $\text{FWHM}(\delta\nu) = \frac{\Delta\nu}{F}$. Here, $F$ is the coefficient of finesse, given by $F = \dfrac{\pi\sqrt{R}}{1 - R}$.

The FSR of the resonator in this case is $1.5 \times 10^8$ Hz. The value of finesse $F$ for the resonator with 99.69% reflectivity is 1011. Hence, the FWHM is $1.48 \times 10^7$ Hz.

**Exercise 9.4.** Determine the $Q$ of the cavity mode, if the output of an indium–gallium–aluminum phosphate diode laser operating at 650 nm is transmitted through the FP interferometer. The interferometer mirrors are placed 10 mm apart and have reflectivities of 99.99%.

**Solution:** The $Q$ of a cavity mode in a cavity is defined as the ratio of the peak frequency $\nu_0$ to the FWHM ($\delta\nu$).

The FWHM of the mode of this cavity $4.78 \times 10^7$ Hz, which is similar to the value given in exercise 9.3. The quality factor of a mode is given by $= \frac{\nu_0}{\delta\nu}$. Assuming that the peak wavelength of 650 nm is transmitted with the obtained value of the FWHM, we get $Q = 9.66 \times 10^8$. (By looking at the next exercise, 9.5, you will notice that the actual wavelength is about 21 pm away from 650 nm. However, that does not influence the value of $Q$ appreciably.)

### 9.2.6.4 Cavity lifetime

In equation (9.8), the expression $\left(\frac{d\sqrt{R}}{c(1-R)}\right)$ has the dimensions of time and is defined as the *cavity lifetime* ($\tau_c$), as follows:

$$\tau_c = \frac{d\sqrt{R}}{c(1-R)}. \tag{9.9}$$

Therefore, in terms of $\tau_c$, we can write $Q$ as

$$Q = \omega_0 \tau_c.$$

For two mirrors with different reflectivity values of $R_1, R_2$, based on equation (9.8), $Q$ can also be written as $\quad Q = \frac{2d\pi\nu_0}{c\left(1 - \sqrt{R_1 R_2}\right)} (R_1 R_2)^{1/4}.$

### 9.2.6.5 Alternate definition of the cavity lifetime

The cavity lifetime is the time spent by a photon in a cavity of length $d$. For $N$ reflections, the cavity lifetime is

$$\tau_c = \frac{Nd}{c}. \tag{9.10}$$

After $N$ reflections with reflectivity $R$, the intensity is given by $I = I_0 R^N$ (where $I_0$ is the initial intensity). Let us assume that after $N$ reflections, the intensity $I$ falls to $1/e$ of its initial value, i.e.

$I_0/e = I_0 R^N$ or $N \ln R = -1$.

Since, for highly reflective mirrors, the quantity $(1–R)$ is small, the above expression can be rewritten as

$$N \ln[1 + \{-(1-R)\}] = -1. \tag{9.11}$$

We now make use of the mathematical inequality $\ln(1+x) = x - \frac{x^2}{2} + \frac{x^3}{3} - \dots$. For small $x$, taking only the first term ($x = \{-(1-R)\}$), equation (9.11) becomes

$$N(-1 + R) \cong -1$$

or

$$N \cong \frac{1}{1 - R}.$$

Now the cavity time in equation (9.10) can be written as $\tau_c \cong \frac{d}{c(1 - R)}$. For large values of $R$ (~1), this is same as that defined above in equation (9.9), except for a factor of $\sqrt{R}$ (which is close to one for a highly reflective cavity).

## 9.3 FP etalon

An important optical element known as the FP etalon is worth introducing here. An FP etalon can be a piece of glass with thickness of a few mm that can transmit a laser line with the required FSR and FWHM. Etalons are extremely useful devices in optics and spectroscopy that are used to reduce the line widths of lasers (♠ see problem 4).

**Exercise 9.5.** If a glass plate of 5 mm is used as an etalon for a semiconductor laser output of 650 nm, what precise wavelength nearest to the laser peak wavelength is transmitted through the etalon ?
    **Solution:** The frequency $\nu$ $(= c/\lambda)$ of the laser is $4.62 \times 10^{14}$ Hz. The FSR for the etalon 5 mm thick is $\Delta\nu = \frac{c}{2d} = 3 \times 10^{10}$ Hz.
    The new frequency that is transmitted, which is nearest to the laser's peak wavelength, is $\nu' = \nu + \Delta\nu$. The new frequency can be converted into a precise wavelength. The transmitted wavelength is 42.3 pm away from the central wavelength of the laser, i.e. 650.0423 nm.

## 9.4 Fresnel number

The Fresnel number used in resonator theory is of practical importance, as it limits the distance between the mirrors of a laser cavity in accordance with their size. For the sake of understanding, let us consider a sound emitter (a speaker, or source) and a receiver (detector) as follows.

Consider the ear of a person detecting the sound produced by the speaker (figure 9.10). Let us assume that the *size* of the source is denoted by $a_1$, and that the size of the detector is denoted by $a_2$. The size $a_1$ can be taken as an aperture through which the sound of wavelength $\lambda$ is diffracted. The phenomenon of diffraction, as we know, is the bending of light around sharp edges. Diffraction is higher for smaller apertures or longer wavelengths. We can now define the 'diffraction angle' and the 'acceptance angle' for these processes while keeping the following points in mind.

    (i) Red light diffracts more for the same aperture as compared to blue light. Similarly, the diffraction angle of the emitted sound is inversely proportional to $a_1$.
    (ii) The diffraction angle is proportional to the wavelength $\lambda$.
    (iii) The acceptance angle of the detector is proportional to $a_2$.

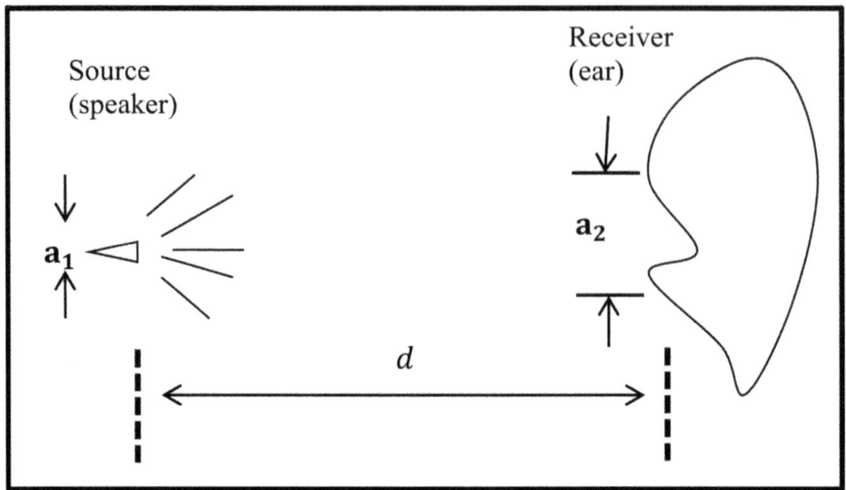


**Figure 9.10.** Schematic illustrating the Fresnel number. The source and receiver, which have aperture sizes of $a_1$ and $a_2$, respectively, are separated by a distance $d$.

(iv) Larger distances between the source and the detector result in smaller acceptance angles.

Based on points (i) and (ii), we can write the diffraction angle as $\propto \frac{\lambda}{a_1}$, and based on points (iii) and (iv), we can define the acceptance angle as $\propto \frac{a_2}{d}$.

For the best signal detection, the acceptance angle has to be larger than the diffraction angle, i.e. or

$$\frac{a_2}{d} > \frac{\lambda}{a_1} \quad \text{or} \quad \frac{a_2 a_2}{\lambda d} > 1. \tag{9.12}$$

The quantity $\frac{a_2 a_2}{\lambda d}$ is also known as the Fresnel number of a resonator for optical waves. Typically, Fresnel numbers of 10, 100, or even higher can be used in lasers. Note that in contrast to their appearance, $a_1$ and $a_2$ are not the radii but the diameters (of circular mirrors) or the sides of square mirrors.

**Exercise 9.6.** Find the Fresnel number of an optical resonator for a laser wavelength of 300 nm. The cavity uses identical mirrors 10 mm in diameter. The separation between the mirrors is 1 m.

**Solution:** The Fresnel number is given by $N = \frac{a_1 a_2}{\lambda d}$. Here, the distance ($d$) between the mirrors is 1 m, and $a_1$, $a_2$ are the diameters of the mirrors. Therefore, for a wavelength of 300 nm, $N = 683$.

## 9.5 Mode pulling

Most of the cavity parameters have been obtained for vacuum, i.e. for a passive cavity. In active cavities, due to the presence of the gain medium, its refractive index has to be considered. As a result, the cavity parameters must be modified for active cavities.

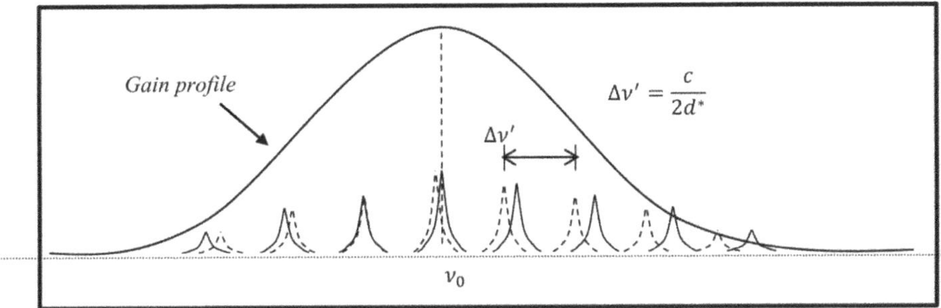


**Figure 9.11.** FSR ($\Delta v'$) of the modes (shown by dashed profiles) for an active resonator for a cavity of length $d$.

For a typical cavity, we define a *round trip time* in the cavity, $\tau_p$. The modes (longitudinal modes) of a passive resonator without any active medium in the cavity have a certain FSR value. We now introduce a gain medium into the cavity so that it becomes an 'active' resonator.

The effective mirror separation $d'$ containing a gain medium of length $L$ with a refractive index $n(\nu)$ is given by

$$d' = (d - L) + n(\nu)L \equiv d + L\{n(\nu) - 1\}.$$

Therefore, there is an effective reduction in the mode spacing, as $\Delta\nu' = c/2d'$. The modes will be pulled towards the central frequency of the Gaussian curve of the grain profile, as shown in figure 9.11. The refractive index $n(\nu)$ of the oscillation mode depends on the frequency of the modes within the gain profile. Hence, this mode pulling is not a linear effect but results in a nonuniform mode distribution. The FSR of the active cavity ($\Delta\nu'$) is inversely proportional to the refractive index of the gain medium.

FP resonator analysis demonstrates that a laser resonator provides well-separated sharp modes with high $Q$ and high finesse. These properties are dependent on the distance between the two mirrors, their degree of reflectivity, and their size. For the visible region, the distances are of the order of few μm to a few meters.

## Questions and problems

1. What is the difference between the phase on reflection and the phase of transmission of a light beam?
2. Relate the expression for finesse to the FWHM of a mode of a cavity in terms of the reflectivity $R$.
3. The identical mirrors of an FP resonator are separated by 100 mm. Take the medium between the mirrors to be air. For mirror reflectivities of 20.0%, 70.0%, and 99.9%, compare the values of the FSR, FWHM, and the quality factor of the modes of passive cavities.
4. In optics, a sheet of glass with a certain thickness (known as an *etalon*, based on the principle of the FP interferometer) is used to further improve the monochromaticity of a light source. For the following FP etalons, calculate the value of the precise wavelength transmitted near the laser line when the

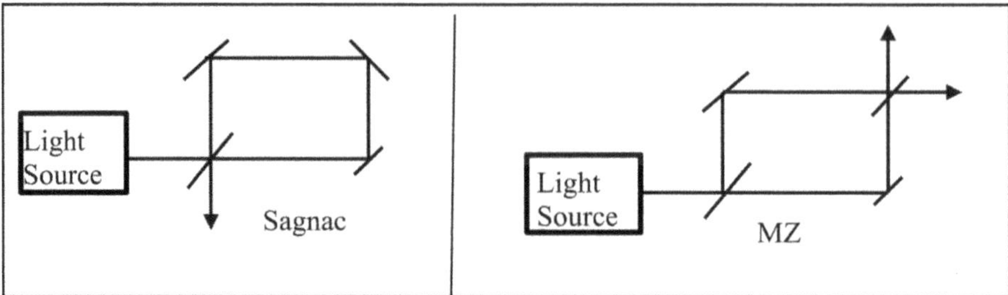


**Figure 9.12.** Schematics of the Sagnac (*left*) and MZ interferometers (*right*). A two-photon interference effect in quantum optics was demonstrated in 1987 by Hong, Ou, and Mandel and is known as the *Hong–Ou–Mandel* effect. This effect occurs when two identical single-photon waves enter a 1:1 beam splitter, one per input port. When the photons are identical, they extinguish each other. If they become more distinguishable, the probability of detection increases. Do you know that another type of interferometer is used for the *detection of gravitational waves*? Can you name a few more interferometers and briefly describe their applications?

etalon is illuminated at the nominal wavelength of (a) a red laser pointer (650 nm) with an etalon thickness of 50 cm; (b) 1064 nm with an etalon thickness of 10 mm.

5. With reference to the Airy function, write the expression $(1 - Re^{i\phi})^2$ in terms of the coefficient of finesse.

6. Write an expression for the Airy function as the ratio of the transmitted intensity to that of the incident intensity ($I_t/I$). Now plot the ratio $I_t/I$ as a function of the phase change by taking the reflection coefficient to have common values for the two mirrors of (i) 0.2 and (ii) 0.9.

7. An optical resonator cavity uses mirrors with an identical intensity coefficient of reflection of 99% at 300 nm. The circular identical mirrors with diameters of 10 mm have radii of curvature of 3 m. For a mirror separation of 1 m, calculate: (i) the photon lifetime in the cavity $\tau_c$ (in seconds); (ii) the resonator cavity's frequency bandwidth $\Delta\nu_c$ in Hertz; (iii) the circular frequency and the quality factor $Q$ of the resonator.

8. What do you understand by the term *mode-pulling* in a passive resonator?

9. In addition to the FP resonator, several interferometers are used for various applications. One of them, the *Mach–Zehnder (MZ) interferometer* (*right-hand panel*, figure 9.12) is a highly configurable instrument. This interferometer is frequently used in the fields of aerodynamics, plasma physics, and heat transfer to measure pressure, density, and temperature changes in gases as well as in quantum optics. Various fiber-optic sensors are based on this interferometer. Another interferometer, known as the *Sagnac type*, is shown in the *left-hand panel* of figure 9.12. This is used, for example, in the figure-of-eight (F8L) laser (♠ see chapter 23).

# Bibliography


[1] Bach R, Pope D, Liou S-H and Batelaan H 2013  Controlled double-slit electron diffraction *New J. Phys.* **15** 33018

[2] Mulligan J F 1998 Who were Fabry and Pérot? *Am. J. Phys.* **66** 797–802

[3] Rayleigh L 1910 CXII. The problem of the whispering gallery *The London, Edinburgh, Dublin Philos. Mag. J. Sci.* **20** 1001–4

[4] Bongu S R *et al* 2014 Influence of localized surface plasmons on Pauli blocking and optical limiting in graphene under femtosecond pumping *J. Appl. Phys.* **116** 073101

[5] Fabry C H and Pérot A 1899 Theorie et applications d'une nouvelle methode de spectroscopic interferentielle *Ann. Chim. Phys.* **16** 115–44

[6] Pérot A and Fabry C 1899 Methodes interferentielles pour la mesure des grandes epaisseurs et la comparaison des longueurs d'onde *Ann. Chim. Phys.* **16** 289–338

**IOP** Publishing

# An Introduction to Photonics and Laser Physics with Applications

**Prem B Bisht**

# Chapter 10

## Basic properties of lasers: directionality, brightness, and coherence

**Summary:** Lasers deliver intense and highly directional light beams. In this chapter, their properties viz. intensity, directionality, monochromaticity, and coherence are compared with those of other light sources. This figure shows a schematic of a Michelson-type interferometer, which is used to explain the term *coherence length* in this chapter. When working with lasers, why do we need to protect our eyes with suitable goggles? The reason is given here.

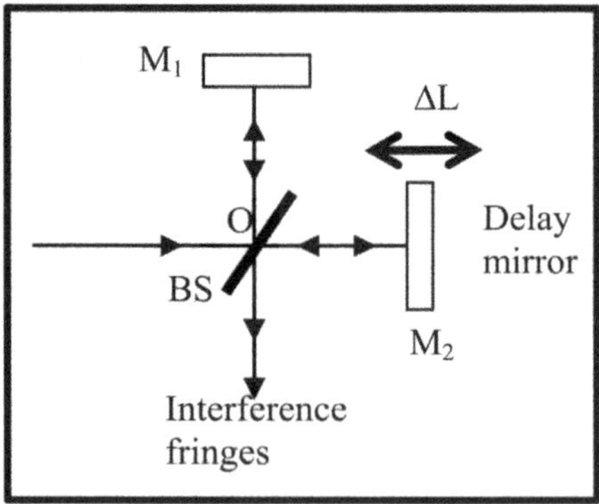


**Learning objectives**
**After reading this chapter, the learner will be able to:**

Recognize the solid angle;
Describe the divergence and brightness of a light source;
Explain the monochromaticity of a light source;

> Identify the coherence length and transverse coherence of a light source;
> Distinguish between temporal and spatial coherence;
> Define the coherence volume.

## 10.1 Directionality of a laser beam

The light produced by conventional light sources such as bulbs, tube lights, or light-emitting diodes (LEDs) spreads in all possible directions. A few of these popular light sources, such as sodium or mercury lamps, have been used in classic groundbreaking science experiments. A beam of light can be obtained from these sources with the help of a slit or an aperture. In addition, a couple of lenses can be used to collimate and focus such light. On the other hand, laser outputs are well collimated, as shown in figure 10.1.

We know that the plane angle $\theta$ can be defined in terms of an arc of radius $r$ as $\theta = \frac{arc}{r}$. Consider a spherical surface of radius $r$ with the unit vector $\hat{e}_r$ in the radial direction. In spherical polar coordinates $(r,\theta,\phi)$, the element of solid angle $(d\Omega)$ subtended by the surface area element $dA$ at the center of a sphere is given by $d\Omega = \frac{\hat{e}_r \cdot d\vec{A}}{r^2} = \frac{\hat{e}_r \cdot rd\theta r\sin\theta d\phi \hat{e}_r}{r^2}$ so that

$$\int d\Omega = \int_0^{\pi} \sin\theta d\theta \int_0^{2\pi} d\phi \equiv 4\pi.$$

The total solid angle $(\Omega)$ around a point source is $4\pi$ steradians (Sr). Knowledge of the solid angle is used to calculate the brightness of a source in the following section.

**Exercise 10.1.** What is the value of the solid angle $(\Omega)$ subtended by the light emanating from an incandescent or a fluorescent lamp?

**Solution:** The light generated by such a lamp spreads in all directions. As shown above, integrating the equation for $\int d\Omega$ yields a net value of $4\pi$. This is also the value of the solid angle subtended by a spherical surface at its center.

♣ With reference to the directional properties of laser beams, we will also discover in chapter 12 (♦ see section 12.4) that the spot size $(2w_0)$ of a laser beam is Gaussian

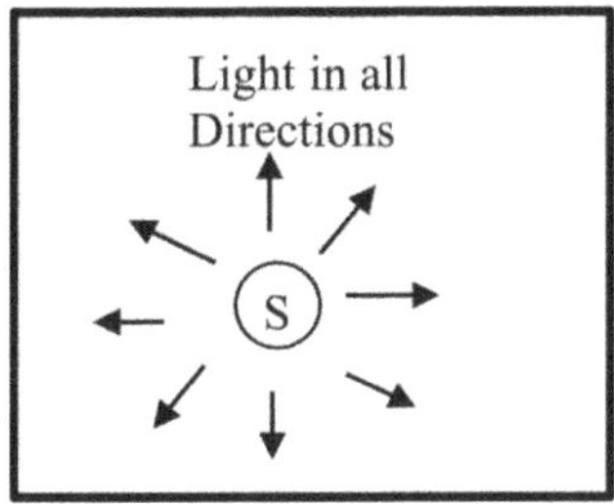


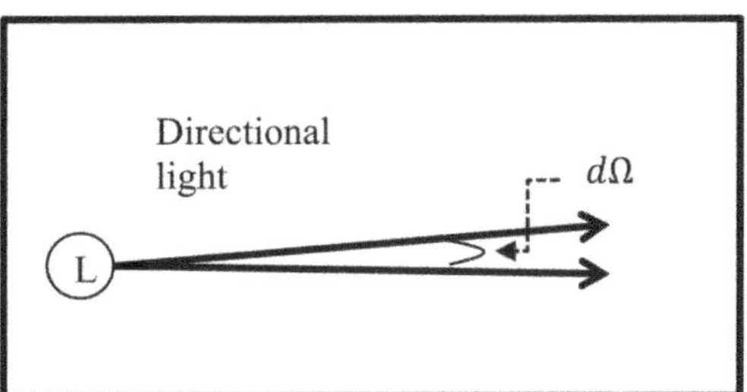


**Figure 10.1.** $S$ is a traditional light source (left). A collimated light beam from a laser ($L$, right); $d\Omega$ is the beam divergence angle, known as the solid angle.

in nature. The relevant radius of the laser beam ($w_0$) is defined as $1/e$ of the value at the center. We will also learn that a beam of wavelength $\lambda$ propagates *almost as* a parallel beam for a distance ($z$) of $\pi w_0^2/\lambda$ (known as the Rayleigh range) before it spreads. The far-field angular spread or divergence angle ($d\Omega$) is approximately given by $\lambda/z$. Lasers are typically found to have a beam divergence of the order of 1 milliradian.

## 10.2 Brightness of a light source

The energy density, i.e. the total energy radiated per unit area for all wavelengths of a thermal source is given by Stefan's law ($\spadesuit$ see section 4.1.3). The 'brightness' of a source is expressed in terms of the radiation intensity in $W\ m^{-2}$ and the solid angle of emitted radiation as follows:

$$\text{Brightness} = \frac{\text{Power/area of the source}}{\text{solid angle of radiation}}. \tag{10.1}$$

The brightness of the Sun is given by

$$\frac{\sigma T^4}{4\pi} = \frac{5.67 \times 10^{-8} \times (6000)^4}{4\pi} = 7.4 \times 10^7\ W/m^2/Sr.$$

Let us now look at the characteristic known as the *spectral brightness* of a light source, for example, a 100 W lamp at a distance of about 1 m. Its light is spread over all the emitted wavelengths of the lamp's spectrum. The human eye is sensitive only in the visible region; with an aperture of about a few mm (the size of the iris of the human eye), the power entering the eye to be focused on the retina is approximately less than a milliwatt.

It is important to mention that at room temperature the absorption spectra of organic molecules are broad and spread over a large number of frequencies. However, absorption at a single frequency is enough to initiate the effect of radiation by optical excitation or any thermal effects that may occur on absorption of the photon. This brings us to the point that one must define the parameter *spectral brightness* as follows:

$$\text{Spectral brightness} = \frac{\text{Brightness}}{\text{spectral purity}}. \tag{10.2}$$

Here, spectral purity refers to the wavelength spread ($\Delta\lambda$) of the light source. For instance, a He–Ne laser line with a line width of 0.001 nm has better spectral purity than the spectral lines of a sodium lamp which have line widths of 0.1 nm. The following subsections examine the spectral brightnesses of a few light sources.

### 10.2.1 Sun

The Sun is necessary for life on earth. It is also a source of photons that have a wide spectral range. Sunlight was also used in a classic experiment that led to a Nobel Prize. The brightness of the Sun, as calculated in the previous section, is $7.4 \times 10^7\ W/m^2/Sr$.

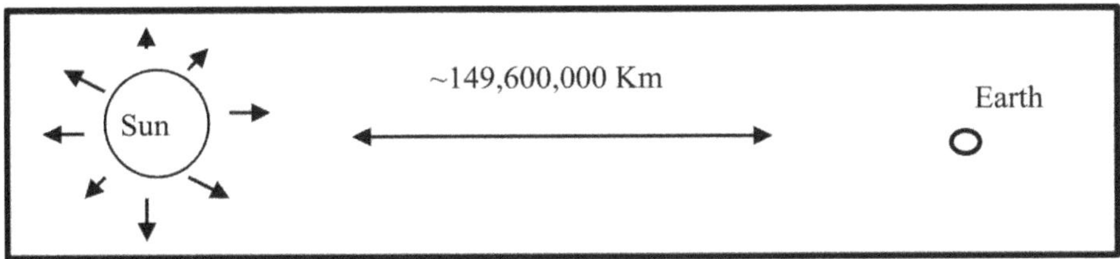


**Figure 10.2.** The brightness of the Sun observed at the Earth's surface is approximately $7.4 \times 10^7$ W/m$^2$/Sr.

♣ Sir C V Raman used sunlight in the discovery of the Raman effect, for which a Nobel Prize was awarded to him in 1928.

♣ Alexander Graham Bell transmitted a telephone signal with his *photophone* a few hundred meters in 1880 using sunlight.

For the visible region of about 400 nm (say 350–750 nm), let us take the brightness for the Sun as $\sim 1 \times 10^7$ W/m$^2$/Sr as received at the surface of the Earth (figure 10.2). We obtain its spectral brightness from equation (10.2), as follows:

$$\frac{10^7 \text{ W/m}^2/\text{St}}{400 \times 10^{-9} \text{ nm}} = 2.5 \times 10^4 \text{ W/m}^2/\text{Sr/nm}.$$

We will compare this value with those of other light sources in the following.

### 10.2.2 Sodium lamp

The sodium lamp[1] has been an essential feature of undergraduate experiments in optics and spectroscopy. It was also used in street lights before LEDs replaced them. From atomic spectroscopy, it is known that the sodium doublet has two close spectral lines viz. 589 and 589.6 nm, the former being the more intense by approximately a factor of two. Assuming that the power is distributed evenly in both lines, an 80 W lamp will deliver a power of 40 W for each of the spectral lines. The spectral purity of the sodium lines is 0.1 nm at room temperature, as mentioned above. As the lamp radiates in all directions, the source divergence in this case is also $4\pi$.

As shown in figure 10.3 (left), the radiating area of the lamp is $100 \times 50$ mm ($= 0.005$ m$^2$). From equations (10.1) and (10.2), its spectral brightness is given by

$$\frac{40 \text{ W}}{(5 \times 10^{-3} \text{ m}^2) \times 4\pi \times 0.1 \text{ nm}} = 3.18 \times 10^3 \text{ W/m}^2/\text{Sr/nm}.$$

This value of spectral brightness is an order less than that of the Sun.

### 10.2.3 A laser—the He–Ne laser

The He–Ne laser, which is generally available in an undergraduate teaching laboratory, has a typical output power in the range of a few mW. As we know

---

[1] Arthur H Compton invented the low-pressure sodium lamp in 1919.

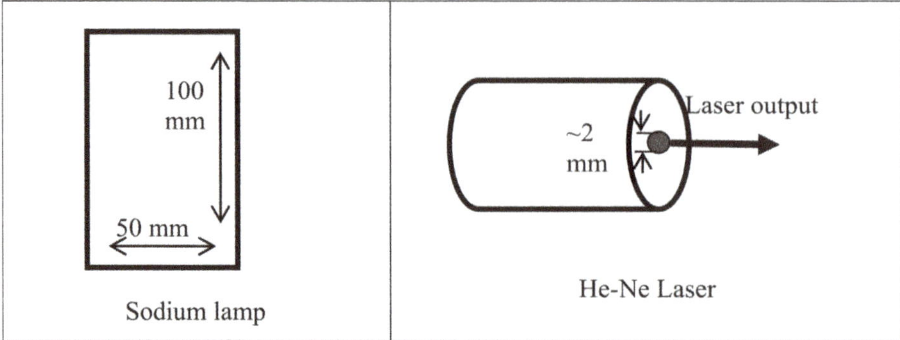



**Figure 10.3.** Typical sizes of a sodium lamp (left) and a He–Ne laser output aperture of a few millimeters (right).

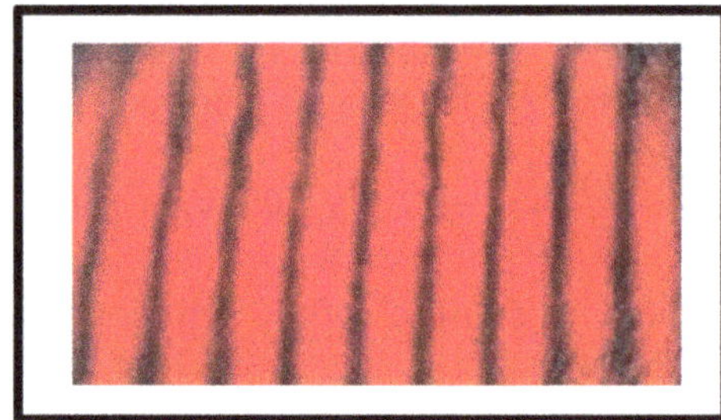

**Figure 10.4.** Typical two-beam interference pattern produced by a low-power He–Ne laser observed on the wall (Recorded at IIT Madras).

from chapter 6 (♠ see section 6.3.2), two of the main He–Ne laser wavelengths are 632.8 and 543.5 nm and the spectral purity is of the order of 0.001 nm. If the aperture of the output mirror is about 2 mm [with an area of $\pi\,(10^{-3})^2 = 3.14 \times 10^{-6}$ m$^2$], the spectral brightness in this case can be calculated as follows:

$$\frac{10^{-3}\,\text{W}}{3.14 \times 10^{-6}\,\text{m}^2 \times 10^{-3}\,\text{St} \times 10^{-3}\,\text{nm}} = 3.18 \times 10^8\ \text{W/m}^2/\text{Sr/nm}.$$

If we compare this value with that of the Sun, we find that the spectral brightness is $10^4$ times higher in the case of the He–Ne laser, even though its power level is measured in mW! This is a reason why, in the laboratory, lasers are kept at a height far below eye level in order to avoid any accidental exposure to the eyes. In addition, in undergraduate laser-related experiments, even interference fringes (♠ see figure 10.4) should not be directly observed using the eyes. Instead, they should be observed on a screen, a piece of card, or the wall.

♣ Consider a person staring at the Sun for a while. Within a few seconds, sunlight makes the eyes uncomfortable. From the above comparisons of sunlight and laser light, we realize that even watching a milliwatt-power laser with the naked eye is extremely dangerous for the eyes. On the lighter side, if one stares at a laser beam once with the naked eye, one may never see it again!

♣ Therefore, when a laser is in operation, special goggles for each laser wavelength region are used to protect the eyes from random reflections from other optics mounted on the table. Lasers are classified based on their output powers. While low-power lasers in the milliwatt range are in classes 1–2, higher-powered lasers fall into class 3 and above; such lasers can cause damage to tissues with direct exposure.

**Exercise 10.2.** **(i)** Calculate the number of photons per second produced by a sodium lamp at a power of 100 W.

**(ii)** Comment on the number of photons emitted by a typical laser with a power of 1 mW, as discussed above.

**Solution: (i)** The energy $U$ of a light source is measured in joules. Light power is expressed in units of watts or joules per second.

The relation between the number of photons ($n$) produced by a source and its energy is given by

$$U = nh\nu.$$

The frequency ($\nu$) corresponding to the lamp wavelength ($\lambda$, assumed to be the average of the sodium doublet, i.e. 589 nm), is

$$\nu = c/\lambda = \frac{3 \times 10^8}{589 \times 10^{-9}} = 5.09 \times 10^{14} \text{ Hz.}$$

The energy of a single photon corresponding to $\lambda$ is $U = hc/\lambda = 6.64 \times 10^{-34} \times 5.09 \times 10^{14}$ J $= 3.38 \times 10^{-19}$ J.

The number of photons produced per second ($n$) is given by

$$n = \frac{\text{power}}{\text{energy of a single photon}} = \frac{100 \text{ J s}^{-1}}{3.38 \times 10^{-19}} = 2.96 \times 10^{20} \text{ photons.}$$

**(ii)** Note that the lamp emits this number of photons over its total area. Compared with a laser of the same power, the number of photons in a laser beam with a radius of 1 mm is greater by a factor of about $10^6$ (check it yourself). However, since its power is smaller by a factor of $10^{-5}$, the number of photons in this case is $2.96 \times 10^{21}$.

## 10.3 Monochromaticity

For the closely spaced lines of atomic sodium vapor (589.0 and 589.6 nm), the spectral purity is 0.1 nm. For practical purposes, we often take the sodium vapor lamp to be a monochromatic light source. In chapter 8, we learnt about natural line broadening and related broadening mechanisms that contribute to the observed spectral line widths ($\delta\nu$) up to the kilohertz range or even higher. We now also know about FP etalons (♠ see section 9.2), which help to reduce line widths further.

Quantitatively, we can define the *degree of non-monochromaticity* ($\epsilon$) of a wave with a central frequency $\nu_0$ as $\frac{\delta\nu}{\nu_0}$. For a resonant frequency with a peak at $\nu_0$, this is equal to the reciprocal of the quality factor. Very often, the term *degree of monochromaticity* is used, which is the inverse of the quantity $\epsilon$. We must understand that monochromaticity is a relative term. For example, lasers are more mono-chromatic than sodium lamps. The output of a highly stable gas laser can have a bandwidth of 500 Hz. No source can have a single frequency; in other words, there is no *ideal* laser light.

♣ You will learn in chapter 17 that ultrafast lasers with pulse durations of the order of fs are not monochromatic. However, this is related to the time–bandwidth product.

## 10.4 Coherence

The term *coherence* denotes the relationship between the present phase of a light wave and its past. It is the degree of ordering of its waveforms, starting from the source. In contrast to *ideal* laser light, the so-called monochromatic light used for practical purposes is a combination of a number of frequencies. These frequencies or modes (longitudinal or transverse modes of the laser) may be 'in phase' up to a certain distance while propagating along their path, but can also be completely 'out of phase'.

For example, figure 10.5 shows waves that are self-coherent and self-incoherent as well as waves that are coherent with other waves. Generally, when the phase difference in a wave packet of a wave slips by more than $\pi$, the wave is categorized as incoherent light. For a source with uncoupled emitters, phase fluctuations decrease the degree of coherence.

As a simple example, a laser output may consist of two slightly different frequencies that are in phase with each other until a certain time $t$, but slowly become out of phase beyond $t$, as shown in figure 10.6. We also understand that there is a corresponding length $\Delta L$, beyond which the two frequencies are completely out of phase.

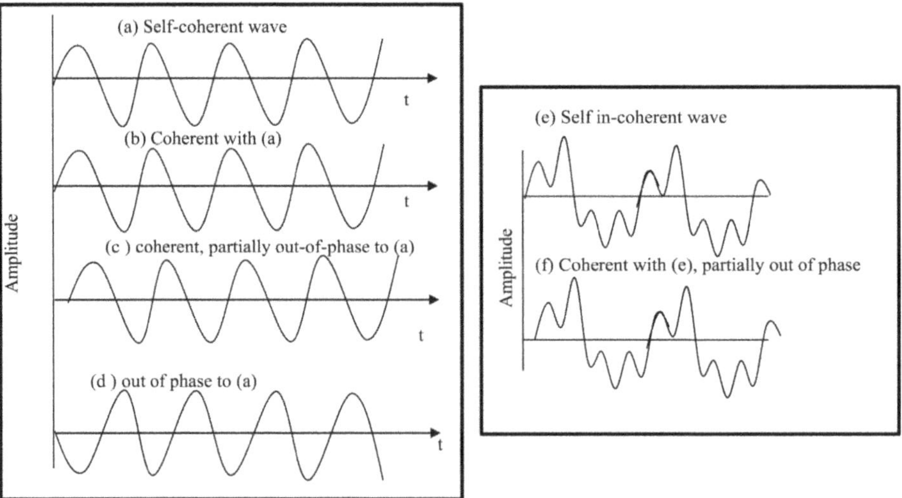


**Figure 10.5.** Schematic of self-coherent and self-incoherent waves with various phase relations to each other.

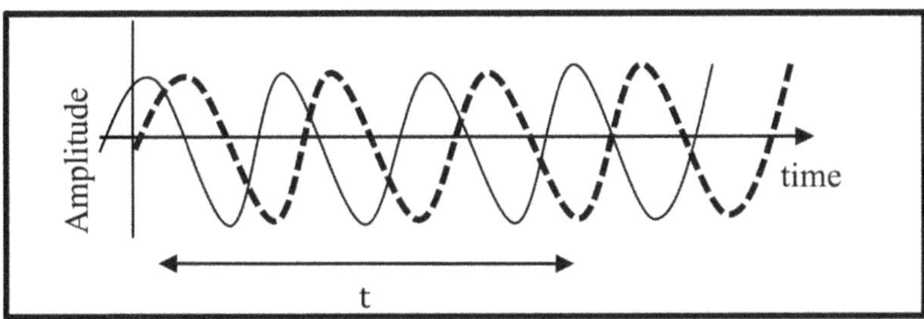

**Figure 10.6.** Propagating similar frequencies with similar phases become out of phase after a time $t$.

### 10.4.1 Visibility or contrast in interference fringes

When the two light waves from a source with a common wavelength are superimposed on a screen (♠ see section 10.4.3), we observe an interference pattern. The bright and dark fringes observed are due to the phase relation between the troughs and crests. For example, two troughs add up when they are completely in phase but they cancel each other when they are out of phase.

Let us assume two electric fields $E_1$ and $E_2$ are spatially close, as follows:

$E_1 = E_1^0 e^{-i(k_1 z - \omega_1 t)}$ and $E_2 = E_2^0 e^{-i(k_2 z - \omega_2 t)}$.

The resultant field $(E)$ is written as

$$E = E_1^0 e^{-i(k_1 z - \omega_1 t)} + E_2^0 e^{-i(k_2 z - \omega_2 t)}.$$

The intensity is given by $I = |E|^2 = |E\,E^*|$

$$I = \left(E_1^0 e^{-i(k_1 z - \omega_1 t)} + E_2^0 e^{-i(k_2 z - \omega_2 t)}\right)\left(E_1^0 e^{i(k_1 z - \omega_1 t)} + E_2^0 e^{i(k_2 z - \omega_2 t)}\right)$$

$$I = \left(E_1^0\right)^2 + \left(E_2^0\right)^2 + 2E_1^0 E_2^0 \cos\phi = I_1 + I_2 + \sqrt{I_1 I_2}\,\cos\phi, \tag{10.3}$$

where $\phi = (k_1 - k_2)z + (\omega_2 - \omega_1)t$ is known as the phase of the electric field. The value of $I$ is maximal ($I_{\max}$) at $\phi = 0°$ and minimal ($I_{\min}$) for $\phi = 180°$. One can also assume $(k_1 z - \omega_1 t) = \phi_1$ and $(k_2 z - \omega_2 t) = \phi_2$ and proceed to get $\phi = \phi_2 - \phi_1$.

$\phi$ can have a value of $0°$ (both waves are in phase), $180°$ (they are completely out of phase) or $90°$ (they are partially in phase). The intensity varies depending upon the value of $\phi$. For example, assuming $\varpi_1 = \varpi_2$ and for $k_1 = k_2$, for two similar frequencies that are in phase, the value of $\phi$ is zero. Mathematically, a parameter which can indicate the degree of coherence of a source is known as the *visibility* ($V$) or the *contrast* of the interference fringes. It is an excellent parameter with which to get an insight into coherence. For the general case, we define the visibility of the interference fringes for varying values of $I$ in equation (10.3) as follows:

$$V = \left(\frac{I_{\max} - I_{\min}}{I_{\max} + I_{\min}}\right). \tag{10.4}$$

When $I_{\min} = 0$, $V = 1$. This means that the brightness of the fringes is dependent on the values of the frequencies and the wave vector of the incident fields. If the sources have distinctly different parameters, the visibility is minimal and hence the sources can no longer be called coherent. It should be mentioned again that as long as the phase difference of the incident fields does not slip by more than $\pi$, the sources can be called coherent. There are two types of coherence characteristics, viz. the temporal and spatial coherences of a light source, as described below.

### 10.4.2 Temporal coherence: the idea of 'coherence length'

If we review figure 10.6, we find that the phases of the two frequencies are initially close to each other. After time $t_1$, on traveling a distance of $\Delta L$, the frequencies become out of phase by $\pi$. In view of this fact, *temporal coherence* can be understood using a typical *Michelson-type interferometer*[2], as shown in figure 10.7.

Here, the light beam is split in two with the help of a beam splitter (BS). The mirrors $M_1$ and $M_2$ are used to reflect the beam collinearly onto a screen. The two beams match spatially at the screen. For a zero path difference, the interference pattern observed has the maximum value of visibility. While the mirror $M_1$ is kept fixed, $M_2$ is now moved away, such that the visibility of the fringes decreases. At a certain distance $\Delta L$, the interferences fringes just disappear (i.e. minimum $V$). Under these conditions, the distance $\Delta L$ is called the *coherence length* of the light source.

On moving by the distance $\Delta L$ shown in figure 10.7, let $\Delta t$ ($= \Delta L/c$) be the corresponding time up to which the fringes are observed. Let $l_c$ be the coherence length of the laser, then from the figure, $OM_1 - OM_2 = \Delta L \equiv l_c$ (say) and $l_c = c\Delta t$. Dimensionally speaking, as the frequency is the inverse of time, we can write it as $l_c = c/\Delta\nu$. For a small difference in frequencies, using the value of $\Delta\nu$ from $c = \nu\lambda$, we obtain

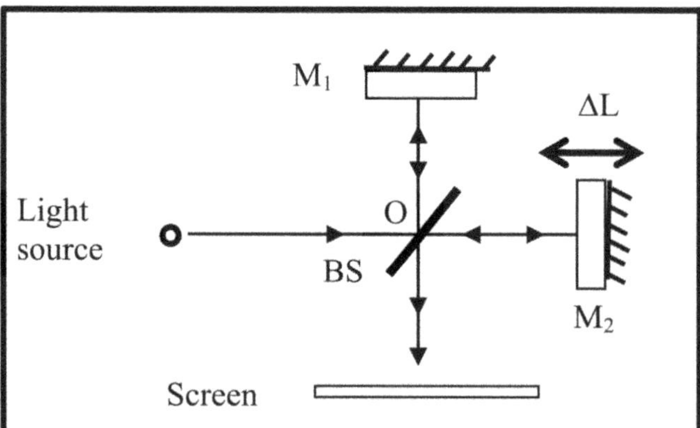



**Figure 10.7.** A Michelson-type interferometer used to measure the temporal coherence of a laser. $M_1$, $M_2$ are mirrors and BS is the beam splitter. The double arrow ($\leftrightarrow$) indicates the translation of mirror $M_2$ by a distance $\Delta L$.

---

[2] In a Michelson interferometer, a glass plate in the transmission path compensates for the path length of the reflected beam in the compact setup. Here, we consider the beam splitter to be very thin and the path compensation takes place at $M_2$.

$$l_c = \left(\frac{\lambda}{\Delta\lambda}\right)\lambda. \tag{10.5}$$

Here, $\lambda$ is the average wavelength and $\Delta\lambda$ is the difference between two similar wavelengths. The SI unit of $l_c$ is m. For an ideal laser light $l_c = \infty$. The coherence length describes the distance up to which the waves are coherent with each other. $l_c$ is significant when the difference in the wavelength $\Delta\lambda \ll \lambda$. The coherence length of lasers in the visible region can typically be of the order of several tens of centimetres.

**Exercise 10.3.** Find out whether the coherence length of a sodium lamp is smaller than that of a He–Ne laser.

**Solution:** Let us take a He–Ne laser wavelength of 543 500 nm.

For a He–Ne laser with $\Delta\lambda = 10^{-3}$ nm, $l_c = \dfrac{\lambda^2}{\Delta\lambda} = \dfrac{\left(543.5 \times 10^{-9}\right)^2}{10^{-3} \times 10^{-9}} = 29.54$ cm.

Similarly, for a sodium lamp with an average $\lambda = 590$ nm and $\Delta\lambda = 0.1$ nm,

$$l_c = \frac{(590 \times 10^{-9})^2}{10^{-1} \times 10^{-9}} = 0.35 \text{ cm}.$$

Therefore, we see that the coherence length of the He–Ne laser is larger by approximately two orders of magnitude.

### 10.4.3 Spatial or transverse coherence

Besides the *degree of monochromaticity*, the size of the light source also determines its coherence properties. For example, we have already seen that the size of a sodium lamp is larger by a factor of $10^3$ than the beam aperture size of a He–Ne laser. The waves emanating from different portions of the light source have a phase relationship. Therefore, it is very relevant to discuss the concept of spatial, lateral, or transverse coherence.

Lateral coherence can best be understood by the Young's classic double slit experiment with light. This leads to the important conclusion that when light waves passing through two slits are separated, they give rise to interference fringes. With reference to figure 10.8, let us assume that the slits (P and Q) are separated from each other by a distance $s$ and a screen is placed at a distance $r$ from them at point R.

The vertical distance ($y$) of the bright fringe in terms of the angle $\theta$ is given by

$$y = r \tan\theta; \text{ and } \sin\theta = \frac{\delta}{s} = \frac{m\lambda}{s}.$$

For small angles, the trigonometric functions tan and sin are equal. Therefore, $\dfrac{y}{r} = \dfrac{m\lambda}{s}$. For $m = 1$,

$$y = \frac{r\lambda}{s} = l_t. \tag{10.6}$$

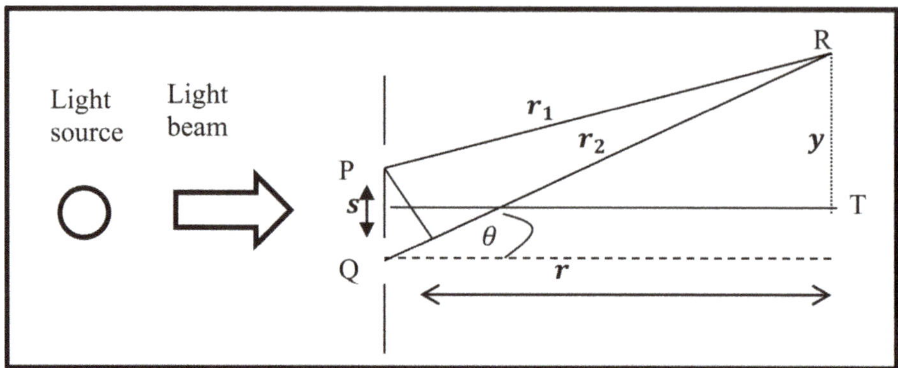



**Figure 10.8.** Young's *double slit-type* configuration used to measure spatial coherence. The slit separation is $s$ and the screen is located at a distance $r$ from the slits. The bright fringe at R is located at a certain distance along the $y$-axis. Point R makes an angle $\theta$ with the horizontal as seen from Q.

From equation (10.6), we see that lateral coherence is therefore proportional to the wavelength of the light source. It is inversely proportional to the slit spacing ($s$).

♣ Young's double slit experiment has been voted the most beautiful experiment in the history of science. Its variants with electrons have also been carried out. This experiment is important to emphasize the idea of wave-particle duality, in other words that photons or electrons can be considered to be particles as well as waves.

♣ Three-slit experiments are important in checking the possibilities of multi-order interference in quantum mechanics.

**Exercise 10.4.** Suitable setups have been designed to detect interference between classical or quantum particles or waves in a similar way to Young's double slit experiment, as schematically shown in the figure (below left). The replaceable source can deliver bullets, water, light, or matter waves (electrons). A suitable movable detector is used in each case to measure the intensities, amplitudes, or probabilities of the events. Either slit or both can be open. The intensity or probability distributions recorded by the detector are shown in panels I and II on the right. Comment on the distributions obtained at the screen for the various sources.

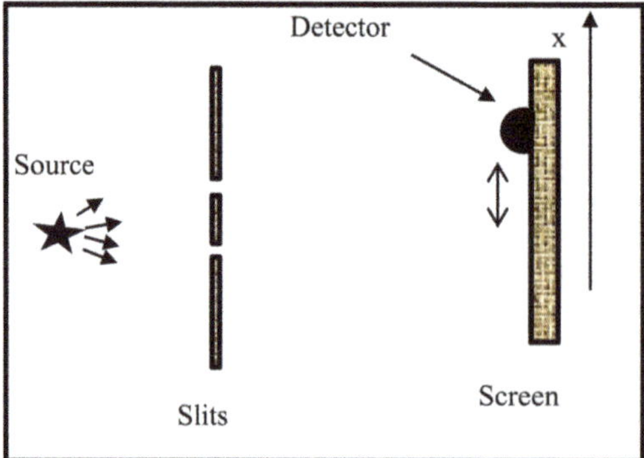



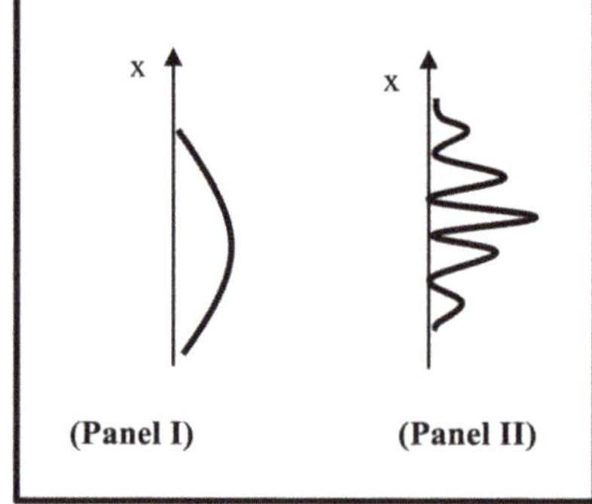

**Solution:** For a single open slit, the probability distributions for water (waves) and electrons are as given by the plot in panel I. However, when both slits are open, the probability distributions are as given by the plot in panel II. Even when both the slits are open, due to their classical nature, the probability distribution for bullets is as given by the plot in panel I. Note that the detector is moving.

### 10.4.4 Coherence surface and coherence volume

One of the applications of lasers with large coherence lengths is holography. Holograms are extremely important in industry, as they are used to authenticate products. In addition, old/mediaeval manuscripts or art of archeological importance can be preserved as three-dimensional images using holography. Digitized images of the holography can also be used in modern 'online museums'.

That is where the term 'coherence volume' becomes relevant. The volume of space occupied by an object that is expected to make a successful hologram is known as the *coherence volume*. It is sometimes also known as *depth of focus*.

While deriving equations (10.5) and (10.6), we encountered the coherence length $l_c = \frac{c}{\Delta \nu}$ and the lateral coherence, $l_t = \frac{r\lambda}{s}$. We can now define the coherence surface when $s$, $r$ and $\lambda$ are measured in meters, as follows:

$$S_c \quad (\text{in m}^2) = (l_t)^2 = \left(\frac{r\lambda}{s}\right)^2. \tag{10.7}$$

The coherence volume is given by

$$V_c = l_c \times S_c = \frac{r^2\lambda^2}{s^2}\frac{c}{\Delta \nu}\text{m}^3. \tag{10.8}$$

## Questions and problems

1. Calculate the divergence angle (in radians) of a laser with a wavelength of 1064 nm at 1 km from the laser.
2. Find the spectral brightness of an argon-ion laser that produces a single line (514.5 nm) with an output power of 78.50 mW. Take the spectral purity to be 0.001 nm, the beam diameter to be 2 mm and the solid angle for the beam divergence to be 0.001 Sr.
3. Compare the magnitude of the spectral brightness of a 10 W sodium lamp with that of a He–Ne laser with an output power of 10 mW. What are the main factors responsible for the difference?
4. Define the term *coherence length*. What is the magnitude of the coherence length of a He–Ne laser?
5. Define the term *visibility* in the context of an interference experiment.

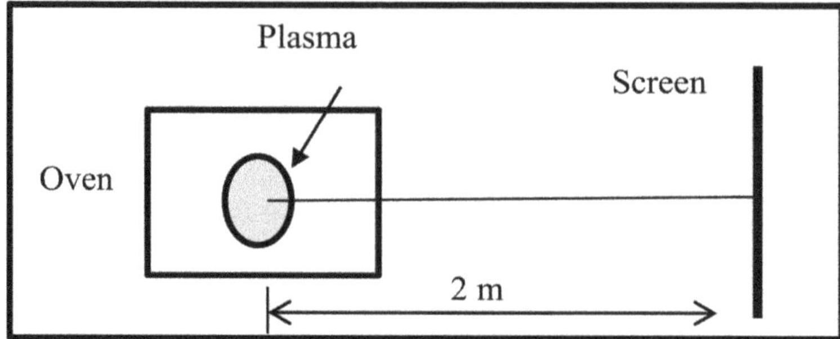


**Figure 10.9.** Diagram for question 6: a plasma that produces interference at the screen.

6. If we place two spherical grapes inside a kitchen microwave oven, a strong plasma can be produced due to strong confinement of the microwaves along the surface of the grapes. The plasma is produced as a result of interference in the strong EM field of the microwaves in the spherical cavity. This effect has a similarity with the *whispering gallery modes* of a microcavity. Assume that the plasma is of a spherical shape and radiates strongly at a wavelength in the far UV (20 nm) with an emission bandwidth of $3 \times 10^{14}$ Hz. A screen is placed 2 m away from the oven (i.e. from the center of the plasma) perpendicular to the axis joining the screen and the plasma (see the schematic in figure 10.9). At the screen, the spatial coherence resulting from the light emitted from opposite sides (*say, A and B*) of the plasma is $5 \times 10^{-5}$ m. Find (i) the diameter and (ii) the coherence length of the plasma.

7. In a Newton's rings experiment, the diameter of a wire, which is expected to be 200 μm, has to be measured. With reference to the coherence length, find out whether a sodium lamp would be suitable for an accurate measurement of the diameter.

8. The spatial coherence ($l_t$) of two sources (open circles in the figure 10.10) is defined as follows. If the two sources (figure 10.10) give rise to interference fringes on the screen kept at a distance $r$, the spatial coherence is the vertical distance from point $a$ to point $b$ at which the two sources cease to show interference effects, show that $l_t = r\lambda/s$.

9. As shown in figure 10.11, in a Young's double slit experiment with a slit separation $s = 0.3$ mm and a screen distance $r = 100$ mm, a parallel beam of light that has a wavelength of 600 nm is incident at a small angle $\theta$. PTX divides the slit's separation into equal parts. Find the value of $\theta$ at which destructive interference takes place on the screen at point R for RT = 11 mm.

10. What is the principle of holography? Calculate the coherence volume of a standard He–Ne laser with an output beam size of 10 mm for an object placed at a distance of 1 m.

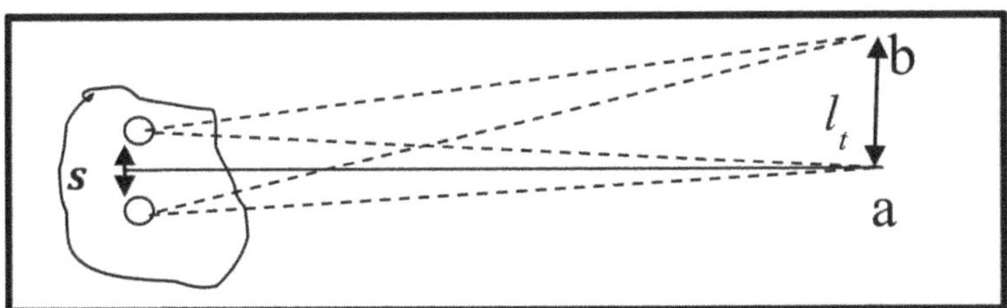

Figure 10.10. Diagram for question 8.

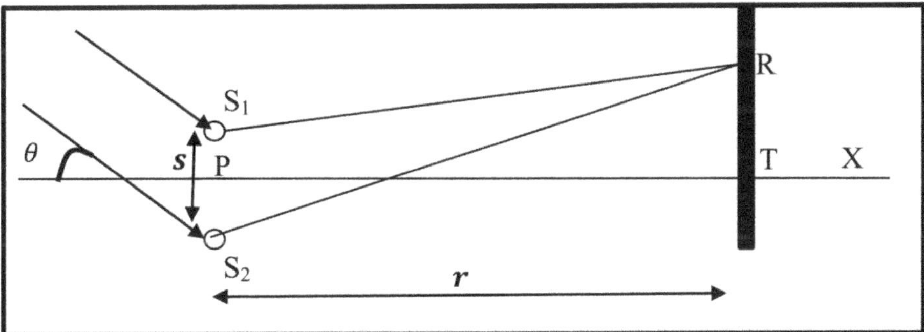

Figure 10.11. Diagram for question 9. The separation between slits $S_1$ and $S_2$ is $s$.

## Bibliography

[1] Crease R P 2002 The most beautiful experiment *Phys. World* **15** 17

[2] Sinha U, Couteau C, Jennewein T, Laflamme R and Weihs G 2010 Ruling out multi-order interference in quantum mechanics *Science* **329** 418–21

**IOP** Publishing

# An Introduction to Photonics and Laser Physics with Applications

**Prem B Bisht**

# Chapter 11

# ABCD matrices and stability diagrams

The properties of the Fabry–Pérot (FP) resonator were described in a previous chapter in terms of the reflectivity of the cavity mirrors and the distance between them. We also know that the Fresnel number must be greater than one to restrict the keep within the laser cavity. In lasers, the important property of directionality requires that the radiation should travel parallel to the cavity's axis during oscillation. In the cavity, the intracavity beam is reflected by one mirror, translated to the other, reflected back, and so on. A matrix representation of the phenomena of translation, reflection, and refraction is a useful tool for modeling this behavior. We can identify a well-defined region in which the laser cavity provides stable laser output. In the figure, the parameters $c_1$ and $c_2$ are functions of (i) the radius of the curvature of the mirrors and (ii) the distance between them. The shaded region near the origin represents *stable* resonance. As well as discussing stable conditions, we will also see that the stability restrictions can be relaxed and that *unstable resonators* can be useful for high-power applications.

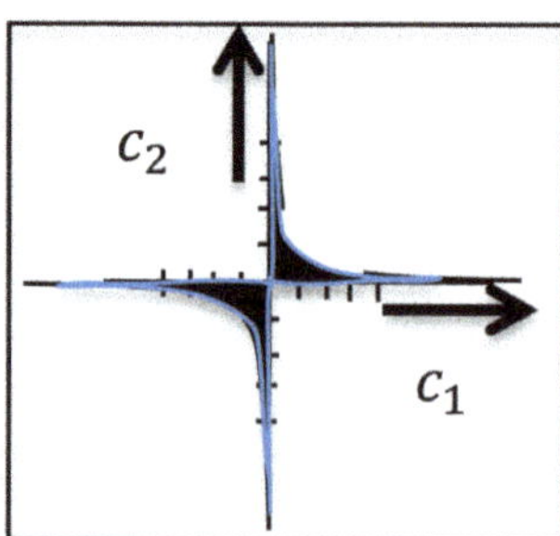

> **Learning objectives**
> **After reading this chapter, the learner will be able to:**
> Describe the processes of reflection and translation in matrix form;
> Construct a matrix for the round trip in a laser cavity;
> Examine the stability diagram;
> Identify the stability condition of the laser cavity;
> Recognize stable and unstable laser cavities.

# 11.1 Geometrical optics and ABCD matrices

The Fabry–Pérot (FP) resonator cavity helps to achieve specific oscillations of the longitudinal modes of a laser. The characteristics of the FP cavity—for example, the quality factor of its modes and the finesse of the cavity, were described in chapter 9. FP resonator theory also indicates the fact that beam oscillation in the cavity is confined to the direction of the central axis of the resonator. This occurs because the resonating field propagates nearly normal to the resonator. In practical terms, within a laser cavity, this is achieved by working within the *Fresnel number* limits (♠ see section 9.4).

During amplification by the gain medium, the intracavity beam experiences a translation of its coordinates from a position $(x_1, y_1)$ on a mirror to a position $(x_2, y_2)$ on the other mirror, as indicated in figure 11.1. Following this translation, there are multiple reflections as well as translations. Let us look at these processes individually and represent each of them as $2\times2$ matrix, known as the *ABCD matrix*, in which the characters A, B, C, and D represent the elements of the matrix. We will see that as well as ray tracing, the matrix representation of light propagation not only enables a better understanding of laser resonators but also provides several useful parameters for laser applications.

## 11.1.1 Translation matrix

Consider figure 11.2 in which a light beam originating from a point $P$ makes a *small* angle $\theta_1$. From the horizontal axis, $P$ is translated to point $Q$. For any optical element present at point $Q$, the beam can move away from the axis, i.e. with $\theta_2 > \theta_1$ or towards the axis, with $\theta_2 < \theta_1$.

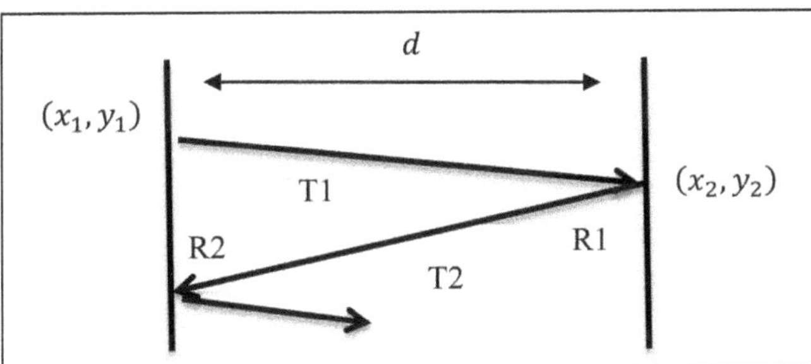


**Figure 11.1.** A laser cavity with cavity spacing $d$. The coordinates of the initial beam and the first reflection are denoted by $(x_1, y_1)$ and $(x_2, y_2)$, respectively. The translations $(T_1, T_2)$ and reflections $(R_1, R_2)$ of light are also indicated in the cavity.

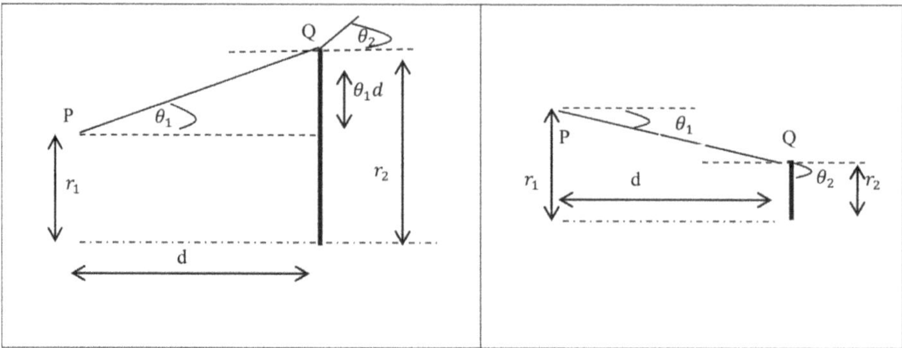

**Figure 11.2.** Point $P$ at height $r_1$ is translated to point $Q$ with height $r_2$ at a distance $d$. Here, $\theta_1$ is the small angle from the horizontal to point $Q$. The angle $\theta_2$ is greater than $\theta_1$ in the left-hand panel, while it is less than $\theta_1$ in the right-hand panel. For a pure translation, i.e. one without reflection or refraction, $\theta_2 = \theta_1$.

For this process of translation, we can write $r_2$ in terms of $r_1$ and the small angle $\theta_1$ as

$$r_2 = r_1 + d \times \theta_1. \tag{11.1}$$

For a translation to point Q without any reflection or refraction, $\theta_2$ in terms of $\theta_1$ is written simply as

$$\theta_2 = 0 + \theta_1. \tag{11.2}$$

Based on equations (11.1) and (11.2), $r_2$ and $\theta_2$ can be written as the product of two matrices in terms of the coefficients,

$$\begin{bmatrix} r_2 \\ \theta_2 \end{bmatrix} = \begin{bmatrix} 1 & d \\ 0 & 1 \end{bmatrix} \begin{bmatrix} r_1 \\ \theta_1 \end{bmatrix} \tag{11.3}$$

so that the translation matrix $[T]$ can be represented as

$$[T] = \begin{bmatrix} 1 & d \\ 0 & 1 \end{bmatrix} \tag{11.4}$$

$[T]$ is known as the ABCD matrix for translation.

### 11.1.2 Reflection matrix

Consider the reflection of a point object $P$ (kept on the $x$-axis in Cartesian coordinates) from a concave mirror, as shown in figure 11.3. The image is formed at point $Q$ after reflection at a height $y_1$ in the mirror. In the ray-diagram optics, the distances of $P$ and $Q$ from the axis of the mirror are $u$ and $v$, respectively.

As the height at the mirror ($y_1$) is the same before and after the reflection, we can write

$$y_2 = y_1 + 0 \tag{11.5}$$

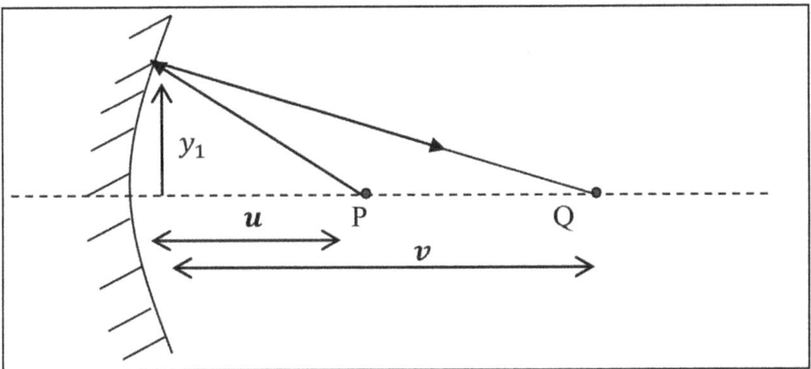

**Figure 11.3.** A point object $P$ at a distance $u$ from the mirror is imaged at point $Q$ at a distance $v$ after reflection from the concave mirror. The ray strikes the mirror at $y_1$ from the axis (dashed line) of the mirror.

The slopes of the incoming $(y'_1)$ and reflected beams $(y'_2)$ are opposite, as follows

$$y_1' = \frac{y_1}{u}$$

$$y_2' = -\frac{y_1}{v}.$$

The difference between the slopes is

$$y_2' - y_1' = \left(\frac{-y_1}{v}\right) - \left(\frac{y_1}{u}\right)$$

or

$$y_2' - y_1' = -y_1\left(\frac{1}{u} + \frac{1}{v}\right).$$

By using the relation for the focal length $(f)$ given by $\frac{1}{f} = \frac{1}{u} + \frac{1}{v}$ for the mirror, we obtain

$$y_2' = y_1\left(-\frac{1}{f}\right) + y_1'. \tag{11.6}$$

Using equations (11.5) and (11.6), the coefficients can be written as the product of two matrices, as follows:

$$\begin{bmatrix} y_2 \\ y_2' \end{bmatrix} = \begin{bmatrix} 1 & 0 \\ -\left(\frac{1}{f}\right) & 1 \end{bmatrix}\begin{bmatrix} y_1 \\ y_1' \end{bmatrix},$$ and we can introduce the reflection matrix $[R']$ as

$$[R'] = \begin{bmatrix} 1 & 0 \\ -\left(\frac{1}{f}\right) & 1 \end{bmatrix} \tag{11.7}$$

For the simple case of a spherical mirror, the focal length is half of the radius of curvature $(R)$; the matrix $[R']$ can also be written as

$$[R'] = \begin{bmatrix} 1 & 0 \\ -\left(\frac{2}{R}\right) & 1 \end{bmatrix}.$$

$[R']$ is known as the ABCD matrix for reflection. Note that a prime is used for the matrix $[R']$ to distinguish it from the radius of curvature $R$.

### 11.1.3 Refraction matrix

As shown in figure 11.4, a light beam is incident at angle $\theta_1$ on the surface $OO'$ separating two media with refractive index values of $n_1$ and $n_2$, respectively. The beam at point $Q$ emerges at an angle $\theta_2$. It is necessary to know the refractive index and distance $(d)$ between the refracting point $P$ and the normal YY' in order to obtain the ABCD matrix for a refraction at $Q$ through small angles $(\theta)$, as $\sin \theta \to \theta$.

We now know that the position of the beam at the point of refraction and also the subsequent translation can be expressed by the translation matrix. The angle in the medium that has a refractive index of $n_2$ changes according to Snell's law as

$$\frac{\sin \theta_1}{\sin \theta_2} = \frac{n_2}{n_1}.$$

Again, for small angles, this can be written as $\theta_1/\theta_2 = n_2/n_1$ or

$$\theta_2 = 0 + \theta_1\left(\frac{n_1}{n_2}\right). \tag{11.8}$$

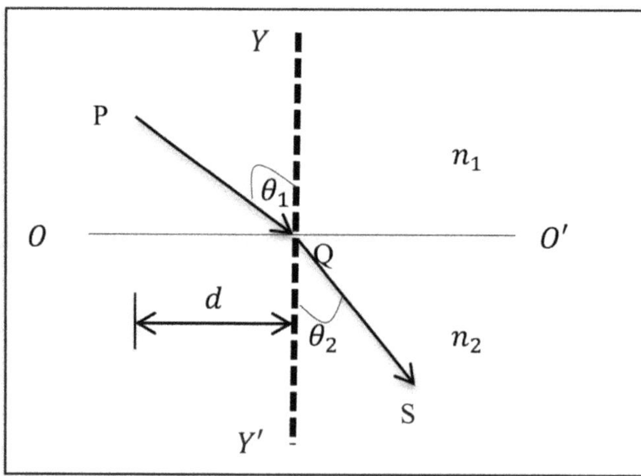


**Figure 11.4.** Light beam PQ strikes the interface $OO'$ separating the two media with refractive indices $n_1$ and $n_2$, respectively, to emerge as a beam QS. $\theta_1$, $\theta_2$ are the incident and refraction angles, respectively, made by the beam relative to the perpendicular $YY'$ to the surface. The projection of point $P$ onto $OO'$ is $d$.

Invoking the case of translation, by taking the perpendicular distance from $P$ to the normal as $r_1$ and that from $S$ as $r_2$, equation (11.1) can be rewritten as

$$r_2 = r_1 + d\theta_1.$$

Therefore, the matrix representation of the equations for $r_2$ and $\theta_2$ is

$$\begin{bmatrix} r_2 \\ \theta_2 \end{bmatrix} = \begin{bmatrix} 1 & d \\ 0 & \dfrac{n_1}{n_2} \end{bmatrix} \begin{bmatrix} r_1 \\ \theta_1 \end{bmatrix}.$$

The ABCD matrix for refraction ($N$) can now be written as

$$[N] = \begin{bmatrix} 1 & d \\ 0 & \dfrac{n_1}{n_2} \end{bmatrix}. \tag{11.9}$$

**Exercise 11.1.** A collinear laser beam strikes a concave mirror that has a focal length of 10 cm. After reflection, it is blocked by a metal sheet at a distance of 50 cm. What ABCD matrix describes the process by which the beam strikes the block?

   **Solution:** There will be three matrices for this process, as follows. For the beam originating from infinity (i.e. a collinear parallel laser beam) equation (11.2) gives

$$\theta_2 = 0 + 0.$$

Let us assume that the height at which it strikes the mirror is $r_2(=r_1)$ (from equation 11.1):

$$r_2 = r_1 + d \times 0.$$

Therefore, one of the matrices describes the translation ($T1$) of the beam from infinity to strike the mirror. $T1 = \begin{bmatrix} 1 & 0 \\ 0 & 0 \end{bmatrix}$. Similarly, after reflection by the mirror, the matrix ($R1$) is as follows:

$[R1] = \begin{bmatrix} 1 & 0 \\ -10 & 1 \end{bmatrix}$; and finally, the second translation matrix ($T2$) required for the beam to arrive at the metal sheet is $[T2] = \begin{bmatrix} 1 & 0.5 \\ 0 & 1 \end{bmatrix}$. The first process of reflection is

$[R1][T1] = \begin{bmatrix} 1 & 0 \\ -10 & 0 \end{bmatrix}$. The matrix for the second process, i.e. that one that allows the beam to arrive at the block is

$$[T2][R1][T1] = \begin{bmatrix} -4 & 0 \\ -10 & 0 \end{bmatrix}.$$

**Exercise 11.2.** Using the ABCD matrix method, find the refraction of a light beam from a curved surface of radius $R$ with its center at $C$ (i.e. into a thick lens).

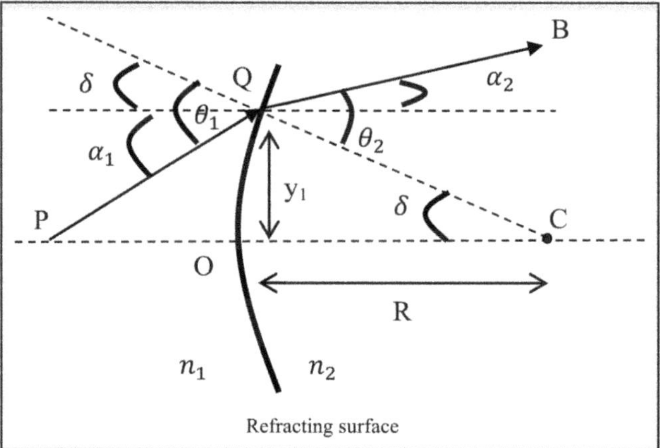


Refracting surface

**Solution**: As shown in the figure, the beam PQ refracts at QB into the refracting surface. The refractive indices of the media are $n_1$ and $n_2$, respectively. Since the height does not change at the refraction point Q, for small refracting angles, we have

$$y_2 = y_1 + 0 \quad \text{(i)}$$

$$n_2\theta_2 = n_1\theta_1 \quad \text{(ii)}.$$

Equation (ii) can be expressed in terms of the other angles indicated in the figure as follows:

$$n_1(\alpha_1 + \delta) = n_2 \quad (\alpha_2 + \delta)$$

$$n_2\alpha_2 = n_1\alpha_1 + \delta(n_1 - n_2)$$

$$\delta = \frac{y_1}{R}$$

$$n_2\alpha_2 = \frac{y_1}{R}\left(n_1 - n_2\right) + n_1\alpha_1$$

$$\alpha_2 = -\frac{y_1}{n_2R}(n_2 - n_1) + \frac{n_1}{n_2}\alpha_1 \text{(iii)}.$$

Using equations (i) and (iii), we can write

$$\begin{bmatrix} y_2 \\ \alpha_2 \end{bmatrix} = \begin{bmatrix} 1 & 0 \\ -\dfrac{(n_2 - n_1)}{n_2\,R} & \dfrac{n_1}{n_2} \end{bmatrix} \begin{bmatrix} y_1 \\ \alpha_1 \end{bmatrix}. \tag{11.10}$$

Therefore, the refraction matrix for a thick lens is $\begin{bmatrix} 1 & 0 \\ -\dfrac{(n_2 - n_1)}{n_2R} & \dfrac{n_1}{n_2} \end{bmatrix}$. You can compare it with the matrix [*N*] described above for the case of a lens with a curved surface radius (*R*) of infinity.

## 11.2 Round trip in a cavity

An active laser cavity is a light oscillator light. While it is being reflected back and forth by the mirrors, light is also amplified in presence of the gain medium. The light beam makes several round trips in the cavity before emerging as a laser beam. In the following subsections, we examine the process by which the beam makes round trips in the cavity with translations and reflections.

### 11.2.1 Light ray from the middle of the cavity

Look at the path of the ray of light in figure 11.5, starting at point $P$, which is at the middle of the cavity. Point $P$ is also denoted by number (1) as the starting point. Let us trace the return path of the beam back to this point after reflection by the mirrors $M_1$ and $M_2$, which have radii of curvature of $R_1$ and $R_2$, respectively. Before the translation denoted by (6), which brings the beam back to the original point $P$, the beam is reflected at $M_1$ (5). To reach the mirror $M_1$, it sequentially undergoes a translation (4) preceded by a reflection (3) at $M_2$ and a translation (2). It should be noted that for translation process (4), the beam travels a horizontal distance $d$, while for processes (2) and (6), it travels a distance of $d/2$.

For the round trip shown in figure 11.5, there are three translations and two reflections. Let us assume that the final matrix is

$$\begin{pmatrix} y_6 \\ y_6' \end{pmatrix} = \begin{pmatrix} A & B \\ C & D \end{pmatrix}\begin{pmatrix} y_1 \\ y_1' \end{pmatrix}. \tag{11.11}$$

For a round trip inside the cavity, the matrices for the various processes can be inserted into equation (11.11) as follows:

$$\begin{pmatrix} y_6 \\ y_6' \end{pmatrix} = [T_3][R'_2][T_2][R'_1][T_1]\begin{pmatrix} y_1 \\ y_1' \end{pmatrix}.$$

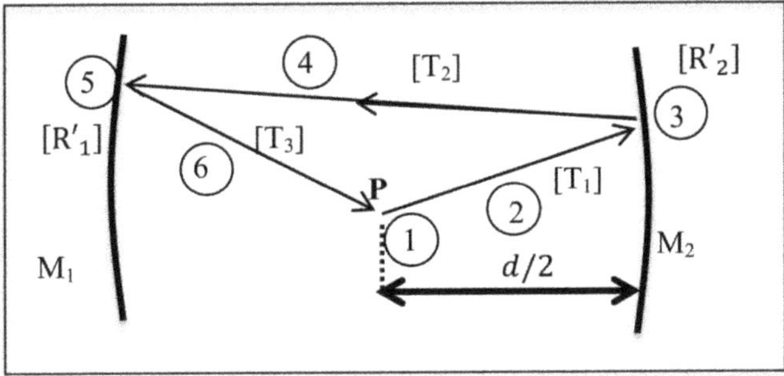


**Figure 11.5.** Light beam translating from point $P$ (at the center) and reflecting from the two mirrors located at a distance $d$ from each other. The corresponding matrices for reflection and translation are denoted by square brackets. The encircled numbers indicate the sequence of processes a light beam undergoes during the trip.

Based on figure 11.5, the matrices for translation are

$$[T_3] = [T_1] = \begin{bmatrix} 1 & d/2 \\ 0 & 1 \end{bmatrix} \quad \text{and} \quad [T_2] = \begin{bmatrix} 1 & d \\ 0 & 1 \end{bmatrix}; \text{ and the reflection matrices are}$$

$$[R'_1] = \begin{bmatrix} 1 & 0 \\ -2/R_1 & 1 \end{bmatrix} \text{ and } [R'_2] = \begin{bmatrix} 1 & 0 \\ -2/R_2 & 1 \end{bmatrix}.$$

Using these values, the elements of the ABCD matrix in equation (11.11) can be found, as follows:

$$\begin{pmatrix} A & B \\ C & D \end{pmatrix} = \begin{pmatrix} 1 - \dfrac{d}{R_1} - \dfrac{3d}{R_2} + \dfrac{2d^2}{R_1 R_2} & -\dfrac{2}{R_1} - \dfrac{2}{R_2} + \dfrac{4d}{R_1 R_2} \\[2ex] -\dfrac{d}{2} - \dfrac{d^2}{2R_1} + \left(1 - \dfrac{d}{R_2}\right)\left(\dfrac{3d}{2} - \dfrac{d^2}{R_1}\right) & -\dfrac{d}{R_1} + \left(1 - \dfrac{d}{R_2}\right)\left(1 - \dfrac{2d}{R_1}\right) \end{pmatrix}. \tag{11.12}$$

This is the matrix for a single round trip when the light originates at the middle of the cavity. In the following, let us take another case, that in which the light originates at one of the mirrors.

**11.2.2 Light initiated at a mirror**

If the light beam originates from point P on a mirror ($M_1$), the round trip can be completed as shown in figure 11.6. There are two translations and two reflections for the round trip in this case.

Using the appropriate values of $[T_1]$, $[T_2]$ and $[R'_1]$, $[R'_2]$, the ABCD matrix for this round trip can be written as

$$\begin{pmatrix} y_4 \\ y_4' \end{pmatrix} = [R'_2][T_2][R'_1][T_1] \begin{pmatrix} y_1 \\ y_1' \end{pmatrix} \tag{11.13}$$

or

$$\begin{pmatrix} y_4 \\ y_4' \end{pmatrix} = \begin{pmatrix} A & B \\ C & D \end{pmatrix}\begin{pmatrix} y_1 \\ y_1' \end{pmatrix}$$

The matrices $[R_1]$ and $[R_2]$ remain similar to the previous case:

$$[R_1] = \begin{bmatrix} 1 & 0 \\ -2/R_1 & 1 \end{bmatrix} \text{ and } [R_2] = \begin{bmatrix} 1 & 0 \\ -2/R_2 & 1 \end{bmatrix}.$$

The values of the translation matrices are equal in this case, as follows:

$$[T_2] = [T_1] = \begin{bmatrix} 1 & d \\ 0 & 1 \end{bmatrix}.$$

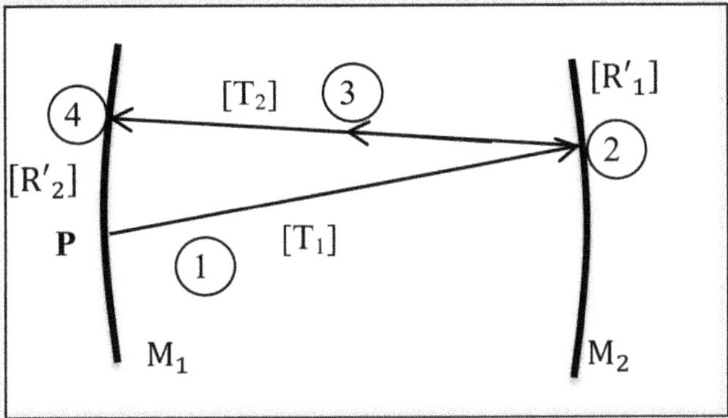


**Figure 11.6.** Light beam making a round trip from a point $P$ on the mirror $M_1$ after reflection by the mirror $M_2$. The numbers indicate the sequence of processes for the light beam during the trip.

Using these values, the elements of the ABCD matrix in equation (11.11) can be written as

$$\begin{pmatrix} A & B \\ C & D \end{pmatrix} = \begin{pmatrix} 1 - \dfrac{2d}{R_2} & 2d - \dfrac{2d^2}{R_2} \\ \dfrac{4d}{R_1 R_2} - \dfrac{2}{R_1} - \dfrac{2}{R_2} & 1 - \dfrac{2d}{R_2} - \dfrac{4d}{R_1} + \dfrac{4d^2}{R_1 R_2} \end{pmatrix}. \tag{11.14}$$

As expected, we see that the final matrix depends on the place of origin of the beam. Photons can originate anywhere in the laser cavity, depending on the type of the gain medium.

## 11.3 Cavity with several round trips

In order for light to be amplified in the cavity, it bounces back and forth between the cavity mirrors and hence passes through the gain medium several times. When the radiation is constrained to oscillate between the cavity mirrors for a large number of round trips, the cavity is said to be stable. For a cavity to be stable over $m$ round trips, the following condition is mandatory:

$$\begin{pmatrix} y_m \\ y'_m \end{pmatrix} = \begin{pmatrix} A & B \\ C & D \end{pmatrix}^m \begin{pmatrix} y_0 \\ y'_0 \end{pmatrix}. \tag{11.15}$$

For the stability criterion of the resonator, it is assumed that after $m$ round trips, the beam should lie collinearly along the axis of the resonator. Therefore, we need the solution of the $m$th power matrix. Its solutions are obtained using Sylvester's theorem [♠ see the reference below]. It is found that the roots $(\lambda_1, \lambda_2)$ are given by equation (11.16):

$$\lambda^2 - \lambda(A + D) + AD - BC = 0 \tag{11.16}$$

♣ A detailed account of the procedure is given by Tovar and Casperson in [3].

### 11.3.1 Stability condition for $m$ round trips

In the ABCD matrices for translation and reflection, i.e. equation (11.16), $AD - BC = 1$. Therefore, for $m$ round trips, equation (11.16) has two solutions:

$$\lambda_{1,\,2} = \frac{A + D}{2} \pm \left[ \left( \frac{A + D}{2} \right)^2 - 1 \right]^{\frac{1}{2}}.$$

If we take $\frac{A+D}{2} = \cos\theta$ then, using Euler's relations, the eigenvalues ($\lambda_1$, $\lambda_2$) can be written as

$$\lambda_{1,2} = \cos \theta \pm i \sin \theta = e^{i\theta}.$$

For real values of $\theta$, $\cos\theta$ must lie between $-1$ and $+1$, i.e.

$$-1 \leqslant \cos\theta \leqslant 1$$

or

$$-1 \leqslant \frac{A + D}{2} \leqslant 1. \tag{11.17}$$

This is the condition under which laser oscillations take place along the axis due to multiple reflections or in $m$ round trips. The values of $A$ and $D$ for a single round trip, for example, from equation (11.12) can be substituted into equation (11.17), so that

$$\frac{A + D}{2} = \frac{1}{2}\left( 1 - \frac{d}{R_1} - \frac{3d}{R_2} + \frac{2d^2}{R_1 R_2} - \frac{d}{R_1} + \left( 1 - \frac{d}{R_2} \right) - \frac{2d}{R_1} + \frac{2d^2}{R_1 R_2} \right)$$

$$= \frac{1}{2}\left( 1 - \frac{d}{R_1} - \frac{3d}{R_2} + \frac{2d^2}{R_1 R_2} + 1 - \frac{3d}{R_1} - \frac{d}{R_2} + \frac{2d^2}{R_1 R_2} \right)$$

$$= 1 - \frac{2d}{R_1} - \frac{2d}{R_2} + \frac{2d^2}{R_1 R_2}.$$

Substituting the value of $\frac{A+D}{2}$ into equation (11.17) yields

$$-1 \leqslant 1 - \frac{2d}{R_1} - \frac{2d}{R_2} + \frac{2d^2}{R_1 R_2} \leqslant 1.$$

By adding $+1$ on all sides and rearranging, we get $0 \leqslant \left( 1 - \frac{d}{R_1} \right)\left( 1 - \frac{d}{R_2} \right) \leqslant 1$. If we take $\left( 1 - \frac{d}{R_1} \right) = c_1$ and $\left( 1 - \frac{d}{R_2} \right) = c_2$, we can write

$$0 \leqslant c_1 c_2 \leqslant 1. \tag{11.18}$$

This inequality is known as the stability condition for the laser cavity and it helps us to prepare a *stability diagram*, as described in the next section.

**Exercise 11.3.** Check whether cavities with the following configurations are stable: (i) two mirrors that have radii of 6 m and 3 m, respectively, separated by 2 m; (ii) both mirrors have a radius of 0.7 m and distance between them is also 0.7 m.

    **Solution:**

    (i). For the given cavity, we look for the value of the product $c_1c_2$ in equation (11.18) and find that it falls between zero and one; i.e. $c_1c_2 = 0.22$. The cavity is stable.

    (ii). Note that this is a confocal cavity with $c_1c_2 = 0$ and is marginally stable.

### 11.3.2 Stability diagram

A simplified picture of the laser cavity is drawn with two plane mirrors, as shown in figure 11.7. Generally speaking, in all laser cavities, mirrors with certain radii of curvature $R$ are separated by a distance $d$. In this section, we learn about the ideas behind the various designs for laser cavities and their consequences with reference to the stability of the laser output.

Not all laser cavities give rise to stable laser operation. The stability of the laser output depends on the values of $R$ and $d$. This is decided by the stability condition given by equation (11.18). If we plot $c_1$ vs $c_2$ with the constraint of equation (11.18), we get a plot known as a stability diagram, as shown in figure 11.8. This diagram has shaded regions in the first and third quadrants.

There are two branches of the diagram, depending on the values of $c_1$ or $c_2$: the positive branch for $c_1c_2 > 0$ and the negative branch for $c_1c_2 < 0$. The resonator cavities that fall in the shaded region are known as stable cavities. In addition, we notice that there are points $c_1c_2 = 0$ and 1 as well. The designs corresponding to these values are shown in figure 11.8. For other specific values of $c_1$ and $c_2$, the positions of the various resonator cavities and their properties are described in the following section.

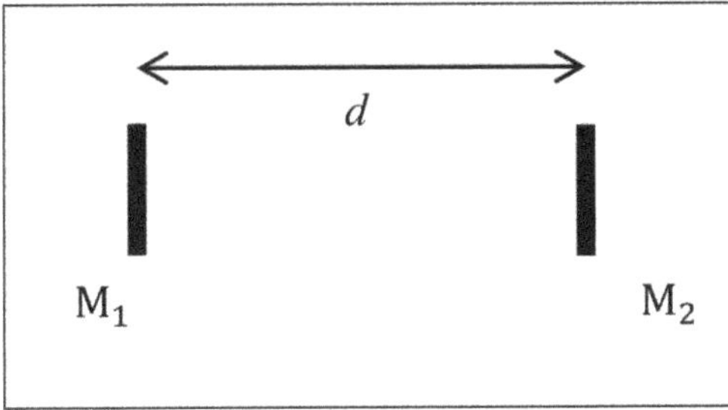


**Figure 11.7.** Schematic of a laser cavity consisting of plane mirrors ($M_1$, $M_2$) separated by a distance $d$.

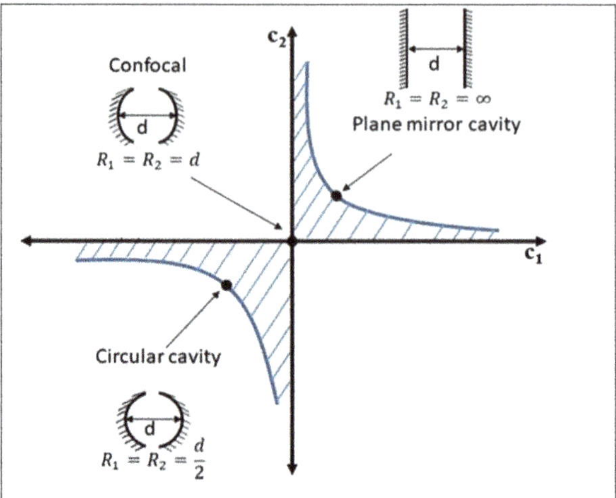


**Figure 11.8.** Stability diagram obtained using equation (11.18). The shaded part is called the stability region. The positions of a few laser cavities are also shown in the diagram.

## 11.4 Nearly stable or marginally stable resonators

To sustain the oscillation and feedback required to deliver a stable output, the geometric parameters of the resonator or the laser cavity should fulfill the conditions shown in the stability diagram (i.e. equation (11.18)). Let us again look at the product $c_1 c_2$ that lies between zero and one and describe some of the resonators.

I. *Planar cavity*: for plane mirrors with a distance $d$ between the mirrors, the radii of curvature are $R_1 = R_2 = \infty$ and $c_1 = c_2 = 1$. This resonator falls on the curve in the first quadrant. This is a *marginally stable* resonator (table 11.1(a)).

II. *Confocal cavity*: one for which $R_1 = R_2 = d$ and $c_1 = c_2 = 0$. This resonator falls at the origin. This is also a *marginally stable* resonator (table 11.1(b)).

III. *Spherical* (or concentric) cavity: one for which $R_1 = R_2 = d/2$ and $c_1 = c_2 = -1$. This resonator falls on the curve in the third quadrant of the stability diagram. This is a *marginally stable* resonator (table 11.1(c)).

IV. *Hemispherical* (hemi-concentric) cavity: one for which $R_1 = d$, $R_2 = \infty$, and $c_1 = 0$, $c_2 = 1$. This resonator also falls on the curve in the first quadrant of the stability diagram. This is a *marginally stable* resonator.

♣ **Spherical microcavity:** With reference to III above, this image shows a spherical cavity at the micrometer scale. The photograph on the right shows a 130 μm-diameter droplet of benzene floating in water. The droplet contains a fluorescent dye. The excitation wavelength was 543.5 nm, which was generated by a He–Ne laser. A long-pass filter at > 590 nm was used to image the droplet. The excitation spot on the right is larger than the one on the left due to cavity lensing effects (also known as a photonic jet). Selective confinement of the light in the cavity can also be seen (recorded by Sandeep and Bisht, IIT Madras).

**Table 11.1.** Some of the common cavities in the stability diagram with beam paths inside the cavity.

| Type | Diagram | Parameters | Remarks |
|---|---|---|---|
| (a). Planar or plane parallel | $R_1$ $R_2$ $d$ | $R_1 = R_2 = \infty,$ $c_1 = c_2 = 1$ | Marginally stable |
| (b). Confocal | $R_1$ $R_2$ $d$ | $R_1 = R_2 = d,$ $c_1 = c_2 = 0$ | Marginally stable |
| (c). Spherical | $R_1$ $R_2$ $d$ | $R_1 = R_2 = d/2,$ $c_1, c_2, = -1$ | Marginally stable |
| (d). Hemi-Concentric | $R_1$ $d$ $R_2$ | $R_1 = d, R_2 = \infty, c_1 = 0,$ $c_2 = 1$ | Marginally stable |
| (e). Semi-confocal | $R_1$ $2d$ $R_2$ | $R_1 = 2d, R_2 = \infty,$ $c_1 = 1/2,$ $c_2 = 1$ | Stable |
| (f). Nearly planar concave (Large radii mirrors) | $R_1$ $d$ $R_2$ | $R_1 \gg d, R_2 \gg d, c_1,$ $c_2 < 1$ | Stable |

## 11.5 Stable resonators

The resonators outlined in the previous section are marginally stable, as they lie on the curves of the stability diagram. It should be expected that to convert these into stable cavities, we have to push the corresponding points in the stable region, i.e. inside the curves. This is simply achieved by changing the values of the radii of curvature or the distance between the mirrors. Examples for some cavities are shown in the following:

   i. **Nearly planar (concave) cavity**: when $R_1$, $R_2 > d$ and $c_1$, $c_2 < 1$. This resonator falls in the *first quadrant*. This is also a stable resonator.
   ii. **Nearly confocal cavity**: when $R_1$, $R_2 > d$ and $g_1$, $g_2 > 0$. This resonator also falls in the *first quadrant*. This is a stable resonator.
   iii. **Nearly spherical cavity**: when $R_1 > d/2$, $R_2 > d/2$ (*or* $d/R_1 < 2$) and $c_1$, $c_2 > -1$. This resonator falls in the *third quadrant*. This is a stable resonator.

Stable resonators are generally used for low-powered lasers that have narrow beam sizes, where the stability of the output beam is the main criterion over several hours of operation.


**Exercise 11.4.** Compare the following resonators with those described in table 11.1.
   (i) A nearly planar (concave) cavity with row (a) of the table
   (ii) A nearly confocal cavity with row (b) of the table.
   (iii) A nearly spherical cavity with row (c) of the table.
   **Solution:** It can be seen that all three cavities fall just inside the stability diagram; hence, they are stable (and not marginally stable).


## 11.6 Unstable resonators

As mentioned earlier, according to equation (11.18), in order to obtain a stable resonator, the product $c_1c_2$ should lie between zero and one. However, certain applications require laser powers of several kW at large beam sizes. In stable cavities, a stringent requirement for stable output is to keep the volume of the gain medium small. A schematic of a limited portion of the amplifying medium can be seen, for example, in table 11.1 for confocal (b) and spherical cavities.

For high-power applications, stability is not the main criterion. In fact, higher powers cannot be obtained from stable configurations. This is mainly because the gain medium is not completely utilized to its full potential. Therefore, it is worth working with the *unstable regime* of the stability diagram. In unstable cavity designs, part of the beam fills the cross-section of the mirror completely and spills around it. This is due to the fact that the higher-order laser modes do not propagate in a well-defined manner and are generally random in nature. Such a beam passes through a larger volume of the gain medium. The leakage from the mirrors is the output beam that is used for the application. In high-power applications, the laser output can be pulsed or may not even need to be stable over several shots. In the following, the principles of the two unstable cavities are described.

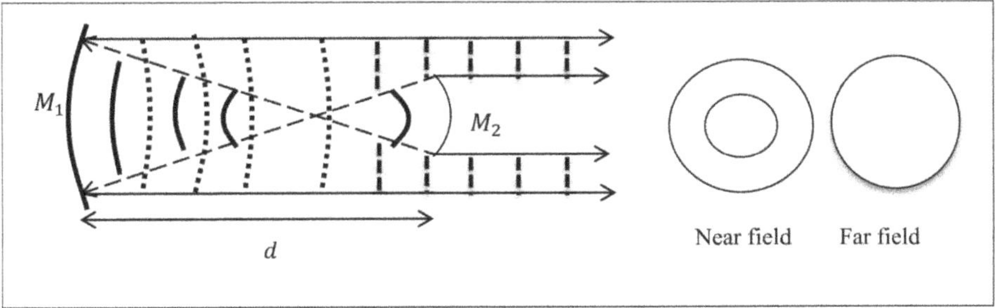


**Figure 11.9.** Negative-branch confocal *unstable* resonator cavity and its output beam shapes in the near field and the far field.

### 11.6.1 Negative-branch confocal unstable cavity

As shown in figure 11.9, in this kind of cavity, the output coupler ($M_2$) is smaller than the large rear reflecting mirror. The radii of curvature of the two mirrors are also different. The radius of the rear reflecting mirror is $R_b$, while that of the output coupler is $R_{oc}$. The beam passes through a larger volume of the gain medium to achieve the high power. Due to spillover beyond the output coupler, the output has a doughnut shape[1] in the near field, while it becomes nearly Gaussian in the far field. We note that in such resonators, both mirrors can have 100% reflectivity.

As the output coupler is small, it is possible to define the magnification ratio ($M$) for the obtained spot size. It is given by the ratio of the curvature of the rear mirror ($R_b$) to that of the output coupler ($R_{oc}$), i.e. $M = R_b/R_{oc}$. The distance ($d$) between the two mirrors obeys the relation $R_b + R_{oc} = 2d$. This resonator falls in the $c_1c_2 < 0$ region, i.e. in the second and fourth quadrants of the stability diagram.

The disadvantage of the negative-branch confocal cavity is that the oscillator beam is focused in the gain medium. This creates difficulties related to damage to the gain medium, which is susceptible to burning.

### 11.6.2 Positive-branch confocal unstable cavity

As mentioned above, in the negative-branch confocal resonator, the gain medium can be damaged due to tight focusing of the light within the cavity. To avoid this problem, another resonator design known as the *positive-branch confocal* cavity is used. This is shown in figure 11.10. This resonator falls in the $c_1c_2 > 1$ region, i.e. in the first and third quadrants of the stability diagram. Here, the radii of curvature of the mirrors are again different. The radius of the rear reflecting mirror is $R_b$, while that of the output coupler is $R_{oc}$.The distance between the two mirrors obeys the relation $R_b + R_{oc} = 2d$.

Large gain volumes can be used in both of these resonators. Therefore, in contrast to stable resonators, large output powers can be achieved for various applications.

---

[1] This is due to a geometrical effect and is entirely different from the orbital angular momentum (OAM) beams referred in chapter 1.

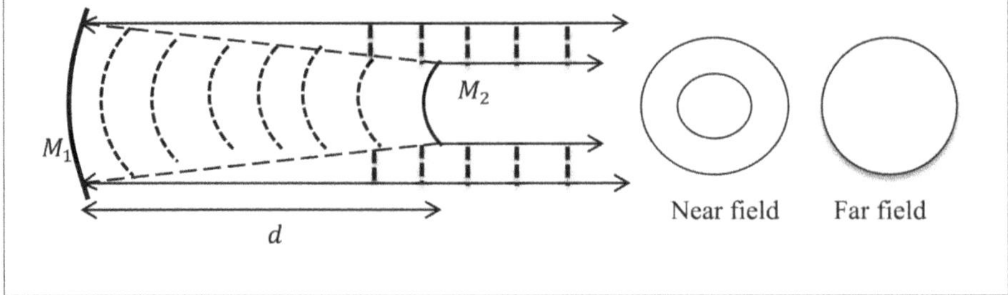



**Figure 11.10.** Positive-branch confocal unstable resonator cavity and its output beam shapes in the near field and the far field.

## Questions and problems

1. Find the ABCD matrix for the refraction of a thin lens of focal length $f$.
2. Prove that the ABCD matrix for a translation followed by a reflection is different from that for a reflection followed by translation in a cavity.
3. Two thin lenses that have focal lengths of $f_1$ and $f_2$ are placed adjacent to each other (with no distance between them) along the axis. Find the ABCD matrix for this lens combination. What is the focal length of the combination?
4. Two mirrors with radii of curvature of $10R$ are separated by a distance of $R$. Write down the ABCD matrix for a light beam that originates at the center of the system and reaches the second mirror after reflection by the first mirror.
5. Write down the ABCD matrix for one round trip through a resonator for the following two cases: (i). a beam originating at the center of the cavity; (ii). a beam originating at one of the mirrors. The radii of curvature of the mirrors are $R$ and $2R$, respectively, and the mirrors are separated by a distance of $2R$.
6. If $R_1$, $R_2$ are the radii of curvature of two mirrors separated by a distance $d$, confirm whether the following resonators fall in the stable or unstable regime:
   (i) $R_1 = R_2 = \infty$, irrespective of the value of $d$; (ii) $R_1 = R_2 = \frac{d}{2}$;
   (iii) $R_1 = R_2 = \frac{d}{3}$; (iv) $R_1 = R_2 = -d$; (v) $R_1 = R_2 = d$;
   (vi) $R_1 = \infty$, $R_2 = 2d$; (vii) $R_1 = \frac{d}{2}$, $R_2 = -d$; (viii) $R_1 = R_2 = -d$;
   (ix) $R_1 = R_2 = 2$ m and $d = 0.5$m.
7. In a resonator configuration, $R_1$, $R_2$ are the radii of curvature and $d$ is the distance between two concave mirrors. If $R_1 > R_2 \sim d$:
   (i) write the name of the resonator (and relate $R_1$, $R_2$ and $d$);
   (ii) draw the beam path; (iii) draw the beam shape of the output at the near and far field.
8. One of the structured beams in optics has a doughnut shape (a doughnut is a kind of bread with a hole at the center that is similar in shape to a snack named *bada* in the Indian subcontinent). What are the ways of producing *bada*-shaped structured beams, such as Laguerre–Gaussian and Bessel–Gaussian beams? What are their applications? (Hint: see, for example,

Allen L *et al* 1992, Orbital angular momentum of light and the transformation of Laguerre–Gaussian laser modes *Phys. Rev.* A **45** 8185–9).

9. Sketch the following optical resonator configurations for two mirrors that have radii of curvature $R_1$ and $R_2$ and a separation $d$ indicating the values of the parameters $R_1$, $R_2$ in terms of $d$:

(i). symmetrical confocal; (ii). symmetrical spherical; (iii). hemispherical.

10. For the positive-branch unstable resonator, using $R_r - R_o = 2d$, prove that $c_1 + c_2 = 2c_1c_2$.

## Bibliography


[1] Eakin D M and Davis S P 1966 An application of matrix optics *Am. J. Phys.* **34** 758–61
[2] Halbach K 1964 Matrix representation of gaussian optics *Am. J. Phys.* **32** 90–108
[3] Tovar A A and Casperson L W 1995 Generalized Sylvester theorems for periodic applications in matrix optics *J. Opt. Soc. Am.* A **12** 578–90
[4] Dantham V R, Bisht P B and Namboodiri C K R 2011 Enhancement of Raman scattering by two orders of magnitude using photonic nanojet of a microsphere *J. Appl. Phys.* **109** 103103

**IOP** Publishing

# An Introduction to Photonics and Laser Physics with Applications

**Prem B Bisht**

# Chapter 12

# Stability conditions according to Gaussian beam analysis

A laser beam can be sent to the Moon and detected back on the Earth within 2.5 s following reflection by the retroreflector placed there by astronauts in 1969[1]. Even a highly directional laser beam undergoes a spread in its wavefront. A beam with a spot size 10 mm in diameter sent from the Earth results in a corresponding diameter of several kilometers at the Moon. The beam diameter obtained at the destination can be controlled by the launch properties. Laser beams propagate as nearly unidirectional waves and have finite cross-sections. We know that: (a) plane wavefronts have infinite cross-sections and (b) a propagating spherical wave diverges rapidly; it is not unidirectional. Thus, plane and spherical waves do not describe laser beams. Laser beams are mathematically described by Gaussian functions. The laser cavity in the figure shows the wavefronts of a Gaussian beam. In this chapter, we will learn that a laser beam at a specific location ($z$) is characterized by its beam waist ($\varpi(z)$) and the curvature of its wavefront ($R(z)$). Furthermore, transverse electromagnetic (TEM) laser modes are also introduced here as a special case of Hermite polynomials.

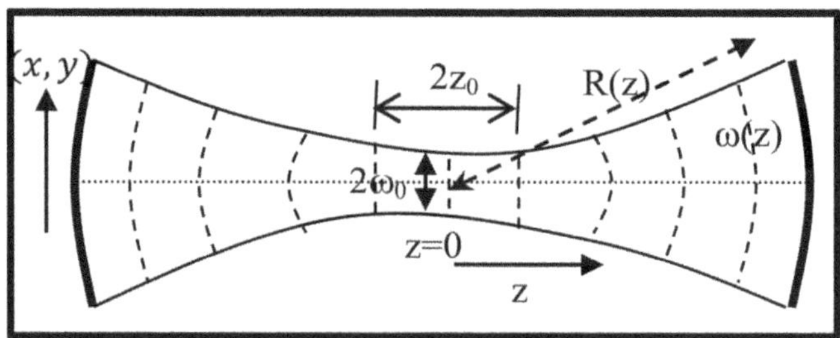


---

[1] A retroreflector was placed on the Moon by astronauts Buzz Aldrin and Neil Armstrong on July 21, 1969.

**Learning objectives**
**After reading this chapter, the learner will be able to:**
Identity the nature of laser light, i.e. a Gaussian beam;
Describe the properties of Gaussian beams;
Outline Kirchhoff's diffraction formula;
Predict the directional properties of laser light;
Compute the Rayleigh range, confocal parameter, and beam waist;
Describe transverse laser cavity modes.

## 12.1 Cavity mirrors as diffracting elements

In chapter 9, we learnt about the Fabry–Pérot (FP) cavity, where light makes round trips in between two partially reflective mirrors placed parallel to each other at a distance. In this configuration, the modes of a 3D cavity are confined to two dimensions. We also learnt that the longitudinal modes of the oscillating light are restricted to remain within the bandwidth of the gain medium. The two mirrors of the FP cavity can be treated as small diffracting elements for light within the limit defined by the Fresnel number. In this chapter, we will learn about the properties of laser beams, such as their size and the curvatures of their wavefronts. The beam parameters are extremely useful while working in the interdisciplinary field of 'laser physics and applications'.

## 12.2 Laser light: a plane or spherical wave?

In the context of the FP resonator, we worked with the plane wavefronts incident on one of the beam splitters. The wavefront strikes the beam splitter from outside the resonator, as seen in figure 9.8. We have described how different longitudinal modes separated by the free spectral range (FSR) are generated within the laser cavity. Instead of plane waves striking the mirror configuration from outside, we will now consider the case in which the light source (i.e. the gain medium) is located within the laser cavity. We will continue to work with mirrors of finite size limited by the Fresnel number.

Table 12.1 presents a schematic that shows how the wavefronts of a spherical wave change while traveling long distances. It can be seen that the spherical

Table 12.1. Schematic of the changes in the wavefront of a spherical wave originating from a source on traveling long distances along the $z$-axis.

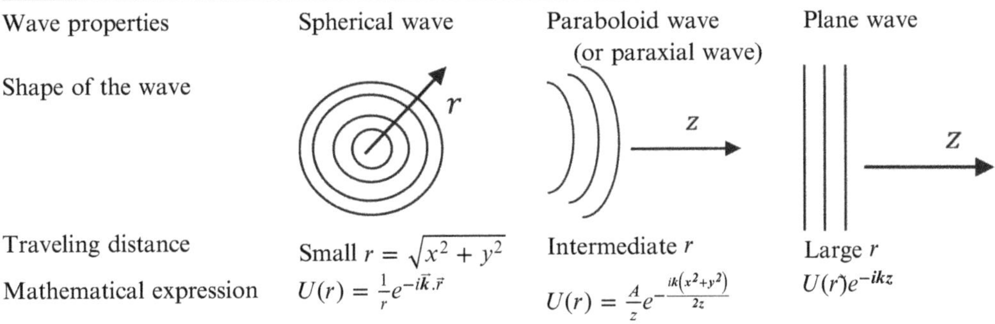

| Wave properties | Spherical wave | Paraboloid wave (or paraxial wave) | Plane wave |
|---|---|---|---|
| Shape of the wave | | | |
| Traveling distance | Small $r = \sqrt{x^2 + y^2}$ | Intermediate $r$ | Large $r$ |
| Mathematical expression | $U(r) = \frac{1}{r}e^{-i\vec{k}\cdot\vec{r}}$ | $U(r) = \frac{A}{z}e^{-\frac{ik(x^2+y^2)}{2z}}$ | $U(r)e^{-ikz}$ |

wavefronts turn into those of a plane wave when the distance traversed approaches infinity. Laser beams are highly directional and can be expected to undergo minute changes in their spot size with distance. Therefore, before we proceed, it is worth mentioning that the wavefronts of the laser beam are neither spherical nor plane in nature. A laser beam is typically a paraboloid wave, also known as paraxial wave—the combination of a plane wave and a spherical wave. A Gaussian expression comes closest to describing the shape of the laser beam spot.

Using the Fresnel approximation, we can describe the propagation of a spherical wave along the $z$-axis for the right-handed Cartesian coordinate $(x, y, z)$ system, as presented in table 12.1. Note that the wavefront spread is given in $x$ and $y$ coordinates while it travels along the $z$-axis in all cases.

## 12.3 Kirchhoff's diffraction

The solution of the FP resonator problem using Kirchhoff's diffraction has been tackled both numerically and analytically. Its full description is quite involved and outside the scope of this book. Therefore, only the salient features of this theory are given here.

### 12.3.1 Single diffraction

In figure 12.1, a wavefront is emanating from a point source $P$ with amplitude $U_0$. The incident wave makes an angle $\theta_1$ with the unit normal $\hat{n}$ of the diffraction surface, as shown in figure 12.1(a), and $\theta_2$ is the angle made by the diffracted beam with the unit normal. $dS$ is the area of the aperture $S$.

The wave amplitude $U_S$ at the aperture $S$ situated at the position vector $\vec{r}'$ is given by

$$U_S = U_0 \frac{1}{r'} e^{i\vec{k}\cdot\vec{r}'}. \tag{12.1}$$

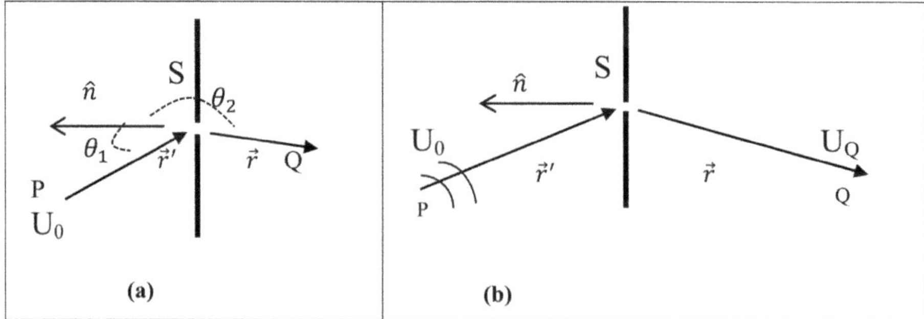


**Figure 12.1.** (a) A light ray with an amplitude of $U_0$ falling on an aperture $S$. (b). Diffraction of a wavefront at $S$ by an area element $dS$; after diffraction, the wavefront reaches point $Q$. $\theta_1$ and $\theta_2$ are the angles between the unit normal $\hat{n}$ and the surface.

Here, $\vec{k}$ is the wave vector. Let $U_Q$ be the amplitude at point $Q$ of all secondary wavelets. The final amplitude at $Q$ is given by Kirchhoff's diffraction formula integrated over the aperture area $dS$:

$$U_Q = \frac{(-ik)}{4\pi} \iint U_0 \left( \frac{e^{ikr'}}{r'} \right) \frac{e^{ikr}}{r} [\cos(\theta_2) - \cos(\theta_1)] dS. \tag{12.2}$$

It should be noted that the factor $(-ik)$ appears due to Huygens' treatment of the diffraction. Here, the factor $[\cos(\theta_2) - \cos(\theta_1)]$, known as the obliquity factor, has to be taken into account. Assuming axial propagation, for the incident beam shown in figure 12.1, the formula in given terms of $U_S$ by equation (12.1) can be rewritten as

$$U_Q = -\frac{ik}{4\pi} \iint U_S \frac{e^{ikr}}{r} [\cos(\theta_2) + 1] dS. \tag{12.3}$$

We note that $U_Q$ gives the amplitude on single diffraction from the aperture. Taking mirrors as diffracting elements, this is a single reflection in a laser cavity. Let us now look at the case of multiple reflections, as follows.

### 12.3.2 Multiple diffractions

The idea of diffraction is based on the Fresnel number, which restricts the sizes of the mirrors that can be used, according to equation (9.12). First, we consider the laser cavity mirrors to be diffracting elements. Second, in the laser cavity, the light reflected from the one of the mirrors falls on the other cavity mirror and this process continues during several round trips. In figure 12.2, the two mirrors S and S′ are shown alternately along with the a simple representation of the light beam for two round trips in a laser cavity. The required condition is that the exit beam remains directed along the axis of the cavity.

This representation is similar to that shown in the following. The distribution of radiation on the primed mirror ($Q$) as a result of the unprimed mirror (P) can be indicated as shown in figure 12.3.

After multiple reflections, the two functions $U(x, y)$ and $U(x', y')$ become identical to a proportionality constant ($\gamma$, say, known as the *eigenvalue*). These are related to the following surface integral:

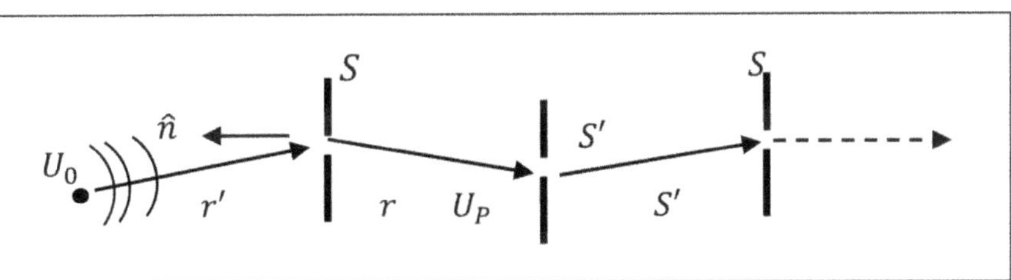


**Figure 12.2.** Representation of laser cavity mirrors (S and S′) as diffracting elements for a light beam passing through an aperture area of $dS$.

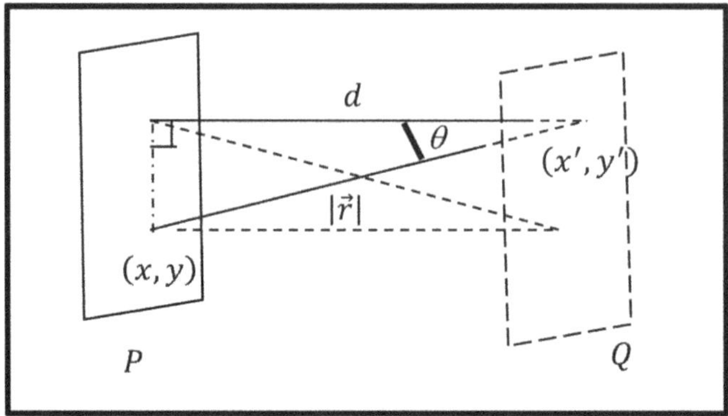

**Figure 12.3.** Re-representation of figure 12.2 as a laser cavity with two mirrors.

$$\iint U(x, y)K(x, y, x', y')dxdy = \gamma \quad U(x', y'). \tag{12.4}$$

At this point, $U$ is unknown. Here, $K$ (known as the *kernel* of the equation) is extracted from equation (12.3) as follows:

$$K(x, y, x', y') = -\frac{ik}{4\pi}\frac{e^{ikr}}{r}[\cos\theta + 1].$$

There are infinite (or, say, $n$) solutions (known as modes) of equation (12.4) represented by $U_n$. The eigenvalue for those solutions is given by $\gamma_n = |\gamma_n| e^{i\phi_n}$. Here, $\phi_n$ is the phase shift associated with the $n$th mode. A simple approximation using the above description (equation 12.4) suggests that functions that are identical under a Fourier transform of themselves are Gaussian functions as well as Hermite polynomials (♠ see section 12.6). While the Gaussian function describes the longitudinal modes, the Hermite polynomials are helpful in describing the *transverse modes* of the resonator. The amplitude of the output of the cavity, which is a result of the radiation diffracting back and forth between the mirrors, can be given by a Gaussian function, i.e. a lowest-order Hermite polynomial, as follows. For $\rho^2 = x^2 + y^2$, the amplitude $U(x, y)$ is

$$U(x, y) \propto e^{-\frac{\rho^2}{w^2}}, \tag{12.5}$$

where $w$ is defined as the beam radius. The intensity ($I$) is given by

$$I = I_0 e^{-\frac{2\rho^2}{w^2}}. \tag{12.6}$$

Here, $I_0$ is a constant. The Gaussian mode with the (0,0) set of the Hermite polynomial is also known as the $\mathrm{TEM}_{0,0}$ mode (♠ see section 12.6). Even though the divergence of a laser beam is extremely small, nevertheless, the value of $w$ changes for long propagation distances. Subsequently, as described in table 1.1, the radius of curvature ($R$) also changes. Thus, we need to find the mathematical relation between

$w(z)$ and $R(z)$ in terms of the wavelength ($\lambda$) and the minimum beam waist ($w_0$) as a function of the propagation distance $z$.

Since they are EM beams, laser beams satisfy Maxwell's equations. Ideal laser light can be described by the monochromatic Helmholtz equation. For a particular polarization propagating in the $z$ direction, we can write a scalar Helmholtz equation in 2D. This is described in the following sections by considering the spatial part of the Helmholtz equation.

## 12.4 Directional properties of laser light

### 12.4.1 Obtaining the Helmholtz equation from the wave equation

The wave equation (1.2) for $U(r, t)$ in space ($r$) and time ($t$) can be written as

$$\left(c^2\nabla^2 - \frac{\partial^2}{\partial t^2}\right)U(r, t) = 0. \tag{12.7}$$

We can perform a separation of variables to get the Helmholtz equation as follows. Let $U(r, t)$ be expressed in terms of its spatial and temporal parts as $U(r, t) = U(r)T(t)$, so that the partial derivatives with respect to $r$ and $t$ can be separated as follows:

$$\nabla^2 U(r, t) = \frac{1}{c^2}\frac{\partial^2}{\partial t^2}U(r)T(t) \equiv (-k^2) \text{ (a constant, say). Therefore, we can write}$$

$$\frac{\nabla^2 U(r)}{U(r)} = (-k^2) \text{ and } \frac{1}{c^2}\frac{1}{T(t)}\frac{\partial^2}{\partial t^2}\frac{T(t)}{} = (-k^2)$$

so that the Helmholtz equation for the spatial part is

$$(\nabla^2 + k^2) \quad U(r) = 0 \tag{12.8}$$

and the temporal part is given by $\left(\frac{\partial^2}{\partial t^2}+c^2k^2\right)T(t) = 0$. Using $\omega = ck$, this can be rewritten as

$$\left(\frac{\partial^2}{\partial t^2} + c^2k^2\right)T(t) = 0.$$

We will now work on the spatial part in order to discuss paraxial waves. A paraxial wave is a complex envelope containing a plane wave ($e^{-ikz}$) modulated by a slowly varying function of position $r$, i.e. $A(r)$. To understand this, a plane wave with a wave function $U(r)$ is considered using

$$U(r) = A(r)e^{-ikz}. \tag{12.9}$$

The whole envelope ($U(r)$) propagates in the the $z$ direction and satisfies the spatial Helmholtz equation (12.8).

### 12.4.2 Slowly varying envelope approximation (SVEA)

The amplitude of the paraxial wave must be modified by the form of a complex envelope $A_1(r)$, which is a slowly varying function of position $r$ given by $A(r) = A_1(r)e^{-ikz}$. The variation of $A_1(r)$ with position $(z)$ must be slow within the distance of a wavelength $\lambda = 2\pi/k$, so that the wave approximately maintains its underlying plane-wave nature, as indicated in figure 12.4.

A slowly varying envelope approximation (SVEA) with a carrier wave, i.e. a plane wave of a certain frequency whose amplitude is modulated by a slowly varying function of $r$, is shown in figure 12.4. As outlined in section 12.2, the wavefronts of a paraxial wave (table 12.1) are neither plane nor spherical, but have a paraboloid curvature (see figure 12.4(b)). A wave is said to be paraxial if its wavefront normals are paraxial rays.

### 12.4.3 Transverse Helmholtz equation

Placing the value of $U(r)$ in the spatial part of the Helmholtz equation within the condition $\Delta z = \lambda$ and using equation (12.9), equation (12.8) can be written in terms of a Laplacian as

$$\left(\frac{\partial^2}{\partial x^2} + \frac{\partial^2}{\partial y^2}\right)A(r)e^{-ikz} + \frac{\partial^2}{\partial z^2}A(r)e^{-ikz} + k^2A(r)e^{-ikz} = 0. \tag{12.10}$$

Here, we define the transverse Laplacian as $\nabla_T^2 = \frac{\partial^2 A}{\partial x^2} + \frac{\partial^2 A}{\partial y^2}$, so that the above equation can be rewritten as

$$\nabla_T^2 A(r)e^{-ikz} + \frac{\partial^2}{\partial z^2}A(r)e^{-ikz} + k^2A(r)e^{-ikz} = 0. \tag{12.11}$$

The value of $\Delta A$ is much smaller than $A$ itself, i.e. $\Delta A \ll A$. This implies that $k\Delta A \ll kA$, so that $\frac{2\pi}{\lambda}\Delta A \ll kA$.

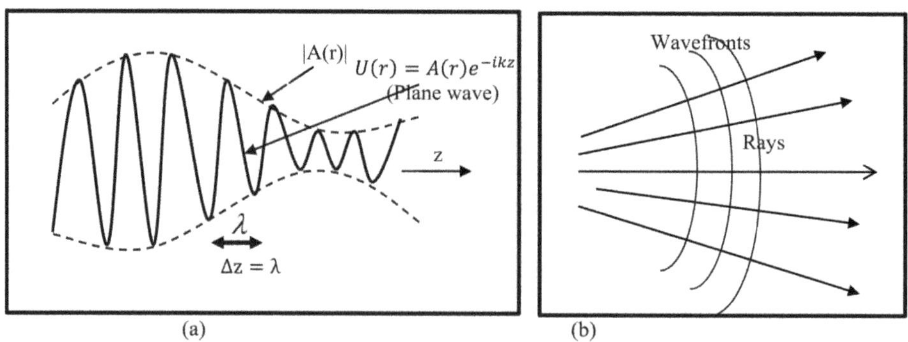


**Figure 12.4.** (a). Slowly varying envelope approximation (SVEA): a carrier wave (plane wave) representing a paraxial wave. The net amplitude of the wave envelope is $A(r) = A_1(r)e^{-ikz}$. (b). The wavefronts of a paraxial wave are nearly paraboloid in nature.

As $\lambda \equiv \Delta z$, we can write

$$\frac{\partial A}{\partial z} \ll kA. \tag{12.12}$$

Accordingly,

$$\frac{\partial^2 A}{\partial z^2} = k\frac{\partial A}{\partial z}. \tag{12.13}$$

Therefore, the value of $\frac{\partial^2 A}{\partial z^2}$ also becomes $\ll k^2 A$. The second term of equation (12.11) can be rewritten as

$$\frac{\partial^2}{\partial z^2}(A(r)e^{-ikz}) = \frac{\partial}{\partial z}\left(\frac{\partial}{\partial z}(A(r)e^{-ikz})\right)$$

$$= \frac{\partial^2 A}{\partial z^2}e^{-ikz} - 2ik\frac{\partial A(r)}{\partial z}e^{-ikz} + (ik)^2(A(r)e^{-ikz}).$$

Equation (12.11) now becomes

$$\left(\nabla_{\mathrm{T}}^2 A(r) + \frac{\partial^2 A}{\partial z^2} - 2ik\frac{\partial A(r)}{\partial z} + (ik)^2 A(r) + k^2 A(r)\right)e^{-ikz} = 0. \tag{12.14}$$

Neglecting the term $\frac{\partial^2 A}{\partial z^2}$ (in equations (12.12) and (12.13)), the equation becomes

$$\nabla_{\mathrm{T}}^2 A(r) - 2ik\frac{\partial A(r)}{\partial z} = 0. \tag{12.15}$$

This is known as the transverse Helmholtz equation. The simplest solution of equation (12.15) is a paraboloid wave, as follows:

$$A(r) = \frac{A_1(r)}{z} \, e^{-\frac{ik\rho^2}{2z}}. \tag{12.16}$$

♣ Assuming a solution for a differential equation is a well-known practice in science: assume a solution and then see whether it works well by inserting it into the differential equation. For an example, see chapter 1 for the solution of a plane-wave equation.

Equation (12.16) is the paraxial approximation of the spherical wave. It is useful to compare this equation with the one obtained from multiple Fresnel diffractions (equation 12.5). We now use this equation to discover the properties of the beam.

### 12.4.4 Radius of curvature of the beam

We introduce a complex number $q(z) = z \mp \zeta$. This is an important quantity, as a laser beam that has a beam waist in the cavity experiences diffraction as it propagates. The spread of the wavefront results in the variation of two quantities, viz. (i). the beam waist itself and (ii). the radius of curvature of the wavefront. In the forthcoming description, we will see that in traveling from one cavity mirror to the other, the wavefronts have various curvatures ranging from nearly spherical to infinite. This introduces an additional phase term known as the Gouy phase. Therefore, instead of $z = 0$, we choose the value of $q(z)$ at $z = +\zeta$ (say) that satisfies another imaginary quantity, $\zeta$. We now define $z_0$ as the *Rayleigh range*, such that $q(z) = z + iz_0$ remains the solution of equation (12.15), as follows:

$$A(r) = \frac{A_1\ (r)}{q(z)} e^{-\frac{ik\rho^2}{2q(z)}}. \tag{12.17}$$

In view of the aforementioned discussion, we can introduce the value of $q(z)$ as follows:

$$\frac{1}{q(z)} = \frac{1}{z + iz_0}. \tag{12.18}$$

In order to separate the amplitude and phase parts of the complex envelope, equation (12.18) can be normalized:

$$\frac{1}{z + iz_0} = \frac{z}{(z^2 + z_0^2)} - \frac{iz_0}{(z^2 + z_0^2)}.$$

Normalization gives $q(z)$ in terms of real and imaginary parts containing $R(z)$ and $z_R(w)$. Here, the radius of the beam has conventionally been denoted by $w$ in optics— not to be confused with the angular frequency, $\omega$:

$$\frac{1}{q(z)} = \frac{1}{R(z)} - i\frac{1}{Z_R(w)}. \tag{12.19}$$

On simplification, the real part of equation (12.19) can be rewritten as

$$\frac{1}{R(z)} = \frac{z}{(z^2 + z_0^2)}$$

or

$$R(z) = z\left[1 + \left(\frac{z_0}{z}\right)^2\right]. \tag{12.20}$$

Equation (12.20) is a very useful equation for finding the radius of curvature $R(z)$ as a function of the distance ($z$) traveled by the beam. Note that this expression includes the Rayleigh range (see section 12.4.6).

**Exercise 12.1.** The radius of curvature ($R(z)$) of a laser resonator is infinity at two points. For which value of $z$ this is true (choose from the following): $z = 0$, $z \to 0$, $z = z_0$, and $z = \infty$?

   **Solution:** The radius of curvature ($R(z)$) is given as a function of the propagation distance $z$ by $R(z) = z\left[1 + \left(\frac{z_0}{z}\right)^2\right]$.

**Table 12.2.** Values of the radius of curvature at different locations.

| Value of $R(z)$ | Position inside or outside the cavity |
| --- | --- |
| $\infty$ for $z = 0$ | At the center of the cavity |
| $\infty$ for $z \ll z_0$ | Within the Rayleigh range |
| $\infty$ for $z \to \infty$ | For large distances |
| $2z_0$ for $z = z_0$ | At the Rayleigh range |

Here, $z_0$ is the Rayleigh range. For $z \to 0$ (or for values less than $z_0$), the expression is governed by the second term, while for $z = \infty$, it is governed by the first term.

In both of these cases, the value of $R(z)$ becomes infinity (see table 12.2). Similarly, it is straightforward to check that the value of $R(z) = 2z_0$ at $z = z_0$.

### 12.4.5 Beam waist

The beam waist ($w(z)$) is the minimum width of the beam at a given location ($z$). Let us compare the argument of the exponential part of equation (12.17) with the Gaussian expression from equation (12.5), so that the quantity ($z_R(w)$) can be obtained in terms of $w(z)$ from equations (12.18) and (12.19):

$$\frac{k\rho^2}{2q(z)} = \frac{\rho^2}{w^2(z)}$$

so that

$$q(z) = \frac{\pi w^2(z)}{\lambda}. \tag{12.21}$$

We identify this quantity $\frac{\pi w^2(z)}{\lambda}$ as $z_R(w)$ in equation (12.19). Therefore, equation (12.19) can now be rewritten as

$$\frac{1}{q(z)} = \frac{1}{R(z)} - \frac{1}{z_R(w)} \text{ with } \frac{1}{z_R(w)} = \frac{z_0}{\left(z^2 + z_0^2\right)}.$$

Using the values of $z_R(w)$ and $z_0$, we obtain the expression for the beam waist ($w(z)$), as follows:

$$w(z) = w_0 \left[\left(1 + \frac{z^2}{z_0^2}\right)\right]^{\frac{1}{2}}. \tag{12.22}$$

Equations (12.20) and (12.22) give the two important parameters, viz. $R(z)$ and $w(z)$ for the cavity in terms of Gaussian wave analysis. We are now in a position to place these parameters in the cavity, as shown in figure 12.5. One of the most important parameters in the diagram is the minimum beam waist ($w_0$). All the other parameters can be obtained once the value of $w_0$ is known.

### 12.4.6 Rayleigh range in terms of the minimum beam waist

The beam waist $w(z)$ is similar to the radius of the laser beam's cross-section, which is known as the beam *spot size*. The minimum spot size at the center of the laser cavity is generally given as twice the value of the minimum beam waist, i.e. $2w_0$.

The Rayleigh range ($z_0$) for the wavelength $\lambda$ is defined in terms of $w_0$ as $z_0 = \pi w_0^2 / \lambda$. This parameter is important in optics and spectroscopy. In practical applications, it suggests that the position of the sample under study should be within a factor of $\sqrt{2}$ from the focused beam waist, $w_0$. The equal distance on the either size of the focus, which is twice the size of the Rayleigh range, is known as the *confocal parameter* ($2z_0$). Sometimes, this is also known as depth of focus.

**Exercise 12.2.** For a laser resonator, the beam waist ($w(z)$) is at a minimum at its center. Find the values of ($w(z)$) at $z = 0$ and $z = z_0$.

**Solution:** The beam waist ($w(z)$) as a function of the propagation distance z is given by $w(z) = w_0\left[\left(1 + \dfrac{z^2}{z_0^2}\right)\right]^{\frac{1}{2}}$.

For $z = 0$, using this expression, the minimum beam waist is $w_0$. The value of $w(z) = w_0\sqrt{2}$ at $z = z_0$. This is also indicated in figure 12.5.

**Exercise 12.3.** For an argon-ion laser, the mirrors have to be designed to give a spot size of 20 mm at a distance of 10 m from the laser. The laser's operating wavelength is 514.5 nm and its cavity length is 1.5 m. What are possible values of the minimum beam waist at the center of the cavity?

**Solution:** (a). The expression for the beam waist $w(z)$ is

$$w(z) = w_0\left[\left(1 + \frac{z^2}{z_0^2}\right)\right]^{\frac{1}{2}}.$$

As the beam waist is half of the spot size, $w(z) = 10^{-2}$ m. We obtain $z$ by adding half of the cavity length to the distance from the output mirror: $z = 10.75$ m. The wavelength $\lambda = 514.5$ nm and the Rayleigh range $z_0 = \dfrac{\pi w_0^2}{\lambda}$.

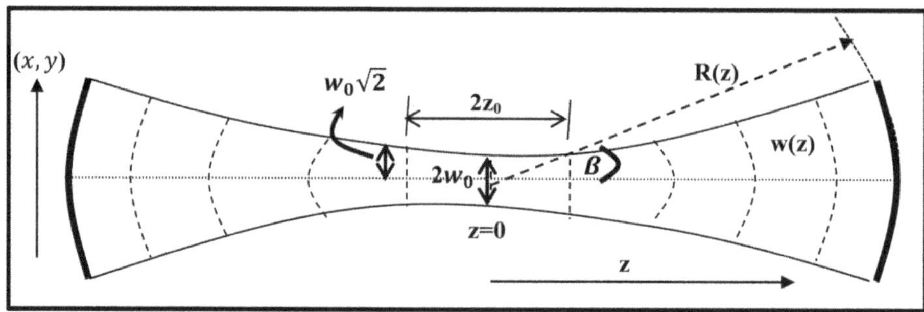


**Figure 12.5.** Schematic representation of the cavity parameters $R(z)$ and $w(z)$. The coordinates $(x, y)$ at the vertical axis indicate the 2D nature of the beam inside the cavity. The beam propagation is in the $z$ direction.

Therefore, by placing the numerical values into $w^2(z) = w_0^2\left[\left(1 + \left(\frac{\lambda z}{\pi w_0^2}\right)^2\right)\right]$, we get

$(10^{-2})^2 = w_0^2\left[\left(1 + \left(\frac{514 \times 10^{-9} \times 10.75}{\pi w_0^2}\right)^2\right)\right]$. After rearrangement, this can be written in the form of a quadratic equation:

$$w_0^4 - (10^{-4})\ w_0^2 + 3.1 \times 10^{-12} = 0.$$

This equation has two solutions, as follows:

$$w_0^2 = \frac{-(-10^{-4}) \pm \sqrt{(10^{-4})^2 - 4\times 3.1 \times 10^{-12}}}{2}.$$

Solving this, we get two values of $w_0^2$, namely $9.99 \times 10^{-5}\,\mathrm{m}^2$ and $3.10 \times 10^{-8}\,\mathrm{m}^2$. Finally, by taking the square roots of these values, the two values of $w_0$ obtained are 9.99 mm and 0.85 mm. Note that the selection of one of these values also depends on other parameters, such as the radii of curvature of the mirrors.

### 12.4.7 Laser beam spot and Gaussian expression

At the beginning of this chapter, in table 12.1, it was observed that there is a transformation between spherical wavefronts and the limiting case of plane wavefronts, which takes place as a function of the propagation distance from the source. In the previous sections, we compared equations (12.5) and (12.17) to deduce the propagation properties of the beam. In this sense, equation 12.5 is, in principle, similar to equation (12.17). Equation (12.5) gives the transverse distribution of the intensity of a Gaussian beam in the form of $I = I_0 e^{-\frac{2\rho^2}{w^2}}$. There are two cases for the full width at half maximum (FWHM) as indicated in figure 12.6, as follows:

    i. For the case of the electric field, $E = E_0 e^{-\frac{\rho}{w}}$ for $\rho = w$, $\frac{E}{E_0} = \frac{1}{e}$, and $1/e \cong 0.4$.

       For practical applications, the value of the FWHM can be roughly taken to be half of $E_0$.

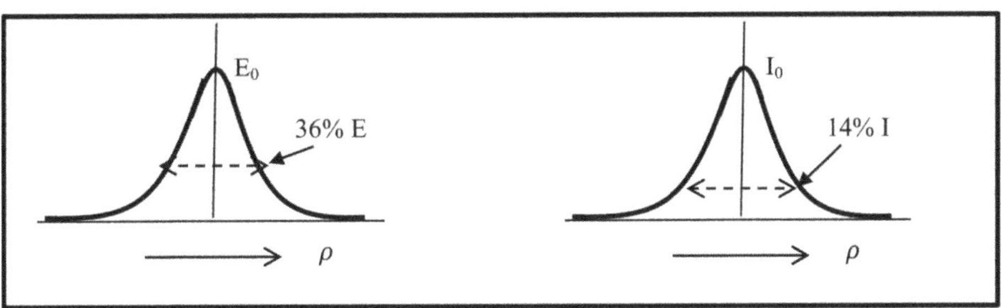


**Figure 12.6.** Electric field (left) and intensity (right) profiles of the Gaussian beam and the corresponding FWHMs (indicated by arrows).

ii. For the case of intensity, at $\rho = w$, $\frac{I}{I_0} = 1/e^2$. As $1/e^2 = 0.14$, this indicates that 86% of the light is within the FWHM of the beam.

### 12.4.8 Angular spread

From figure 12.5, the angular spread (i.e. divergence) of the beam can be obtained as follows:

$$\tan \beta = \lim_{z \to \infty} \frac{w(z)}{z} \equiv \frac{\lambda}{\pi w_0}, \tag{12.23}$$

where $w(z)$ is given by equation (12.22). Therefore, the angular spread is given by

$$2\beta = 2\tan^{-1}\left(\frac{\lambda}{\pi w_0}\right).$$

It can be seen that the angular spread is proportional to the wavelength of the laser beam but inversely proportional to its spot size.

### 12.4.9 Special cases of $R(z)$

From equation (12.20), we can find the radius of curvature ($R(z)$) of the Gaussian beam at various locations inside and outside the cavity. The values of $R(z)$ are given in table 12.2.

For higher values of $z$ ($>z_0$), the value of $R(z)$) can be approximated by $z$. Figure 12.7 gives the plot of $R(z)$ as a function of $z$. Inside the cavity, it can be seen

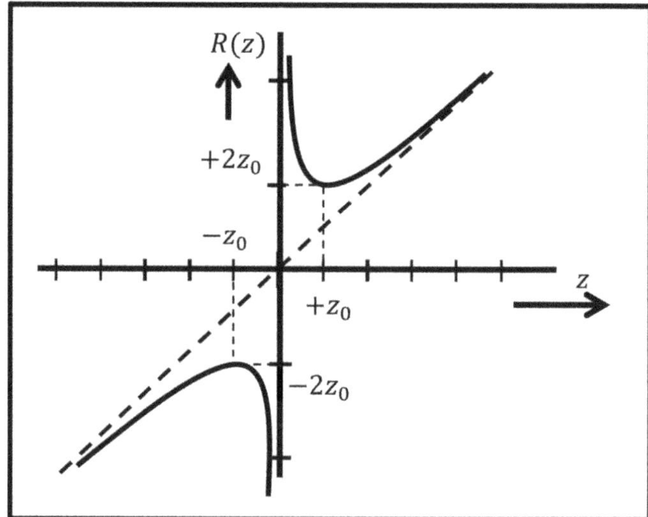


**Figure 12.7.** Plot of the radius of curvature of the wavefront $R(z)$ as a function of the propagation distance for a Gaussian beam; $z_0$ is the Rayleigh range. The dashed line passing through the origin indicates the case for the spherical wavefront.

from the plots that for the Gaussian beam, $R(0) = \infty$. As the beam propagates outward in the cavity along $z$, the curvature of $R(z)$ falls to certain values. However, for the higher values of $z$, we see that it follows a straight line, i.e. $R(z) \cong z$. This is the case of a spherical wavefront. This essentially reveals the fact that at $z = \infty$, the wavefront curvatures again become infinite.

### 12.4.10 Appearance of the Gouy phase

As we see from figures 12.5 and 12.7, a laser cavity has a minimum beam waist, $w_0$. The beam converges to the position of the beam waist (at the center in figure 12.8), and thereafter diverges to the right. This happens due to the diffraction of light, similar to the diffraction that occurs when light beams are forced to pass through small apertures ($w_0$ in this case). Any convergent beam acquires a phase change of $\pi$ when traveling from $-\infty$ to $+\infty$. In its propagation direction, a Gaussian beam also acquires a phase shift which differs from that for a plane wave of the same optical frequency. This effect was experimentally and theoretically explained by Gouy in the late nineteenth century. As mentioned at the beginning, a Gaussian beam can be considered to be a superposition of plane waves with different propagation directions. The plane wave components that have propagation directions different from the beam axis experience phase shifts as they propagate along $z$. In the case of paraxial beams, the overall phase shift $\psi(z)$ that arises is given by a superposition of all these components. It is popularly known as the Gouy phase and is described by

$$\psi(z) = \tan^{-1}(z/z_0). \tag{12.24}$$

Based on the aforementioned discussion, the amplitude $U(r)$ of the paraxial wave for $z \gg z_R$ in equation (12.9) can now be written as follows. Using the value of $A(r)$ from equation (12.17), $q(z)$ from equation (12.19), and the Gouy phase ($\psi(z)$) from

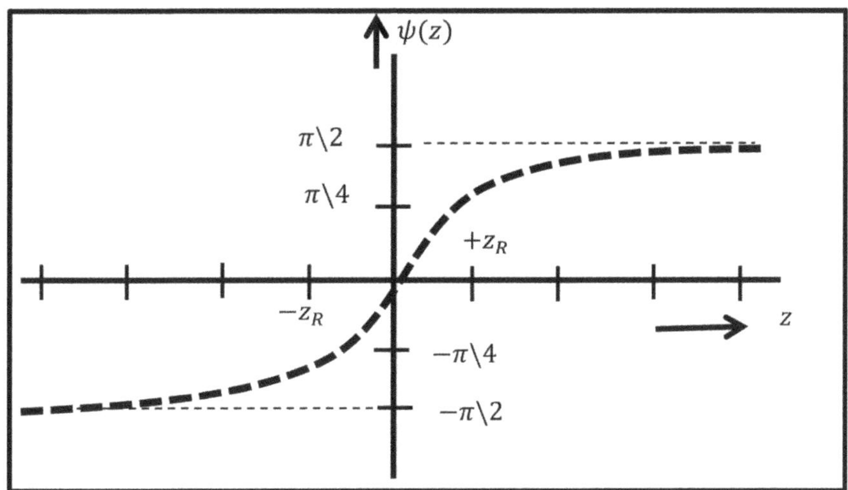


**Figure 12.8.** Variation of the Gouy phase ($\psi(z)$) as a function of propagation distance ($z$).

equation (12.24), the compete solution for the paraxial Gaussian beam can be written as

$$U(r) = A(r) \quad e^{-(\rho^2/w^2(z))} \quad e^{-ik(\rho^2/2R(z))}e^{-ikz}e^{-i\psi(z)}. \tag{12.25}$$

## 12.5 Stability condition and Gaussian wave analysis

We now know that laser beams are Gaussian in nature. The important property of a Gaussian beam is that it remains Gaussian at every point in space during its travel until it encounters a distorting medium. More importantly, the Fourier transform of a Gaussian function is itself a Gaussian function (♠ see question 11). Therefore, when such beams are focused by a lens or a concave mirror, they preserve their Gaussian nature.

The stability condition obtained via Gaussian wave analysis states that the calculated value of $R(z)$ for the required values of $w(z)$ must be obtained from equation (12.20). We also know that when calculated value of $R(z)$ from equation (12.20) matches the radius of curvature of the mirror in the laser cavity, the wavefront's paraxial rays are perpendicular to the full mirror surface, resulting in perfect reflection. This stability condition goes in parallel with the stability condition (equation (11.18)) discussed in the previous chapter in the context of the geometrical optics and stability diagram (♠ see section 11.3.2).

## 12.6 TEM modes

We know that the light inside the laser cavity undergoes multiple reflections, as outlined in section 12.3.2. These reflections occur in such a way that the standing wave along the axis gives an output beam. However, within the limitation of the Fresnel number, the beams can also make multiple reflections in round trips in a 2D cavity away from the cavity axis[2]. These reflections continue to keep the light beams within the cavity, even after several reflections. Part of the light passes through the output coupler per round trip in the form of the laser's output. Although the part of the output beam that does not travel along the axis as a single spot (known as a $\text{TEM}_{00}$ mode) is undesirable, it is nevertheless important for academic purposes and applications in other research fields, such as in holey beams.

The mathematical expression for the output is expressed in form of a polynomial known as a Hermite polynomial. In general, we can write the output as the product of a Hermite polynomial and a Gaussian distribution, as follows:

$$U_{mn}(x, y) = H_m\left(\frac{x\sqrt{2}}{w}\right)H_n\left(\frac{y\sqrt{2}}{w}\right) \times e^{-\frac{\rho^2}{w^2}}. \tag{12.26}$$

---

[2] This is similar to some types of art (calligraphy); for example, drawing an asterisk (*) on paper without lifting the pen. ♣ Chinese/Japanese dictionaries are based on 'the number of times the pen is lifted while making a character'. ♣ *Kolam*—a cultural morning ritual art made in front of the house in south India with rice powder.

Here, for 2D mirrors, the integers $m$ and $n$ denote the order of the Hermite polynomial. Therefore, every set of $(m, n)$ represents a distribution of the amplitude $U_{mn}(x, y)$ at the mirror. Each of these distributions is known as the TEM ($TEM_{m, n}$) mode of the cavity.

**Exercise 12.4.** In equation (12.26), let us assume a generalized Hermite polynomial $H_p(u)$ with order $p$ and variable $u$, as follows: $H_p(u) = (-1)^p e^{u^2} \times \frac{d^p}{du^p}(e^{-u^2})$, where $u$ denotes either $\left(\frac{x\sqrt{2}}{w}\right)$ or $\left(\frac{y\sqrt{2}}{w}\right)$. What are the various orders of the polynomial?

   **Solution:** In this case, the various orders of the polynomial are
$H_0(u) = 1; H_1(u) = 2u; H_2(u) = 2(2u^2 - 1)$.
   Note that Gaussian functions are zeroth-order functions of Hermite polynomials.
   Some of the TEM modes are described in the following. It should be noted that the longitudinal mode number (which is a large number) is not written in the description.

   i. **TEM$_{0,0}$ mode:** The TEM$_{0,0}$ mode is a perfect Gaussian mode ($H_0(u) = 1$) and is generally a property of stable resonators at their best alignment along the laser axis. While propagating along the axis of the cavity, if the light repeatedly bounces between only two points on the mirrors, this results in a TEM$_{0,0}$ mode (see figure 12.9). This is known as the fundamental or axial mode of the laser.

   ii. **TEM$_{1,0}$ mode:** There is a possibility that a round trip is completed in such a way that the light reflects at two points on each mirror along the horizontal or the vertical axis of the mirror. As shown in figure 12.10, this behavior results in a mode other than TEM$_{0,0}$. The nomenclature for a TEM mode is based on the number of minima the mode has along the horizontal or vertical axes. For example, for the mode shown in figure 12.10, there is a single minimum in the vertical direction, and the mode is therefore known as the TEM$_{1,0}$ mode. Similarly, a mode with a minimum between two lobes on the horizontal axis would be known as a TEM$_{0,1}$ mode.

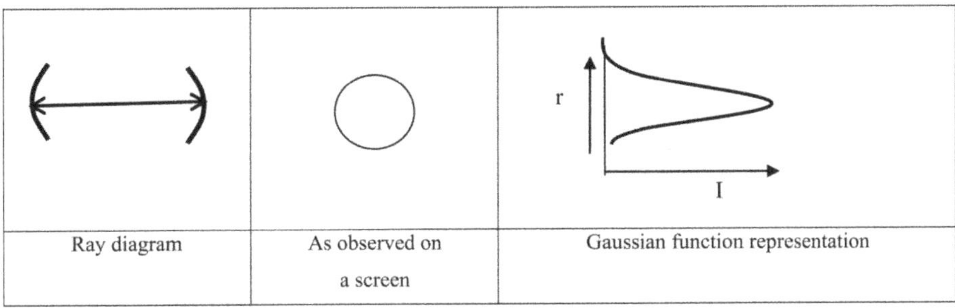



**Figure 12.9.** Depiction of the TEM$_{0,0}$ mode in terms of a ray diagram in a cavity, a spot on a screen, and its Gaussian representation in terms of its intensity ($I$) and its position ($r$).

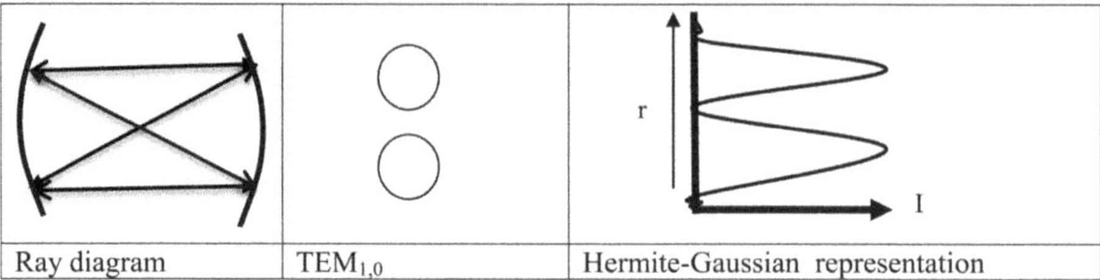

| Ray diagram | $TEM_{1,0}$ | Hermite-Gaussian representation |

**Figure 12.10.** Schematic view of the $TEM_{1,0}$ mode in terms of a ray diagram in a cavity, spots on a screen, and a Gaussian representation.

**Table 12.3.** Higher-order Hermite–Gaussian (or other TEM) mode representations of lasers as observed on the screen.

| $TEM_{0,2}$ | $TEM_{1,1}$ | |
|---|---|---|
| ○  ○  ○ | ○  ○<br>○  ○ | |
| $TEM_{0,3}$ | $TEM_{3,1}$ | $TEM_{1,3}$ |
| ○  ○  ○  ○ | ○ ○<br>○ ○<br>○ ○<br>○ ○ | 0 0 0 0<br>0 0 0 0 |
| $TEM_{1,2}$ | $TEM_{2,2}$ | $TEM_{3,3}$ |
| 0 0 0<br>0 0 0 | 0 0 0<br>0 0 0<br>0 0 0 | 0 0 0 0<br>0 0 0 0<br>0 0 0 0<br>0 0 0 0 |

The cavity mirrors have a 2D shape, therefore the higher-order modes appear in both dimensions. Depending upon the various round trips that can take place in the possible geometries, we can observer higher-order modes $TEM_{m,\,n}$, as shown in table 12.3.

## Questions and problems

1. In the mathematical expressions for a (i) plane wave, (ii) a spherical wave, and (iii) a paraboloid wave, find the similarities at extreme propagation distances.
2. Find the required cavity parameters for a laser operating at a wavelength $\lambda$ that produces a beam of diameter $2\omega$ at a distance $d$ from its output mirror. Assume that the laser has a cavity gain length of $d/2$.

3. Use the information given in problem 2 to solve the following:
    (i) Find the parameters for a He–Ne laser with a required spot size ($w$) of 5 mm at a distance of 1 m from the output mirror. Take the cavity length of the laser to be 0.5 m.
    (ii) The minimum spot size of the beam inside the cavity of a Nd: yttrium aluminum garnet (YAG) laser is 0.5 mm. The cavity length of the laser is 2 m. The YAG rod is 100 mm long and it is placed at the center of the cavity. Take the refractive index of the YAG rod to be 1.80. Find the spot size of the beam at a distance of 5 m from the output coupler.
    (iii) Find the minimum beam waist for a $CO_2$ laser with a confocal cavity that has a length of 2 m.
4. Find the value of the radius of curvature ($R(z)$) for the obtained values of the minimum beam waist ($\omega_0$) in exercise 12.2. Comment on the obtained values of $R(z)$.
5. A laser is beamed through a transmitting telescope to hit an orbiting satellite 36 000 km away from the Earth. A retroreflecting mirror surface on the satellite equal to the beam spread of 10 m is just fully illuminated, as sensed by detectors. Assume the Earth to be stationary during the experiment and that both signals are required for estimation. (i) Assuming the orbiting path to be straight for 2 km, what curvature of the mirrors of the laser cavity allows the beam to continue to be detected by the satellite? (ii) What is the time taken by the light to return to the Earth after reflection? (iii) What is the size of the beam that arrives at the Earth after reflection from the satellite?
6. The spot size obtained with the help of a lens of size $D$ and focal length $f$ at the wavelength $\lambda$ is $\frac{2f\lambda}{D}$. Can you approximate this from the beam waist of the Gaussian beam analysis?
7. Using approximate expressions for a beam waist inside a cavity, show that the divergence of a laser beam is inversely proportional to its initial spot size.
8. Comment on the statement 'the radius of curvature of the Gaussian beam is infinite at a point inside a resonator'.
9. Define a paraxial beam.
10. Gaussian functions are associated with Hermite polynomials. Express various TEM modes in the form of Hermite polynomials.
    i. What order of Hermite polynomial is the Gaussian mode?
    ii. Take an appropriate Gaussian function and show that its Fourier transform is also a Gaussian function.
    iii. For a cavity with square mirrors, draw the appearance of the $TEM_{2,3}$ mode on a screen.
11. Two short laser pulses of the same wavelength ($\lambda$) at a certain repetition rate are focused by a thin lens at the sample, as shown in the figure. (*This experiment detects the change in the probe intensity due to absorption by the sample*).

The detected probe intensity varies if the pump pulse strikes the sample before the probe. The focal length of the lens is $f$. The diameters of the pump ($D_P$) and probe ($D_D$) beams are larger than those of the circular variable slits placed in front of the lens. The spot sizes of the pump and probe beams at the sample are $w_p$ and $w_D$, respectively.

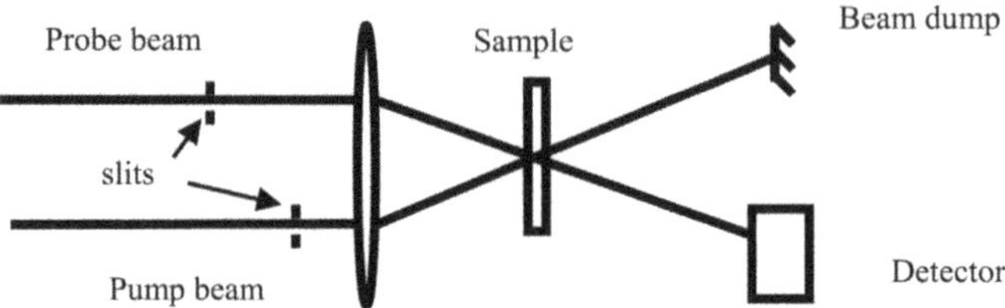


i. Give the spot sizes of the two beams at the sample in terms of $f$, $D_P$, $D_D$ and $\lambda$? How can the sizes of the focused spots at the sample be controlled while keeping the same lens?

ii. Which of the following conditions can be used to get the best signal-to-noise (S/N) ratio in the detection of the probe pulse: (i). $w_p > w_D$, (ii). $w_p < w_D$, or (iii). $w_p = w_D$?

iii. Modify the diagram by sketching the minimum addition required to control the time of arrival of the pump pulse at the sample.

# Bibliography


[1] Turyshev S (ed) 2009 *From Quantum to Cosmos* (Singapore: World Scientific)
[2] Siegman A E 1986 *Lasers* (Mill Valley, CA: University Science)
[3] Fowles G R 1975 *Introduction to Modern Optics* 2nd edn (New York: Dover Publications)
[4] Feng S and Winful H G 2001 Physical origin of the Gouy phase shift *Opt. Lett.* **26** 485–7
[5] Gouy L G 1890 Sur une propriete nouvelle des ondes lumineuses *C. R. Acad. Sci., Paris* **110** 1251
[6] Rubinowicz A 1938 On the anomalous propagation of phase in the focus *Phys. Rev.* **54** 931
[7] Gouy L G 1890 Sur la propagation anomale des ondes *Compt. Rendue Acad. Sci. Paris* **111** 33

# Part II

Pulsed lasers and nonlinear optical applications

**IOP** Publishing

# An Introduction to Photonics and Laser Physics with Applications

**Prem B Bisht**

# Chapter 13

# Laser spiking and $Q$-switching

In the history of measuring fast events, two popular experiments have taken place in the last two centuries, separated by a gap of almost 75 years. The first was a galloping horse photographed by Muybridge in 1878. This was a classic experiment that paved the way for the modern movie industry. The second was the imaging of an apple shot by a bullet in 1964. Both of these experiments provided details of events that took place on shorter timescales than those discernible by the human eye. The first experiment used the shutters of 12 cameras clicked sequentially by strings attached to a running horse, while the second used a faster imaging camera available at the time. In these experiments, the timescales were of the order of milliseconds ($10^{-3}$ s) to microseconds ($10^{-6}$ s), respectively. In order to optically record events that have even shorter timescales, the time resolution needs to be improved further. To achieve this, we need pulsed laser excitation sources with faster pulse durations. The figure shows the generation of nanosecond-scale ($10^{-9}$ s) laser pulses by manipulating the loss in the cavity—the details of which are provided in this chapter. Another technique used to generate laser pulses of the order of femtoseconds ($10^{-15}$ s) is described in chapter 17, following the introduction of nonlinear optics in subsequent chapters.

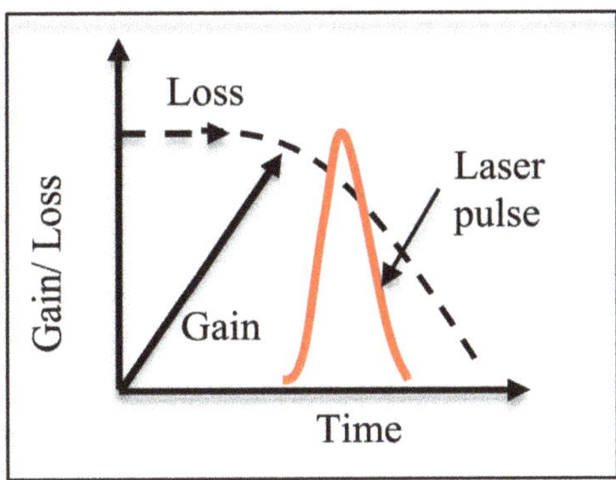

**Learning Objectives**
**After reading this chapter, the learner will be able to:**
Identify the requirements of pulsed lasers;
Describe the phenomenon of spiking;
Outline the conditions required for $Q$-switching;
Classify the methods of $Q$-switching;
Understand the electro-optic and acousto-optic methods.

# 13.1 Pulsed light sources

Pulsed light can be obtained from a light source by modulating the light source externally. In addition, there are methods of generating optical pulses by internal modulation of the laser cavity parameters.

### 13.1.1 External modulation

The most direct way of achieving external modulation is to use a light chopper in front of a continuous-wave (CW) laser, as shown in figure 13.1 (top frame). The

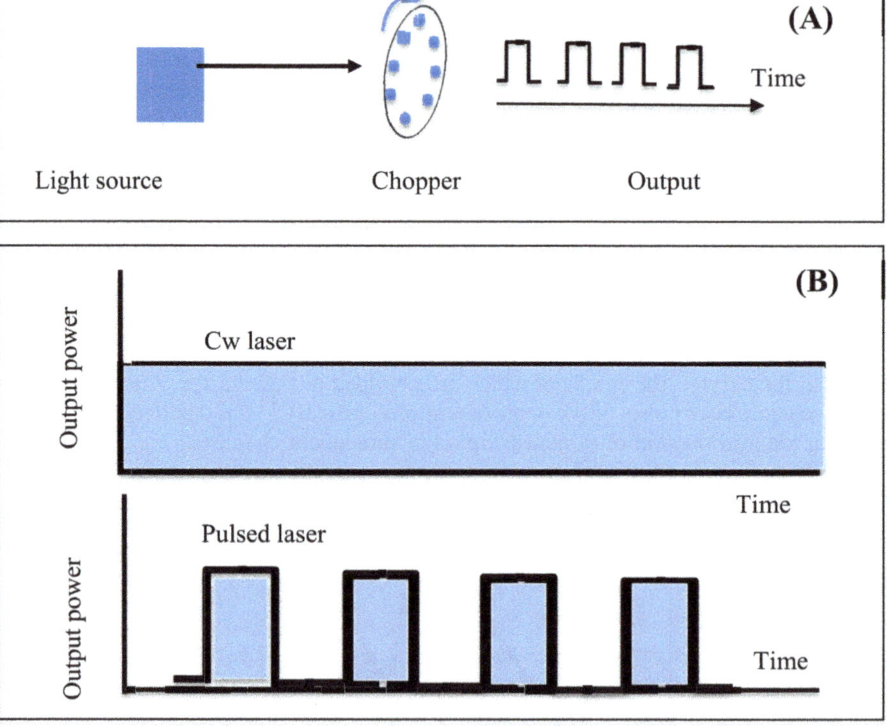


**Figure 13.1.** A rotating chopper in front of a laser can produce pulses of light (panel (A)). In panel (B), the shaded portion indicates the CW laser output (top). The shutter allows the power within the shaded area to be transmitted during its 'on' time (bottom) in the form of pulses. Square pulses are shown for illustration purposes.

chopper permits the light to pass through in the form of light pulses, as shown in the top panel of figure 13.1. The repetition rate of the laser is decided by the number of openings, their positioning on the chopper, and its rotational speed.

The chopper can be controlled by an external switch or a modulator. It transmits light only during the time window selected by the repetition rate of the opening, or the 'on' time. However, it has the following two disadvantages: (i) the EM energy of the pulse train is wasted during the opaque portion of the chopper, i.e. during its 'off' time; (ii) the peak power of the pulse cannot exceed the steady-state value of the CW laser.

Therefore, an efficient scheme is necessary by which EM energy can be stored during the 'off' time and released only during the 'on' time. This can be achieved by internal modulation, in which light is periodically allowed to escape from the cavity at high peak power. Here, is it worth defining a few terms related to pulsed lasers. The *average power* of a pulsed laser (in units of J s or watts) is defined as the product of the *Energy of the pulse* and its *repetition rate*, and the peak power of each pulse is given by the ratio of the *Energy of the pulse* to its *pulse duration*. As the energy of an optical pulse is confined by the pulse shape, the peak power is higher than the average power measured by a power meter.

**Exercise 13.1.** A laser delivers 10 ns pulses at a repetition rate of 10 Hz. The average power of the laser as measured by a power meter is 100 mW. What is the peak power of each pulse? Distinguish the obtained power from that of a 100 mW CW laser.

**Solution**: The average power of the laser is $100 \times 10^{-3}$ J s$^{-1}$ at a 10 Hz repetition rate. This means that the energy per pulse $= \frac{1}{10}(100 \times 10^{-3})$ J $= 10 \times 10^{-3}$ J. The laser is delivering each pulse with an energy of 10 mJ.

The peak power of the laser pulse $= \dfrac{\text{(energy of single pulse)}}{\text{pulse duration}} = \dfrac{10^{-2}}{10 \times 10^{-9}} = 10^6$ W. In this example, we can distinguish between a CW laser and a pulsed laser with the same average power. It should be noted that even though a CW laser and a pulsed laser have the same average powers of 100 mW, as recorded by a power meter, the laser pulses exhibit peak powers that are higher by a factor of $10^7$. Such lasers with high peak powers are useful in nonlinear optical applications and micromachining. On the other hand, high-power CW lasers are used for heat-generating applications such as welding and cutting metals.

### 13.1.2 Intra-cavity modulation

The methods used to generate pulses by internal modulation of the laser cavity are: gain switching, cavity dumping, $Q$-switching, and mode locking. Gain switching is a direct method of controlling the gain. It involves turning the laser pumping 'on' and 'off'. For example, in case of a pulsed ruby laser, the flash lamp is periodically switched on for brief periods of time. In this way, sequences of electric pulses are used for gain switching. During the *'switch on'* time, the gain coefficient exceeds the losses and laser output is obtained. Cavity dumping is used in mode-locked lasers (♠ see chapter 17) to reducing the repetition rate or to select a pulse train produced by the laser cavity.

## 13.2 The spiking phenomenon

As mentioned above, laser pulses can be generated by pumping the gain medium in pulsed mode, as shown in figure 13.2. A flash lamp suitably located in the laser cavity pumps the gain medium at a certain repetition rate to obtain a pulsed output. The parameters that should be taken into account are:

    (i) the duration of the pumping flash $(\tau_p)$,
    (ii) the cavity round-trip time $(\tau_{rt})$, and

the lifetime of the upper state $(\tau_f)$ that participates in the lasing process.

### 13.2.1 Pump flash duration and population inversion

Let us consider the duration $(\tau_p)$ of the pump flash. Flash lamps have typical flash duration of about 1 ms. Figure 13.3 shows a portion of a flash produced by a lamp with its full time duration, $\tau_p$. The behavior of the net population inversion in the gain medium $(\Delta N)$ and its threshold value $(\Delta N_{\text{th}})$ are also shown for one round trip $(\tau_r)$ of the cavity. Let $\tau_s$ be the time taken for the gain to exceed the loss. For a sufficiently large value of $\Delta N$ (i.e. $\geq \Delta N_{\text{th}}$), we observe a spike in the output over a single round trip in the cavity. If the refractive index of the gain medium is $n_L$ and that of the vacuum is $n_c$, the round-trip time in the cavity $(\tau_{rt})$ of length $d$ is given by

$$\tau_{rt} = \frac{2}{c}[Ln_L + (d - L)n_c].$$

Here, $L$ is the length of the gain medium.

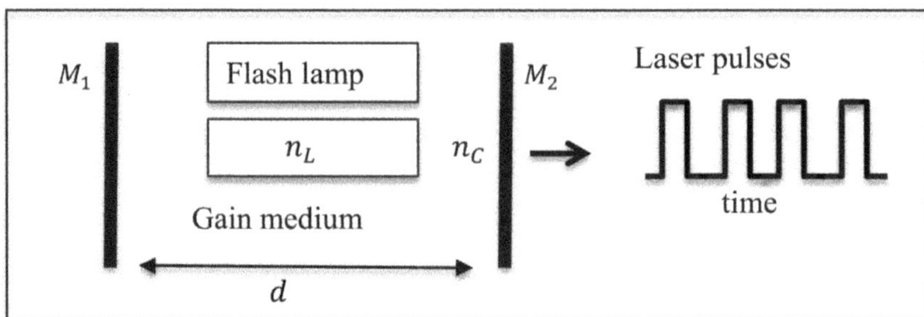


**Figure 13.2.** Schematic of a typical pulsed laser. Here, $n_L$ and $n_C$ are the refractive indices of the gain medium and the cavity, respectively. $M_1$, $M_2$ are cavity mirrors separated by a distance $d$.

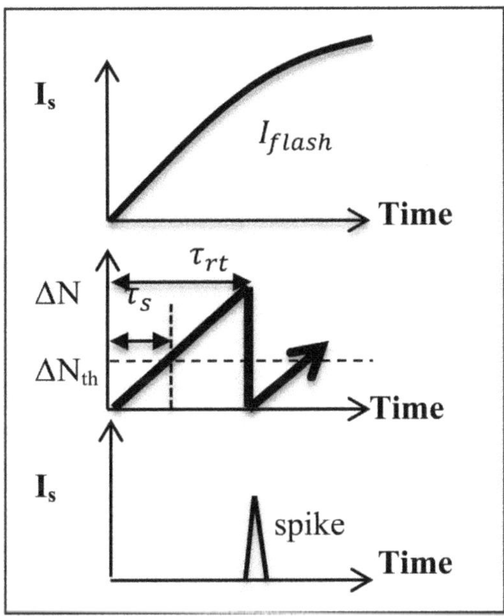


**Figure 13.3.** Formation of a laser spike (bottom) after a round-trip time ($\tau_{rt}$) in the cavity during the initial portion of the excitation ($I_{\text{flash}}$) by a flash (top). The middle panel shows the behavior of the excited population, in which the dashed line indicates the threshold value at which population inversion occurs ($\Delta N_{\text{th}}$).

### 13.2.2 Laser spiking

On every round trip, as the laser beam is coupled out by the mirror ($M_2$), no gain is left behind in the cavity. The time required to reach gain equilibrium in several ($m$, say) passes in a cavity of length $L$ is

$$\tau_m = \frac{m}{c}[Ln_L + (d - L)n_c]. \tag{13.1}$$

If the intensity of the flash is strong enough to maintain $\Delta N$ above its threshold value for the next round trip, there will be a subsequent output pulse, as indicated in figure 13.4. Similarly, in later round trips, if $\Delta N$ is restored to a sufficient level during $\tau_m$, a series of light spikes will be observed as long as $\Delta N \geqslant \Delta N_{th}$. For every flash of the pumping source, we obtain several peaks in the output of the laser with varying intensities.

Typically, for a pump pulse of $\tau_p$ of 1 ms, 1000 spikes can be observed for a pulse duration of 1 µs. For every cavity round trip, this process of gain depletion accompanied by an output burst during a single pump flash is known as *laser spiking*. As the output is a function of the round-trip time as opposed to a single output pulse per pump, several spike pulses are obtained. Therefore, laser spiking can be considered to be a waste of energy for a meaningful output. To obtain a single output pulse, the $Q$-switch mechanism is employed, as described in following sections.

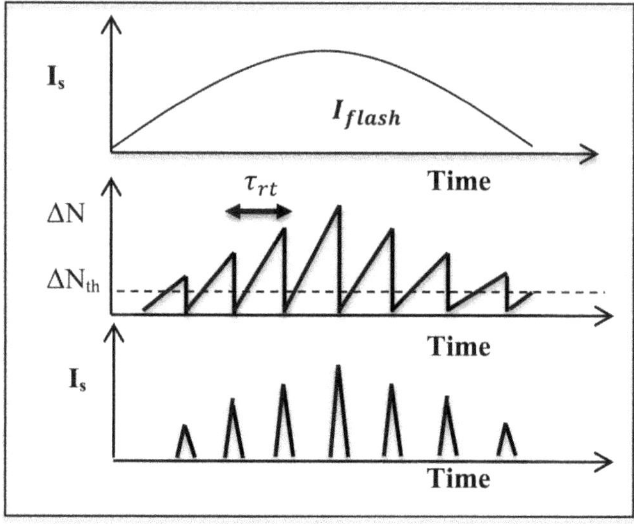


**Figure 13.4.** Laser cavity events during a complete flash of the lamp (top) showing the phenomenon of spiking (bottom). As in figure 13.3, the dashed line (middle panel) indicates the threshold value of population inversion.

## 13.3 The $Q$-switching phenomenon

In a similar way to that in which we defined the quality factor ($Q$) of a cavity mode (see chapter 9), here we define the $Q$ value of a laser cavity as

$$Q = \frac{\text{energy stored in the cavity}}{\text{energy dissipated per cycle}}.$$

Thus, we now describe a mechanism used to obtain laser output by cyclically switching the $Q$ of the cavity. During the $Q$-switch cycle of the laser, its output is actually turned off by increasing the resonator losses. However, at the same time, the gain medium is allowed to be pumped by the excitation source, with the result that the excited states continue to be populated. We now know that the time required to reach gain equilibrium in $m$ passes is given by equation 13.1. The $Q$-switch cycle completes when the $Q$-factor of the resonator is changed suddenly. In this process, the largely accumulated population inversion between energy states is released, generating a short pulse of laser light with a high peak power. This is done by extracting the maximum energy from the cavity in the form of a single output pulse; in this process, the cavity gain is depleted once. To do this, a few conditions must be met, as described in the following section.

### 13.3.1 Conditions for $Q$-switching

The parameters of the gain medium and the laser cavity are important when considering a laser for $Q$-switching. The critical variables that must be considered to fulfill the conditions of this mechanism are: the round-trip cavity time $\tau_{rt}$, the upper-state lifetime ($\tau_0$), and the duration of the pump flash $\tau_p$. As described above, $\tau_m$ is the time taken for the gain to exceed the loss in $m$ round trips. Let us consider the following two cases:

(a) (i) $\tau_0 > \tau_m$: this is one of the main requirements of $Q$-switching—this ensures that the upper state gets sufficient time for 'population inversion' due to pumping; (ii) the pump flux duration $\tau_p$ must be longer than $\tau_m$ and preferably greater than or equal to $\tau_0$.

(b) $\tau_0 < \tau_m$: when $\tau_0$ is smaller than $\tau_m$, there is no build-up of gain in the cavity. Hence, this condition is redundant for the $Q$-factor change.

For case (a) above, to achieve sufficient population inversion, the initial cavity losses have to be kept high during the pumping process. This is achieved by simply blocking one of the cavity mirrors during $\tau_m$ *to initially avoid laser oscillation*. This is a condition in which the initial cavity losses are kept greater than the gain of the amplifier. To get an output laser pulse, the cavity losses are reduced instantaneously. Here, $Q$-switching is achieved by suddenly switching the cavity back to its original state, i.e. by unblocking the cavity mirror.

The phenomenon of $Q$-switching is schematically presented in the following diagrams. Generally, the pump flash duration is much larger than $\tau_{rt}$. Figure 13.5 shows the situation in the cavity on initial blocking of one of the mirrors (panels A and B). During this period, the gain medium is pumped to achieve the maximum gain, i.e. until gain saturation during the pump flash ($\tau_p$). As indicated in panel C, bringing the cavity back into place suddenly reduces the $Q$ of the cavity. As a result, the photon density ($\phi(t)$) increases and a giant pulse of nanosecond duration is obtained as the output of the laser.

A more detailed picture of the build-up of the gain and its depletion during the process of $Q$-switching is shown in figure 13.6, which displays the gain switching along with its subsequent effects in terms of the populations of the lower ($N_L$) and upper ($N_U$) states. At time $t < 0$, the population difference between the upper and the lower states $N(t) = N_L$. Therefore, as the population has a value below the threshold, no oscillation occurs, indicating $(t) = 0$. This situation continues until time $t_1$, as indicated in the lower panel of figure 13.6, even when the pump is turned on at $t = 0$.

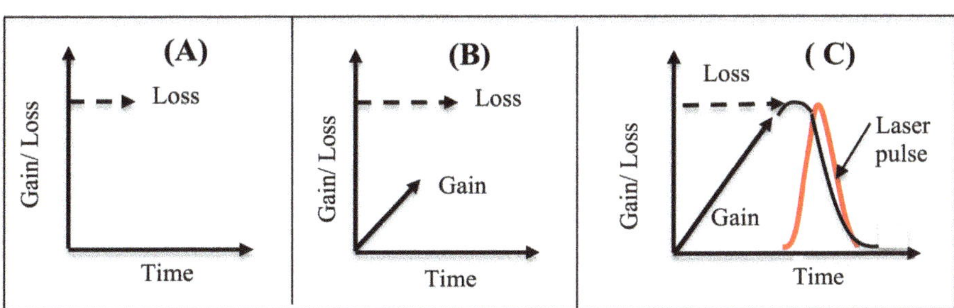


**Figure 13.5.** Schematic of the process of $Q$-switching in a laser cavity. Initially (panel A), the cavity has high loss, as one of the mirrors ($M_1$ in figure 13.2) is blocked. The gain builds up during the pump flash ($\tau_p$) (panel B). The loss is suddenly reduced when the mirror is unblocked (panel C).

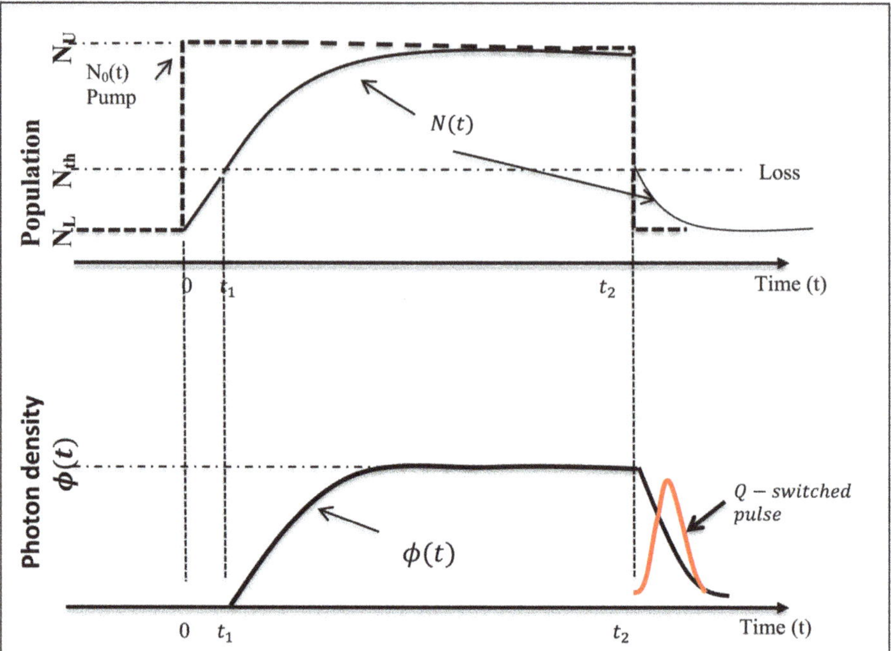


**Figure 13.6.** (b). Photon number density $\phi(t)$ as a result of the population difference with time $t$ in $Q$-switching. $N_U$ and $N_L$ are the population densities in the lower and upper states, respectively. $N_{th}$ is the population at the laser threshold. The lower panel shows the generation of the $Q$-switched pulse when the laser cavity is brought back into operation at time $t = t_2$.

However, after the pump is turned on, $N(t)$ grows exponentially until $t_2$, despite the aforementioned loss mechanism, due to the laser cavity. At time $t = t_2$, the cavity is brought back into operation by unblocking the mirror. As shown in the bottom panel of figure 13.6, the photon density $\phi(t)$ attains a value between $t_1$ and $t_2$ and a $Q$-switched pulse appears after $t_2$.

### 13.3.2 Methods of $Q$-switching

As mentioned above, in $Q$-switching, one of the cavity mirrors initially needs to be blocked during the gain build-up. The mirror has to be brought back to its original position to obtain the $Q$-switched pulses on the nanosecond timescale. There are a few ways of blocking and unblocking the cavity mirror within the required time-scale, as described below.

#### 13.3.2.1 Mechanical methods

In these methods, one of the cavity mirrors is repetitively blocked or removed by a mechanical device, such as a chopper wheel or a rotating hexagonal mirror assembly. Such mechanical procedures were used at the beginning of 1960s to achieve $Q$-switching in ruby lasers. However, high chopping speeds lead to vibration

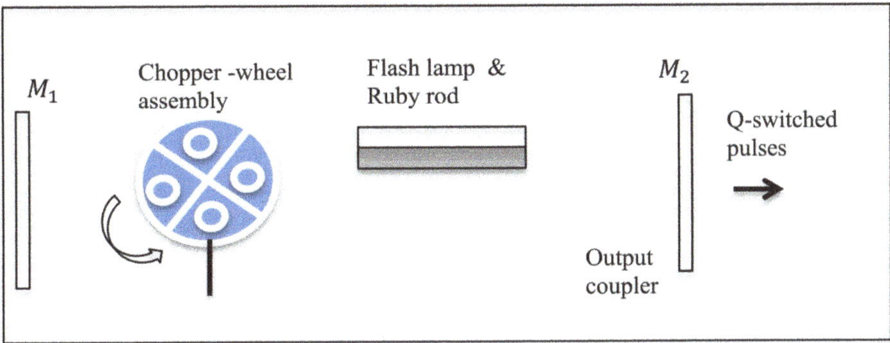



**Figure 13.7.** Chopper-wheel method of $Q$-switching. Here, the output coupler ($M_2$) is fixed, while the laser light aligns with one of the holes of the chopper and is reflected by the mirror $M_1$.

and misalignment during the operation. This is the disadvantage of this method, as seen in the following exercise.

**Exercise 13.2.** One of the methods used to eliminate the phenomenon of spiking in laser output is shown in figure 13.7. As shown schematically, a rotating chopper-wheel assembly is used with a ruby laser cavity. The holes of the chopper can align with the cavity during rotation. What is the repetition rate of the pulsed laser output for a chopper with four holes?

**Solution**: The excited-state lifetime of ruby is 3 ms. Therefore, in order to synchronize the repetition rate with the upper-state lifetime of the gain medium, one-quarter of the rotation should take place in 3 ms. This means that the chopper should complete a full rotation in 12 ms, corresponding to a rotational speed of 5000 revolutions per minute (rpm). This is a high rpm value, which contributes to the instability of the laser. In this case, the repetition rate of the laser is one-third of a millisecond, or ~333 Hz.

### 13.3.2.2 Electro-optic method

In this method, the 'blocking' of the cavity mirror for the desired duration is achieved using an electro-optic (EO) shutter. The EO shutter is made of a birefringent material, such as a quartz block. A birefringent material exhibits variation in its refractive index on the application of a DC electric field, due to the Pockels effect or the Kerr effect, as described in chapter 3 (♠ see section 3.7.1). Due to the EO effect, a phase change is introduced that helps to rotate the polarization of the propagating beam.

Consider figure 13.8, in which a linear polarizer and an EO shutter are kept inside a cavity. A voltage is applied across the crystal. The thickness of the quartz block in the EO shutter is selected in such a way that it rotates the polarization of the light by 45° when the electric field is applied. After the light is reflected by the mirror and

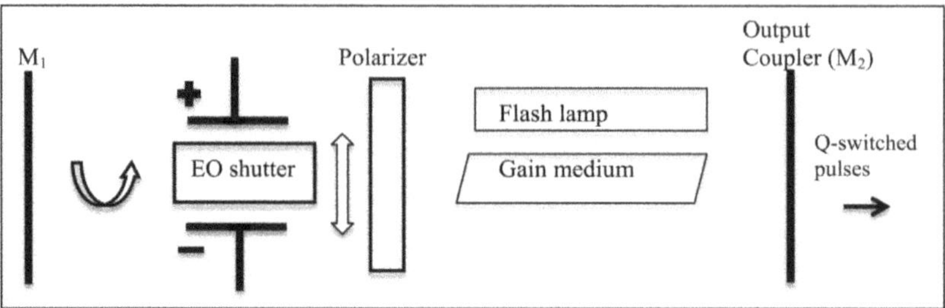


**Figure 13.8.** Electro-optic (EO) method of $Q$-switching. The arrows indicate that the linearly polarized light ($\Leftrightarrow$) becomes circularly polarized (bent arrow) after passing though the EO shutter.

passes through the quartz block again (on its return path), the polarization is rotated by another 45°, making it perpendicular to the incident beam. As this point, the light cannot pass through the polarizer located in the cavity. The repetition rate of the EO shutter is chosen such that it synchronizes with the other parameters of the $Q$-switch cavity.

**Exercise 13.3.** If, on application of a DC voltage to the EO crystal, the crystal produces a phase change equal to adding a $\lambda/4$ plate, what is the total phase change for a double pass of the same light? For $Q$-switching, both an EO crystal and a linear polarizer are placed in the cavity, as seen in figure 13.8. Can the light pass through the linear polarizer on its return path?

    **Solution:** The net phase change for a crystal length of $L$ is

$$\Delta\phi = kL = \frac{2\pi}{\lambda}\left(\frac{\lambda}{4}\right) \times 2 = \pi.$$

The total phase change is $\pi$. This means that the light polarization is rotated by 90°. Thus, it will not pass through the polarizer when the voltage is applied.

*13.3.2.3 Acousto-optic method*

In acousto-optics, the refractive index of an optical medium is altered by the presence of sound waves produced by a radio-frequency (RF) transducer, i.e. an ultrasound source (figure 13.9). Sound is a mechanical dynamic strain involving molecular vibrations that results in longitudinal waves depending on the characteristics of the medium. Acoustic waves create a perturbation of the refractive indices of materials such as quartz crystals. The strain $S(x, t)$, expressed as a function of time $t$ and position $x$ from the source of the acoustic wave, is given by $S(x, t) = S_0\cos(\Omega t - kx)$. Here, $S_0$ is a constant, $\Omega$ is the angular frequency of the acoustic wave of wavelength $\Lambda$, and k is the wave vector ($2\pi/\Lambda$). The change of refractive index ($\Delta n$) at the grating in the $x$ direction (say) is given by $\Delta n(x, t) = -\frac{1}{2}pn^3 S(x, t)$, where $p$ is the strain-optic constant. In the presence of

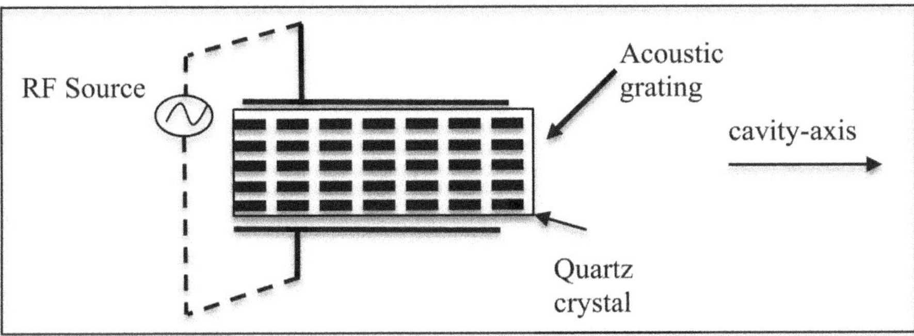



**Figure 13.9.** Principle of the acousto-optic modulator. In the presence of an RF pulse, an acoustic grating is created across the crystal that selectively operates the cavity.

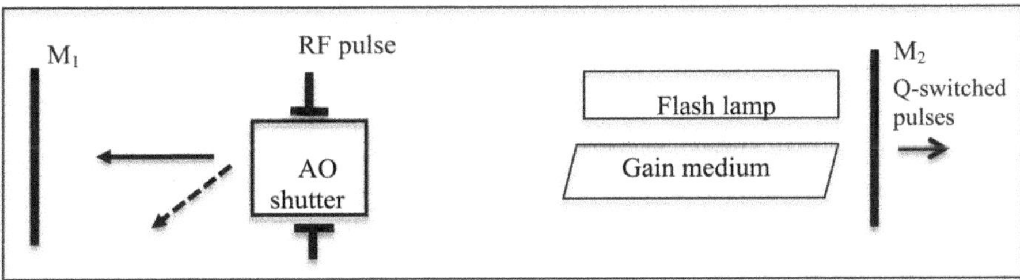



**Figure 13.10.** Acousto-optic (AO) method of Q-switching. The dashed arrow indicates the diffraction of the light that is sent out of the cavity during the RF pulse.

the acoustic waves, the crystal behaves as a graded index medium. This can be regarded as an acoustic grating in which photon–phonon interaction takes place.

The grating in the quartz crystal diffracts light away from the cavity axis for the duration of the applied RF pulse. Therefore, during the RF pulse, since one of the cavity mirrors does not receive light along the cavity axis, the gain builds up in the cavity. When the RF pulse is turned off, the cavity mirrors are in place, resulting in Q-switched output. Typically, a RF wave in frequency range of 25–50 MHz is applied across the crystal, at which point an acoustic grating is created with a grating constant of the order of micrometers. This results in Bragg diffraction of the light away from the cavity axis for the duration of the applied frequency (figure 13.10).

### 13.3.2.4 Saturable absorber (SA) method

In this method, an absorbing medium known as a saturable absorber (SA) is placed in the cavity. The SA modulates the losses in the cavity as follows. For lower incident EM fields, the SA strongly absorbs the light pulses. Conversely, at higher energies, it transmits the pulses uninterrupted. The Q-switching method based on saturation of absorption is also known as *passive* Q-switching. The process of saturable absorption is described later in this book under applications of nonlinear optics (see section 16.12). The necessary criterion for the use of an SA material located in the beam path away from the amplifier is that the absorption cross-section of the SA material should be reasonably high to absorb the lower-power laser modes.

## Questions and problems

1. A laser delivers 30 ps pulses at a rate of 10 Hz. If the peak power of the pulse is 1 GW, what is the energy per pulse and the average power of the laser?

2. Define the term 'laser spiking'. A gain medium that has a refractive index of one uniformly fills a cavity 0.5 m long. Calculate the minimum number of passes required to obtain a maximum-energy $Q$-switched pulse if the gain medium has an upper-level lifetime of 1 ms.

3. A hexagonal mirror fixed on a rotating stage is used to reflect the light in a $Q$-switch cavity. For a Nd:yttrium aluminum garnet (YAG) laser, if one-sixth of a turn for one of the faces of the hexagonal mirror coincides with the gain saturation time, what is the rotation speed of the stage?

4. A circuit consisting of a capacitor (10 mF) and a resistor (1 K$\Omega$) is used to operate a shutter in a cavity to generate optical pulses from a CW laser. What is the minimum time limit for the pulse obtained in this case? Is the average power of the pulsed laser the same as that of the CW in this case?

5. An electro-optic crystal is placed near the end mirror in a laser cavity. A linear polarizer is inserted in front of the crystal towards the output coupler. DC power is applied across the crystal so that the beam of light of wavelength $\lambda$ passes through the polarizer on its return path. What is the total phase change in this process?

6. Can you differentiate between a Pockels cell and a Kerr cell, as used in the generation of pulsed laser beams? (Hint: see section 3.5.1.)

7. A frequency range of 20–100 KHz is used by bats and dolphins for communication and navigation. You may know that the human ear cannot hear sound frequencies beyond 20 KHz (ultrasonic). In medicine, the frequency range of 1–20 MHz, referred to as 'ultrasound' is used for imaging at a micrometer resolution. Assuming that the speed of sound is 330 m s$^{-1}$, what is the wavelength of a 20 MHz source? What is the use of the RF pulse in the acousto-optic method of $Q$-switching?

8. Calculate the electric field of a 10 ns $Q$-switched Nd:YAG laser pulse that has 1 mJ of energy and a focused spot diameter of 2 mm.

9. Calculate the number of Nd$^{+3}$ ions in the upper (lasing) state just before a Nd:YAG laser delivers a $Q$-switched pulse of energy of 500 mJ. Assume that each ion emits once during the pulse and that the net population inversion between the two levels is also equal to the number of ions in the upper state.

10. If a saturable absorber is inserted into the cavity of a $Q$-switched laser so that the effective cavity length remains the same, what are its effects in terms of the output power and the pulse duration? (Hint: see chapter 16.)

# Bibliography


[1] Muybridge E photographer. (ca. 1878) The Horse in motion. 'Sallie Gardner,' owned by Leland Stanford; running at a 1:40 gait over the Palo Alto track, 19th June/ Muybridge. California Palo Alto, ca. 1878. [Photograph] Retrieved from the Library of Congress, https://loc.gov/item/97502309/

[2] Edgerton H E Bullet through Apple, 1964, printed 1984, dye transfer print, Smithsonian American

[3] Collins R J and Kisliuk P 1962 Control of population inversion in pulsed optical masers by feedback modulation *J. Appl. Phys.* **33** 2009–11

[4] Maydan D and Chesler R B 1971 *Q*-switching and cavity dumping of Nd:YAG lasers *J. Appl. Phys.* **42** 1031–4

[5] McClung F J and Hellwarth R W 1962 Giant optical pulsations from ruby *J. Appl. Phys.* **33** 828

[6] Wagner W G and Lengyel B A 1963 Evolution of the giant pulse in a laser *J. Appl. Phys.* **34** 2040

**IOP** Publishing

# An Introduction to Photonics and Laser Physics with Applications

**Prem B Bisht**

# Chapter 14

# Introduction to nonlinear optical phenomena

The previous section introduced the Q-switching technique for pulse generation. Such short-duration, intense laser pulses can have an electric field strength of the order of intra-atomic fields. Under such incident fields, the polarization induced in a material can exhibit nonlinear behavior, as indicated in the figure. This brings us to a new field of *nonlinear optics* (NLO). This chapter introduces the subject of NLO to the beginner, before continuing with a detailed description of related phenomena in the next chapters. The connection between linear dielectrics and higher-order optical non-linearities is discussed here. As an example, two nonlinear optical processes, viz. second-harmonic generation (SHG) and optical rectification (OR) are introduced in this chapter along with the idea of virtual levels. The differences between linear and nonlinear optics are also highlighted.

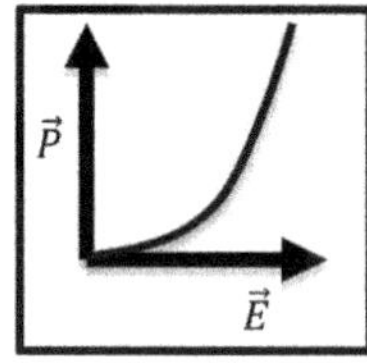

**Learning objectives**
**After reading this chapter, the learner will be able to:**
Describe the refractive index and dielectric constant of a medium;
Relate the interatomic field strength with the EM field of a laser;
Identify higher-order optical nonlinearities, together with their units and dimensions;
Estimate the order of magnitude of optical nonlinearities;
Define the phenomenon of optical rectification;
Identify the virtual levels;
Differentiate between linear and nonlinear optics.

doi:10.1088/978-0-7503-5226-0ch14          © IOP Publishing Ltd 2022

## 14.1 Review of linear dielectrics

From electrostatics, we know that the polarization $(\vec{P})$ induced by an incident electric field $(\vec{E})$ in a dielectric is given by

$$\vec{P} = \epsilon_0 \chi_e \vec{E}. \tag{14.1}$$

Here, $\chi_e$ is the susceptibility of the dielectric medium. It is a dimensionless quantity, a tensor[1] of rank 2. It should be noted that $\epsilon_0$, the permittivity of free space has units of $C^2 Nm^{-2}$. Here, it is included in the equation to balance the dimensions on the right-hand side.

Introduced in section 1.4.5, the dielectric constant $(K)$ of a material is given by the ratio of the permittivity of the medium $(\epsilon)$ to that of free space. It is also known as relative permittivity $(\epsilon_r)$.

$$K \text{ or, } \epsilon_r = 1 + \chi_e. \tag{14.2}$$

We know that the dielectric constants of most of materials are in the range from 1 to 100. Therefore, their susceptibilities are also of the same order. The speed of light $(c)$ is defined as $c = \dfrac{1}{\sqrt{\epsilon_0 \mu_0}}$, where $\mu_0$ is the permeability of the medium. As mentioned in section 1.4.7, the complex refractive index $(\tilde{n})$ of a medium can be written as

$$\tilde{n} = n' + in''. \tag{14.3}$$

Here, $n'$ is the real part of the refractive index, and $n''$ is the imaginary part appearing due to absorption. $n''$ is related to the absorption coefficient $(\alpha_0)$ and the magnitude of the wave vector $(k)$, as follows: $n'' = \alpha_0/2k$ .

We now introduce the polarizability $(\delta)$ of an atom or a molecule as a function of the external electric field. It is related to the dipole moment $(\vec{p})$ as follows: $\vec{p} = \delta \vec{E}$. If $i, j$ denote the Cartesian components of the polarisability, the tensor form of $\delta_{ij}$ can be written as follows:

$$\delta_{ij} = \begin{vmatrix} \delta_{11} & \delta_{12} & \delta_{13} \\ \delta_{16} & \delta_{22} & \delta_{23} \\ \delta_{15} & \delta_{24} & \delta_{33} \end{vmatrix}. \tag{14.4}$$

In general, the EM fields corresponding to ordinary light (such as lamps) are considered to be weak. Equation (14.1) is valid for linear, isotropic, homogeneous dielectric media for weak incident electric fields. As described in section 1.3, the magnetic component of EM radiation is the smaller by a factor of $c$; the major effects are due to its electric component.

---

[1] A scalar is a tensor of rank zero and a vector is a tensor of rank one that has three components. A tensor of rank two has $3^2$ components, and so on. Two vectors are related by a tensor one rank higher.

## 14.2 Wave equation in nonlinear optics

The 3D wave equation was derived from Maxwell's equations for an arbitrary homogeneous dielectric medium in chapter 1. When the polarization is independent of time, the 3D wave equation is given by equation ((1.2). Using equation (14.1), the partial differential equation for time-dependent polarization can be written as follows:

$$\nabla^2 \vec{E} - \frac{1}{c^2}\frac{\partial^2 \vec{E}}{\partial t^2} = \mu_0 \frac{\partial^2 \vec{P}}{\partial t^2}. \tag{14.5}$$

For nonzero values of the factor $\frac{\partial^2 \vec{P}}{\partial t^2}$ ($\neq 0$), charges are being accelerated. The time derivative of nonlinear polarization plays a major role in our understanding of nonlinear optical processes.

### 14.2.1 Interatomic field strength

To understand the nonlinear optical response, let us visualize an electron attached to a crystal lattice by a spring (figure 14.1). In the Lorentz model of the electron, under the influence of a normal electric field, the electron oscillates with the same frequency as that of the incident field, and radiates. We know that for small oscillations, the displacement ($x$) is proportional to the applied force $F$ ($\spadesuit$ also see section 8.4). This force between the nucleus and the electron of an hydrogen atom can be estimated using Coulomb's law. For a typical distance of 0.05 nm (the Bohr radius), the electric field experienced by an electron is found to be ~$10^{12}$ V m$^{-1}$.

**Exercise 14.1.** Considering the Lorentz model of atom, calculate the value of the interatomic field for the hydrogen atom.

**Solution:** For a positive charge $q$ centered at the nucleus, the magnitude of the electric field $|\vec{E}|$ can be written as $|\vec{E}| = \frac{q}{4\pi\epsilon_0 r^2}$. For the Bohr radius of 0.05 nm, the magnitude of the electric field is $|\vec{E}| = \frac{9\times10^9 \times 1.6 \times 10^{-19}}{\left(5\times10^{-11}\right)^2} = 6.34 \times 10^{11}$ V m$^{-1}$. It should be noted that we do not encounter such high electric field strengths (~$10^{11}$ V m$^{-1}$) in routine undergraduate laboratory classes.

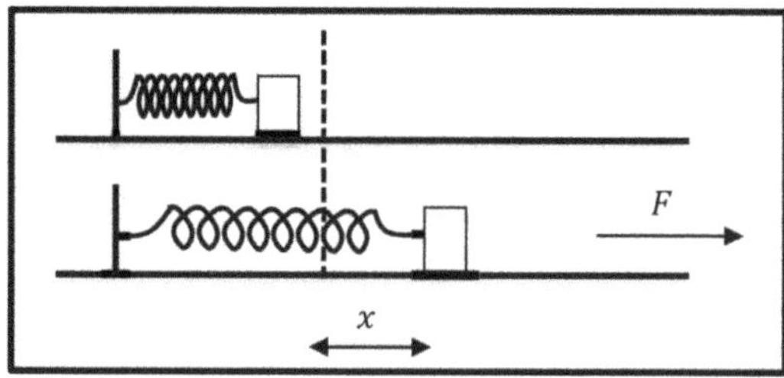


**Figure 14.1.** In the Lorentz model, the motion of an electron in an atom is considered to be similar to the simple harmonic motion of a mass attached to a spring. The displacement ($x$) of the spring is measured from the equilibrium position indicated by the vertical dashed line.

For molecular systems, the situation could be complex, as they are more easily polarized in one direction than another. Hence, the components of the polarizability tensor become useful. Under the influence of stronger electric fields, the electronic distribution can be perturbed further and the oscillations can no longer be described by simple harmonics. Under stronger incident fields, higher-order terms appear in Hook-type equations, for example $F = -y_1 x - y_2 x^2 - y_3 x^3$ and so on, where $y_1$, $y_2$, and $y_3$ are constants of appropriate dimensions.

### 14.2.2 Higher terms of polarization

In recent years, higher electric field strengths have readily been available from ultrafast lasers[2]. For instance, laser pulse durations of 100 fs at a 1 KHz repetition rate with an average power of 100 mW correspond to a peak power of 1 GW per pulse (♠ see section 13.1.1). As a result, at the large incident electric field strengths corresponding to such powerful laser pulses, the polarization of the medium ($P$) contains higher terms in the electric field. Under such circumstances, the medium is said to be *nonlinear* and equation (14.1) can be rewritten as

$$P = \varepsilon_0[\chi_e E + \chi^{(2)} \quad E^2 + \chi^{(3)} \quad E^3 + ....]. \tag{14.6}$$

The total polarization ($P_{\text{total}}$), for example, in terms of the first-, second-, and third-order polarizations $P_1(t)$, $P_1(t)$, and $P_1(t)$ respectively, can be written as

$$P_{\text{total}} = P^{(1)}(t) + P^{(2)}(t) + P^{(3)}(t) + ... \tag{14.7}$$

♣ We note that equation (14.6) is in its scalar form. In fact, in contrast to equation (14.1), equation (14.6) can no longer be written in its vector form. This is the case because the electric fields on the right-hand side must be identically expressed by either the dot product or the cross products. Such equations are generally written in their tensor form.

We rewrite equation (14.6) for a component of the polarization tensor in index form as

$$P_i = \varepsilon_0\left[\chi_{ij} E_j + \chi^{(2)}_{ijk} E_j E_k + \chi^{(3)}_{ijkl} E_j E_k E_l + ...\right]. \tag{14.8}$$

Here, $\chi^{(m)}_{ijk....}$ ($m = 2, 3,...$) is the $m$th-order susceptibility tensor of rank ($m + 1$). The subscripts $i$, $j$, $k$, and $l$ represent the Cartesian components of the vector fields. We know that the matrix form of linear susceptibility ($\chi_e$) is written as a tensor of rank two, as follows.

$$\chi_{ij} = \begin{vmatrix} \chi_{11} & \chi_{12} & \chi_{13} \\ \chi_{16} & \chi_{22} & \chi_{23} \\ \chi_{15} & \chi_{24} & \chi_{33} \end{vmatrix}. \tag{14.9}$$

---

[2] Electric fields are available from lasers up to the petawatt scale (see chapter 24). However, for nonlinear optical applications in free space, peak powers in the range of megawatts to gigawatts are sufficient.

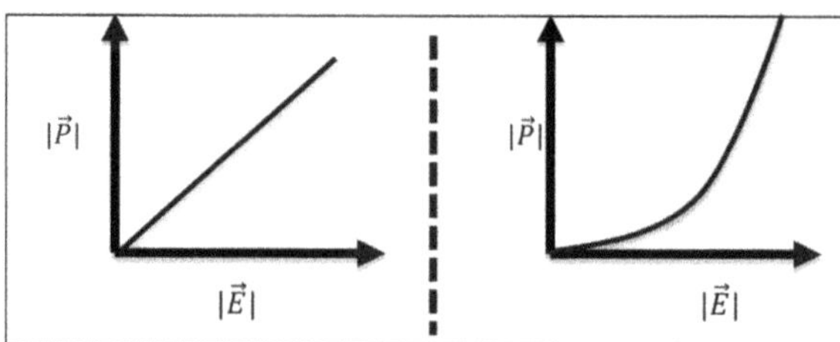

**Figure 14.2.** Schematic description of induced polarization $|\vec{P}|$ in a dielectric for low values of the incident electric field $|\vec{E}|$ (left-hand panel), and the nonlinear behavior of the light–matter interaction for higher incident fields (right-hand panel).

$\chi_{ij}$ has nine components. The linear susceptibility is related to the refractive index ($n$) and the absorption coefficient ($\alpha_0$) as introduced above, in section 14.1.

Second-order nonlinear susceptibility ($\chi^{(2)}_{ijk}$) is responsible for the generation of second-harmonic, sum, and difference frequencies. Third-order susceptibility ($\chi^{(3)}_{ijkl}$) is involved in phenomena such as third-harmonic generation. The various phenomena associated with these nonlinear susceptibilities will be explained in subsequent sections/chapters.

Based on the above information, we now can propose the variations in the induced polarization under the influence of incident electric fields as indicated in the plots of $|\vec{P}|$ versus $|\vec{E}|$ in figure 14.2. We can see that at lower values of incident fields, the plot is linear, and that it becomes nonlinear at higher electric field strengths. The light–matter interaction under high-peak-power optical pulses falls under a subarea of photonics known as NLO.

## 14.3 Units and estimates of susceptibilities

The second term $P_2(t)$ in equation (14.7) can be compared with $P^{(1)}(t)$. Using equations (14.6) and (14.7), we can write

$$P^{(2)}(t) = \varepsilon_0(\chi^{(2)}E)E \equiv (\Delta\chi)E.$$

Here, $(\Delta\chi)$ is the change in the susceptibility, which is equal to $(\chi^{(2)}E)$. Therefore, $\chi^{(2)}$ now has the dimensions of the inverse of E, and its units are m V$^{-1}$. Similarly, for the case of third-order optical nonlinearity, when we write $P^{(3)}(t) = (\Delta\chi)E$, the value of $(\Delta\chi)$ becomes $(\chi^{(3)}E^2)$, in which the units of $\chi^{(3)}$ are m$^2$ V$^{-2}$, and so on.

**Exercise 14.2.** Estimate the orders of magnitude of the nonlinear optical suscepti-bilities $\chi^{(2)}$ and $\chi^{(3)}$.

**Solution:** The previous discussion ascertained that the orders of magnitude for the higher-order susceptibilities differ by a factor of the magnitude of $E$. For a rough estimate, let us take the magnitude of $E$ to be of the order of $10^{10}$ V m$^{-1}$.

Let us work with $P^{(2)}(t) = \varepsilon_0(\chi^{(2)}E)E \equiv (\Delta\chi)E$. For $\chi^{(2)}$ to have an appreciable effect on the system, the value of $\Delta\chi$ must be comparable with the first term of

**Table 14.1.** Estimated values and units of nonlinear optical susceptibilities.

| Susceptibility | Order | Units |
| --- | --- | --- |
| $\chi_e$ | 1 | Dimensionless quantity |
| $\chi^{(2)}$ | $\sim 10^{-10}$ | $(\text{mV}^{-1})$ |
| $\chi^{(3)}$ | $\sim 10^{-20}$ | $(\text{mV})^{-2}$ |
| $\chi^{(q)}$ | $\sim 10^{-10(q-1)}$ | $(\text{mV})^{-1(q-1)}$ |

equation (14.6). It follows that the minimum value of $(\chi^{(2)}E)$ should be equal to unity. Therefore, the estimated value of $\chi^{(2)}$ must be equal to the inverse of the magnitude of E i.e. $10^{-10}\,\text{mV}^{-1}$.

Similarly, the estimated value of $\chi^{(3)}$ is $10^{-20}\,\text{mV}^{-1}$. The estimated values are summarized in table 14.1.

## 14.4 Characteristics of second-order susceptibility

We now turn to the second term of the nonlinear polarization equation (14.6):

$$P^{(2)}(t) = \varepsilon_0 \chi^{(2)} E^2.$$

We take a typical EM field $(E)$ that has an oscillation frequency of $(\omega)$, given by $E = E_0 \cos \omega t$. By substituting this value of E into equation (14.6) and rearranging, we obtain

$$P^{(2)}(t) = \frac{\varepsilon_0 \chi^{(2)} E_0^2}{2} + \left( \frac{\varepsilon_0 \chi^{(2)} E_0 \cos 2\omega t}{2} \right) E_0. \tag{14.10}$$

The first term of equation (14.10) gives a DC type component (no oscillation as in alternating current (AC)) and the second is an oscillating term with a frequency of $2\omega$. While the DC term is known to give rise to the effect of optical rectification, the second term is responsible for generating the second harmonic $(2\omega)$ of the fundamental frequency $(\omega)$. These effects are briefly described in the next sections.

### 14.4.1 Optical rectification (OR)

Optical rectification (OR) was observed for the first time in 1962. When laser pulses pass through a nonlinear optical medium that has second-order nonlinear susceptibility $\chi^{(2)}$, a DC field is induced across the medium. This phenomenon is analogous to the process of electrical rectification. Figure 14.3 shows a schematic representation of the OR process. In this optical rectification process, no new oscillating field is generated across the $\chi^{(2)}$ material except for a DC voltage. The generation of terahertz frequencies is an important application of this process.

As outlined at the beginning, we know that the high-frequency dielectric constant (or the refractive index) is given by $n^2 = 1 + \chi_e$. Therefore, the change of refractive index $\Delta n$ is related to the change of susceptibility $(\Delta\chi)$ as follows:

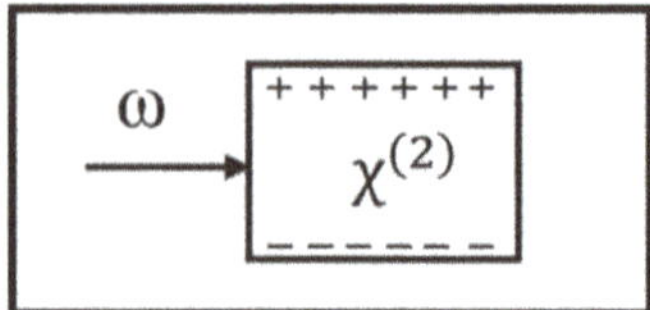

**Figure 14.3.** Schematic representation of the optical rectification (OR) process. An incident field of frequency $\omega$ generates a DC field across the $\chi^{(2)}$ material.

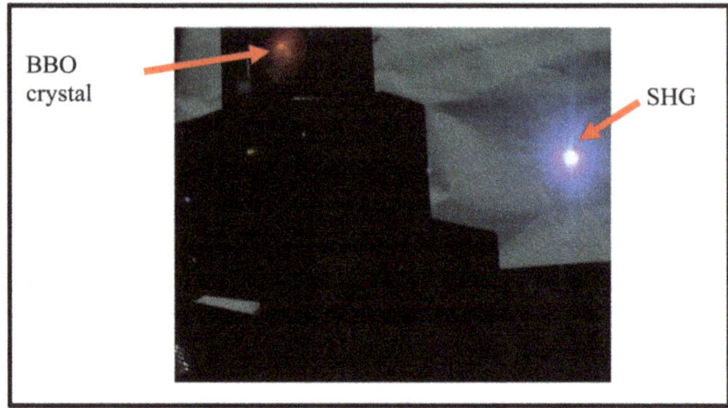


**Figure 14.4.** Photograph of the generation of the second harmonic of Ti:sapphire laser output by a $\beta$-barium borate (BBO) crystal. The bright bluish spot (388 nm) is the second harmonic of the incident IR (776 nm) light, which is visible as a red spot on the BBO crystal) (Sailaja and Bisht, IIT Madras).

$$\Delta n = \frac{\Delta \chi}{2n}. \tag{14.11}$$

The DC component does not lead to any new radiation, but results in a change of refractive index of the $\chi^{(2)}$ material, as indicated by equation (14.11). This is similar to the Pockels effect observed when a DC voltage is applied across birefringent crystals ($\spadesuit$ see section 3.5.1).

### 14.4.2 Second-harmonic generation

From exercise 14.1 and table 14.1, it can be seen that the product of the field amplitude $E$ and $\chi^{(2)}$ becomes comparable to the electrical susceptibility, $\chi_e$ in its magnitude and dimensions. Thus, the second term in equation (14.10) can be compared with equation (14.6). We see that this term contains the equivalent quantity to $\Delta \chi$, i.e. $\frac{\chi^{(2)} E_0 \cos 2\omega t}{2}$. Here, the $\cos(2\omega t)$ factor represents the oscillating term at twice the incident frequency, or SHG. A photograph of SHG due to fundamental femtosecond pulses of a Ti:Sapphire laser generated by a nonlinear optical crystal is shown in figure 14.4. SHG was first reported by Franken and coworkers in 1961.

SHG can be visualized as a light–matter interaction process in terms of an exchange of photons between the frequency components of the field, as shown in figure 14.5.

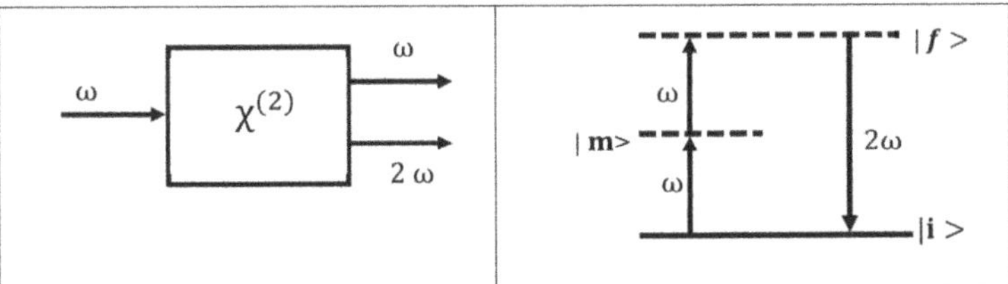



**Figure 14.5.** (left-hand panel): SHG process in a $\chi^{(2)}$ crystal; $\omega$ is the fundamental frequency. The right-hand panel contains an energy-level diagram with an initial level |i> and two virtual levels, namely the intermediate level (|m>) and the final level (|f>) (dashed lines).

## 14.5 Virtual levels

According to the photon picture illustrated in the right-hand panel of figure 14.5, two photons of frequency $\omega$ are destroyed. They simultaneously create a photon of frequency $2\omega$ in a single quantum-mechanical process. The dashed horizontal lines in the figure represent the *virtual levels,* as explained below.

On interaction with the incident photon, the electrons in an atom can oscillate about their equilibrium position. This process does not involve the absorption of photons or a change to the more excited levels. The incident photon can therefore prepare a molecule that can act as a dipole and oscillate with a tiny change of its polarizability. Due to this change of polarizability, other events can take place in the ground state, such as Rayleigh/Raman scattering. These processes are said to be taking place through *virtual levels*. The virtual levels are not the eigenstates of an atom alone, i.e. they correspond to no specific $n$ or $l$ states of an atom (♠ see chapter 2). These states are the eigenstates of the combined system, i.e. the atom and the photon together. Virtual states are also understood to be imaginary states for which the excited state lifetime is negligible in contrast to the case for eigenstates. In other words, the closer the virtual state is to an actual eigenstate (or a real energy level), the longer the lifetime of the virtual state.

## 14.6 Linear and nonlinear optics

With the introduction of pulsed lasers into optics, electric fields of the order of gigawatts or beyond became accessible. In fact, we now have laser pulses at petawatt power levels. Such incident EM fields are capable of interacting with the electronic and nuclear configurations of atoms and molecules and of imparting perturbations such that light can control matter in a disciplined manner.

The exciting field of nonlinear optics opens up the area of 'modern optics'. The rest of the chapters in this section are devoted to this exotic field. In summary, some of the differences between *linear* and *nonlinear* optics are listed in table 14.2.

### Questions and problems

1. The intensity ($I$) or the power radiated per unit area of a light source is given by $= \frac{1}{2}\epsilon_0 E^2 c$. A laser has power of 30 mW and a beam cross-section of 10 mm$^2$. Calculate the magnitude of $E$ corresponding to the laser beam.

**Table 14.2.** Differences between *linear* and *nonlinear* optics.

| Linear optics | Nonlinear optics |
|---|---|
| **(i)** The refractive index ($n$) and the absorption coefficient ($\alpha$) are independent of light intensity | **(i)** $n$ and $\alpha$ can change with light intensity |
| **(ii)** The principle of superposition holds for light beams. | **(ii)** The principle of superposition is not followed. |
| **(iii)** The frequency of light cannot be altered by the medium. | **(iii)** Light can alter its frequency as it passes through a medium. |
| **(iv)** Light cannot interact with light. Two beams crossing each other cannot have any effect on each other in a medium— i.e. light cannot control light. | **(iv)** Photons do interact with each other. Light can control light. |

2. A laser delivers pulses of 100 fs duration at a repetition rate of 1 KHz. The average power measured by a power meter is 100 mW. What is the peak power of the laser pulse?

3. A dielectric can behave as a nonlinear optical medium, provided that the field strengths are higher than the interatomic fields. Write down the equation for the induced polarization ($P$) up to first two terms in tensor form as a function of the incident electric field ($E$).

4. Electrical susceptibility $\chi_e$ is a *dimensionless* quantity. What are the dimensions of $\chi^{(2)}$ and $\chi^{(3)}$?

5. For systems with inversion symmetry, $\chi^{(2)}$ is identically zero. Comment on this statement.

6. **(i)** How do you relate the Pockels effect to the phenomenon of optical rectification? **(ii)** What is the mechanism by which the polarization of incident light is changed using a Pockels cell?

7. Derive equation (14.5) by using Maxwell's equations in a medium.

# Bibliography


[1] Shen Y R 2003 *The Principles of Nonlinear Optics* (New York: Wiley Interscience)

[2] Boyd R W 2003 *Nonlinear Optics* 2nd edn (New York: Academic)

[3] Bass M, Franken P A, Ward J F and Weinreich G 1962 Optical rectification *Phys. Rev. Lett.* **9** 446–8

[4] Bloembergen N, Chang R K, Jha S S and Lee C H 1968 Optical second-harmonic generation in reflection from media with inversion symmetry *Phys. Rev.* **174** 813–22

[5] Burresi M, van Oosten D, Kampfrath T, Schoenmaker H, Heideman R, Leinse A and Kuipers L 2009 Probing the magnetic field of light at optical frequencies *Science* **326** 550–3

[6] Franken P A, Hill A E, Peters C W and Weinreich G 1961 Generation of optical harmonics *Phys. Rev. Lett.* **7** 118–9

[7] Tomasino A, Parisi A, Stivala S, Livreri P, Cino A C, Busacca A C and Morandotti R 2013 Wideband THz time domain spectroscopy based on optical rectification and electro-optic sampling *Sci. Rep.* **3** 3116

**IOP** Publishing

# An Introduction to Photonics and Laser Physics with Applications

**Prem B Bisht**

# Chapter 15

# Second-order susceptibility, phase matching, and applications

We saw in the previous chapter how the intense EM field associated with lasers can invoke optical nonlinearity in a material. In this chapter, second-order nonlinearity is described in detail by highlighting its applications. The techniques used to generate tunable coherent radiation covering the UV–vis–IR region, based on the $\chi^{(2)}$ phenomenon, are: optical parametric oscillation (OPO) and optical parametric amplification (OPA). The principles of optical parametric generation (OPG), super-fluorescence, and conical emission (CE) are also described in this chapter. The figure shows the generation of entangled photons ($\omega_s$ and $\omega_i$) by pumping a $\chi^{(2)}$ material with photons $\omega_p$ —the details are covered in this chapter.

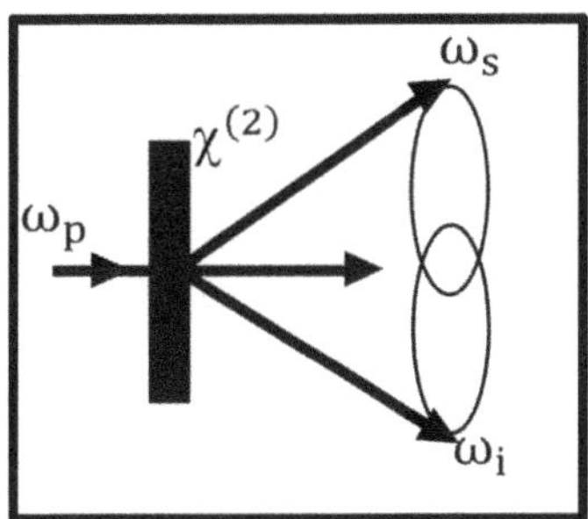


**Learning objectives**
**After reading this chapter, the learner will be able to:**
Describe the three-wave mixing processes and the matrix representation of $\chi^{(2)}$;
Define signal and idler photons;

doi:10.1088/978-0-7503-5226-0ch15



© IOP Publishing Ltd 2022

Define the phase-matching conditions;
Distinguish between angle-tuned, temperature-tuned, and quasi phase-matching conditions;
Understand the phenomenon of parametric amplification;
Assemble an optical parametric amplifier (OPA);
Explain the phenomenon of superfluorescence;
Recognize the phenomenon of photon entanglement.

## 15.1 Sum- and difference-frequency generation

Optical nonlinearity in a material is invoked by the EM fields of pulsed lasers with pulse durations on the nanosecond to femtosecond timescale. Several nonlinear optical (NLO) materials exist, such as $\beta-$ barium borate (BBO) crystal and potassium trihydrogen phosphate (KTP) crystal, which have high $\chi^{(2)}$ values. On order to obtain a glimpse of the exotic phenomenon of the nonlinear interaction of light with NLO crystals, we take the incident electric field expression in Euler's form. Using $E(t) = E_0 e^{-i\omega t} + E_0^* e^{i\omega t}$, the second-order nonlinear polarization term in equation (14.6) i.e. $P^{(2)}(t) = \varepsilon_0 \chi^{(2)} E^2(t)$ gives

$$P^{(2)}(t) = \varepsilon_0 \chi^{(2)} E_0 E_0^* + \varepsilon_0 \chi^{(2)} E_0^2 e^{-i(2\omega t)} + \text{Complex conjugate (CC) term}$$

Here, the first term is a DC term, and the second term is responsible for second-harmonic generation (SHG). Let us assume that two frequencies $\omega_1$ and $\omega_2$ are incident on the NLO crystal at the same time and the same place (i.e. with a temporal and spatial match). For the two electric fields $E_1$ and $E_2$ incident on the crystal, the net incident field is given by

$$E(t) = (E_1 e^{-i\omega_1 t} + E_1^* e^{i\omega_1 t}) + (E_2 e^{-i\omega_2 t} + E_2^* e^{i\omega_2 t}).$$

As a result of the interaction between this field and the crystal dimensions (and also as a function of position ($\vec{r}$)), we can expect to observe the SHG of both of the fundamental frequencies ($\omega_1$ and $\omega_2$); we can write $P^{(2)}(r, t)$ as

$$P^{(2)}(r, t) = \varepsilon_0 \chi^{(2)} \left[ \left( 2E_1 E_1^* + 2E_2 E_2^* \right) + \left( E_1^2 e^{-2i\omega_1 t} + E_1^{*2} e^{2i\omega_1 t} + E_2^2 e^{-2i\omega_2 t} + E_2^{*2} e^{2i\omega_2 t} \right) \right.$$
$$\left. + \left( 2E_1 E_2 e^{-i(\omega_1+\omega_2)t} + 2E_1^* E_2^* e^{i(\omega_1+\omega_2)t} \right) + \left( 2E_1 E_2^* e^{-i(\omega_1-\omega_2)t} + 2E_1^* E_2 e^{i(\omega_1-\omega_2)t} \right) \right] \tag{15.1}$$

It can be seen that the first two terms $(2E_1 E_1^* + 2E_2 E_2^*)$ are again the DC terms. These correspond to the process of optical rectification (OR), which gives rise to charges on the surface of the $\chi^{(2)}$ material.

The third and fifth terms, respectively, are the second harmonics of the two fundamental frequencies ($2\omega_1$ and $2\omega_2$, respectively). In addition, we also have the sum of the two frequencies (in the seventh term), known as the sum-frequency generation (SFG) process. The ninth term is the difference between the two frequencies and is therefore called the difference-frequency generation (DFG) process. These processes are schematically shown in figure 15.1.

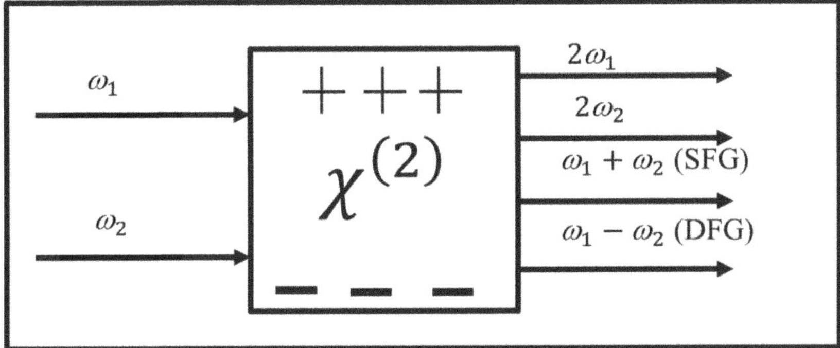


**Figure 15.1.** Various nonlinear optical processes can take place when two frequencies $\omega_1$ and $\omega_2$ interact with a $\chi^{(2)}$ material. The positive and negative charges on the surface of the material are due to the OR process.

♣ The OR process was outlined in the previous chapter (♠ section 14.6). It is the working principle of the AC Pockels effect used in electro-optic beam deflection, which has applications in industry in mode lockers and optical switches.

### 15.1.1 Three-wave mixing processes

It should be noted that the generated frequencies depend on the phase-matching condition (♠ see section 15.4). As there are two input beams along with a third beam generated for a particular orientation of a nonlinear optical material, it is also known as *three-wave mixing* process. The processes based on $\chi^{(2)}$ are as follows: OR, SHG, and SFG, optical parametric generation (OPG), optical parametric oscillation (OPO), and optical parametric amplification (OPA). The first three processes were described in the previous section, and the other processes are briefly described in forthcoming sections of this chapter.

**Exercise 15.1.** Differentiate between the processes of SFG and SHG. How do you generate the third harmonic using a $\chi^{(2)}$ material?

**Solution:** The SFG is nonlinear optical process that takes place when two frequencies $\omega_1$ and $\omega_2$ give rise to the sum of the two frequencies ($\omega_1 + \omega_2$) as indicated in figure 15.1. As a special case, when $\omega_1 = \omega_2$ ($= \omega$, when the two beams are derived from the same laser), the generated signal is known as SHG and has a frequency of $2\omega$.

## 15.2 Signal and idler photons

From the point of view of energy conservation, in the SFG process, the two incident photons with frequencies $\omega_1$ and $\omega_2$ are destroyed and a photon with a frequency ($\omega_3$) equal to the sum of these frequencies ($=\omega_1 + \omega_2$) is created (see figure 15.2). This corresponds to the seventh term in equation (15.1).

On the other hand, to obtain the difference between the frequencies (the ninth term in equation 15.1), the pump photon, which has a frequency of $\omega_1$, is destroyed to create two photons that have frequencies $\omega_2$ and $\omega_3$ such that the difference

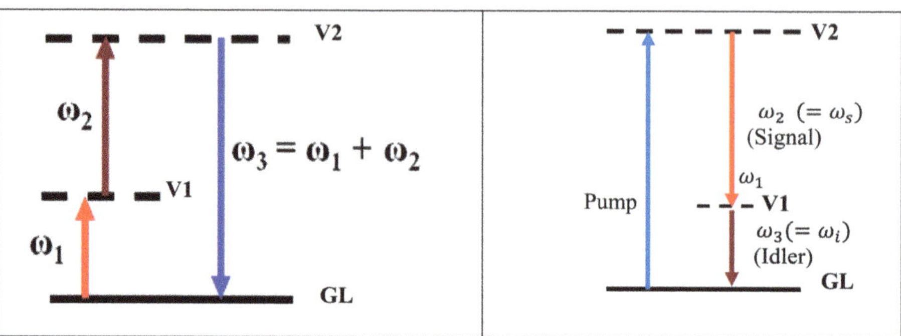

**Figure 15.2.** The processes of SFG (left panel) and DFG (right panel) with two incident photons that have frequencies of $\omega_1$ and $\omega_2$. V1 and V2 are virtual levels and GL is the ground state.

frequency $(\omega_1 - \omega_2)$ is equal to the new frequency $\omega_3$. Conventionally, in the DFG process, the higher-frequency created photon is known as the *signal* photon (or $\omega_s$) and that associated with the lower frequency (or the higher wavelength) is known as the *idler* photon (or $\omega_i$). This process of the creation of signal and idler photons of lower frequencies from a single pump photon is also known as *parametric downconversion* and is used in quantum optics[1] for studies of the entanglement process ( ♠ see section 15.7).

♣ As an aside, if the frequency $\omega_s$ is chosen as one the incident frequencies (= $\omega_2$), there are interesting consequences, as discussed in the next section (15.6).

## 15.3 Properties of, and contracted notation for, $\chi^{(2)}$

In the expression for the second-order polarization $P_i^{(2)}(r, t) = \varepsilon_0 \chi_{ijk}^{(2)} \, E_j E_k$, the quantity $\chi_{ijk}^{(2)}$ is a third-rank tensor with 27 components.

♣ A third-rank tensor has to be written as a triad in three 'pages' of 3×3 matrices in a cuboid appearance, in which the index $k$ appears respectively in each 'page' as 1, 2, or 3 as the third index of an element, while the indices $i$ and $j$ respectively represent the rows and columns of each 'page' as the first and the second indexes of the all the elements in all the pages. In the following, the first three columns represent 'page 1' of the triad, and so on.

$$\begin{bmatrix} \chi_{111}^{(2)} & \chi_{121}^{(2)} & \chi_{131}^{(2)} & \chi_{112}^{(2)} & \chi_{122}^{(2)} & \chi_{132}^{(2)} & \chi_{113}^{(2)} & \chi_{123}^{(2)} & \chi_{133}^{(2)} \\ \chi_{211}^{(2)} & \chi_{221}^{(2)} & \chi_{231}^{(2)} & \chi_{212}^{(2)} & \chi_{222}^{(2)} & \chi_{232}^{(2)} & \chi_{213}^{(2)} & \chi_{223}^{(2)} & \chi_{233}^{(2)} \\ \chi_{311}^{(2)} & \chi_{321}^{(2)} & \chi_{331}^{(2)} & \chi_{312}^{(2)} & \chi_{322}^{(2)} & \chi_{332}^{(2)} & \chi_{313}^{(2)} & \chi_{323}^{(2)} & \chi_{333}^{(2)} \end{bmatrix}. \tag{15.2}$$

---

[1] To prove Bell's inequalities, see [9].

♣ You might also come across the above matrix in the following form:

$$\begin{bmatrix} \chi^{(2)}_{111} & \chi^{(2)}_{121} & \chi^{(2)}_{131} \\ \chi^{(2)}_{211} & \chi^{(2)}_{221} & \chi^{(2)}_{231} \\ .. & .. & .. \\ \chi^{(2)}_{213} & \chi^{(2)}_{223} & \chi^{(2)}_{233} \\ \chi^{(2)}_{313} & \chi^{(2)}_{323} & \chi^{(2)}_{333} \end{bmatrix}.$$

In index notations, we can choose to define the components of $\chi^{(2)}$ as a function of the incident frequencies for three-wave mixing. For example, for the case of SFG, the sum frequency ($\omega_3$) of $\omega_1$ and $\omega_2$ from equation (14.8), we can write a typical second-order nonlinear polarization term, as follows:

$$P_i^{(2)}(\omega_3) = \epsilon_0 \sum_j \sum_k (\chi^{(2)}_{ijk}(\omega_3; \omega_1, \omega_2) E_j(\omega_1) E_k(\omega_2)).$$

The subscripts $ijk$ represent the Cartesian components of the vector fields. It should be noted that here, the quantity $\chi^{(2)}_{ijk}(\omega_3; \omega_1, \omega_2)$ also includes $\chi^{(2)}_{ijk}(\omega_3; \omega_2, \omega_1)$ by exchanging the incident frequencies, even though the result is unchanged.

♣ You can easily write the above equation for the case of second-harmonic generation.

Consider the SFG and DFG, as follows:

$$\omega_3 = \omega_2 + \omega_1;\ \omega_1 = \omega_3 - \omega_2,\ \omega_2 = \omega_3 - \omega_1.$$

The six tensors of $\chi^{(2)}_{ijk}$ are

$$\chi^{(2)}_{ijk}(\omega_3; \omega_1, \omega_2),\ \chi^{(2)}_{ijk}(\omega_3; \omega_2, \omega_1),\ \chi^{(2)}_{ijk}(\omega_1; \omega_3, -\omega_2),\ \chi^{(2)}_{ijk}(\omega_1; -\omega_2, \omega_3),\ \chi^{(2)}_{ijk}(\omega_2; \omega_3, -\omega_1),$$

and $\chi^{(2)}_{ijk}(\omega_2; -\omega_1, \omega_3)$.

Six more tensors arise from their complex conjugates. Each of these tensors has 27 components, making a total of 324 components. Therefore, it is timely to discuss the following properties and the reducible representation of $\chi^{(2)}_{ijk}$ in general.

    (i) Reality of fields

As $P_i^{(2)}(\omega_3)$ is a physically measurable quantity, it is real. Using a similar approach to that used for the symmetry of linear susceptibility $\chi_{ij}$, we can express the reality of fields condition as $\chi^{(2)*}_{ijk}(\omega_3; \omega_1, \omega_2) = \chi^{(2)}_{ijk}(-\omega_3; -\omega_1, -\omega_2)$. This is also true for SHG, as $\chi^{(2)*}_{ijk}(2\omega; \omega, \omega) = \chi^{(2)}_{ijk}(-2\omega; -\omega, -\omega)$.

    (ii) Intrinsic permutation symmetry

The positive and negative components are related by $P_i^{(2)}(-\omega_3) = P_i^{(2)*}(\omega_3)$; this relation also hold true for the positive and negative components of the incident fields $E_j(-\omega_1) = E_j(\omega_1)^*$ and $E_k(-\omega_2) = E_k(\omega_2)^*$. $P_i^{(2)}(\omega_3)$ is the product $\chi^{(2)}_{ijk}(\omega_3; \omega_1, \omega_2) \times E_j(\omega_1) E_k(\omega_2)$ and $j, k$ 1, 2, 3, respectively, are dummy indices.

Therefore, for convenience, we can write the produce by interchanging $j$ and $k$ followed by the simultaneous interchange of one and two, as follows: $\chi_{ijk}^{(2)}(\omega_3; \omega_1, \omega_2) \times E_j(\omega_1)E_k(\omega_2)$ and $\chi_{ikj}^{(2)}(\omega_3; \omega_2, \omega_1) \times E_k(\omega_2)E_j(\omega_1)$. As a consequence, the latter two products are numerically equal and the action sequence is immaterial.

(iii) Whole permutation symmetry

For non-resonant low-frequency ($\omega_0$) excitation, the material response is instantaneous and does not involve any energy exchange between the light and the media (such as an absorption process). Here, the Cartesian indices are to be interchanged freely with a condition that the corresponding frequencies are permuted. For instance, for the case of $\chi_{ijk}^{(2)}$,

$$\chi_{ijk}^{(2)}(\omega_3 = \omega_1 + \omega_2) = \chi_{jki}^{(2)}(-\omega_1 = \omega_2 - \omega_3).$$

The complex conjugate of RHS can now be written as $\chi_{jki}^{(2)*}(\omega_1 = -\omega_2 + \omega_3) = \chi_{jki}^{(2)}(-\omega_1 = \omega_2 - \omega_3)$.

Therefore, using condition (i), we can also write this for lossless media as

$$\chi_{ijk}^{(2)}(\omega_3 = \omega_1 + \omega_2) = \chi_{jki}^{(2)}(-\omega_1 = \omega_2 - \omega_3).$$

The first frequency argument is the sum of the latter two, therefore when interchanging the Cartesian indices, the sign of the frequency must also be taken care of.

**Exercise 15.2** Show that $\chi_{ijk}^{(2)}(\omega_3 = \omega_1 + \omega_2) = \chi_{kij}^{(2)}(\omega_2 = \omega_3 - \omega_1)$.

**Solution:** We know that in order to change the frequency components, we need to exchange the corresponding indices by taking care of the sign, so

$$\chi_{ijk}^{(2)}(\omega_3 = \omega_1 + \omega_2) = \chi_{kij}^{(2)}(-\omega_2 = \omega_1 - \omega_3)$$

$$= \chi_{kij}^{(2)*}(\omega_2 = -\omega_1 + \omega_3) \text{ (Complex conjugate)}$$

$$= \chi_{kij}^{(2)}(\omega_2 = \omega_3 - \omega_1) \text{ (Reality of fields for lossless media).}$$

(iv) Kleinman conjecture

If the nonlinear optical effects are the result of electron displacement only, without other relaxation effects due to molecules, ions, or nuclei, then for such a lossless media, the dispersion of $\chi^{(2)}$ can be ignored. In this case, all the components of the susceptibility tensor are taken to be real. In addition to conditions (i) and (ii), according to this symmetry condition, if the corresponding Cartesian indices associated with the frequency components are interchanged simultaneously, all the frequency arguments of $\chi^{(2)}$ can be freely interchanged while the values remain constant.

$$\chi_{ijk}^{(2)}(\omega_3 = \omega_1 + \omega_2) = \chi_{jik}^{(2)}(\omega_3 = \omega_1 + \omega_2) = \chi_{kij}^{(2)}(\omega_3 = \omega_1 + \omega_2).$$

This leads to the conclusion that even the $n$th-order nonlinear susceptibility $\chi^{(n)}$ tensors become independent of the frequency. Under such conditions, since the medium is lossless, the nonlinear optical medium is said to follow the full permutation symmetry of Kleinman symmetry.

Under the Kleinman conjecture, the number of components (27) of the $\chi^{(2)}$ tensor can be reduced, as described below. To do so, let us work with the $\chi_{ijk}^{(2)}$ matrix components given above as equation (15.2). Here, let us introduce new matrix elements $d_{il}$ such that $d_{il} \equiv d_{ijk} = \frac{1}{2}\epsilon_0\chi_{ijk}^{(2)}$. This newly introduced matrix element $d_{il}$ is such that the index $i$ is taken as 1, 2, 3, and the values of indices $jk$ are taken as 11, 22, 33, 23, 32, 31, 13, and 12, 21. If we now take the new index $l = 1, 2, 3, 4, 5, 6$; a $3 \times 6$ matrix is obtained, as shown in equation (15.3). We write $d_{il}$ with 18 components as follows:

$$d_{il} = \begin{vmatrix} d11 & d12 & d13 & d14 & d15 & d16 \\ d16 & d22 & d23 & d24 & d25 & d26 \\ d15 & d24 & d33 & d23 & d13 & d36 \end{vmatrix}. \tag{15.3}$$

As can be seen, here $d12 = d122 = d212 = d26$ and $D14 = d123 = d213 = d25$. Using this, the second-order polarization expression, for example, can be written as

$$\begin{bmatrix} P_x(2\omega) \\ P_y(2\omega) \\ P_z(2\omega) \end{bmatrix} = \begin{vmatrix} d11 & d12 & d13 & d14 & d15 & d16 \\ d16 & d22 & d23 & d24 & d14 & d12 \\ d15 & d24 & d33 & d23 & d13 & d14 \end{vmatrix} \begin{bmatrix} E_x(\omega)^2 \\ E_y(\omega)^2 \\ E_z(\omega)^2 \\ 2E_y(\omega)E_z(\omega) \\ 2E_x(\omega)E_z(\omega) \\ 2E_x(\omega)E_y(\omega) \end{bmatrix}.$$

By applying the argument of the Kleinman conjecture, we find that the matrix $d_{il}$ in equation (15.3) has only ten independent components, namely d11, d22, and d33 and seven other components that appear twice.

## 15.4 Conditions for refractive-index matching

According to Cauchy's formula (♠ see section 1.4.6), the refractive index of a material varies with wavelength. As discussed in chapter 3 (♠ see section 3.4.5, figure 3.14), in birefringent materials there are, in general, two mutually perpendicular modes of propagation for each frequency. One of the modes (the o-wave) experiences a constant value of the refractive index known as the ordinary refractive index ($n_0$) and is represented as a circle. The other mode (the e-wave) also experiences a variation in the refractive index with respect to the optic axis (as a function of an angle $\theta$ with the optic axis). The latter is known as the extraordinary refractive index ($n_e$) and is represented as an ellipse. The wavelength dependences of both indices are shown in figure 15.3.

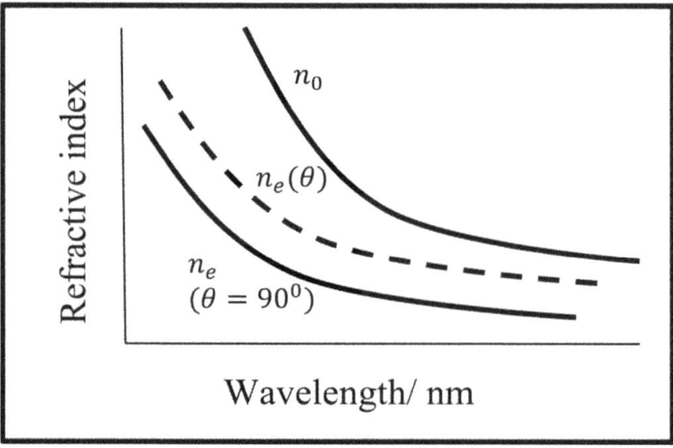


**Figure 15.3.** Representative diagram of the wavelength dependence of $n_0$ and $n_e$ for a negative uniaxial crystal. The values of $n_e$ vary with the angle ($\theta$) made with the optic axis.

### 15.4.1 Angle dependence of the extraordinary refractive index

Crystalline materials are characterized by a particular direction called the $C$-axis, also known as the optic axis (OA). Let us take an example of a positive uniaxial crystal in which the o-wave is polarized perpendicular to the OA in the $x$–$y$ plane. The e-wave is perpendicular to the incident wave, but makes an angle with the optic axis. The angle dependence of $n_e$ as a function of the phase-matching angle ($\theta$) with the $C$-axis, in terms of the index ellipsoid (or indicatrix) symmetric about the $C$-axis, is given by

$$\frac{1}{n_e^2(\omega,\,\theta)} = \frac{\cos^2\theta}{n_o^2(\omega)} + \frac{\sin^2\theta}{n_e^2(\omega)}. \tag{15.4}$$

From equation (15.4), we can see that for $\theta = 0°$, $n_e(\theta) = n_o$ and for $\theta = 90°$, $n_e(\theta) = n_e$. The required phase-matching angle ($\theta_{pm}$) for SHG ($2\omega$) for a positive uniaxial crystal for which $n_e > n_0$ can be obtained as follows (♠ see question 7):

$$\sin^2\theta_{pm} = \frac{n_o^{-2}(\omega) - n_o^{-2}(2\omega)}{n_0^{-2}(2\omega) - n_e^{-2}(\omega)}. \tag{15.5a}$$

For a negative uniaxial crystal, $\theta_{pm}$ is given by

$$\sin^2\theta_{pm} = \frac{n_o^{-2}(\omega) - n_o^{-2}(2\omega)}{n_e^{-2}(2\omega) - n_o^{-2}(2\omega)}. \tag{15.5b}$$

### 15.4.2 Phase matching for SHG in a crystal

Let us take the well-known case of the SFG of $\omega_1$ and $\omega_2$, which generates a third frequency $\omega_3$ (see figure 15.2). The momenta ($k_1$, $k_2$, and $k_3$) associated with the

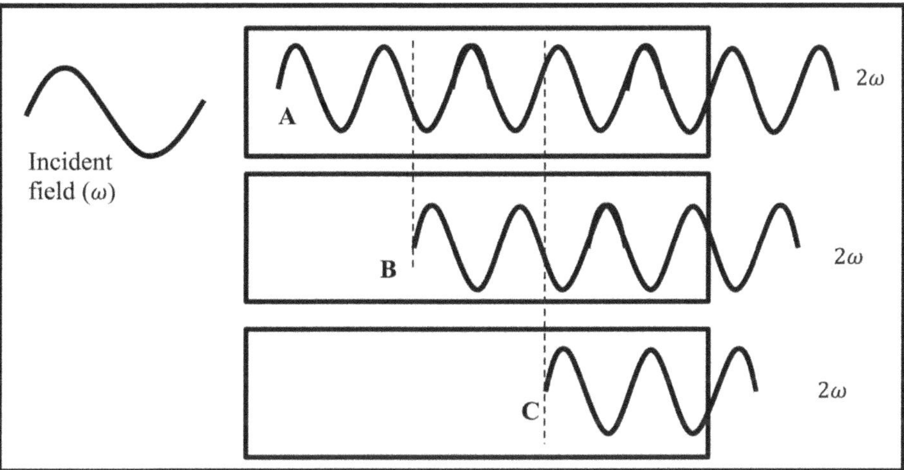



**Figure 15.4.** The second harmonic ($2\omega$) is generated at three points A, B, and C inside an NLO crystal on propagation of the incident field ( $\omega$). The vertical dotted lines indicate the locations of the waves inside the crystal.

corresponding frequencies ($\omega_1$, $\omega_2$, and $\omega_3$) give the following relation under the conservation of momentum for *three-wave mixing*:

$$k_1 + k_2 = k_3; \ \text{ or } \ \Delta k = 0.$$

Using $k = \frac{\omega}{c}$, this can be rewritten as $\frac{\omega_1 n_1}{c} + \frac{\omega_2 n_2}{c} = \frac{\omega_3 n_3}{c}$. When both incident frequencies are same, i.e. $\omega_1 = \omega_2$ (say $= \omega_1$), we can write a relation for SHG as follows:

$$\frac{2\omega_1 n_1}{c} = \frac{\omega_3 n_3}{c}.$$

Considering the second harmonic of frequency $\omega_1$ to be $\omega_3$ ($= 2\omega_1$), the above equation implies that $n_1 = n_3$. This indicates that for SHG, the refractive index of the fundamental (or pump) frequency and that of the generated frequency (the second harmonic) are the same.

The incident fundamental field ($\omega$) propagating inside an NLO crystal is shown at three points A, B, and C in figure 15.4. At points A and C, the generated SHG ($2\omega$) signal is in phase, while at point B it is out of phase. However, to obtain good efficiency, the second harmonics generated throughout the crystal must be in phase with each other as well as with the incident fundamental. This condition can be met when the speeds of both frequencies are identical inside the crystal. This is only possible when (i) the refractive index of the fundamental ($n_0(\omega)$) is equal to that of the SHG $n_e(2\omega)$, so that (ii) the frequencies $\omega$ and $2\omega$ have polarizations mutually perpendicular to each other. This explains why the polarization of the generated SHG in such a case is perpendicular to that of the incident fundamental beam.

As a consequence, the speeds of the two waves inside the NLO medium are equal. The condition of the identical value of the refractive index is satisfied in birefringent

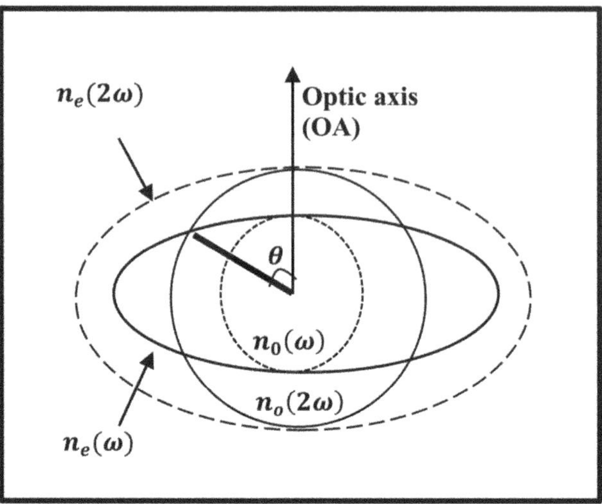


**Figure 15.5.** The phase-matching condition $n_0(2\omega) = n_e(\omega)$ is satisfied at an angle of $\theta$ with the OA where the circle and the ellipse intersect each other.

materials when the $n_0$ value of the fundamental beam is equal to that of $n_e(\theta)$, as shown in figure 15.5.

### 15.4.3 Effects of the length and area of nonlinear crystals

Estimating the efficiency of SHG of a crystal is important. It depends on geometrical parameters such as the length and the cross-section of the crystal as well as on the incident beam parameters.

The second-harmonic power is found to be proportional to square of the length ($L$) of the crystal. It is also inversely proportional to the cross-sectional area ($A$) of the fundamental beam incident on the crystal. From equation (15.1), we see that the second term in the polarization equation is proportional to the square of the electric field. This clearly indicates that the power of SHG is proportional to the square of the fundamental power as well as that of $\chi^{(2)}$.

Let us consider a crystal of thickness $z$ with an element $dz$ only along the propagation length, as shown in figure 15.6. As $E \propto \exp(i(k_1 z - \omega t))$, the power corresponding to the SHG for an element $dz$ can be written as

$$P_{\text{SHG}} \, dz \propto \exp(2i(k_1 z - \omega t))dz.$$

For the second-harmonic frequency and a wave vector $k_2$, the electric field due to SHG (in terms of the wave vector $k_1$) is given by

$$dE_z^{(2)} \propto \exp(2i(k_1 z - \omega t))dz. \tag{15.6}$$

To find the SHG radiation with wave vector $k_2$ produced at the exit surface of the crystal (with thickness elements $dz$) we can express $dE_L^{(2)}$ (at $z' = L - z$) as

$$dE_L^{(2)} \propto dE_z^{(2)} \exp(i(k_2(L - z))dz.$$

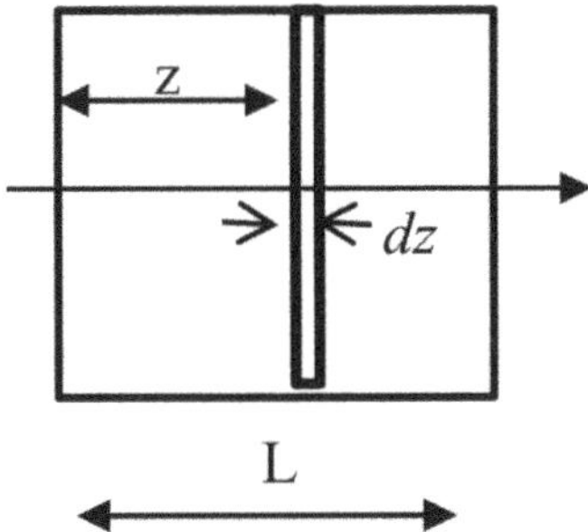

**Figure 15.6.** Schematic of a nonlinear optical crystal of length $L$ showing the element $dz$ contributing to SHG.

Using the value of $dE_z^{(2)}$ from equation (15.6) and assuming that the propagation power does not change inside the crystal,

$$dE_L^{(2)} \propto \exp(2i(k_1 z - \omega t))dz \quad \exp(i(k_2(L - z)))$$

$$\propto \exp(i(2k_1 - k_2))z \times \exp(i(k_2 L - 2\omega t))dz$$

or

$$E_L^{(2)} \propto \int_0^L \exp(i(2k_1 - k_2))z \times \exp(i(k_2 L - 2\omega t))dz.$$

As the first factor is a constant, we can find

$$E_L^{(2)} \propto \frac{(e^{i(2k_1 - k_2))L}) - 1}{i(2k_1 - k_2)}.$$

On simplification and taking the real part of the proportionality factor,

$$E_L^{(2)} \propto \frac{\sin\left(\frac{2k_1 - k_2}{2}\right)L)}{\frac{2k_1 - k_2}{2}L} \equiv \frac{\sin\left(\frac{\Delta k L}{2}\right)}{\left(\frac{\Delta k L}{2}\right)}. \tag{15.7}$$

Here, we have replaced $2k_1 - k_2$ with $\Delta k$. Similarly, we can write the equations for the SFG process. Equation (15.7) shows that the SHG field will be maximal when

$$L = \frac{\pi}{2} \times \left(\frac{2}{\Delta k}\right) = \frac{\pi}{\Delta k}. \tag{15.8}$$

This length value, $\left(\frac{\pi}{\Delta k}\right)$, is known as the *coherence length* $(L_c)$ for the SHG. Within $L_c$, the crystal provides the maximum conversion efficiency. Increasing the length beyond this value does *not* increase the SHG intensity.

♣ The electric field corresponding to the phase-matched power of the second harmonic is given by $E_z^{(2)} \propto \sqrt{P_{\text{SHG}}}\, dz \propto \int_0^L e^{-i\Delta k}dz$ ; and $\int_0^L e^{-i\Delta k}dz = \frac{e^{-i\Delta k L} - 1}{i\Delta k}$.

The phase-matching factor for the intensity (and the power) is given by

$$P_{\text{SHG}} \propto \left| \left[ \frac{L^2}{L^2} \frac{(e^{-i\Delta kL} - 1)(e^{i\Delta kL} - 1)}{(\Delta k)^2} \right] \right| \propto L^2 \frac{\sin^2\left(\frac{\Delta kL}{2}\right)}{\left(\frac{\Delta kL}{2}\right)^2}. \tag{15.9}$$

For $\Delta kL \to 0$, the expression will tend to one. Here, length cannot be zero, hence $\Delta k = 0$. The final expression for the obtained SHG power in terms of the power of the fundamental beam ($P_f$) along with the length and cross-sectional area of the crystal is given by

$$P_{\text{SHG}} \approx \frac{L^2 P_f^2}{A} \, \{\chi^{(2)}\}^2 \, \left( \frac{\sin^2\Delta\phi}{\Delta\phi^2} \right). \tag{15.10}$$

Here, $\Delta\phi = \frac{\Delta kL}{2}$ and the term in the bracket is known as the *phase-mismatch factor*. The power is maximal when $\Delta\phi = \frac{\pi}{2}$. Figure 15.6 gives a plot of this expression. The efficiency of the SHG is given as follows:

$$\eta = \frac{P_{\text{SHG}}}{P_f} \approx \frac{I_{\text{SHG}}}{I_f} = \frac{I_f}{A} \, \{\chi_{ij}^{(2)}\}^2 \, L^2 \left( \frac{\sin^2\Delta\phi}{\Delta\phi^2} \right).$$

We remind the reader here that $\chi_{ij}^{(2)}$ is a quantity that has units of mV$^{-1}$.

We know that due to high reflectivity requirements, the output coupler of a laser transmits only a fraction of the power available in the laser cavity. The intra-cavity power in the laser cavity is large compared to the output power of any laser. Therefore, to obtain better efficiency, second-harmonic-generating crystals are kept inside the laser cavity. For example, in an argon-ion laser, UV light at 229 nm can be obtained in this manner with high efficiency. Similarly, in green laser pointers, the NLO crystal is installed inside the tiny laser cavity of a Nd:yttrium aluminum garnet (YAG) laser.

As can be seen from equations (15.9) and (15.10), the conversion efficiency of a nonlinear process (e.g. SHG) is highly dependent on the phase-mismatch factor. It can be seen from figure 15.7 that the efficiency of the nonlinear interaction is maximized when $\Delta k$ is equal to zero. This is known as the phase-matching condition, which literally means '*equal phase velocities of interacting waves in a nonlinear medium*'. The fields emitted by each dipole match in phase and add up inside the crystal as indicated in figure 15.4 above.

**Exercise 15.3.** In SHG generation, how does the efficiency of SHG changes in following cases: (i). if only the radius of the incident laser beam is doubled? (ii) if only the length of the crystal is doubled?
    **Solution:**
      (i) From equation (15.10), it can be seen that the SHG power is inversely proportional to the area of the fundamental laser beam. Therefore, keeping

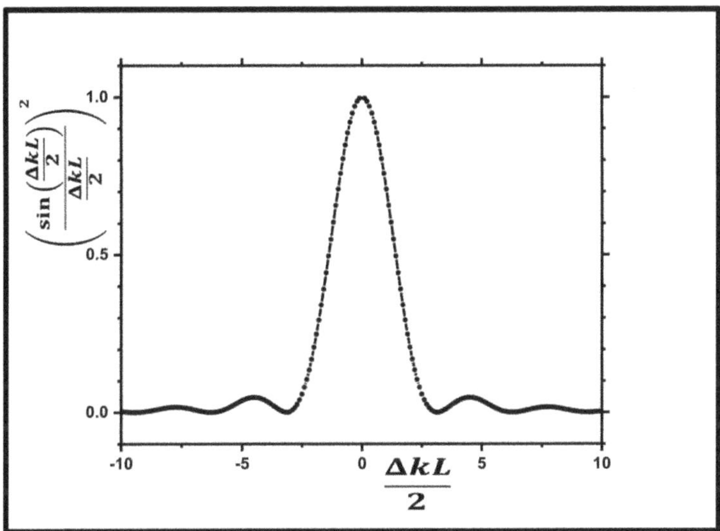

**Figure 15.7.** Plot of the phase-mismatch factor $\left(\sin\left(\frac{\Delta kL}{2}\right)\big/\left(\frac{\Delta kL}{2}\right)\right)^2$ versus $(\Delta kL/2)$ .

all other parameters constant, when the beam radius is doubled, the SHG power decreases by a factor of four.

(ii) Similarly, the SHG efficiency is increased by a factor of four by doubling the length of the crystal.

### 15.4.4 Types of phase-matching interaction

The phase-matching condition can be achieved using birefringent materials, whose refractive indices are direction dependent (see chapter 3). Birefringent phase matching can be achieved in two ways, viz. by (i) angle tuning and (ii) temperature tuning of nonlinear optical materials.

#### 15.4.4.1 Angle-tuned phase matching

There are two types of angle-tuned phase-matching interaction, depending on the polarization of the incident beams. If the incident beams have a common polarization direction then the phase matching is known to undergo *type-I* interaction. On the other hand, for beams with mutually perpendicular polarization, the phase matching is known as *type-II* interaction. The polarization direction of the output beam depends on the nature of the crystal. In the nonlinear process, the highest-frequency beam is polarized in the direction that chooses the lower of the two possible refractive indices. The phase-matching interactions in negative and positive uniaxial crystals are summarized in table 15.1.

For angle tuning (figure 15.8), the crystals are cut at a particular angle such that the speeds of the propagating frequencies are same inside the crystal. For values of $\theta_{pm}$ other than 0° or 90°, the Poynting vector of the extraordinary wave deviates from its direction of propagation. This causes a walk-off effect between the interacting beams and the conversion efficiency becomes low due to this limitation.

**Table 15.1.** Types of phase-matching interaction in negative and positive uniaxial crystals. The wave vectors $k_1$ and $k_2$ are incident on the crystal and generate $k_3$. The superscripts $o$ and $e$ represent the ordinary and extraordinary polarizations, respectively.

| Phase-matching interactions | Negative uniaxial crystals | Positive uniaxial crystals |
|---|---|---|
| **Type I  Type II** | $\vec{k}_1^{o} + \vec{k}_2^{o} = \vec{k}_3^{e}(\theta_{\text{pm}})\ \ \vec{k}_1^{o} + \vec{k}_2^{e}(\theta_{\text{pm}}) = \vec{k}_3^{e}(\theta_{\text{pm}})$ | $\vec{k}_1^{e}(\theta_{\text{pm}}) + \vec{k}_2^{e}(\theta_{\text{pm}}) = \vec{k}_3^{0}\ \ \vec{k}_1^{o} + \vec{k}_2^{e}(\theta_{\text{pm}}) = \vec{k}_3^{e}(\theta_{\text{pm}})$ |

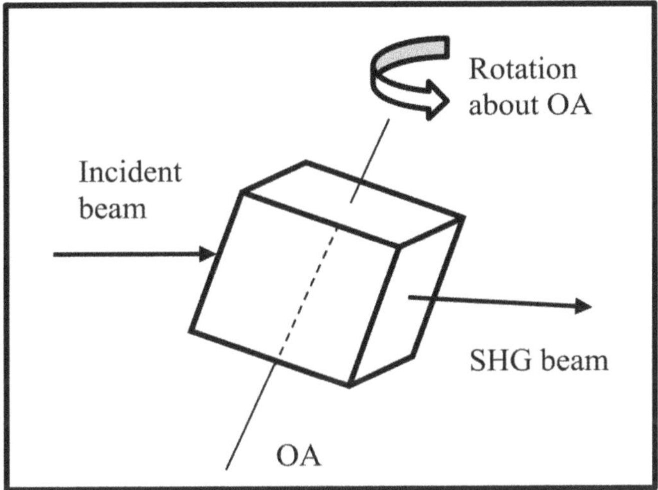


**Figure 15.8.** Schematic of angle-tuned phase matching.

The angle tuning is dependent on the thickness of the crystal, as the SHG signal has to pass through the crystal while selecting the angle. Angle-tuned phase-matching is also known as critical-phase matching.

♣ The phase-matching angles for SFG (with a refractive index of $n_3$) can be derived for a fundamental beam with wave vectors $k_1$ and $k_2$ corresponding to the wavelengths $\lambda_1$ and $\lambda_2$, respectively, as follows (♠ see question 7):

$$\sin^2 \theta_{\mathrm{pm}} = \frac{(n_3^e)^2}{\left[(\lambda_3/\lambda_1)n_1^0 + (\lambda_3/\lambda_2)n_2^0\right]^2} \times \frac{\left((n_3^0)^2 - \left[(\lambda_3/\lambda_1)n_1^0 + (\lambda_3/\lambda_2)n_2^0\right]^2\right]}{(n_3^0)^2 - (n_3^e)^2}.$$

For a DFG process with idler ($\lambda_i$) and signal ($\lambda_s$) wavelengths resulting from the pump wavelength ($\lambda_p$), the phase-matching angle is given by

$$\sin^2 \theta_{\mathrm{pm}} = \frac{(n_i^e)^2}{(n_i^e)^2 + (n_i^o)^2} \times \frac{\left[\left[n_p^0 - (\lambda_p/\lambda_s)n_s^0\right]^2 - (\lambda_p/\lambda_i)^2(n_i^o)^2\right]}{\left[n_p^0 - (\lambda_p/\lambda_s)n_s^0\right]^2},$$

where $n_p$, $n_s$, and $n_i$ are the refractive indices of the pump, signal, and idler beams, respectively. The superscripts $o$ and $e$ represent the ordinary and extraordinary polarizations, respectively.

♣ Using the above equations, computer programs can be written that calculate the phase-matching angles ($\theta_{\mathrm{pm}}$) for various crystals and phase-matching conditions (see, for example, http://www.as-photonics.com/snlo). Some of the angles for **BBO** are given in table 15.2.


**Exercise 15.4.** Describe the type of phase matching required in order for a **BBO** crystal to obtain a 355 nm output from $p$-polarized 1064 nm laser pulses.

**Table 15.2.** Phase-matching angles for nonlinear optical crystals for some popular nonlinear processes.

| Nonlinear optical process | Input wavelength(s) (nm) | Crystal | Output | $\theta_{pm}$ (in degrees) | Type of phase matching |
|---|---|---|---|---|---|
| SHG, Nd:YAG laser | 1064 $(n_0)$, $n_0 + n_0 \rightarrow n_e$ | BBO | 532 nm $(n_e)$ | 22.5 | I |
| SHG, Nd:YAG laser | 532 $(n_e)$, $n_e + n_e \rightarrow n_o$ | BBO | 256 nm $(n_0)$ | 47.0 | I |
| SHG, Ti:sapphire laser | 800 $(n_0)$ | BBO | 400 nm $(n_e)$ | 29.6 | I |
| SHG, argon-ion laser | 488 $(n_0)$ | Urea | 244 nm $(n_e)$ | 65.0 | I |
| SHG, argon-ion laser | 458 $(n_0)$ | KDP | 229 nm $(n_e)$ | 65.0 | I |
| THG, Nd:YAG laser | 1064 $(n_0)$+532 $(n_e)$ | BBO | 355 $(n_e)$ | 59.8 | II |
| SHG, 1500 nm | 1500 $(n_0)$ | BBO | 750 $(n_e)$ | 19.8 | I |
| OPA (collinear) | Pump 532, signal 800 | BBO | 800 nm 2100 nm | 21.34 | I |
| OPA (collinear) | Pump 400, signal 600 | BBO | 600 nm 1400 nm | 26.79 | I |
| OPA (noncollinear angle 2.2°) | Pump 532, signal 700 | BBO | 700 nm 2200 nm | 22.65 | I |
| Parametric up/downconversion process | 800 $(n_e)$ + 800 $(n_o)$ $\rightarrow$ 400$(n_e)$ | BBO | 400$(n_e)$ | 42.4 | II |

**Solution:** To obtain 355 nm, we first need the second harmonic of the Nd:YAG laser. From the above table, we can see that we need the following phase-matching condition from the type-I crystal.

$$n_0(1064\text{nm}) + n_0(1064\text{nm}) \rightarrow n_e(532\text{nm}).$$

Furthermore, to obtain the third harmonic, we need to work with the following phase-matching condition:

$n_0(1064\text{nm}) + n_e(532\text{nm}) \rightarrow n_e(355\text{nm})$. It should be noted that we will need a type-II phase-matching crystal in this case.

### 15.4.4.2 Temperature-tuned phase matching

To meet the requirements for high-efficiency SHG, temperature tuning of a thick nonlinear optical crystal is useful for phase matching. This approach is also known as non-critical-phase matching, in which the phase-matching angle is kept at a fixed position and the temperature of the crystal is varied (to vary the birefringence) to achieve the desired phase-matching condition. For example, the second harmonic of the fundamental wavelength of a Nd:YAG laser can be generated by keeping a lithium triborate (LBO) crystal at a constant temperature of 148 °C.

### 15.4.4.3 Quasi phase matching (QPM)

The limitation of angle- and temperature-tuned phase matching is that these methods work when the medium exhibits birefringence. In contrast, the quasi-phase-matching (QPM) technique is useful for cases in which birefringent phase-matching techniques are impossible (e.g. for isotropic materials). QPM was first described by Armstrong in 1962. It can be achieved in a crystal material by arranging a stacked structure of crystal in a periodic manner known as *periodic poling*. Such a structure can be prepared by slicing the crystal into a number of segments, as follows: the length of each segment is kept equal to the coherence length ($L_C$) given by equation (15.8), and the $C$-axis (optic axis) of every alternative segment is inverted, as schematically shown in figure 15.9.

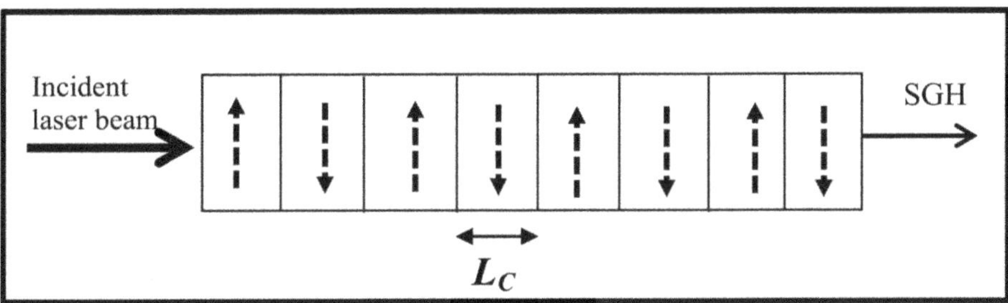


**Figure 15.9.** Schematic representation of a periodically poled nonlinear medium. The dashed arrows pointing in alternately opposing directions indicate the optic axes of the crystal segments stacked together. $L_c$ is the coherence length (♠ see equation (15.8)).

The nonlinear optical coefficient ($d_{il}$ in equation (15.3)) of such a periodically poled crystal becomes negative when its crystalline axis is inverted. The wave-vector mismatch that occurs when the beam propagates inside the crystal is compensated by varying the sign of $d_{il}$. Therefore, the generated signal continues to grow throughout the medium.

♣ The process that occurs within a periodic poled structure in lithium niobate as a result of applying a static electric field that surrounds the crystal has been described in some publications (♠ See [6]). The conversion efficiency of the SHG process ($\eta_{2\omega}$) in QPM is proportional to the number of slices $N$, according to $\eta_{2\omega} \propto (8\pi^2 d_{il}^2 (NL_c)^2 I_\omega)/\ \epsilon_0 n_\omega^2 n_{2\omega} c \lambda_\omega^2$ .

## 15.5 Parametric oscillation and amplification

A popular example of a parametric process is a child on a swing. The child can amplify the amplitude of the swinging motion by adjusting the position of her legs. Similarly, OPA is the stimulation of an optical frequency in the presence of one of the frequencies of the system (known as the signal frequency). It works on the principle of DFG.

As seen in fig. 2a, when two beams, viz. a strong *pump* beam and a weak *signal* beam (known as the seed) are incident on a crystal, a third beam (the idler) is generated. At the same time, the signal is coherently amplified while propagating through the nonlinear crystal. In OPA, the power of the seed is much lower than that of the pump beam. In contrast, in the case of DFG, the two incident beams are of comparable magnitudes. Two-photon emission from the virtual level is known as parametric fluorescence, which is weak (♠ see figure 15.14). Several processes, namely SFG, SHG, DFG, OPO, and OPA fall into the category of *parametric processes* and the efficiency of each process is controlled by its phase-matching condition. The *non-parametric* nonlinear optical processes, which involve real energy levels, are described in chapter 16.

### 15.5.1 Optical parametric oscillation

As shown in figure 15.10, a pump frequency ($\omega_P$) is incident on a nonlinear optical crystal kept inside a cavity. If the crystal is phase matched for DFG, the frequency $\omega_s$ (*signal*) and $\omega_i$ (*idler*) fields can build up in the cavity. This process is known as OPO. The OPO can be said to be either *singly resonant* (if the output consists of *either* the signal or the idler) or *doubly resonant* (if both outputs are produced).

The net gain is increased by placing the nonlinear crystal in between the cavity mirrors. If the parametric gain exceeds the optical loss then oscillation takes place in the cavity. The end mirrors of the cavity are designed for low reflectivity at the pump wavelength but high reflectivity at the signal or idler wavelengths. The output power of the selected wavelengths is increased by multiple reflections at the end mirrors. OPO was first reported in 1965 by Giodmaine and Miller. An OPO is not a laser but a coherent oscillator that produces laser light.

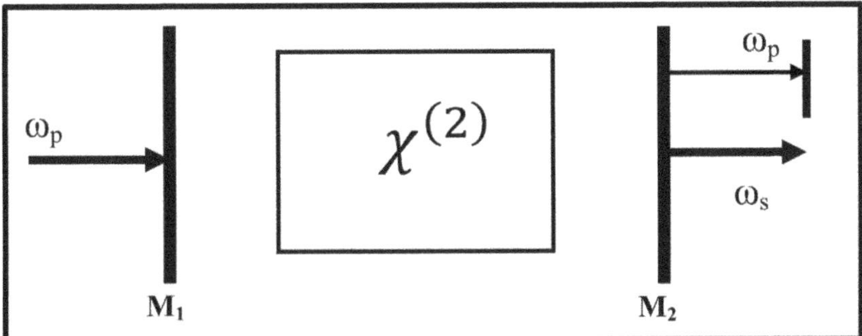

**Figure 15.10.** Schematic representation of the OPO process in a cavity made of two mirrors, $M_1$ and $M_2$. A nonlinear crystal that has second-order optical nonlinearity ($\chi^{(2)}$) is pumped with $\omega_p$. The output of this singly resonant case is the *signal* ($\omega_s$) for an appropriate selection of cavity mirrors.

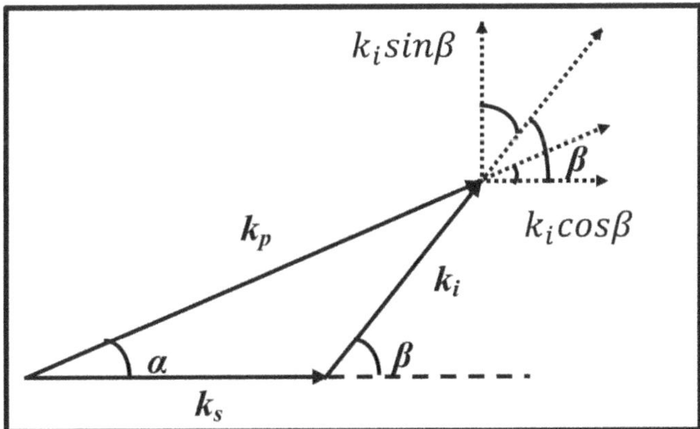

**Figure 15.11.** Schematic representation of the noncollinear interaction of pump and signal beams at an angle $\alpha$. $\beta$ is the angle between the signal and idler beams; $k_p$, $k_s$, and $k_i$ are the propagation vectors of the pump, signal, and idler beams, respectively.

### 15.5.2 Noncollinear geometries of parametric amplification

In the DFG process, if a weak beam with frequency $\omega_2$ is also incident on the system, it can be amplified. Here, the generated signal photons ($\omega_s$) from the DFG processes are added to the incident photons, as shown in figure 15.2. This process is known as optical parametric amplification (OPA), in contrast to OPO which requires a cavity. OPA pumped by the fundamental or the second harmonic of Ti:sapphire or Nd: YAG lasers can provide pulses from the visible region to the near-IR region.

An additional degree of freedom can be obtained using a noncollinear OPA geometry. As shown in figure 15.11, the signal and pump beams in OPA make an angle, $\alpha$, while the idler beam is emitted at an angle of $\beta$ with respect to the pump beam.

The phase-mismatch factor ($\Delta k$) in the parallel ($\Delta k_{II}$) and perpendicular ($\Delta k_{\perp}$) directions is given by

$$\Delta k_{II} = k_p \cos\alpha - k_z - k_i \cos\beta \tag{15.11}$$

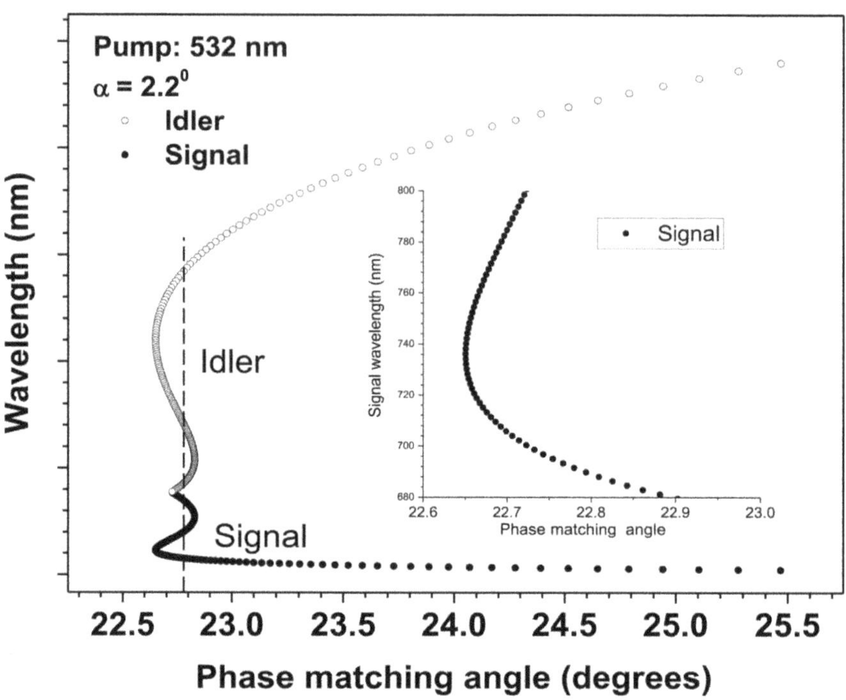



**Figure 15.12(A).** The angle-tuning curve of BBO (*type I*) for parametric oscillations at a pump wavelength of 532 nm for the noncollinear angle α = 2.2°. The inset shows broadband phase matching near 700 nm for a phase-matching angle of the BBO crystal of ~ 22.65° [13].

$$\Delta k_\perp = k_p \sin\alpha - k_z - k_i \sin\beta. \tag{15.12}$$

A plot of the signal and the idler as a function of the phase-matching angle for a BBO crystal is shown in figure 15.12(A). One of the interesting aspects of this area of study is to obtain more than one signal as well as idler wavelengths at a single phase-matching angle. As can be seen, at a phase-matching angle of 22.8°, more than one pair can theoretically be generated, a fact that was verified experimentally (figure 15.12(B)). Simulation programs that perform calculations of the phase-matching angles can be useful in identifying the required phase-matching angles[2].

The group velocity mismatch between the interacting pulses can be reduced by choosing appropriate noncollinear angles. This results in a broad tunability range, a high conversion efficiency, and short-duration output pulses.

From figure 15.11, it can be seen that to achieve broad phase matching using the group velocity ($V_g$), the phase-matching condition can be written as

$$V_{gs} = V_{gi} \cos\beta. \tag{15.13}$$

Here, the subscripts $s$ and $i$ again denote the signal and the idler beams, respectively. Several researchers have demonstrated noncollinear phase-matched OPA (NOPA)

---

[2] One such program is at: https://www.as-photonics.com/snlo.

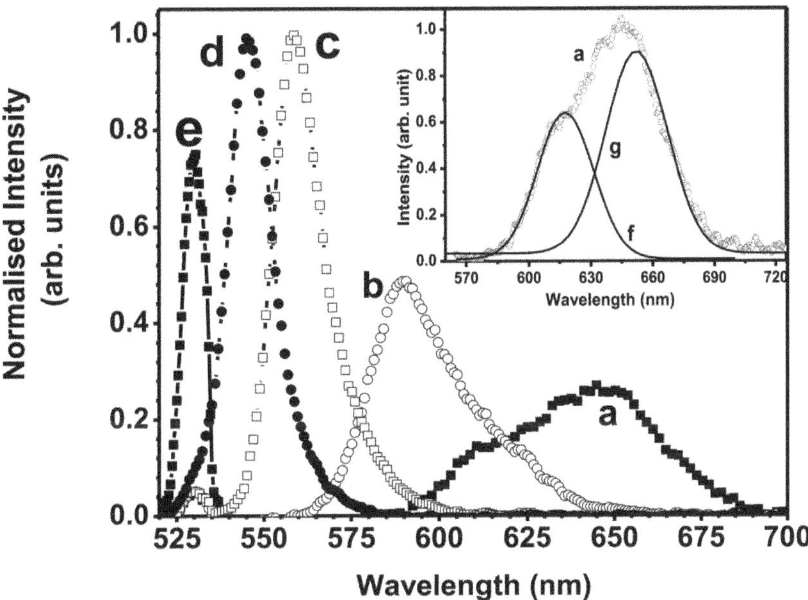


**Figure 15.12(B).** The line emission from a BBO crystal (*type I*) at tilt angles of (a) 1.5°, (b) 3°, (c) 4°, and (d) 5°. The inset shows the sum of two Gaussians for curve (a) [13].

using a BBO crystal and achieved tunable pulses of the order of 10 fs in duration in the visible region with 3–8 μJ pulse energies. Tunable fs pulses from 250 to 335 nm have been generated by frequency doubling [4] of high-power NOPA.

**Exercise 15.5.** An OPO is degenerate when its *signal* and *idler* beams are of the same wavelength. Find the wavelength of the OPO pumped by (i) a $Nd^{+3}$: YAG laser; (ii) the second harmonic generated by the $Nd^{+3}$: YAG laser.

**Solution:** In OPO, the pump photons with frequency $\omega_p$ are split into two frequencies known as the signal $(\omega_s)$ and the idler $(\omega_i)$ as follows: $\omega_p = \omega_s + \omega_i$. For degenerate OPO, the corresponding wavelength requirement is $\lambda_s = \lambda_i$. Therefore, expressing the frequencies in terms of the wavelengths, we get $\frac{1}{\lambda_p} = \frac{1}{\lambda_s} + \frac{1}{\lambda_i}$ or $\lambda_p = \frac{\lambda_s \lambda_i}{\lambda_s + \lambda_i}$. For $\lambda_s = \lambda_i (=\lambda)$, we get, $\lambda_p = \frac{\lambda}{2}$. We use this relation to look at these two special cases.

(i) For OPO pumped by $Nd^{+3}$: YAG, we have $\lambda_p = 1064$ nm; we obtain a degenerate OPO value of 2 $\lambda_p$ or 2128 nm.

(ii) Similarly, for $\lambda_p = 532$ nm, we get a degenerate OPO wavelength of 1064 nm.

## 15.6 Superfluorescence

In section 6.3.4, the phenomenon of superradiance was introduced in terms of a cooperative emission process performed by a population $(N)$ in a critical volume $(V)$ with emitting dipoles aligned in a single direction. In contrast, superfluorescence is the parametric amplification of quantum noise (with dipoles aligned in random directions). It is the radiation emitted by nonlinear crystals due to the spontaneous decay of pump photons into signal and idler photons.

It is a process similar to DFG, in which the available noise frequencies play the role of the signal. Harris *et al* were the first to observe the parametric luminescence known as superfluorescence. It is used as one of the tunable frequency source for parametric amplification in collinear as well as in noncollinear geometries of OPA. The observation of conical emission (see next section) in nonlinear optical crystals under femtosecond pumping in the visible region is also a good example of superfluorescence.

Although only a quantum mechanical treatment can account for this effect, classically it can be understood as follows. Generally, it is emitted by a suitable nonlinear crystal due to the spontaneous decay of $\omega_3$ photons into $\omega_1$ and $\omega_2$ ($\omega_3 = \omega_1 - \omega_2$). The situation is somewhat similar to that depicted in figure 15.2. Amplification occurs at those wavelengths for which the parametric interaction is phase matched, i.e. $k_3 = k_1 - k_2$. For a given orientation of the nonlinear optical crystal, numerous photons simultaneously satisfy the energy and momentum conservation conditions, Therefore, the spectral width of the superfluorescence is broad (figure 15.13).

Considering it to be a kind of spontaneous emission, the characteristic features of superfluorescence are as follows:

(a) The beam width of the superfluorescence depends upon (i) the crystal, (ii) the pump wavelength, and (iii) the phase-matching angle.
(b) For a fixed pump direction, the emission frequency changes with the direction of the emission. The frequency changes depend upon the crystal and its orientation.
(c) The spectral width $\Delta\omega_s$ of the superfluorescence also changes continuously with the direction of emission.

The use of parametric superfluorescence in tunable sources is advantageous for achieving large amplification and substantial seed energy. Large bandwidths in stimulated parametric fluorescence emission tunable over the range of 960 nm to 1160 nm have been obtained using a *barium sodium niobate* crystal pumped by the second harmonic of a $Nd^{3+}$:glass laser. In view of the above, table 15.3 gives the differences between various emission processes, viz. spontaneous emission,

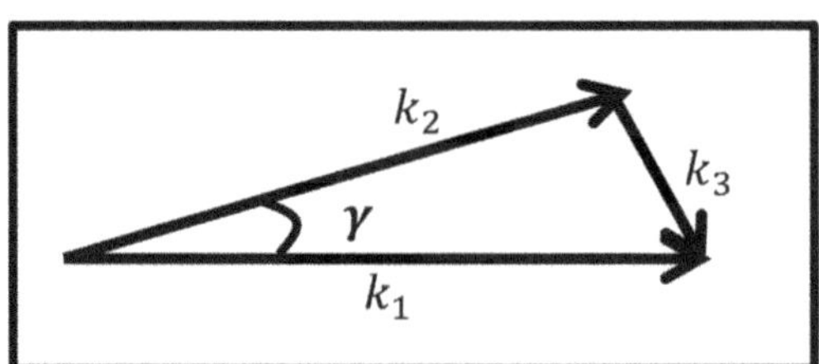

**Figure 15.13.** $\bar{k}$-vector phase matching: the signal wave ($k_2$) makes an angle $\gamma$ with the pump ($k_1$) to generate $k_3$.

**Table 15.3.** Identification of emission processes related to lasers and nonlinear optics.

| Emission process | Directionality | Spectral width | Phase-matching requirement | Examples |
|---|---|---|---|---|
| Spontaneous emission | Random | Broad | No | All luminescent materials |
| Superradiance | Directional | Sharp | No | $N_2$ laser |
| Superfluorescence | Conical | Broad | Yes | Nonlinear optical crystals |
| Amplified spontaneous emission (ASE) | Directional | Sharp | No (emission is amplified as it travels in the gain medium) | All lasing media as unwanted radiation |

superfluorescence, superradiance, and amplified spontaneous emission. Except in the case of spontaneous emission, the lifetime of the emission is faster, as described by the time–bandwidth relation, i.e. the broader the spectral width, the shorter the lifetime of the emission process.

Superradiance is directional (as outlined for the case of the nitrogen laser, ♠ see section 6.3.4). However, superfluorescence is often observed in the form of a ring in noncollinear optical amplifiers. From the point of view of decay times, the distinction between superfluorescence and superradiance is as follows: the peak of cooperative emission in superradiance decay is close to the time zero, while that of superfluorescence has a time lag such that phase matching is achieved. Some of the subtle differences between the similar-looking phenomena of superradiance, superfluorescence, and amplified spontaneous emission (ASE) are given in table 15.3.

## 15.7 Generation of polarization-entangled photons

The generated signal and idler photons in the parametric process possess mutually perpendicular polarizations. It is known (table 15.1) that these photons are generated in the type-II spontaneous parametric *downconversion* process (figure 15.14). In this case, the signal and idler photons can also appear as *entangled photons*. The downconversion efficiency is kept small, such that the generated photons can be described as individual single photons. High-power downconversion processes have also been reported for applications in distant places, such as in space communications.

As shown in figure 15.14, on interaction with a type-II nonlinear crystal, the pump beam generates new low-frequency signal and idler photons with mutually perpendicular polarizations. The two downconverted photons are emitted in two cones because of transverse momentum conservation. Therefore, polarization-entangled states are formed at the points of interaction of the two cones. Such

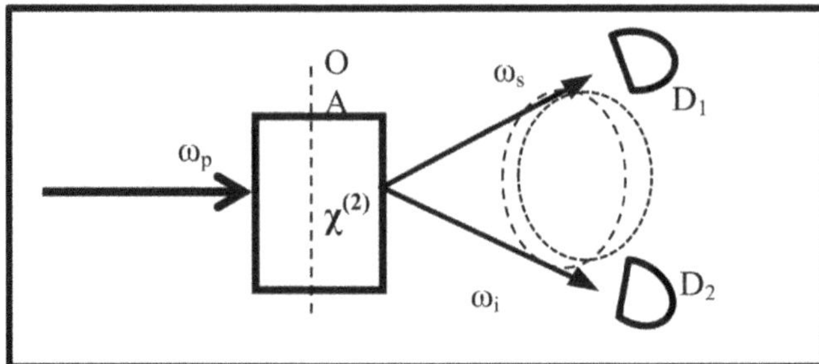


**Figure 15.14.** Schematic representation of the type-II spontaneous parametric downconversion process. OA is the optic axis of the crystal with $\chi^{(2)}$; $\omega_p$, $\omega_s$, and $\omega_i$ are the pump, signal, and idler frequencies, respectively, with phase-matched wave vectors. D1 and D2 are detectors.

observations are useful for basic undergraduate-level experiments in quantum optics, e.g. to verify Bell's inequality.

The two downconverted photons can be detected by the two detectors. One of the detectors can be used to predict the properties (such as the frequency and polarization) of the photon falling on the other detector. Such experiments have applications in quantum lithography, quantum metrology, quantum imaging, quantum teleportation, and quantum computation. A squeezed state of light can also be created by these processes, which has applications in optical communication and gravitational wave detection. Polarization-entangled photons have various applications in quantum information processing.

**Exercise 15.6.** As we can see from figure 15.1, when two frequencies interact in a crystal, in addition to SFG and SHG, difference-frequency generation (DFG) also takes place. OPA works on the principle of DFG by using a portion of the white-light continuum (WLC) (♠ see chapter 21) as the signal frequency ($\omega_s$) and the second harmonic generated by a pump laser as the pump frequency $\omega_p$. For these parameters, give the schematic for a noncollinear optical parametric amplifier (NOPA) that can be used for the amplification of signal frequencies in the visible–near-IR region.

**Solution:** This is a good application of DFG and a simple method of generating ultrafast tunable laser pulses. For the noncollinear interaction of the signal and idler beams, let us select the angle of 2.2° as given in the simulation shown in figure 15.12. For a pump wavelength of 532 nm, a wavelength range of 200 nm with a maximum at 700 nm can be obtained with the crystal that has a phase-matching angle of

22.7°. The required crystal is known as an NOPA crystal. We would also need another SHG crystal to generate 532 nm (the pump frequency) from the fundamental beam. The source of the signal frequency to be amplified, as mentioned in the question, is WLC, but it can be any weak source of photons. The design is given in figure 15.15.

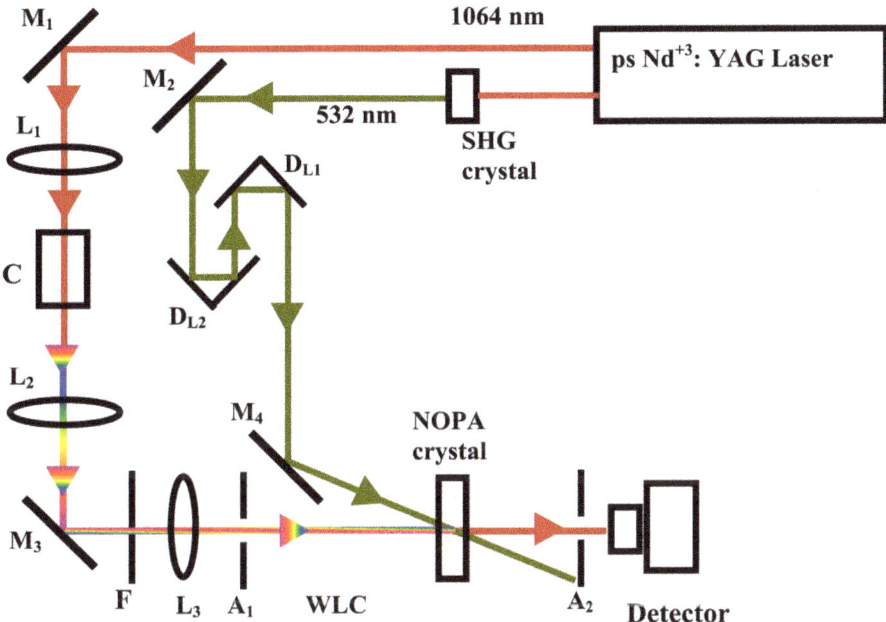


**Figure 15.15.** Schematic of NOPA with a picosecond Nd:YAG laser as the pump source. $L_1$-$L_3$: lenses, $M_1$-$M_3$: plane mirrors, $M_4$: concave mirror, $D_{L1}$ -$D_{L2}$: delay stages, $A_1$ and $A_2$: apertures. The pump wavelength of 532 nm is sent to the NOPA crystal at a noncollinear angle of 2.2°. The white light is obtained by pumping a water/$D_2O$ mixture (C); a 1064 nm filter (F) blocks the remaining 1064 nm light [13].

♣ NOPA is a useful and popular source for tunable ultrafast laser pulses in the visible–near-IR region; it uses a Ti:sapphire laser and BBO as NOPA and SHG crystals. (♠ See the details in [11]).

## Questions and problems

1. What is the difference between virtual and electronic energy levels?
2. A *p*-polarized laser beam that has a frequency of $\omega$ is incident on a nonlinear optical crystal. What is the polarization of the generated second-harmonic ($2\omega$) beam?
3. Describe the phase-mismatch factor. If the other parameters are kept constant, by what factors does the efficiency of SHG change in the following two cases: if (i) the fundamental pulse duration is doubled? (ii) the power of the fundamental beam is doubled?
4. Can the efficiency of SHG be increased by increasing the length of a crystal beyond a certain value? Generally, when a large crystal is required for the SHG of high-power lasers, what kind of phase-matching technique should be preferred?

5. With regard to the properties of the second-order nonlinear susceptibility ($\chi_{ijk}^{(2)}$), answer the following:

   (i) What is the rank of this tensor and how many elements does it have?

   (ii) Using Kleinman symmetry, can we reduce the number of elements? The second-order nonlinear term is written in matrix form for the SHG ($2\omega$) of the incident frequency $\omega$ in equation (15.3). Compare it with the equation for SFG ($\omega_3$) for the two incident frequencies $\omega_1$ and $\omega_2$ .

6. Potassium dihydrogen phosphate (KDP) is a uniaxial birefringent crystal. In KDP, the only nonzero components of the susceptibility tensor are: $\chi_{xyz}^{(2)} = \chi_{yxz}^{(2)}$ and $\chi_{zxy}^{(2)}$. An incident wave that has the frequency ($\omega$) travels with its polarization vector $\vec{E}$ and a wave vector $\vec{k} = \left(k_x, k_y, 0\right)$ normal to the optic axis, which is along the $z$-axis of the KDP. Describe the magnitude and direction of the nonzero components of P ($2\,\omega$).

7. For a negative uniaxial crystal, assume that the optic axis is along the $y$-axis. Using symmetry, the $x$ component of the refractive index is equal to the $y$ component. Using the indicatrix formula $\dfrac{x^2}{n_0^2(2\omega)} + \dfrac{z^2}{n_e^2(2\omega)} = 1$, find the phase-matching angle.

8. For a nonlinear optical crystal, the values of $n_0(\omega)$, $n_0(2\omega)$, $n_e(\omega)$, $n_e(2\omega)$ are given below. What is the value of the phase-matching angle for SHG?

| $n_e(\omega)$ | $n_e(2\omega)$ | $n_0(2\omega)$ | $n_0(\omega)$ |
|---|---|---|---|
| 1.54 | 1.554 | 1.674 | 1.654 |

(Hint: check whether the crystal is positive uniaxial or negative uniaxial.)

9. Differentiate between OPO and OPA. What is NOPA? From figure 15.12, find the phase-matching angle for **BBO** for NOPA that is suitable for generating signal wavelengths in the visible region.

10. Give one example each of the processes in which one can observe superradiance and superfluorescence?

11. How is the entanglement of photons described in terms of difference-frequency generation? Write the phase-matching condition for this process. Suggest an experiment in which you can verify the quantum entanglement of the photons (Hint: ♠ see [9].

# Bibliography

[1] Boyd R W 2003 *Nonlinear Optics* 2nd edn (New York: Academic)

[2] Sutherland R and Thompson B 2003 *Handbook of Nonlinear Optics* (Boca Raton, FL: CRC Press)

[3] Giordmaine J A and Miller R C 1965 Tunable coherent parametric oscillation in LiNbO$_3$ at optical frequencies *Phys. Rev. Lett.* **14** 973–6

[4] Hong C K, Ou Z Y and Mandel L 1987 Measurement of subpicosecond time intervals between two photons by interference *Phys. Rev. Lett.* **59** 2044–6

[5] Halbout J, Blit S, Donaldson W and Chung T Oct. 1979 Efficient phase-matched second-harmonic generation and sum-frequency mixing in urea *IEEE J. Quantum Electron* **15** 1176–80

[6] Yamada M, Nada N, Saitoh M and Watanabe K 1993 First-order quasi-phase matched LiNbO$_3$ waveguide periodically poled by applying an external field for efficient blue second-harmonic generation *Appl. Phys. Lett.* **62** 435–6

[7] Ukachi T, Lane R J, Bosenberg W R and Tang C I 1992 Phase-matched second-harmonic generation and growth of a *J. Opt. Soc. Am.* B **9** 1128–33

[8] Franken P A, Hill A E, Peters C W and Weinreich G 1961 Generation of optical harmonics *Phys. Rev. Lett.* **7** 118–9

[9] Dehlinger D and Mitchell M W 2002 Entangled photon apparatus for the undergraduate laboratory *Am. J. Phys.* **70** 898–902

[10] Armstrong J A, Bloembergen N, Ducuing J and Pershan PS 1962 Interactions between light waves in a nonlinear dielectric *Phys. Rev.* **127** 1918–37

[11] Kim B-J, Choi H–J and Cha M 2012 Angle-tuned second-harmonic generation in periodically-poled lithium niobate *Appl. Phys.* B **107** 349–53

[12] Wilhelm T, Piel J and Riedle E 1997 Sub-20-fs pulses tunable across the visible from a blue-pumped single-pass noncollinear parametric converter *Opt. Lett.* **22** 1494–6

[13] Nautiyal A 2009 *PhD Thesis* (IIT Madras)

[14] Abramczyk H 2005 *Introduction to Laser Spectroscopy* (Amsterdam: Elsevier)

**IOP** Publishing

# An Introduction to Photonics and Laser Physics with Applications

**Prem B Bisht**

# Chapter 16

## Third-order nonlinear optical processes

Nonlinear optical interaction between three incident electric fields in a material can result in a fourth wave. This phenomenon is known as four-wave mixing (FWM). The third-order nonlinear optical susceptibility ($\chi^{(3)}$) of the material is responsible for this effect. Some of the other phenomena related to $\chi^{(3)}$ introduced in this chapter are: the optical Kerr effect (OKE), self-phase modulation (SPM), and multiphoton absorption (MPA). The figure indicates that on an increase in the incident intensity, the absorption coefficient ($\alpha$) of a material decreases and it exhibits *saturable absorption*, SA. At even higher incident intensities, the absorption coefficient increases further, resulting in the effect of *reverse* saturable absorption (RSA). The phenomenon of SA becomes useful in optical switching applications when at least two laser beams are used. You will also see in this chapter how a photonic circuit can be realized in practice.

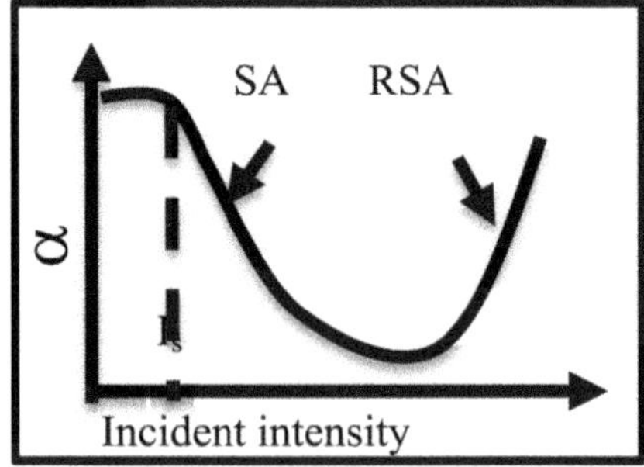


**Learning objectives**
**After reading this chapter, the learner will be able to:**

Differentiate between parametric and nonparametric processes;
Understand degenerate four-wave mixing (DFWM);

doi:10.1088/978-0-7503-5226-0ch16



© IOP Publishing Ltd 2022

Recognize the symmetry properties of $\chi^{(3)}$;
Identify the optical processes due to $\chi^{(3)}$;
Interpret the transverse optical Kerr effect in terms of self-focusing and self-defocusing;
Apply the longitudinal optical Kerr effect in the phenomenon of self-phase modulation;
Illustrate the saturable absorption (SA) effect;
Construct photonic logic gates.

## 16.1 Parametric and nonparametric processes

The concept of the virtual level was introduced in chapter 14 (section 14.5). The virtual level facilitates population exchange between two energy levels within a short time span of $\hbar/\delta E$, where $\delta E$ is the energy gap between levels. This is the signature of a *parametric process* and is responsible for the real part of the susceptibility. In contrast, in the presence of pre-existing excited levels, available pathways of relaxation dissipate the energy such that the photon energies are not conserved. Such processes are known as *nonparametric processes*. In the latter case, some molecular energy is lost during the nonradiative processes, for example, by the generation of heat or the relaxation of the environment around the excited dipole. As we deal with molecular systems with real excited states, this information will be helpful in the forthcoming sections of this chapter.

## 16.2 Third-order nonlinear optical susceptibility

In the third term of the nonlinear polarization equation $P^{(3)}(t) = \epsilon_0\chi^{(3)}E^3(t)$, $\chi^{(3)}$ is the third-order nonlinear optical susceptibility. The resultant polarization is known to be due to the process of four-wave mixing (FWM), since the fourth wave ($P^{(3)}(t)$) is generated as a result of the interaction of three incident fields ($E^3(t)$). For example, if we take the incident electric field $E(t) = E_0\cos\omega t$, we can write

$$P^{(3)}(t) = \epsilon_0\chi^{(3)}(E_0^3\cos^3 \omega t) \tag{16.1}$$

$$= \frac{\epsilon_0\chi^{(3)}E_0^2}{4}\left[E_0\cos^3 \omega t + 3E_0\cos\omega t\right]$$

$$= \left(\frac{\epsilon_0\chi^{(3)}I_0}{4}\right)E_0 \cos^3\omega t + \left(\frac{3\epsilon_0\chi^{(3)} \quad I_0}{4}\right)E(t). \tag{16.2}$$

Here, $I_0 = E_0^2\cos^2 \omega t$. The first term of the above equation corresponds to the third-harmonic generation (THG) process, while the second term varies as a function of the original field intensity ($I_0$). It should be noted that the quantity in parentheses in the second term is the change in susceptibility ($\Delta\chi$). The components of the third-order polarization can also be written in index notation as follows:

$$P_i^{(3)}(t) = \varepsilon_0\chi_{ijkl}^{(3)}E_jE_kE_l. \tag{16.3}$$

It should be noted that the efficiency of higher-order nonlinear optical processes reduces drastically. Although it is difficult to observe experimentally, the seventh harmonic of a krypton fluoride excimer laser (35.4 nm) has been reported. To obtain even higher-order frequencies with good efficiency, an entirely new technique of '*high harmonic generation*' has been implemented (♠ see chapter 24).

## 16.3 Symmetry properties of the susceptibility tensor

Before we talk about the various processes due to $\chi^{(3)}$ of a material, it is useful to briefly recall the reduction method of the components of the $\chi^{(3)}_{ijkl}$ tensor that appears in equation (16.3). This reduction takes advantage of the symmetry conditions of

(i) **The reality of fields:** The complex conjugate of the tensor is equal its complex conjugate, i.e.:

$$\chi^{(3)*}_{ijkl}(-\omega_4; \omega_1, \omega_2, \omega_3) = \chi^{(3)}_{ijkl}(\omega_4; -\omega_1, -\omega_2, -\omega_3).$$

(ii) **Intrinsic permutation symmetry:** We can write $\chi^{(3)}_{ijkl}$ by exchanging the sequence of incident frequency arguments and the corresponding dummy indices, similarly to the method used for the case of $\chi^{(2)}_{ijk}$, as follows:

$$\chi^{(3)}_{ijkl}(\omega_4; \omega_1, \omega_2, \omega_3) = \chi^{(3)}_{ijlk}(\omega_4; \omega_1, \omega_3, \omega_2) = \chi^{(3)}_{iljk}(\omega_4; \omega_3, \omega_1, \omega_2) = \ldots$$

(iii) **Full permutation symmetry with Kleinman symmetry:**
Away from resonant frequencies, for lossless media and pure electronic contributions in which the dispersion effects can be ignored, nonlinear polarization becomes independent of frequency. Therefore, the subscripts of $\chi^{(3)}_{ijkl}$ can be freely exchanged, as shown below

$$\chi^{(3)}_{ijkl}(\omega_4; \omega_1, \omega_2, \omega_3) = \chi^{(3)}_{ijlk}(\omega_4; \omega_1, \omega_2, \omega_3) = \chi^{(3)}_{iljk}(\omega_4; \omega_1, \omega_2, \omega_3) = \ldots$$

$\chi^{(3)}_{ijkl}$ is a fourth-rank tensor containing $3^4$ or 81 elements. For isotropic samples (such as liquids), 60 elements of this tensor vanish due to inversion symmetry. This is due to the fact that the elements must possess an even number of similar Cartesian indices. The $\chi^{(n)}$ tensors become independent of the frequency under the Kleinman conjecture and the associated symmetry properties, as discussed above. As a result, the remaining 21 elements of $\chi^{(3)}_{ijkl}$ can only have four different values, as given below

$$\left.\begin{aligned}
\chi^{(3)}_{1111} &= \chi^{(3)}_{2222} = \chi^{(3)}_{3333} \\
\chi^{(3)}_{1122} &= \chi^{(3)}_{1133} = \chi^{(3)}_{2211} = \chi^{(3)}_{2233} = \chi^{(3)}_{3311} = \chi^{(3)}_{3322} \\
\chi^{(3)}_{1212} &= \chi^{(3)}_{1313} = \chi^{(3)}_{2121} = \chi^{(3)}_{2323} = \chi^{(3)}_{3131} = \chi^{(3)}_{3232} \\
\chi^{(3)}_{1221} &= \chi^{(3)}_{1331} = \chi^{(3)}_{2112} = \chi^{(3)}_{2332} = \chi^{(3)}_{3113} = \chi^{(3)}_{3223}
\end{aligned}\right\}. \tag{16.4}$$

Out of 21 elements, only three are independent; they are related by the following equation:

$$\chi^{(3)}_{1111} = \chi^{(3)}_{1122} + \chi^{(3)}_{1212} + \chi^{(3)}_{1221}. \tag{16.5}$$

## 16.4 Four-wave mixing due to $\chi^{(3)}$

In equation (16.3), three electric fields interact through a $\chi^{(3)}$ material to give rise to the fourth field. If the incident frequencies are identical, the interaction process is known as DFWM. Let us take a non-degenerate case with fields $E_1$, $E_2$, and $E_3$ and their associated frequencies $\omega_1$, $\omega_2$, and $\omega_3$, respectively, as follows:

$E_1(t) = (E_1 e^{-i\omega_1 t} + cc)$; $E_2(t) = (E_2 e^{-i\omega_1 t} + cc)$; $E_3(t) = (E_3 e^{-i\omega_1 t} + cc)$. Here, c.c. is the complex conjugate. The resultant field is given by

$$E_4(t) = E_1 + E_2 + E_3$$

$$E_4(t) = (E_1 e^{-i\omega_1 t} + E_1^* e^{i\omega_1 t}) + (E_2 e^{-i\omega_2 t} + E_2^* e^{i\omega_2 t}) + (E_3 e^{-i\omega_2 t} + E_3^* e^{i\omega_2 t}).$$

For efficient coupling between four interacting beams, both momentum and energy must be conserved. Figure 16.1 shows a typical schematic representation of a phase-matching diagram with the corresponding wave vectors $k_1$, $k_1$, $k_3$, and $k_4$ of the FWM process.

For the resultant third-order polarization term, we have 44 different frequency components (22 original components and their negative frequencies) as given in table 16.1.

From the table, it can be seen that in addition to the well-known process of THG of every incident field, various combinations of frequencies can be obtained. As an example, the following are some of the terms in the polarization equation whose corresponding frequencies appear in table 16.1.

$$P(3\omega_1) \propto \chi^{(3)} E_1^3;$$

$$P(\omega_1 + \omega_2 + \omega_3) \propto 6\chi^{(3)} E_1 E_2 E_3;$$

$$P(2\omega_1 + \omega_2) \propto 3\chi^{(3)} E_1^2 E_2.$$

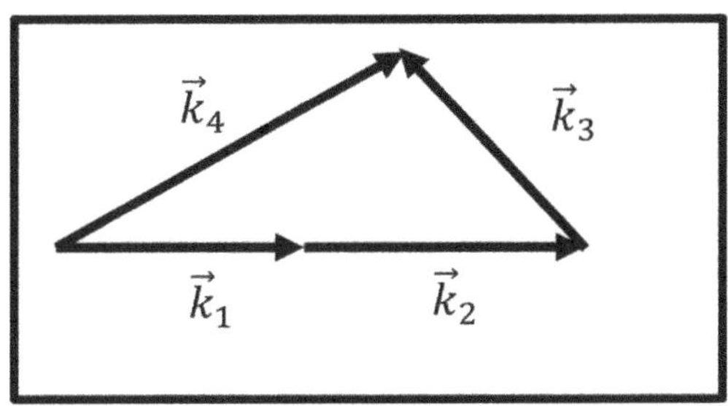


**Figure 16.1.** Schematic representation of a phase-matching diagram of an FWM process. The phase-matching condition is given by $\Delta\vec{k} = \vec{k}_1 + \vec{k}_2 + \vec{k}_3 + \vec{k}_4 = 0$.

**Table 16.1.** Frequencies that can be generated when three frequencies are incident on a $\chi^{(3)}$ material.

| Incident frequencies | Generated frequencies | | |
|---|---|---|---|
| $\omega_1$ | $3\omega_1$ | $\omega_1+\omega_2+\omega_3$ | $2\omega_1 \pm \omega_2$ |
| $\omega_2$ | $3\omega_2$ | $\omega_1+ \omega_2-\omega_3$ | $2\omega_2 \pm \omega_1$ |
| $\omega_3$ | $3\omega_3$ | $\omega_1+\omega_3 - \omega_2$ | $2\omega_1 \pm \omega_3$ |
| | | $\omega_2 + \omega_3 - \omega_1$ | $2\omega_2 \pm \omega_3$ |
| | | | $2\omega_3 \pm \omega_1$ |
| | | | $2\omega_3 \pm \omega_2$ |

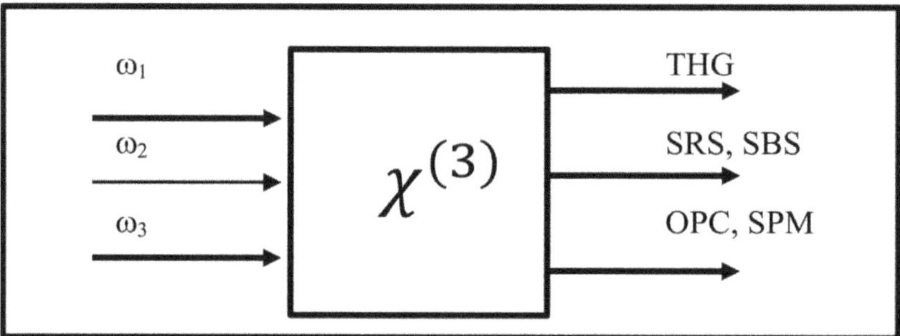


**Figure 16.2.** Phenomena that can take place when frequencies interact with a $\chi^{(3)}$ material.

Figure 16.2 schematically shows the various processes that can take place due to $\chi^{(3)}$. These include: the optical Kerr effect (OKE), self-focusing, self-phase modulation (SPM), soliton formation, stimulated Raman scattering (SRS), stimulated Brillouin scattering (SBS) and optical phase conjugation (OPC). These processes are described in the following sections.

## 16.5 Third-harmonic generation

When the incident field strengths are high enough to invoke the $\chi^{(3)}$ process, the first term in equation (16.2) is responsible for the THG process, which is parametric. The energy-level diagram for the generalized sum-frequency generation (SFG) of three fields is indicated in figure 16.3. For $\omega_1 = \omega_2 = \omega_3 = \omega$, this will give rise to THG ($3\omega$). It should be noted that the efficiency of THG is lower than that of SFG.

**Exercise 16.1.** In commercial lasers, the third harmonic is usually generated using a non-centrosymmetric crystal with $\chi^{(2)}$. In view of third-harmonic generation by $\chi^{(3)}$ materials, how do you justify this?

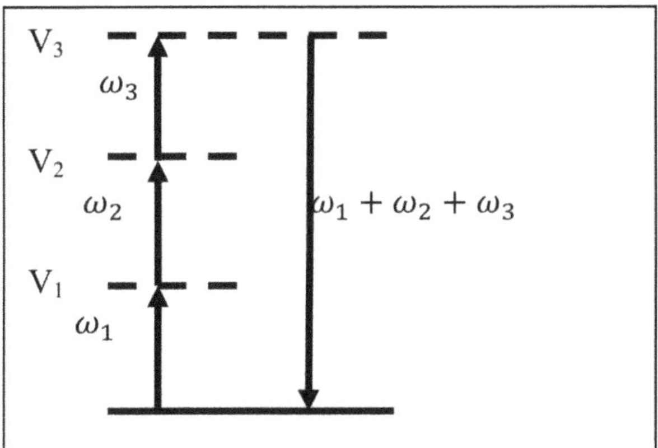

Figure 16.3. Schematic of the sum-frequency generation (SFG) process for three incident fields with frequencies of $\omega_1$, $\omega_2$, and $\omega_3$. $V_1$, $V_2$, and $V_3$ are the virtual levels.

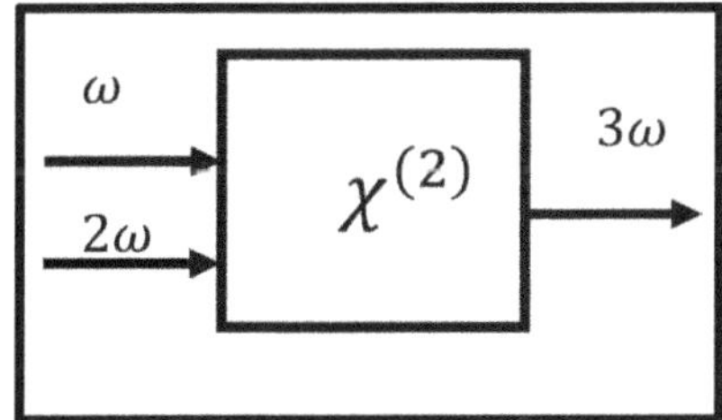

**Solution:** It is interesting to note that it is far easier and more efficient to obtain the THG ($3\omega$) using the fundamental frequency ($\omega$) and its second harmonic ($2\omega$). Once we have $\omega$ and $2\omega$, using sum-frequency generation in a type II $\chi^{(2)}$ crystal allows the third harmonic to be obtained, as shown in the figure. (♠ see table 15.2 for the crystal selection, for example, for the case of a Nd:yttrium aluminum garnet (YAG) laser).

## 16.6 Optical Kerr effect

The net refractive index along with the change of susceptibility ($\Delta\chi$) due to the $\chi^{(3)}$ process can be written in a similar way to that used for equation 14.8,

$$n = \sqrt{(1+\chi) + \quad \Delta\chi} = n_0 \left(1 + \frac{\Delta\chi}{n_0^2}\right)^{1/2}.$$

Here, $n_0$ is the refractive index of the medium that contributes to the linear part of the susceptibility. Let us assume that the nonlinear term is small compared to the linear term. In this case, the above expression can be expended using the Taylor series, such that

$$n = n_0 + \frac{\Delta\chi}{2n_0}.$$

As described by equation (16.2), the value of $\Delta\chi$ is $\frac{3\epsilon_0\chi^{(3)}}{4}I_0$. It follows from the above equation that the second term is the change of the refractive index ($\Delta n$) given by $\Delta n = \frac{3\epsilon_0\chi^{(3)}}{8n_0}I_0$. We can simplify this to

$$n = n_0 + \Delta n.$$

By introducing the *intensity-dependent refractive index*, $n_2(I) = \frac{3\epsilon_0\chi^{(3)}}{8n_0}$, we obtain the following celebrated equation for the refractive index $n(I)$:

$$n(I) = n_0 + n_2(I)I. \tag{16.6}$$

This equation shows the quadratic dependence of the refractive index on the incident optical field, which is similar to that of its counterpart in the DC field (♠ see chapter 3).

**Exercise 16.2.** (i). The AC (optical) Kerr effect takes place when a laser pulse interacts with a medium. (i). How do you differentiate it from its DC counterpart? **(ii).** Does the magnetic field component of the laser pulse contribute to the optical Kerr effect?

    **Solution:**

      (i) The DC Kerr effect was explained in section 3.7. The magnitude of the birefringence at a wavelength $\lambda$ is proportional to the square of the applied DC electric field. This birefringence is due to induced changes in the refractive index. The optical Kerr effect (OKE) operates in a similar way. Here, we can define the AC Kerr constant as $K' = \frac{3\epsilon_0\chi^{(3)}}{8\lambda n_0}$.

      (ii) In contrast to the DC magneto-optics effect, here, the magnetic effects of the laser pulse are not prominent. This is due to the fact that the magnetic field of an EM wave is weaker than the electric field by a factor of $c$.

The materials which exhibit this effect are known as Kerr media. Equation (16.6) is important and has profound applications in nonlinear optics and laser technology. Some of these applications based on the transverse or longitudinal Kerr effect are described in the following sections.

## 16.6.1 Transverse OKE

When a high-intensity Gaussian laser beam profile interacts with a Kerr medium, the refractive index of the medium changes according to the spatial profile of the input beam. The medium experiences an intensity pattern in its transverse plane according to equation (16.6) and acts as a graded-index material. The refractive index is maximized at the center of the medium and decreases on both sides of the medium due to the Gaussian nature of the beam profile, as expected from equation (16.6). The effects of *self-focusing* and *self-defocusing* can be observed as described in the following.

If the medium exhibits a positive sign for $n_2$, it acts like a converging lens and the incident laser beam is focused while propagating in the medium. This is known as

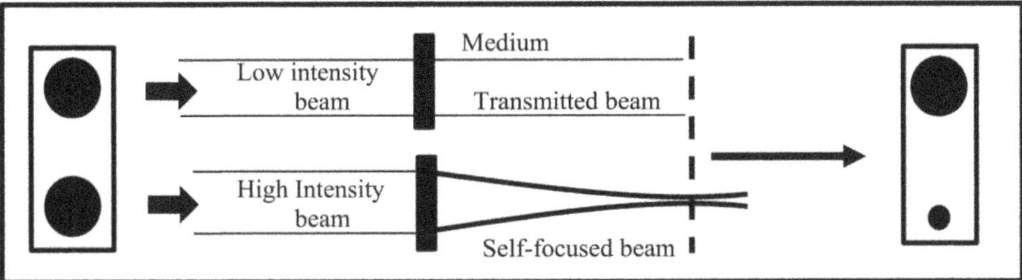


**Figure 16.4.** Self-focusing of a high-intensity beam (bottom) by a medium as compared to that of a lower-intensity beam (top). The black spots indicate the cross-sections of the beams before (left) and as seen at the vertical dashed line after passing through the medium (right).

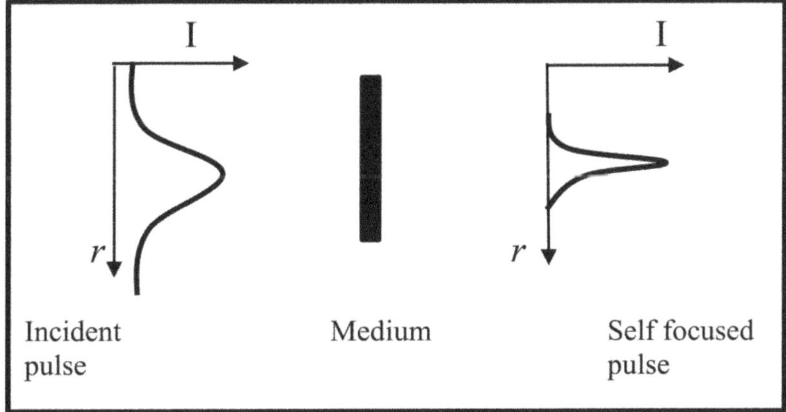


**Figure 16.5.** Representation of a Gaussian input beam undergoing the phenomenon of *self-focusing*.

self-focusing effect. On the other hand, if the medium exhibits a negative sign for $n_2$, then it acts like a diverging lens and the laser beam is defocussed. This is known as the self-defocusing effect. Figure 16.4 gives the schematic representation of the self-focusing effect. The self-focusing or self-defocusing strengths of the medium indicate the amount of nonlinear refraction (NLR) a sample can possess. Figure 16.5 shows a representation of the self-focusing effect for a Gaussian-shaped pulse.

One of the important applications of this effect is Kerr lens mode locking (♠ see chapter 17). Another popular technique frequently used by researchers in nonlinear optics to obtain the sign and magnitude of $\chi^{(3)}$ is known as the *z-scan* (♠ see [3]).

### 16.6.2 Longitudinal OKE: the effect of self-phase modulation

Self-phase modulation (SPM) is a third-order *parametric process*. The intensity-dependent refractive index (equation 16.6) is responsible for SPM. The SPM process is divided into two: spatial and temporal SPM. The transverse OKE results in the phenomenon of self-focusing due to spatial modulation of the input beam by the medium. However, the longitudinal OKE (LOKE) gives rise to additional frequencies (i.e. spectral broadening and white-light continuum generation) due to temporal modulation of the incident beam. In the case of pulsed ultrafast lasers, due to their high peak power, this effect is easily observed even in the air.

The angular frequency ($\omega$) of a wave is related to the rate of change of its phase $\left(\frac{d\phi}{dt}\right)$. The action of the intensity-dependent refractive index (equation 16.6) along the length of the medium is responsible for LOKE. In this process, additional bandwidth is added to the phase, thus changing $\omega_0$ to $\omega_0 \pm \Delta\omega$ as described below.

$$\phi = kL \cong \left(\frac{\omega_0}{c}\right)nL$$

$$\phi_{total} = \phi_{linear} + \phi_{nl}$$

The value of $\phi_{nl}$ is equal to $kL \times n_2(I)$, where $n_2(I)$ is given in equation (16.6). As we are also dealing with frequency and the refractive index is a function of the frequency as well, equation (16.6) can now be written as

$$n(I, \omega) = n_0(\omega) + n_2(I, \omega)I.$$

The total frequency change $\Delta\omega$ due to nonlinearity can be also written as a time derivative of the phase and is given by $\Delta\omega = -\frac{d\phi_{nl}}{dt}$; thus,

$$\Delta\omega = -\left(\frac{\omega_0}{c}\right)L\left(\frac{dn_2(I, \omega)}{dt}\right). \tag{16.7}$$

Here, $\omega_0$ is the frequency of the incident laser beam.

As can be seen from figure 16.6, for a Gaussian beam, the new frequencies ($\omega_0 \pm \Delta\omega$) appear in the detected output of the laser pulse. This effect is known as

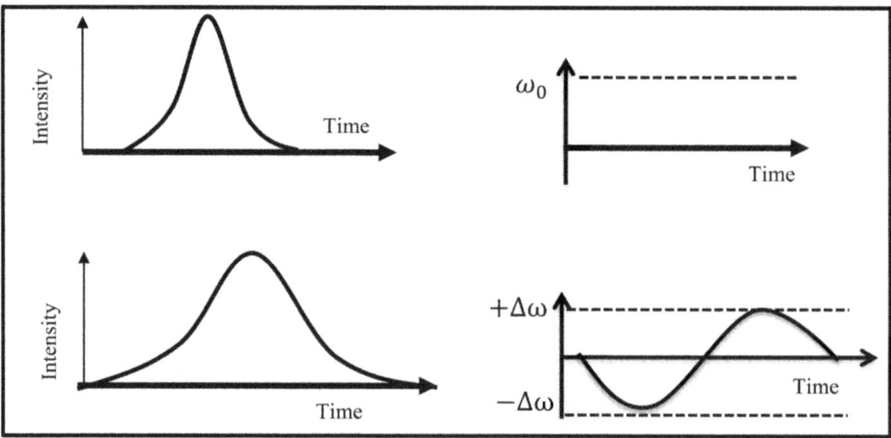


**Figure 16.6.** Schematic representation of pulse chirping due to the phenomenon of SPM. A Gaussian pulse with frequency ($\omega_0$) is shown before chirping (top panel), and after passing through the medium (bottom panel) with its frequency spread ($\Delta\omega$) around the central frequency.

SPM. This happens because $\frac{d}{dt}(n_2(I, \omega))$ is positive on the rising edge, while it is negative along the trailing edge. At the beginning of the laser pulse, the phase front is delayed due to the positive temporal derivation. This results in a reduction of the frequency of the incident laser pulse. On the other hand, for the remaining part of the pulse, the time derivative of the intensity is negative, which increases its frequency. Due to the rising edge, more frequencies arrive at the detector and vice versa for the trailing edge. This spectral broadening of the pulse after propagation through a dispersive medium due to SPM is known as pulse chirping. Additionally, the phenomenon of self-focusing (i.e. spatial SPM) increases the beam intensity inside the medium and contributes to even stronger temporal SPM. Pure temporal SPM can be realized in thin media, where the effect of self-focusing can be neglected. Plasma-induced SPM can also be observed in gas-filled hollow-core photonic crystal fibers.

**Exercise 16.3.** Find the expression for the frequency spread if the incident beam is Gaussian in nature.

**Solution:** Let us take the expression for the intensity of the Gaussian beam as $I = I_0 e^{\left(-\frac{t}{\tau}\right)^2}$, where $\tau$ = pulse duration and $I_0$ is the peak intensity. Equation (16.6) is written as $n(I) = n_0 + n_2 I$.

$$\frac{dn(I)}{dt} = n_2 \frac{dI}{dt} = -n_2 I_0 \times \quad -2t/\tau^2 e^{-\left(\frac{t}{\tau}\right)^2}$$

$\phi(t) = \omega_0 t - kx = \omega_0 t - \left(\frac{2\pi}{\lambda_0}\right) \times n(I)L$, where $L$ is the pulse propagation distance.

$$\omega(t) = \frac{d\phi(t)}{dt} = \omega_0 - \left(\frac{2\pi}{\lambda_0}\right) L \times \frac{dn(I)}{dt}$$

or

$$\omega(t) = \omega_0 - \left(\frac{2\pi}{\lambda_0}\right) L \times -n_2 I_0 \times \quad -2t/\tau^2 e^{-\left(\frac{t}{\tau}\right)^2}$$

$$\omega(t) = \omega_0 + \left(\frac{4\pi L n_2 I_0}{\lambda_0 \tau^2}\right) t e^{-\left(\frac{t}{\tau}\right)^2}.$$

**Exercise 16.4.** Do you know that 'pulse chirping' and 'Kerr lens mode locking' are two sides of the principle responsible for the optical Kerr effect?

**Solution:** Pulse chirping, as explained above, is due to the longitudinal OKE. On the other hand, the transverse OKE is responsible for Kerr lens mode locking (♠ see chapter 17). The effect of SPM and frequency chirping with ultrashort lasers is so obvious and effective that under suitable modulation of the refractive index of a medium, a wide spectrum of radiation covering the visible and near IR region can be obtained. Such effects fall under the topic of white-light continuum generation, which is described in chapter 21.

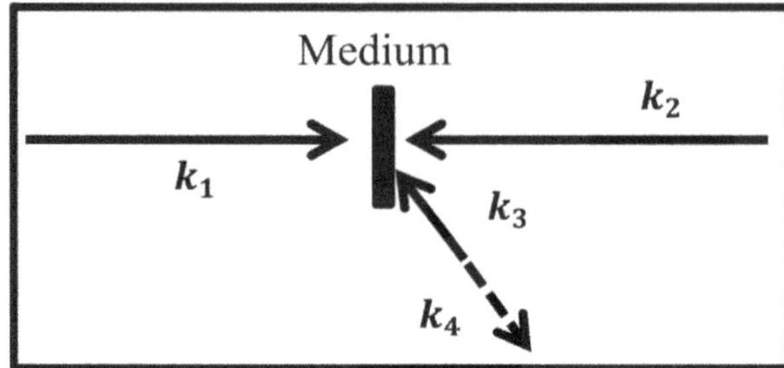


**Figure 16.7.** Schematic of the optical phase conjugation phenomenon. In presence of the incident fields with wave vectors $k_1$ and $k_2$, the nonlinear optical medium generates a beam with wave vector $k_4$ in the direction opposite to the new incident field $k_3$.

## 16.7 Optical phase conjugation

In this configuration of FWM, two counter-propagating pump beams called the forward pump ($k_1$) and the backward pump ($k_2$), along with a third beam known as a probe beam ($k_3$), are incident on a nonlinear medium. Invoking the $\chi^{(3)}$ of the medium, a fourth beam ($k_4$) is created and is referred to as the phase-conjugate beam (figure 16.7). Further details of OPC are given in chapter 19.

## 16.8 Stimulated Raman and Brillouin scattering

Raman scattering gives rise to important characteristic feature of a material in the form of new observed transitions. As outlined in chapter 2, when a photon of frequency ($\omega_1$) is incident on a material, two new frequencies are scattered. The frequencies of the scattered radiation are either smaller or larger than $\omega_1$ and are known as Stokes ($\omega_s$) and anti-Stokes ($\omega_{as}$) frequencies, respectively. When materials are pumped by intense lasers, the phenomenon of stimulated Raman scattering (SRS) takes place, producing intense coherent photons. This is a spatial type of degenerate FWM process in which the vibrational levels of the molecules are involved. An energy-level diagram and a wave-vector diagram of the SRS process are shown in figure 16.8.

When intense laser photons along with Stokes photons are incident on the medium, it leads to stimulation of additional Stokes photons. The generated Stokes photons are coherent with the incident Stokes photons, causing amplification in the form of Stokes lines. This process is known as stimulated Raman scattering (SRS). Panel B of figure 16.8 shows the wave-vector representation of the SRS process in terms of the pump ($k_1$), Stokes ($k_s$), and anti-Stokes ($k_{as}$) vectors.

Stimulated Brillouin scattering (SBS) can be considered to be similar to SRS. However, SRS deals with the involvement of optical phonons, while SBS takes place through acoustic phonons. Therefore, SBS occurs only in the backward direction, while SRS can occur in both directions (♠see section 19.4.1).

**Exercise 16.5.** In an experimental demonstration, Nd:YAG laser (fundamental wavelength) pulses of 50 ps that have an energy per pulse of 20 mJ are focused onto a

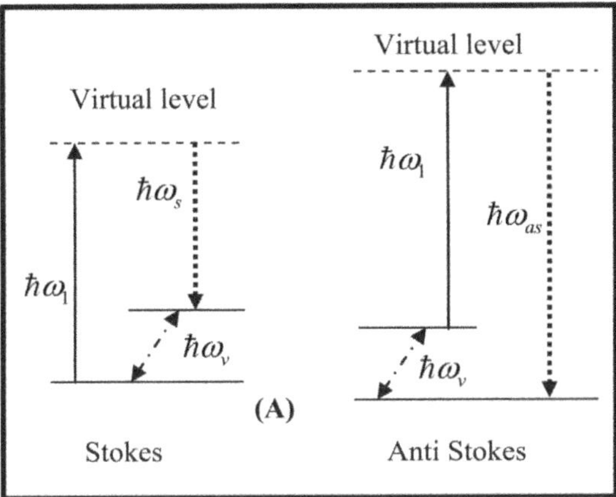


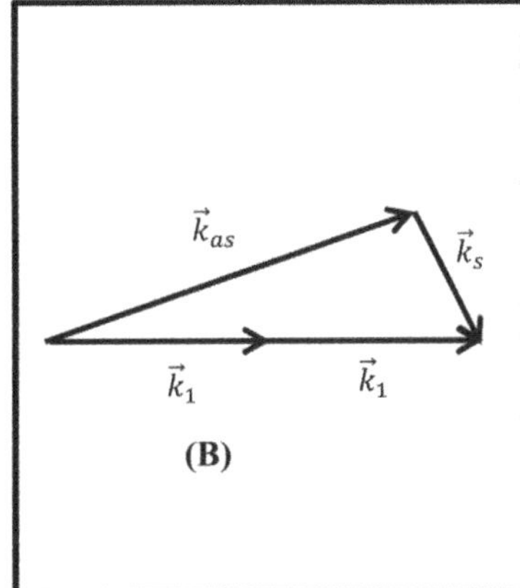


**Figure 16.8.** Energy-level diagram of SRS (left-hand panel). $\omega_l$ and $\omega_v$ refer to the frequencies of the incident photon and the ground-state vibration, respectively. $\omega_s$ is the Stokes frequency and $\omega_{as}$ is the anti-Stokes Raman frequency. The solid and dotted lines represent the real and virtual states, respectively. The phase-matching condition for the wave vectors of the anti-Stokes ($k_{as}$), pump ($k_1$), and Stokes ($k_s$) beams is shown in the right-hand panel.

**Table 16.2.** SRS frequencies of water observed on pumping with the fundamental wavelength of a Nd:YAG laser.

| Stokes shift (cm$^{-1}$) | Frequency ($\nu$) | Wavelength of ring (nm) |
| --- | --- | --- |
| 3000 | $\nu$ (-OH stretch) | 807 |
| 6000 | $2\nu$ (first overtone) | 650 |
| 9000 | $3\nu$ (second overtone) | 544 |
| 12 000 | $4\nu$ (third overtone) | 468 |

10 cm long water (H$_2$O) cell along its length. Red and green rings are obtained on the screen. Assuming that this effect is due to SRS involving the -OH Raman frequency, how do you understand the colored rings? Take the frequency of the active vibration in water (the symmetric stretch of OH) to be ~3000 cm$^{-1}$.

**Solution:** The fundamental wavelength of a Nd:YAG laser (1064 nm) corresponds to a frequency of 9398 cm$^{-1}$. Using an -OH frequency of 3000 cm$^{-1}$ and its second and third overtone Stokes' frequencies $2\nu$ and $3\nu$, the observed wavelengths are ~650 nm and ~544 nm. As the beam is focussed, it will diverge on the screen to appear as rings. Table 16.2 gives the possible Stokes shifts of the obtained rings from the pump wavelength of 1064 nm.

♣ As one might expect, heavy water ($D_2O$) has a lower frequency corresponding to OD vibration than that of OH. A mixture of $H_2O/D_2O$ is often used to obtain a white-light continuum via the SPM process (♠ See chapter 21). Acetone is another good solvent with which to observe these colored rings due to SRS.

## 16.9 Four-photon parametric generation

Similarly to the generation of entangled photons, four-photon parametric generation[1] occurs simultaneously with the SPM process. Photons at the laser frequency ($\omega_1$) parametrically generate photons that are emitted at the Stokes ($\omega_s$) and anti-Stokes ($\omega_a$) frequencies in an angular pattern due to the required phase-matching condition $k_{as} = 2k_1 - k_s$.

## 16.10 Cross-phase modulation

When two or more laser beams interact in a nonlinear medium, each beam experiences a change in phase due to $n_2(I)$. The change in the refractive index due to one beam affects the phase of the other beam. This process is known as cross-phase modulation (XPM). The application of XPM can be found, for example, in the OPA process (♠ see section 15.6), where a part of the signal pulse overlaps with the front part of the pump beam. Due to the SPM effect, this results in a decrease of its frequency relative to the rest of the beam. This process leads to frequency chirping of the signal beam. The XPM process contributes to the generation of a white-light continuum (WLC) in transparent media.


**Exercise 16.6.** In nonlinear optics laboratory demonstrations (say, A and B), two different laser pulses with identical repetition rates and average powers are used. The pulse durations used in the experiments are 10 ns (experiment A) and 10 fs (experiment B). The laser pulses are focused by a lens of focal length $f$ in each case.
   (i) Do the pulse durations change if focusing is performed by a thicker lens? Give reasons.
   (ii) If a sample with a nonzero $\chi^{(3)}$ is placed at the focus, describe the effects for the case of (a). thin and (b). thick samples in both experiments.
   **Solution:**
   (i) The pulse duration after the pulse has passed through a thick lens increases somewhat in experiment (B), as the effect of chirping is significant. The increase in the pulse duration depends on the thickness of the lens. In experiment (A), however, no measurable change is observed in the pulse duration.
   (ii) When a $\chi^{(3)}$ material is placed at the focus in experiment (A), it may generate a detectable third harmonic, as well as possible signals of SRS and SBS, depending on the ground-state vibrational frequencies of the material. The thicker sample also may show similar effects, provided that there is no loss of the incident laser beam due to absorption. On the other hand, in the case of experiment (B), in addition to all these effects, one of the major

---

[1] See, for exercise [6].

differences would be the observation of the phenomenon of SPM-induced white-light continuum generation (♠ wait for chapter 21 for details). The efficiency and quality of the continuum generation may vary with the thickness of the sample.

## 16.11 Self-steepening

Self-steepening is another higher-order nonlinear optical effects caused by the propagation of an intense laser beam through a nonlinear medium. It is the result of intensity-dependent group velocity ($v_g$). Self-steepening has the prominent effect on spectral broadening due to SPM under high-intensity femtosecond pulse excitation. When the group velocity depends not only on the frequency but also on incident intensity (♠ see section 23.4), short laser pulses undergo a process of *self-steepening*. This results in the center part of the pulse being slower than the wings of the pulse. As a result, the peak of the pulse is shifted towards the trailing edge of the pulse. If pulse becomes infinitely steep, it leads to the formation of optical shock waves. Recently, modulation instability and the self-steepening effect have been reported in negative-refractive-index materials.

## 16.12 Saturable absorption

As outlined at the beginning of this chapter, when a process involves transition between real energy levels, it is said to be a *nonparametric process*. This is in contrast to the case of *parametric processes* such as SHG and THG, in which the transitions take place via virtual levels. Some examples of nonparametric processes are saturable absorption (SA) and multiphoton absorption (MPA, ♠ see chapter 20).

In nonlinear optics, when light interacts with matter, not only the refractive index but also the absorption coefficient of the material become intensity dependent, affecting the light transmitted through it. As an example, a description of the SA process in a molecular system is given in figure 16.9. The molecules, initially in the ground state $E_0$, are excited to a non-equilibrium (Franck–Condon) excited state. After excited-state relaxation, $E_1$ is populated by $N_1$ molecules.

If the input irradiance is sufficiently high (but below the damage threshold of the material), the rates of absorption and emission become equal, which gives a saturation effect by transmitting the rest of the laser irradiance (i.e. $I_T = I$, where $I_T$ is the transmitted irradiance and $I$ is the input irradiance). The transmitted irradiance can be expressed in a similar way to equation (6.2) as

$$I_T = I \exp\left[ \sigma_{em}\left(1 - \frac{2N_0}{N}\right)LN \right],  \tag{16.8}$$

where $\sigma_{em}$ is the emission cross-section and $L$ is the thickness of the cell and $N$ is the total population (i.e. $N_0 + N_1$). If, at a given instance, $N_0$ is half of the total population, $I_T$ will be positive, giving rise to the SA effect.

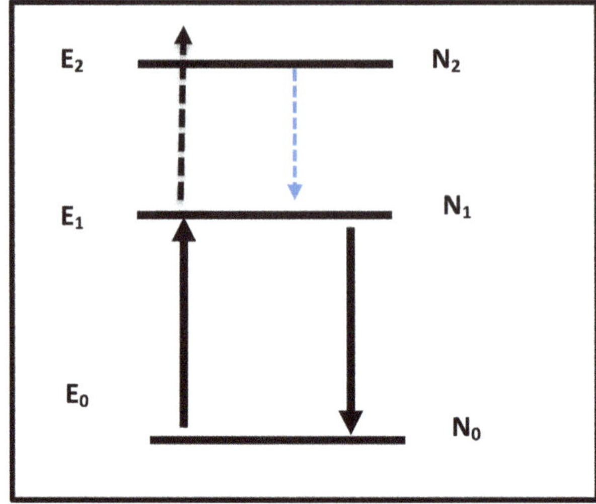

**Figure 16.9.** Three-level energy diagram for a saturable absorber. $E_0$ and $E_1$ are the ground and first excited singlet states with populations $N_0$ and $N_1$, respectively. If the number of incident photons is high, there is the further possibility of excitation into higher states (such as $E_2$) from $E_1$. The dashed downward arrow from $E_2$ to $E_1$ indicates nonradiative transition.

The absorption coefficient of the medium depends on the input laser intensity, resulting in the phenomenon of nonlinear absorption (NLA). In a nutshell, NLA phenomena can be classified as follows:

(i) *Saturable absorption (SA)*: at higher values of incident intensity, the absorption coefficient saturates, hence it decreases for an increase in the incident intensity. This process increases the transmittance of the sample with the input laser intensity.

(ii) *Reverse saturable absorption (RSA)*: on an increase in the incident intensity, the absorption coefficient increases further, resulting in a transmittance decrease with increasing input laser intensity. It is understood that further excitation to the $E_2$ level is necessary to avoid the limitation of two levels, a process known as excited-state absorption (ESA).

Figure 16.10 gives a schematic of the SA and RSA processes as a function of the incident photon energies. SA and RSA are nonparametric processes, as they also involve the real energy levels of the excited states. In NLA (due to processes (i) and (ii)), the change in the absorption coefficient is mainly due to the imaginary part of the refractive index. The NLA coefficient ($\alpha$) for input irradiance ($I$) follows the expression

$$\alpha = \alpha_0 \ / (1 + I/I_s). \tag{16.9}$$

Here, $I_s$ is the saturation intensity and $\alpha_0$ is the linear absorption coefficient. The processes of ESA and MPA are responsible for the phenomenon of RSA. A detailed description of the MPA process is given in chapter 20.

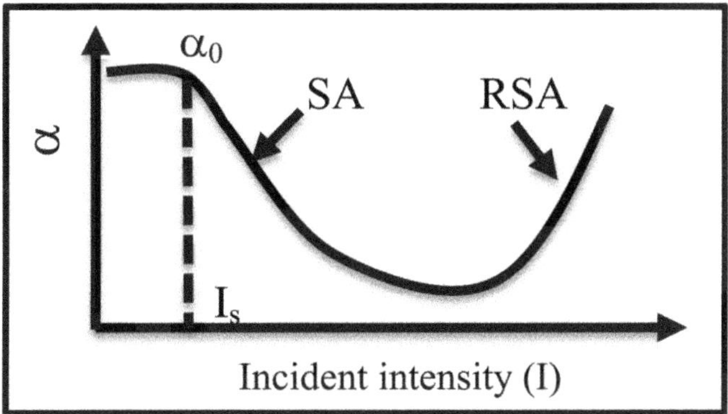



**Figure 16.10.** The absorption coefficient ($\alpha$) as a function of incident intensity ($I$). Linear absorption up to the value of $I_s$ (marked with a dashed vertical line) changes to SA and turns into RSA at higher incident intensities, as shown.

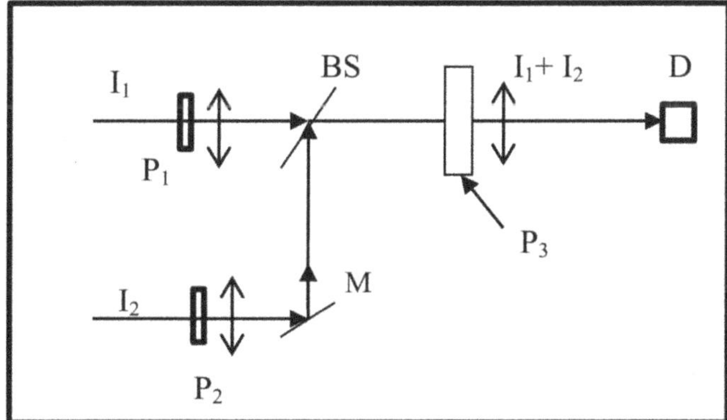



**Figure 16.11.** Schematic representation of a photonic circuit containing three polarizers (P1, P2, and P3). Two polarized light beams with intensities $I_1$ and $I_2$ are combined at the beam splitter (BS). M is the mirror and D is the detector. Double arrows indicate the state of polarization (SoP) of the light, which can be varied (see text for details).

♣ In a standard SA material (such as the dye IR 26), while the dynamics at 1064 nm is explained by two-photon absorption (2PA), at 532 nm it is dominated by the ESA process.

## 16.13 Photonic circuit based on the SA effect

As mentioned in section 1.5, a simple electronic circuit may consist of a resistor, a capacitor, a battery, and a voltmeter. Similarly, a simple *'photonic circuit'* can be made using some optical components such as polarizers, mirrors, beam-splitters, and a detector, as shown in figure 16.11. Here, the detected signal corresponds to the intensities and the states of polarization (SOP) of light beams $I_1$ and $I_2$.

♣ For this photonic circuit, a truth table similar to that for an electronic gate can be made, as shown in table 16.3.

**Table 16.3.** Truth table for a photonic *AND* gate. Here, $2I$ is the resultant maximum intensity. The orientations of the polarizers are indicated by (1) for parallel and (0) for perpendicular to the SoP of the incident beam.

| Orientation of $P_1$ | Orientation of $P_2$ | Orientation of $P_3$ | Detected output |
| --- | --- | --- | --- |
| 1 | 1 | 1 | $2I$ |
| 0 | 0 | 0 | $2I$ |
| 1 | 1 | 0 | 0 |
| 1 | 0 | 0 | $I$ |
| 0 | 1 | 0 | $I$ |
| 0 | 0 | 1 | 0 |
| 1 | 0 | 1 | $I$ |
| 0 | 1 | 1 | $I$ |

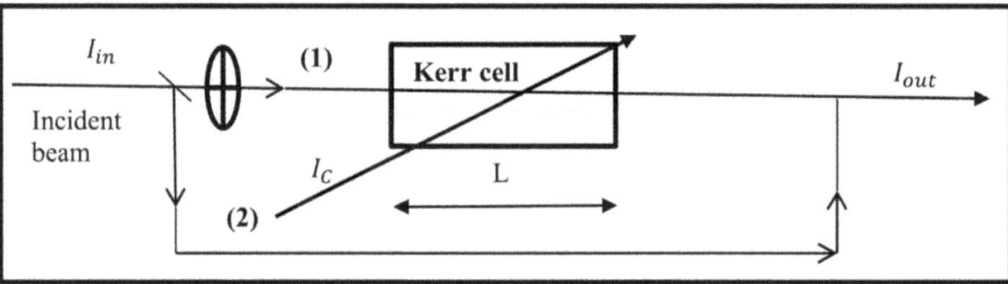


**Figure 16.12.** Kerr cell of length $L$ in a photonic circuit after the incident beam is split. In case (1), a shutter is used to control the output $I_{out}$. In the second case (2), the control light $I_C$ changes the phase of the light in the Kerr cell to influence the output without the use of a shutter.

♣ Similarly, other gates can be realized using photonic circuits and nonlinear optics.

Any SA material with a fast response can be used as an excellent element of a photonic circuit. Consider the sample cell with an SA material of length $L$ (in this case known as a Kerr cell) shown in figure 16.12. The incident light ($I_{in}$) is split into two replica beams. There are two cases shown in the figure. In one of them, the shutter (1) is used to control the output based on the phase difference created by the Kerr cell. When the shutter is on (i.e. the light passes through), the interference between the direct path and the other path can act as a switch. In the second case, a control light passes through an SA material to drive the Kerr cell above or below the SA threshold ($I_S$).

Figure 16.13 shows the transmission response of the SA material for case 2. The following three steps help in the understanding of this photonic system:

    i. Using the input laser beam, keep the system just below $I_S$ (i.e. at $I_L$).
    ii. Flip the system to $> I_S$ by introducing a control light ($I_C$).
    iii. Bring the system back below $I_L$ by controlling the intensity of $I_C$, when required.

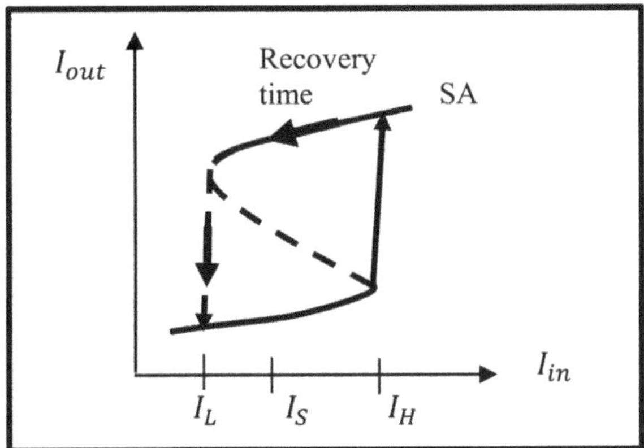

**Figure 16.13.** Response of an SA material ($I_{out}$) as a function of incident intensity ($I_{in}$). A hysteresis-type curve is observed when an SA material is subjected to a control light ($I_C$), as shown in figure 16.12. The value of $I_C$ can lie either below $I_L$, in the $I_S$ region, or above $I_H$.

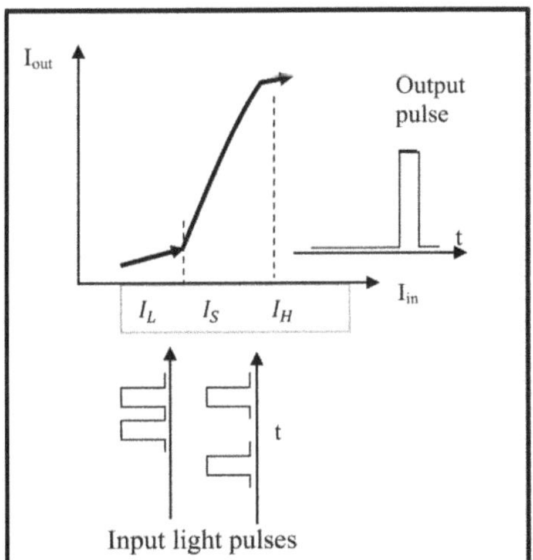

| Truth Table | | | |
|---|---|---|---|
| Conditions | 1/0 | 1/0 | Output |
| Input on, (Control blocked) | 1 | 0 | No |
| Input blocked, ( control on) | 0 | 1 | No |
| Both on | 1 | 1 | Yes |
| Both blocked | 0 | 0 | No |

**Figure 16.14.** Realization of an AND gate using the Kerr cell of figure 16.12. A series of input pulses is used to drive the Kerr cell above or below $I_S$ to get the required output (left panel). Note that there is no output for the cases in which either the pulse or the control light is absent. The truth table for the *AND* gate is given in the right-hand panel.

**Exercise 16.7.** Using the phenomenon of SA in the linear region, show how a photonic AND gate can be realized.

**Solution:** Using the phenomenon of SA, a photonic gate can be realized as shown in figure 16.14. Let us take two light pulses, shown as square pulses. It can be seen that only when both light pulses are present in the system is an output pulse observed. This is due to the fact that at low intensity, the input pulses are absorbed by the system. However, the control light can drive the system into the SA region that transmits the input pulses. This demonstration is the basis of an 'all-optical network' for telecommunications. The truth table for the AND photonic gate is also given.

## Questions and problems

1. What happens when strong laser radiation interacts with complex many-body systems? (Ignore thermal effects).
2. The DC Kerr effect is observed when a DC electric field ($E$) is applied across a material such as quartz. Explain how ultrafast electro-optic birefringence is responsible for the *optical* Kerr effect.
3. THG using the $\chi^{(3)}$ process is not efficient. How one can obtain THG from a $Nd^{+3}$: YAG laser by using an $\chi^{(2)}$ material?
4. For a centrosymmetric material with a refractive index of $n_0$, obtain the equation for the intensity-dependent refractive index $n(I)$.
5. Identify the nonlinear optical processes to which temporal and spatial SPM contribute.
6. With reference to the self-focusing and optical rectification phenomena, can a material that shows self-focusing behavior be used to achieve optical rectification? (Hint: use the symmetry properties of the susceptibilities).
7. Self-phase modulation (SPM) leads to 'frequency broadening (or chirping)' of an ultrashort pulse. For a Gaussian beam, the intensity ($I(t)$) as a function of time is given by $I(t) = I_0 e^{(-\frac{t}{\tau})^2}$, where $\tau$ is the pulse duration and $I_0$ is the peak intensity. Find
   (i) the expression for the corresponding frequency spread. (ii) An ultrafast laser is rated to deliver pulses of a 5 fs duration. If the measured pulse duration is 10 fs at the edge of an optical table that has a length of 2 m, give your comments.
8. Identify (and give reasons) the order of optical nonlinearity in which the following phenomena are invoked: (i) stimulated Raman scattering; (ii) stimulated Brillouin scattering; (iii) difference frequency generation; (iii) optical phase conjugation.
9. Illustrate the distinction between saturable absorption (SA) and reverse saturable absorption (RSA) using energy-level diagrams.
10. Make a photonic circuit using polarizers and write down the truth table for the resulting optical gate.
11. A geometry for four-wave mixing (with $\omega_1$, $\omega_2$, $\omega_3$, and $\omega_4$,) is shown in figure 16.15 (left-hand panel). In the parametric process shown in the right-hand panel, what is the value of frequency $\omega_4$?
12. Draw the phase-matching diagram (momentum conservation vector diagram) for the transitions that result from six-wave mixing for the frequencies ($\omega_1$ to $\omega_5$) for the nonparametric process shown in figure 16.16. Note that $\omega_1 = \omega_2 = \omega_5$ are *pump* frequencies from the same laser source. Another beam with frequency $\omega_3 < \omega_1$ is incident on the sample. $\omega_4$ can be called the *dump* frequency. (Hint: here, $\omega_6$ is the final signal frequency.

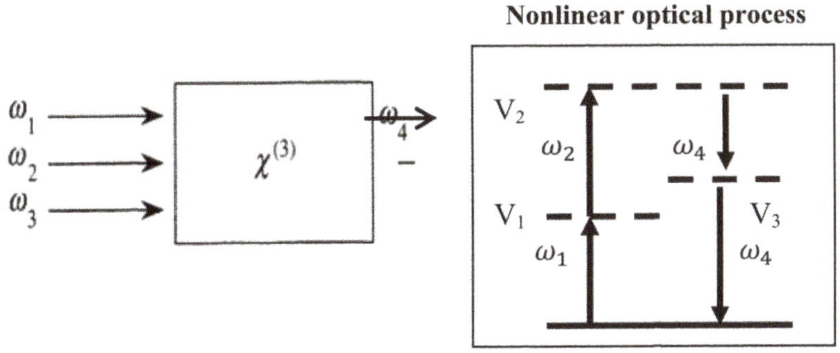

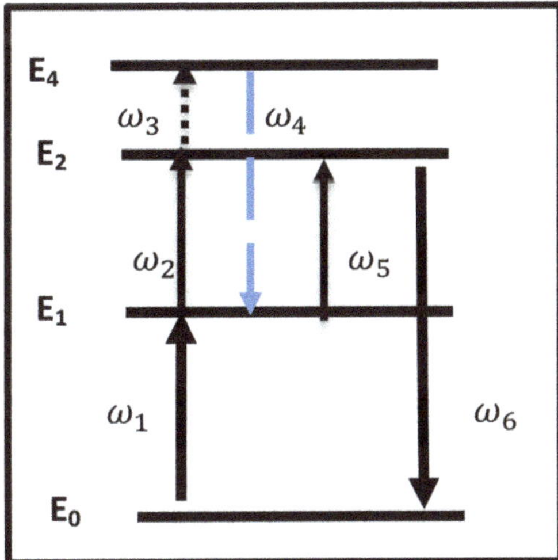



**Figure 16.15.** Figure for question 11.



**Figure 16.16.** Pump transitions ($\omega_1$ to $\omega_5$) and signal frequency ($\omega_6$) for question 12.

To draw the momentum conservation diagram, you will have to take $k_1$, $k_2$, and $k_5$ as collinear in the forward direction along with $k_3$ at an angle to $k_1$. The wave vector $k_4$ should be incident in the backward geometry.)

## Bibliography


[1] Bokor J, Bucksbaum P H and Freeman R R 1983 Generation of 35.5-nm coherent radiation *Opt. Lett.* **8** 217–9

[2] Deeg F W and Fayer M D 1989 Analysis of complex molecular dynamics in an organic liquid by polarization selective subpicosecond transient grating experiments *J. Chem. Phys.* **91** 2269–79

[3] Sheik-Bahae M *et al* 1990 Sensitive measurement of optical nonlinearities using a single beam *IEEE J. Quantum Electron.* **26** 760–9

[4] Klewitz S *et al* 1996 Tunable stimulated Raman scattering by pumping with Bessel beams *Opt. Lett.* **21** 248–50

[5] Sailaja R, Sreeja V and Bisht PB 2005 Studies of self-phase modulation under cw and picosecond laser pumping: White light continuum generation in water *Indian J. Phys.* **79** 1299–304

[6] Xing Q, Yoo K M and Alfano R R Apr. 1993 Conical emission by four-photon parametric generation by using femtosecond laser pulses *Appl. Opt.* **32** 2087–9

[7] DeMartini F *et al* 1967 Self-steepening of light pulses *Phys. Rev.* **164** 312–23

[8] Kranitzky W *et al* 1981 A new infrared laser dye of superior photostability tunable to 1.24 µm with picosecond excitation *Opt. Commun.* **36** 149–52

**IOP** Publishing

An Introduction to Photonics and Laser Physics
with Applications

**Prem B Bisht**

# Chapter 17

## Mode locking

The technique of Q-switching (♠ see chapter 13) generates optical pulses that have nanosecond durations. To generate laser pulses of even shorter durations, the method of *mode locking* is used. In general, continuous-wave (CW) lasers are known to be monochromatic and have laser line widths of the order of nanometers to picometers in the visible region. This is not so for the case of ultrafast lasers! The spectral width of an ultrafast laser pulse is inversely proportional to its pulse duration. An example with a broad spectral profile is shown in the figure, the details of which are given in this chapter. While *active* mode locking is based on fast shutters driven by radio-frequency (RF) waves, the principles of nonlinear optics are used in *passive* mode-locking techniques. The all-optical important technique of *Kerr-lens mode locking (KLM)* works by applying $\chi^{(3)}$. This is the reason that this chapter on *ultrafast pulse generation* appears after the introduction to nonlinear optics in the previous chapters.

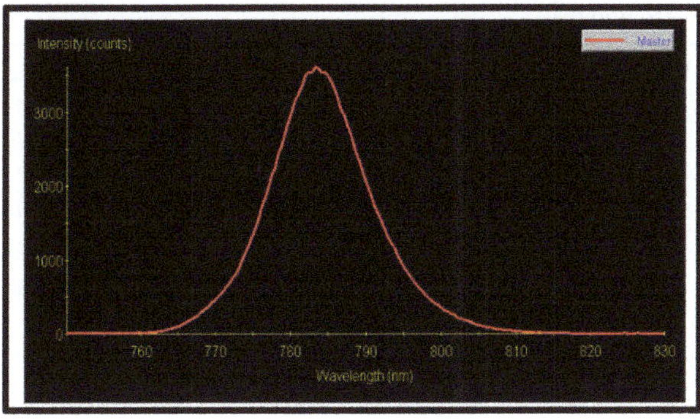

**Learning objectives**
**After reading this chapter, the learner will be able to:**
Identify the requirement for ultrafast pulses;
Describe the general principle of mode locking;
List and classify the methods of mode locking;
Know the techniques of active, passive, and synchronous mode locking;
Explain passive mode locking due to the effect of saturable absorption (SA);
Demonstrate pulse shortening due to effect of SA;
Define Kerr-lens mode locking;
Differentiate between the spectra of CW and pulsed lasers.

## 17.1 The requirement for short-duration optical pulses

We previously introduced Q-switching, a technique used to generate laser pulses of the order of ~10 ns. We also know that it uses a flash lamp (with an electronic control) for optical pumping of the gain medium. Just for an estimate of the speed of electronics, let us take an example of its basic components. For an electronic circuit containing a typical capacitor value (C) in nF and that of a resistor (R) in k $\Omega$, the time constant (RC) is on the microsecond scale. This timescale can be reduced to a few nanoseconds by making use of the stray capacitance (pF) between two wires or by the use of even faster electronic chips. A typical teaching laboratory oscilloscope has a frequency bandwidth of 15–100 MHz and can resolve timescales of the order of nanoseconds, as can be seen from table 17.1. Modern oscilloscopes can have a resolution of the order of 10 ps.

However, the fundamental photonic processes in nature, such as photosynthesis in plants or optical processes in retinal protein, take place on the femtosecond timescale. Therefore, to gain control over nature, it is essential to have faster time resolution in the excitation as well as in the detection response of photonic devices. Therefore, in the following section, we will focus on the methods used to generate ultrafast pulses.

**Table 17.1.** The frequency bandwidths of electronic devices and corresponding estimates of their time resolution.

| Frequency bandwidth | Time-resolution estimate |
| --- | --- |
| 15 MHz | 66 667 ps (66.7 ns) |
| 60 MHz | 16 667 ps (16.7 ns) |
| 100 MHz | 10 000 ps (10 ns) |
| 500 MHz | 2000 ps (2 ns) |
| 1000 MHz (1 GHz) | 1000 ps (1 ns) |
| 5000 MHz (5 GHz) | 200 ps |
| 10 000 MHz (10 GHz) | 100 ps |
| 100 000 MHz (100 GHz) | 10 ps |

♣ The methods used to detect ultrafast pulses will be described in the next chapter. To appreciate these concepts better, you need to understand polarization and the nonlinear optics introduced in chapters 3 and 14, respectively.

## 17.2 Mode locking of lasers

Resonance is known to occur in systems such as two pendula connected by a spring or organ pipes with slightly different frequencies kept close to each other. Mode locking is one such phenomenon, in which the longitudinal modes of a laser are synchronized. It is one of the important methods of ultrafast pulse generation. As described in chapter 9 (♠ see section 9.2.6), the longitudinal modes of the laser oscillate at frequencies equally separated by the intermodal frequency, c/2d where d is the distance between the two mirrors. The corresponding time 2d/c is known as the cavity round-trip time ($\tau_{rt}$.) These modes normally oscillate independently and are called *free-running modes*.

### 17.2.1 Longitudinal modes with random phases

Let the amplitude of the $n$th longitudinal mode of the laser be expressed as $E(t) = E_n e^{i(\omega_n t - \phi_n)}$. Here, $\omega_n$ is the frequency and $\phi_n$ is the phase of the mode. Let there be $N$ modes of equal amplitude ($E_0$) oscillating simultaneously in the cavity. There are two cases in which we can observe resultant output from the cavity, viz. (i) modes oscillating in random phases, and (ii) modes oscillating with a common phase.

For the electric field amplitude $E(t) = E_0 e^{i(\omega_0 t + \phi_0)}$, the resultant field intensity can be given by

$$I(t) = |E(t)E^*(t)| \equiv E_0^2.$$

The intensity for $N$ modes with electric fields $E(t) = E_0 e^{i(\omega_n t + \phi_n)}$ can be expressed as

$$I(t) = E_0^2 \left| \sum_{n=1}^{N} e^{i(\omega_n t + \phi_n)} \right| \equiv NE_0^2. \tag{17.1}$$

Equation (17.1) shows that the intensity resulting from the modes is proportional to $N$, as also indicated in figure 17.1.

### 17.2.2 Longitudinal modes with identical phases

The amplitude resulting from $N$ longitudinal modes (from zero to $N - 1$) can be rewritten as

$$E(t) = E_0 \sum_{n=0}^{N-1} e^{i(\omega_n t + \phi_n)}.$$

Let us take an identical phase, i.e. $\phi_n = \phi_0$ for all $n$. Equating various phase values to a common value ($\phi_0$ here) is also known as freezing the phases. The phase term in the expression of $E(t)$ is taken out of the summation as follows:

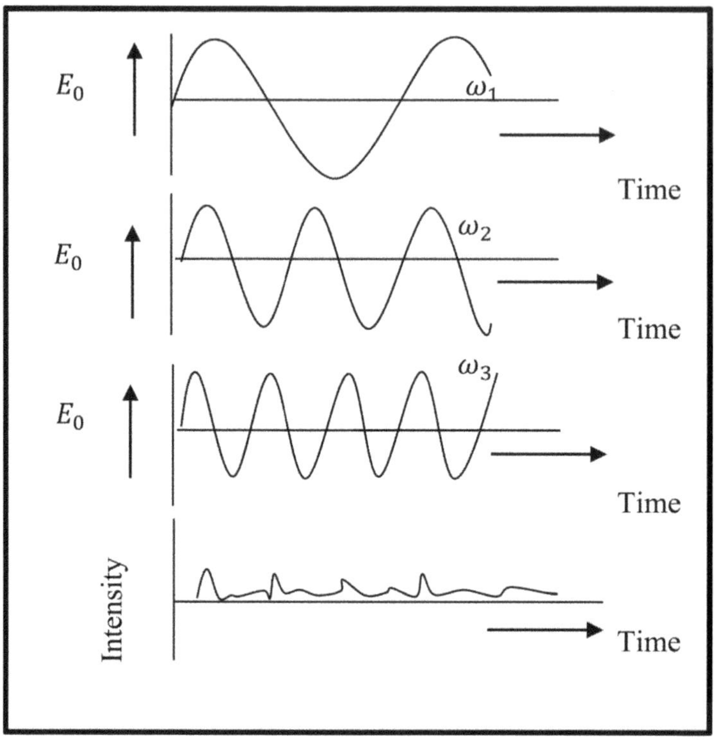

**Figure 17.1.** Three frequencies $\omega_1$, $\omega_2$, and $\omega_3$ of equal amplitude $E_0$ and with random phases result in an intensity pattern given by equation (17.1).

$$E(t) = E_0 e^{i\phi_0} \sum_{n=0}^{N-1} e^{i\omega_n t}.$$

If we define $\omega_n = \omega_{N-1} - n\Delta\omega$, the resultant electric field can be written as

$$E(t) = E_0 e^{i\phi_0} \sum_{n=0}^{N-1} e^{i(\omega_{N-1} - n\Delta\omega)t}$$

or

$$E(t) = E_0 e^{i(\omega_{N-1} t + \phi_0)} \sum_{n=0}^{N-1} e^{i(n\Delta\omega t)}.$$

This can be expressed as a mathematical series as follows:

$$E(t) = E_0 e^{i(\omega_{N-1} t + \phi_n)}[1 + e^{i\Delta\omega t} + e^{i2\Delta\omega t} + e^{i3\Delta\omega t} + \ldots\ldots\ldots e^{i(N-1)\Delta\omega t}]$$

$$= E_0 e^{i(\omega_{N-1} t + \phi_n)} \left[ \frac{\left(1 - e^{iN\Delta\omega t}\right)}{\left(1 - e^{i\Delta\omega t}\right)} \right].$$

The intensity is then given by

$$I(t) = E_0^2 \left| \left[ \frac{\left(1 - e^{iN\Delta\omega t}\right)}{\left(1 - e^{i\Delta\omega t}\right)} \right] \right|^2 = E_0^2 \frac{\sin^2\left(\frac{N\Delta\omega t}{2}\right)}{\sin^2\left(\frac{\Delta\omega t}{2}\right)}. \tag{17.2}$$

Here, the difference in the angular frequency ($\Delta\omega$) is given by $\Delta\omega = \omega_{n+1} - \omega_n$. This can equivalently be written as $2\pi\Delta\nu = \frac{2\pi c}{nd}$. The expression given in equation 17.2 varies with time $t$. The function $I(t)$ reaches a *maximum* for the minimum value of $\sin^2\left(\frac{\Delta\omega t}{2}\right)$, i.e. for $\frac{\Delta\omega t}{2} = 0, \pi, 2\pi, 3\pi, ....n\pi$. Furthermore, $I(t)$ reaches a *minimum* for the minimum value of the numerator, i.e. $\frac{N\Delta\omega t}{2} = p\pi$. Here, $n$ and $p$ are positive integers. As we can see, the $I(t)$ function of a phase-locked laser has maxima and minima that are functions of time, i.e. we get pulsed output.

### 17.2.2.1 Separation of maxima

The repetition rate of the laser pulses can be found by obtaining the difference between consecutive maxima. For a limiting value of $\frac{\Delta\omega t}{2} \to 0$,

$$I = N^2 E_0^2. \tag{17.3}$$

In contrast to the case of modes with random phases (equation 17.1), it can be observed here that the resultant intensity is $N^2$ times the intensity of the single mode when the phases are locked together. This shows that the greater the number of oscillating modes, the better the output intensity of the mode-locked pulses.

♣ This explains the fact that gain media with broad emission bandwidths are good candidates for use in mode-locked lasers. For example, Ti:sapphire is capable of generating tunable femtosecond-duration pulses over a large spectral range of 700–1100 nm.

In equation (17.2), the maxima occur at $\frac{\Delta\omega t_n}{2} = n\pi$ and $\frac{\Delta\omega t_{n+1}}{2} = (n + 1)\pi$. This implies that $t_n = \frac{2n\pi}{\Delta\omega}$ and $t_{n+1} = \frac{2(n + 1)\pi}{\Delta\omega}$. Therefore, the maxima are separated by

$$\Delta t_{\text{sep}} = t_{n+1} - t_n = \frac{2(n + 1)\pi}{\Delta\omega} - \frac{2n\pi}{\Delta\omega} = \frac{1}{\Delta\nu}. \tag{17.4}$$

Pulses separated by $\Delta t_{\text{sep}}$ are shown in figure 17.2.

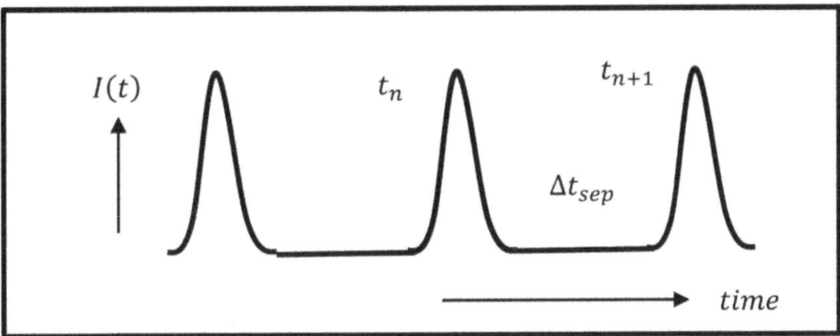


**Figure 17.2.** Under phase-locked conditions, separation of the intensity maxima yields the pulse repetition rate of the laser.

♣ Knowledge about the phase of a well-separated train of pulses is used in the optical frequency comb technique. Theodor Hänsch and John Hall were awarded the Nobel Prize in 2005 for the development of laser-based precision spectroscopy based on this technique. The frequency comb is known as an optical ruler, as it allows the precise determination of an unknown frequency by measuring the beats.

### 17.2.2.2 Separation of minima

The pulse duration is related to the difference between two minima. Its function reaches a minimum value for the minimum numerator. Minima occur at $\frac{N\Delta\omega t_n}{2} = p\pi$ and $\frac{N\Delta\omega t_{n+1}}{2} = (p+1)\pi$. These values give the corresponding times of $t'_n = \frac{2p\pi}{N\Delta\omega}$ and

$$t'_{n+1} = \frac{2(p+1)\pi}{N\Delta\omega}.$$

Therefore, the shortest pulse duration ($\tau_p$) is

$$\tau_p = \Delta t'_{limit} = t'_{n+1} - t_n = \frac{2(p+1)\pi}{N\Delta\omega} - \frac{2p\pi}{N\Delta\omega} = \frac{1}{N\Delta\nu}. \tag{17.5}$$

If we compare this result with the bandwidth of a gain medium (e.g. see the time–bandwidth relation, section 8.5.2.1), we can see that $N\Delta\nu$ is nothing but the entire gain bandwidth. Therefore, if the total gain bandwidth is used by a single mode, the shortest pulse duration limit can be equal to $\frac{1}{\Delta\nu}$. For example, if we consider the visible region (say 400–700 nm) as the entire frequency bandwidth of a gain medium, the corresponding $\tau_p$ is ~3 fs. (♠ Also see question 6.)

**Exercise 17.1.** What can be the fastest pulse that can be obtained from a laser gain medium that has a bandwidth of 50 nm at 500 nm?

**Solution:** Let us take the gain width of 50 nm within 475–525 nm. The corresponding frequency bandwidth is $6.02 \times 10^{13}$ Hz. From the time–bandwidth relation, this gives an estimate of the minimum pulse duration of 16.6 fs.

**Exercise 17.2.** The length of the $Nd^{+3}$: YAG rod (i.e. the YAG rod) is 25 mm in a mode-locked YAG laser. The refractive index of the YAG rod is 1.80. The gain bandwidth is 0.117 THz. Assuming that the laser mirrors are close to the end faces of the rod, find the following:

(i) the effective length ($d$) of the laser cavity, (ii) the longitudinal mode separation, (iii) the number of oscillating modes, (iv) the pulse-to-pulse separation, (v) the pulse duration

**Solution:** The wavelength of a $Nd^{+3}$: YAG laser is 1064 nm. The gain bandwidth is given here as 0.117 THz.

    (i) The effective length ($d$) of the laser cavity = cavity length × refractive index = 0.36 mm.

(ii) The longitudinal mode separation or free spectral range (FSR) $= c/2d = 3.33 \times 10^8$ Hz.

(iii) The number of oscillating modes is given by the bandwidth divided by the FSR, which equals $117 \times 10^9/3.33 \times 10^8 = 350$.

(iv) The pulse-to-pulse separation is $1/FSR = 3$ ns.

(v) The pulse duration of the mode-locked pulse $= 8.6$ ps.

## 17.3 Methods of mode locking

There are several ways to mode lock a laser. The various principles of mode locking are based on the properties of the gain media as well as on the required pulse duration. For instance, the electro-optic shutter-based techniques (known as *active* mode locking) are easy to implement but can only produce pulses of the order of picoseconds.

On the other hand, femtosecond-duration pulses can be obtained by implementing the nonlinear optical method (known as *passive* mode locking) as well as methods based on the optical Kerr effect (♣ see section 16.6). Either a single technique or a combination of these are used for the mode locking of commercial lasers. The chart given in figure 17.3 gives a schematic outline of the various mode-locking techniques.

### 17.3.1 Active mode locking: the acousto-optic method

In active mode locking, a fast shutter is introduced into the cavity to couple the longitudinal modes by locking their phases together to obtain a periodic pulse train.

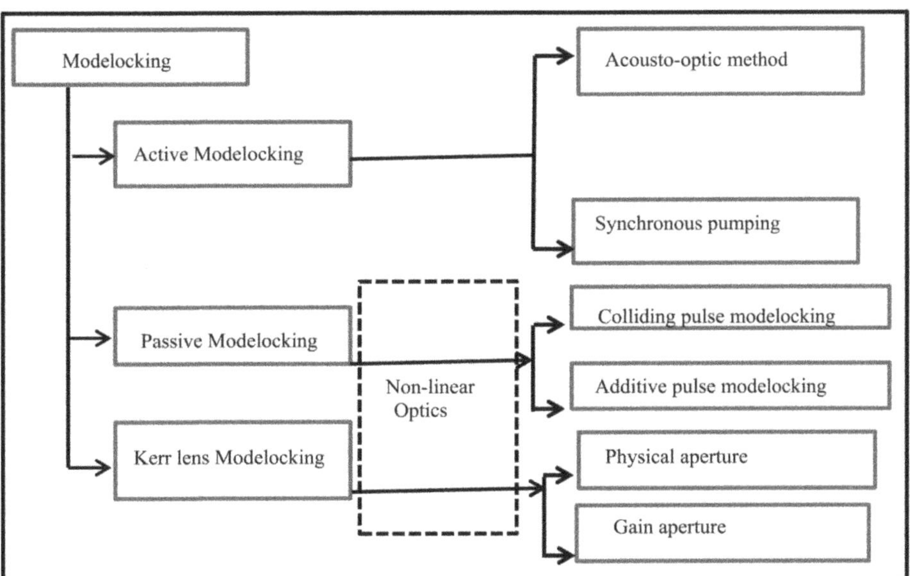


**Figure 17.3.** A classification of mode-locking techniques. The dashed box indicates the techniques that involve *nonlinear optical phenomena.*

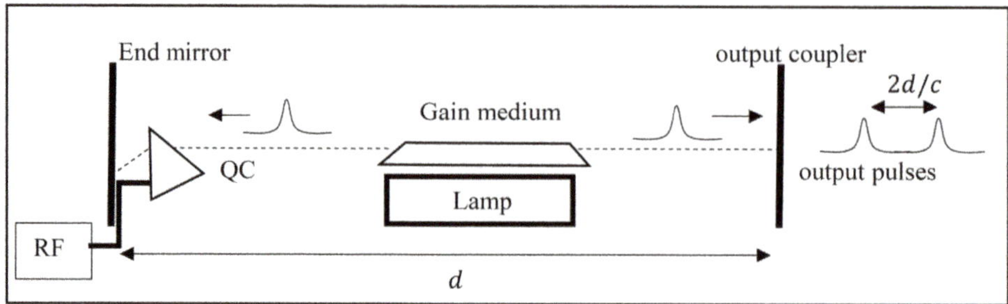


**Figure 17.4.** The principle of active mode locking. An RF is applied to the quartz crystal (ML) that acts as a fast optical switch.

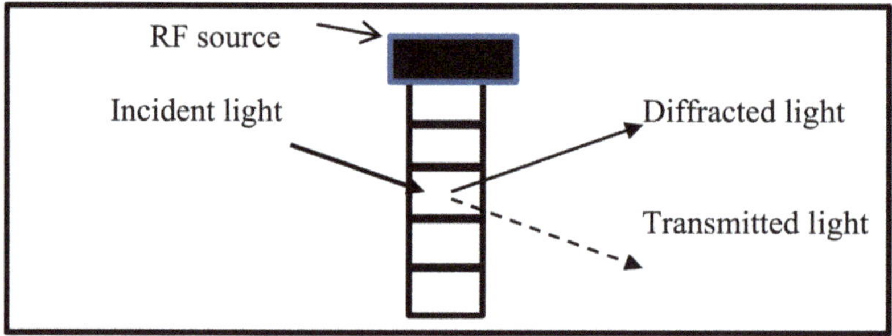


**Figure 17.5.** The quartz crystal diffracts light at the density grating created by the RF.

These shutters, based on a combination of electronics and optics, create small losses in the cavity that are sufficient to produce a pulsed output from low-power lasers at high repetition rates, as described in the following. Figure 17.4 gives a schematic diagram of one of the active mode-locking methods. Here, a quartz crystal is used as the acousto-optic (AO) material. The crystal acts as an optical switch when placed inside the laser cavity, close to one of the mirrors. The switch is used to allow oscillation in the cavity during every cavity round trip. When the switch blocks the cavity, no oscillation can take place. If the switch is opened, the electric fields of all the available modes are simultaneously permitted to be in phase. The switching action is achieved by applying powerful RF waves at a high frequency ($\omega \sim 20$ MHz) obtained from a piezo-electric transducer (♠ see also section 13.3.2.2). Due to photo-elasticity, the RF waves create a density grating in the form of rarefactions and compressions of the material. Thus, an acoustic grating pattern is formed in the quartz crystal, as shown in figure 17.5.

The acoustic grating establishes a standing-wave pattern across the crystal. The grating can diffract the intracavity beam away from the oscillation path and acts as an AO shutter (♠ see question 9). Two pulses simultaneously pass through the cavity when the switch is opened. Thus, the repetition rate of the laser is twice the the RF.

An important point is that the length of the cavity is controlled in such a way that the cavity round-trip frequency is close to the applied RF. Fine tuning of the length of the cavity is necessary to match the value of the RF used for the shutter. When the cavity length is adjusted to the cavity round-trip time $\tau_{rp}$, as each mode develops, it generates sidebands that drive the adjacent modes in phase. In summary, active mode locking involves several procedures, as follows:

    (i) A switch or a shutter is used to simultaneously grow all modes (i.e. all modes develop simultaneously and hence are in phase).

    (ii) The frequency of the shutter is tuned to $\tau_{rp}$. This is achieved by matching the cavity length ($d$) with $\tau_{rp}(=2d/c)$ to match the shutter frequency.

♣ Since the late 1970s, commercial Nd: YAG lasers with low pulse energies have used active mode locking to deliver pulses of about ~100 ps at a repetition rate of ~80 MHz.

♣ If the cavity length is not matched with the RF value, there is a possibility of occurrence of self-Q-switching. This occurs when the energy stored in the laser cavity overcomes the losses when the cavity length is not matched for mode locking. The resulting uncontrolled giant Q-switched pulse can damage the optics.

♣ An RF pulse is used to activate the AO shutter at $\tau_{rp}$ to diffract the laser modes at the end mirror. This method is used for low-power lasers. In absence of an RF pulse, the shutter passes two counter-propagating pulses, as shown in figure 17.4. Therefore, the repetition rate of the mode-locked output is double that of the applied RF frequency.

### 17.3.2 Synchronous pumping

This method is used when a pre-mode-locked laser is used as a pump for another non-mode-locked laser (figure 17.6). The only important aspect in achieving synchronous mode locking is that the cavity round-trip time of the pump laser cavity must equal that of the laser to be mode locked, such as a dye laser. This is

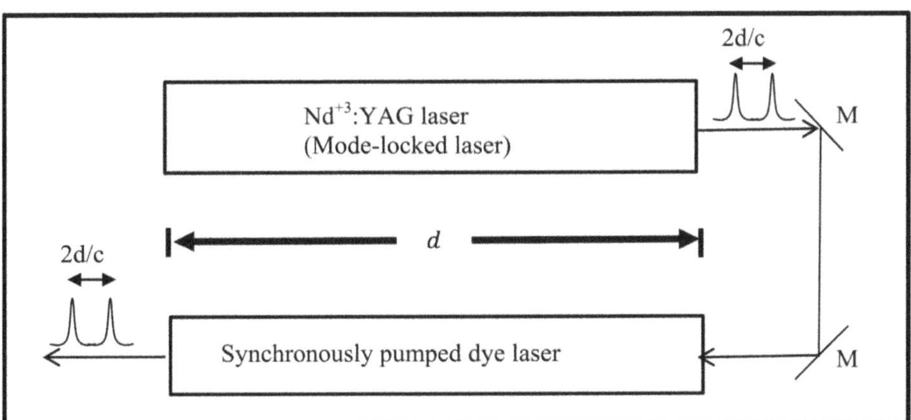


**Figure 17.6.** Synchronous pumping of a dye laser by a mode-locked Nd$^{+3}$:YAG laser. Beam-folding mirrors are denoted by M. The distances ($d$) between the cavity mirrors of both lasers are the same, resulting in an output pulse separation of $2d/c$.

done by simply matching the length of the dye-laser cavity with that of the pump laser. This is due to the fact that the pulses produced at the repetition rate of the pump laser are coupled out to the cavity of the dye laser. As a side effect, this also shortens the pulse duration due to the effect of saturable absorption (♠ see section 17.4).

♣ Synchronously pumped dye lasers deliver approximately picosecond-duration pulses at the repetition rate of the pump laser. Cavity dumping using another acoustic modulator reduces the repetition rate.

### 17.3.3 Passive mode locking

Passive mode locking in ultrafast lasers is based on the phenomenon of saturable absorption (SA, ♠ see section 16.12). Therefore, the materials used in passive mode locking are known as saturable absorbers. Several organic compounds exhibit the SA phenomenon. Organic dyes (such as IR 26) have been used in commercial lasers as passive mode lockers. The stability of the SA material remains a cause of concern, as the material in the laser cavity is exposed to high intracavity power. Frequent replacement of the dye solution is needed to avoid the effects of bleaching. Therefore, semiconductor-based SA materials (SESAMs), which are new materials, are under development for commercial applications even at high power levels.

As outlined in section 16.12, in passive mode locking, as the absorption is a function of the incident intensity, weak modes are absorbed by the SA material, while intense modes are transmitted and subsequently amplified in the cavity. This is shown in figure 17.7.

♣ Recent studies of standard SA materials have indicated the predominantly complex nature of $\chi^{(3)}$, which has a dominant contribution from the imaginary part of the refractive index. The imaginary part belongs to the absorptive type of nonlinearity. Hence, strongly absorbing materials that also have the ability to amplify the signal at nearly the same wavelength are excellent passive mode lockers.

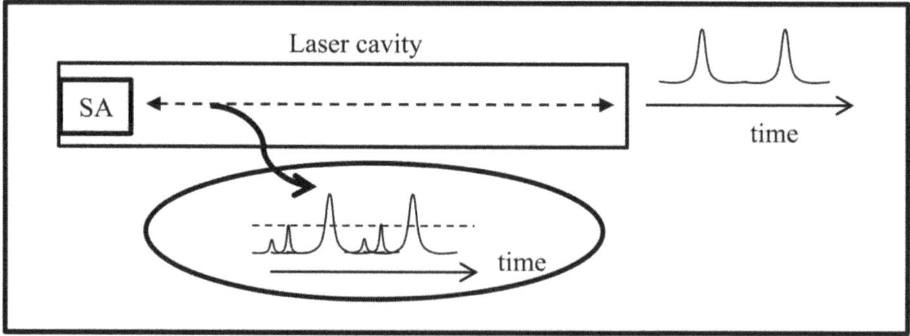


**Figure 17.7.** Intense modes are transmitted from the laser cavity after amplification by the gain medium (top). This happens as the weak modes (shown in the bottom enclosure) are absorbed, since they fall below the absorption threshold (dashed line) of the SA material.

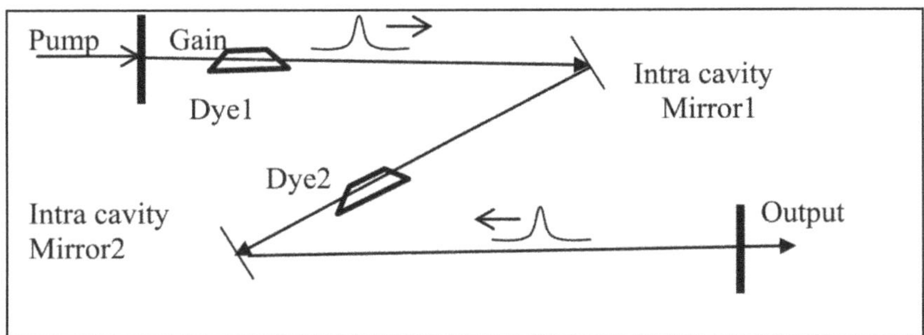


**Figure 17.8.** Colliding-pulse mode locking: Dye1, pumped by the external laser, acts as the gain medium, while Dye2 is an SA material.

### 17.3.3.1 Colliding-pulse mode locking (CPM)

This technique is also based on the principle of SA. Here, in addition to the gain medium (commonly synchronously pumped), a thin SA material (generally a dye solution in the form of a jet) is inserted into the cavity[1]. The design of CPM makes use of carefully aligned intracavity mirrors, as indicated in figure 17.8.

The two counter-propagating pulses meet at the SA material, combining their intensities (figure 17.8). This helps to saturate the absorption; the pulse is transmitted due to the transient bleaching effect of the SA material. The interference pattern at the thin absorber further increases the bleaching effect. This technique also shortens the pulse (♠ see section 17.4).

♣ The insertion of the SA jet adds extra length to the cavity. Therefore, fine adjustment of the cavity is required for synchronous pumping.

♣ Another term known as 'cavity dumping' is useful here. The sequence of pulses obtained by active mode locking is passed through a Pockels cell (♠ see chapter 3) in the dye-laser cavity. This is done to rotate its polarization of the part of the pulse train traveling in the cavity with a timing such that the half of the pulse train is blocked with the help of a quarter-wave plate. Cavity dumping helps in the selection of a particular pulse to be injected into the amplifier. Can you identify the similarity between cavity dumping and the principle of Q-switching? Note that the timescales are controlled here by the Pockels effect.

♣ Dye solutions are also used in low-repetition-rate lasers, in combination with active mode locking. As an excellent example, Nd:YAG (10 Hz) laser mode locking using a combination of an SA dye and another SA material (a layered material made of zinc oxide and gold nanoparticles) is shown in figure 17.9. In this laser, the separation of the actively mode-locked pulse train is controlled by an RF pulse (as described in section 17.3.1) for each pump flash. On introducing a combination of SA materials, a single pulse is obtained, as the weaker modes are absorbed by the SA materials.

---

[1] These lasers include both media (gain and SA) in the form of dye jets circulated by pumps.

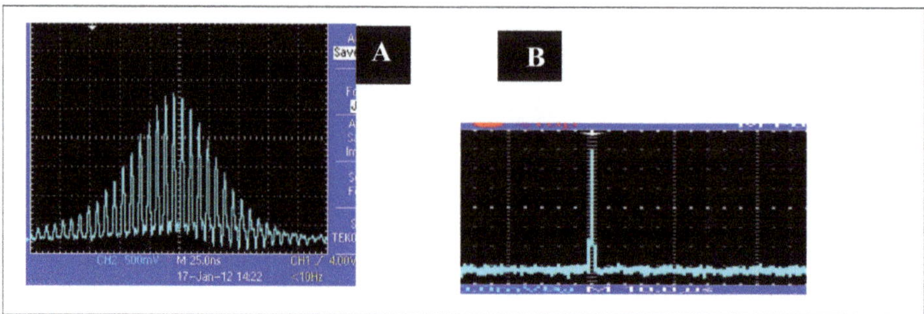

**Figure 17.9.** The actively mode-locked pulse train obtained from the oscillator (panel A). After inserting the passive mode locker, a single pulse is obtained after cavity dumping and amplification, as shown in panel B [7].

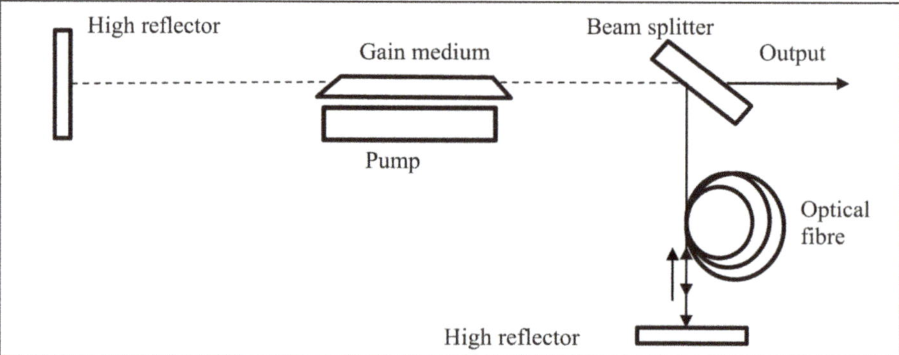


**Figure 17.10.** Additive-pulse mode locking. Part of the beam is sent to an optical fiber for chirping and is then recombined with the original pulse in the cavity.

### 17.3.3.2 Additive-pulse mode locking

In this technique, one of the portions of the output pulse is propagated in an optical fiber and undergoes self-phase modulation (SPM) (♠ see section 16.6.2). As a result of SPM, the pulse in the fiber has a broadened frequency spectrum. The frequency-chirped pulse returns and is injected back into the cavity as shown in figure 17.10. The timing of its injection into the cavity is carefully matched in such a way that its leading edge combines in phase with the pulse in the cavity, while its trailing edge combines with a phase shift of $\pi$ (i.e. out of phase). After several round trips, this process results in a reduction of the pulse duration. The resultant pulse approaches the bandwidth limit of the gain medium.

### 17.3.3.3 Kerr-lens mode locking

This method is an application of $\chi^{(3)}$ for 'all-optical' mode locking - thereby eliminating electronic components. The phenomenon of self-focusing (the longitudinal optical Kerr effect (LOKE), ♠ see section 16.6) contributes to this method of passive mode locking; hence, it is also known as Kerr-lens mode locking (KLM).

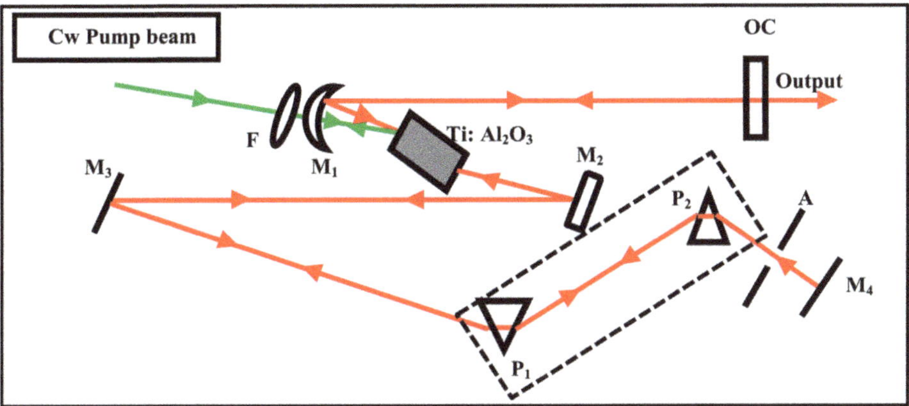


**Figure 17.11.** Schematic layout of a Kerr-lens mode-locked Ti:sapphire oscillator. M1–M4: mirrors, F: convex lens, P1, P2: Brewster-angled prisms, A: aperture, and OC: output coupler. As indicated, the pump beam is derived from a CW laser.

Figure 17.11 is a schematic diagram of a Ti:sapphire oscillator that makes use of the phenomenon of KLM. We can see that there is no component that is controlled by electronics, except for the electronics of the pump source.

Here, an aperture $A$ is located near the end mirror $M_4$. The cavity is defined by the output coupler (OC) and $M_4$. In practical terms, a laser spike is initiated either by tapping the optical table or by blocking/unblocking the pump beam. Due to the effect of self-focusing in the gain medium (which also acts as the Kerr medium), a lensing mechanism is invoked. This lensing mechanism slightly focusses the beam within the material. Due to this self-focusing, the beam size of the pulsed mode becomes smaller than that of the CW mode. The round trips of the light in the oscillator cavity help to further self-focus the intracavity beam. The repetition rate of the laser is decided by the value of $\tau_{rp}$ ($=c/2d$ in the oscillator, where $d$ is cavity length and $c$ the speed of light).

The key point is to select the pulsed mode rather than a CW mode to sustain mode-locked operation. This is done with the help of the aperture (i.e. physical) or the gain, or both. Generally, KLM can make use of either or both of the two following variations: (i) self-focusing of the mode; (ii) aperture control of the mode. In the aperture control system, the aperture is placed between the laser gain medium and the end mirror, as shown in figure 17.12. The aperture is small enough in diameter to provide a relatively high loss to the CW modes and none to their pulsed counterpart. A prism pair shown in a dotted box provides the pulse compression (♠ see chapter 23).

KLM also provides group-velocity dispersion (GVD) within the crystal during self-focusing. This produces the broad frequency spectrum of the pulse. The broad bandwidth thus provides scope for further pulse shortening by compression techniques (♠ see section 18.7).

For picosecond operation, the Ti:sapphire oscillator is also mode locked in the hybrid manner. In the hybrid method, an external device (e.g. an acousto-optic

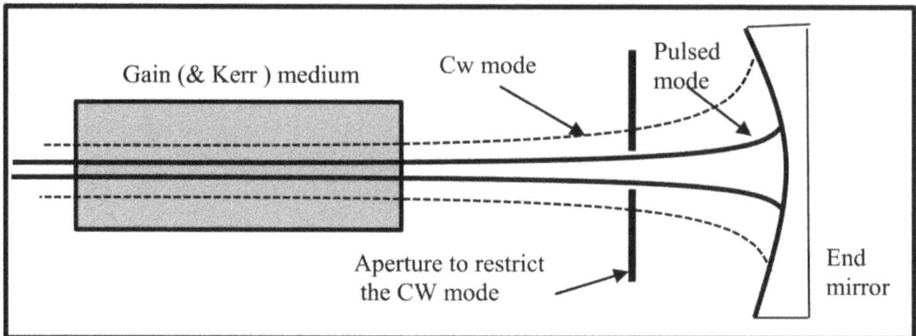


**Figure 17.12.** Schematic for the placement of a physical aperture in the cavity to select the pulsed mode for KLM.

device— ♠ see section 17.3.1) generates a pulsed beam. In the pulsed mode of operation, the self-focusing effect within the gain medium enhances the SPM.

♣ It is particularly interesting to note that unlike synchronous pumping, here, the pump laser operates in CW mode, while the KLM in the Ti:sapphire laser produces pulsed output at a repetition rate controlled by its cavity length.

**Exercise 17.3.** What do you understand by a Kerr lens ?

**Explanation:** In nonlinear optics, a flat piece of material (with no radius of curvature) can exhibit a lensing effect due to the self-focussing phenomenon. This is in contrast to conventional optics, in which a lens is used to focus a light beam. The nonlinear optical interaction between an intense optical field and the $\chi^{(3)}$ of the material is responsible for the so-called Kerr lens. This interaction is also known as the LOKE and is used in KLM as explained above.

## 17.4 Shortening of pulse length

When an ultrafast pulse with a broad spectrum passes through an SA material, the leading and trailing edges (known as the wings) of the pulse are absorbed, as shown in figure 17.13. After the saturation of absorption occurs, the trailing edge passes through the material unaffected (except for the dispersion experienced by the entire pulse envelope). The resultant pulse duration ($\Delta t'$) is shorter than the original one ($\Delta t$). It should be noted that amplification of the pulse also takes place due to the stimulated emission property of the SA material.

## 17.5 Spectra of mode-locked laser pulses

We know that ideal laser light is monochromatic. Even with the presence of a limited number of modes in the observed laser profile and the concomitant line-broadening mechanism, CW laser beams exhibit a line spectrum. Compared to other light

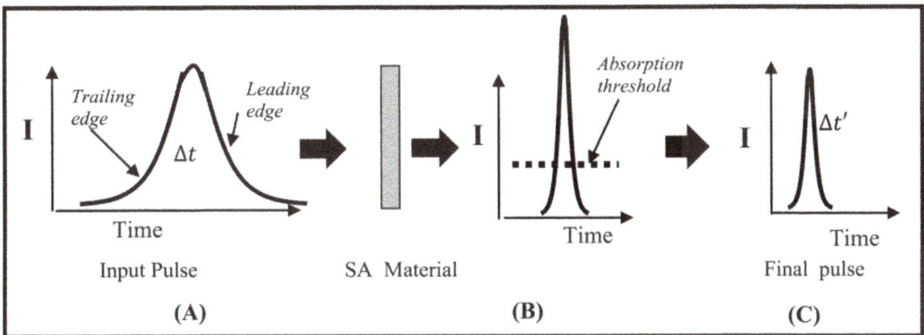



**Figure 17.13.** Effect of SA on the duration ($\Delta t$) of the propagating pulse. The thick arrow shows the progress of the pulse shape as it passes through the SA material. The dotted line in the middle panel (B) shows the absorption threshold of the SA material. The final pulse duration ($\Delta t'$) in (C) has become smaller than the incident pulse duration of panel (A).

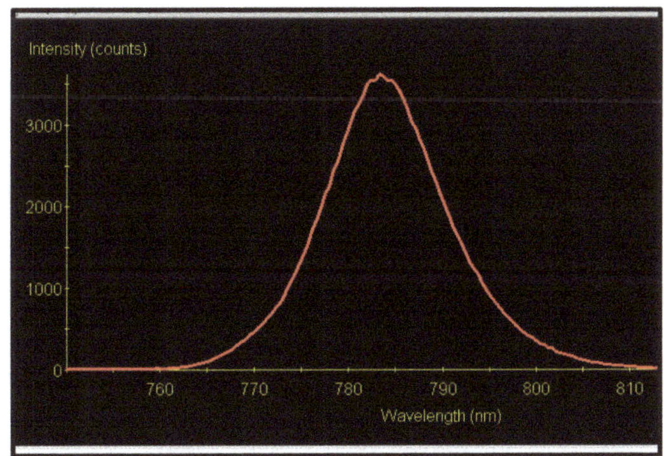



**Figure 17.14.** Screen shot of the spectrum of the fundamental pulse of a Ti:sapphire oscillator; the peak wavelength of ~785 nm was recorded by a fiber-optic spectrometer (HR2000). Note the broad spectral range (~40 nm) of the laser pulse.

sources, lasers are considered to be coherent and monochromatic sources (♠ see chapter 10).

However, as can be seen from equation (17.5), mode-locked laser pulses are expected to show a broad spectrum. The equation indicates that the shortest pulse has a reasonable frequency bandwidth due to the locking together of a large number of modes. A typical spectrum of a mode-locked pulse obtained at the fundamental wavelength of a Ti:sapphire laser is shown in figure 17.14.

It is worth looking at the gain spectrum of a Ti:sapphire crystal. As can be seen from figure 17.15, it has a broad spectral range of 600–1100 nm. This is a clear indication that on mode locking this laser, *ultrafast as well as tunable* pulses can be obtained over a wide spectral range at fundamental wavelengths. Furthermore, it

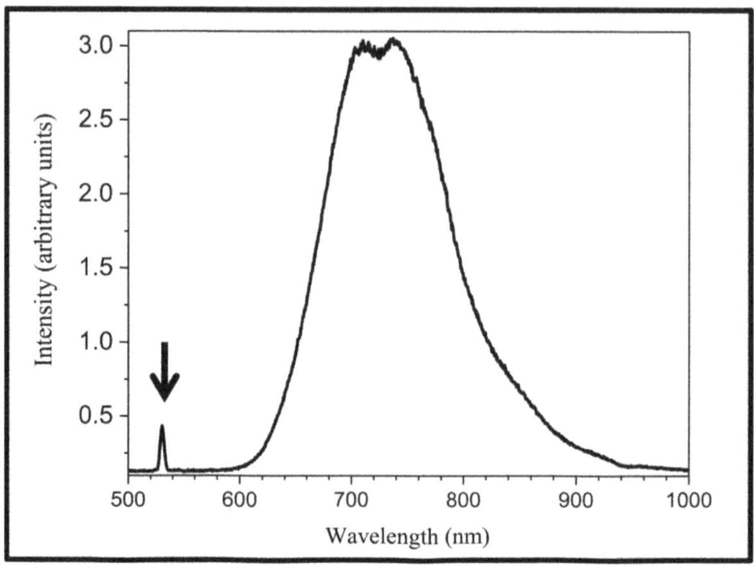

**Figure 17.15.** Photoluminescence spectrum of a Ti:sapphire crystal. The sharp peak indicated by an arrow is the pump wavelength (532 nm).

should be noted that second- and third-harmonic generation can easily provide the near-UV–visible wavelengths.

## Questions and problems

1. Which methods are used to generate ultrafast laser pulses?
2. Classify the mode-locking methods of lasers into active and passive schemes.
3. With reference to the optical Kerr effect (OKE),
   (i) differentiate between the transverse and longitudinal AC Kerr effect;
   (ii) in Kerr-lens mode locking (KLM), describe the effect of the length of the gain medium on the pulse duration.
4. What is cavity dumping? Identify its differences and similarities with the Q-switching method.
5. If there are N modes in the gain profile of a laser, what is the proportionality factor of the intensity of the mode-locked laser pulse in terms of N?
6. Find the shortest pulse duration for a $Ti^{+3}$: sapphire laser that has a gain bandwidth of 150 GHz.
7. A diode laser that has a wavelength of 900 nm and a gain bandwidth of 10 nm is to be mode locked . What is the minimum pulse duration that can be obtained from this laser? If the laser cavity of 0.2 mm is also equal to the width of the gain medium, find the separation between mode-locked pulses (take the refractive index of the semiconductor to be 3.0).
8. What is the pulse duration and the pulse repetition rate of the following mode-locked lasers, each of which has a laser cavity length of 1.5 m?

> (i) A He–Ne laser that has a wavelength of 543 nm (take the refractive index of the gas mixture equal to be one);
> (ii) A $Ti^{+3}$: sapphire laser operating over a gain width of 750–1100 nm and which has a rod length of 20 mm. The refractive index of the $Ti^{+3}$: sapphire is 1.76;
> (iii) A dye laser operating over a gain width of 550–600 nm and which has a jet thickness of 0.1 mm. The refractive index of the dissolving solvent is 1.33;
> (iv) an argon-ion laser at a wavelength of 488 nm (take the refractive index of the gas to be one).

9. An acousto-optic switch made of an optical material that has a refractive index of 1.5 is used to mode lock a He–Ne laser that has a wavelength of 632.8 nm. An RF pulse at 30 MHz is applied across the optical material. Taking the velocity of sound to be 3000 m s$^{-1}$, calculate the diffraction angle of the laser beam inside the switch. (Hint: use Bragg diffraction for the first order.)

10. Is it possible to generate attosecond-duration ($10^{-18}$ s) pulses in the visible region? Give your reasons.

11. Laser pulses 100 fs in duration from an Er-doped fiber laser (1550 nm) propagate through a silica fiber that has a refractive index of 1.45.
> (i) Find the spatial length of the pulse.
> (ii) After propagating through a fiber of length $l$, the spatial length of the pulse increases by a factor of ten. What is the value of $l$ if the linear dispersion value is taken to be $10^5$ m$^{-1}$?

# Bibliography


[1] Smith P W *et al* 1974 Mode-locking of lasers *Prog. Quantum Electron.* **3** 107–229

[2] Haus H A 2000 Mode-locking of lasers *IEEE J. Sel. Top. Quantum Electron.* **6** 1173–85

[3] Benfey D P *et al* 1992 Diode-pumped dye laser analysis and design *Appl. Opt.* **31** 7034–41

[4] Kranitzky W *et al* 1981 A new infrared laser dye of superior photostability tunable to 1.24 μm with picosecond excitation *Opt. Commun.* **36** 149–52

[5] Keller U *et al* 1996 Semiconductor saturable absorber mirrors (SESAM's) for femtosecond to nanosecond pulse generation in solid-state lasers *IEEE J. Sel. Top. Quantum Electron.* **2** 435–53

[6] Saraceno C J *et al* 2012 SESAMs for high-power oscillators: design guidelines and damage thresholds *IEEE J. Sel. Top. Quantum Electron.* **18** 29–41

[7] Ali S A 2013 *Ph D Thesis* IIT Madras

[8] Ell R *et al* 2001 Generation of 5-fs pulses and octave-spanning spectra directly from a Ti: sapphire laser *Opt. Lett.* **26** 373–5

[9] Bausa L E, Vergara I, Jaque F and Garcia Sole J 1990 Ultraviolet laser excited luminescence of Ti-sapphire *J. Phys. Condens. Matter* **2** 9919

[10] Eckstein J N, Ferguson A I and Hänsch T W 1978 High-resolution two-photon spectroscopy with picosecond light pulses *Phys. Rev. Lett.* **40** 847

**IOP** Publishing

# An Introduction to Photonics and Laser Physics with Applications

**Prem B Bisht**

# Chapter 18

# Characterization of ultrafast laser pulses

Phosphorescence (♠ see section 2.9.7.6) decay times[1] can be of the order of seconds and hence can be estimated even with the help of a wrist watch. To measure the fluorescence decay time, however, we need faster time resolution, which is available using electronics. We have learnt two techniques for generating ultrafast laser pulses in chapters 13 and 17. To characterize pulse durations on the picosecond or femtosecond timescales, in addition to electronics, we need to apply the principles of nonlinear optics. This chapter discusses the techniques used to characterize short-duration optical events in the visible and IR regions. The figure shows the principle of one of the pulse characterization techniques: *spectral phase interferometry of direct electric field reconstruction* (SPIDER). It should be noted that the same principles are used to characterize attosecond pulses in the extreme UV and x-ray regions.

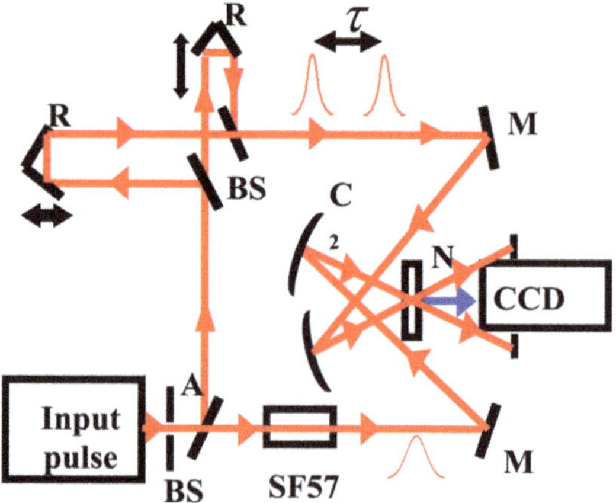


---

[1] The dynamic range is the time window (e.g. 10–200 ps) of an instrument, beyond which it cannot measure.

> **Learning objectives**
> **After reading this chapter, the learner will be able to:**
> Identify the problems in measuring ultrafast events;
> List the various techniques used to characterize ultrafast laser pulses;
> Understand the autocorrelation function;
> Relate the interferometric autocorrelation with the Michelson interferometer;
> Discover the techniques used to characterize the phases of ultrafast pulses;
> Categorize the methods used for frequency up- and downconversion;
> Demonstrate the problem of dispersion and its compensation;
> Identify the need for dispersion-free autocorrelation;
> Illustrate the chirped pulse amplification used to obtain petawatt laser pulses.

# 18.1 Introduction

A cursory search of a teaching or research laboratory will locate an oscilloscope with a frequency bandwidth of 50~60 MHz or even higher. The bandwidth of an oscilloscope reveals the time resolution it can provide. The time resolution provided by a detector (a photodiode or a photomultiplier) depends on the values of the time constant of the electronics involved in its circuit. To obtain a detectable value of the signal voltage, for the smallest values of a resistor ($R \sim 1$ K $\Omega$) and a capacitor ($C \sim 1$ pF), the time constant (RC) of the circuit is $\sim 1$ ns. Therefore, electronic circuits fail to detect short-duration pulses, even though expensive oscilloscopes with higher bandwidths (in GHz) are available on the market.

One of the techniques introduced in the 1980s was the streak camera, which is based upon the principle of converting phosphorescent images on timescales of up to 1 ps. However, the dynamic range[2] of a streak camera is limited. On the other hand, photomultiplier tubes (PMTs) that have several dynodes are susceptible to transit-time spread and color effects. The transit time is the difference in the arrival times of electrons at the anode through dynodes with finite area. In addition, different spectral components of the light travel at different speeds, resulting in 'color effects'. These two effects in PMTs take place on the order of nanoseconds. In microchannel-plate PMTs, these effects are reduced but are not completely removed. Photodiodes

**Table 18.1.** Time-resolution limits of various experimental techniques.

| Technique | Time resolution | Limitation |
| --- | --- | --- |
| PMTs and photodiodes | 100 ps | Transit-time spread |
| Streak camera | 10 ps | Dynamic range is small |
| Autocorrelation | ~1 fs or less | Dispersion effects limit the resolution |

---

[2] Dynamic range is the time window (e.g., 10–200 ps) of an instrument, beyond which it cannot measure.

can measure pulses of the order of hundreds of picoseconds at reduced sensitivity. Table 18.1 lists various techniques and their time-resolution limits.

From table 18.1, we see that one of the economic and accurate ways to measure short-timescale events of the order of picoseconds and femtoseconds is autocorrelation, which is described in the following sections.

## 18.2 Autocorrelators

An autocorrelator measures the temporal correlation between the field amplitudes of two replica pulses derived from a laser source. These correlations are mathematically expressed by normalized correlation functions of the first or second order. In general, the correlation function $G(\tau)$ is found to be between the intensity of a shorter reference pulse ($I_{\mathrm{ref}}$) and the sample pulse, i.e. the pulse to be characterized ($I_{\mathrm{pulse}}$). Since a short reference pulse is generally unavailable, in *autocorrelation,* the sample pulse itself is used as a reference. This is done by splitting the pulse into two replica pulses, one of which is identified as the reference pulse.

♣ The reason for naming this technique *autocorrelation* is also due to the fact that the laser pulse to be characterized is used in characterizing it.

For a reference field amplitude of $E_{\mathrm{ref}}(t)$ and a pulse amplitude that is delayed by $\tau$, $E_{\mathrm{pulse}}(t - \tau)$, the first-order autocorrelation (also known as the field correlation) is given by $G(t) \propto \int_{-\infty}^{+\infty} \left| E_{\mathrm{ref}}(t) + E_{\mathrm{pulse}}(t - \tau) \right|^2 dt$. For the same amplitude pair, the second-order autocorrelation is $G_{(2)}(t) \propto \int_{-\infty}^{+\infty} \left| \left| E_{\mathrm{ref}}(t) + E_{\mathrm{pulse}}(t - \tau) \right|^2 \right|^2 dt$, the third-order autocorrelation is given by $G_{(3)}(t) \propto \int_{-\infty}^{+\infty} \left| \left| E_{\mathrm{ref}}(t) + E_{\mathrm{pulse}}(t - \tau) \right|^3 \right|^2$, and so on. (We are aware of the Michelson interferometer (♠ see chapter 10), which deals with first-order autocorrelation.) Since ultrafast lasers are capable of invoking optical nonlinearity, applications in ultrafast optics often make use of the principles of nonlinear optics. Depending on the application, optical nonlinearities of the second or third orders are used, which address higher-order autocorrelations.

### 18.2.1 Intensity autocorrelation

Let $E(t)$ be the slowly varying envelope function of the electric field, and $I(t)$ be the intensity profile of the laser pulse. The interference between two laser beams with intensities of $I_1$ and $I_2$ follows equation (10.3), i.e. $I_1 + I_2 \pm \sqrt{(I_1 I_2}\cos\phi$, where $\phi$ is the phase difference between the two beams. The detection of this interference pattern by a linear detector (such as a photodiode) results in a background signal, which provides no estimate of the incident pulse duration.

As mentioned in the previous section, we use nonlinear optics in autocorrelation. From equation (15.10), we know that the power of the sum-frequency generation (SFG), is proportional to the square of the power of the fundamental laser beam.

♣ It should be noted that equation (15.10) is dimensionally balanced, as the quantity $\chi^{(2)}$ has dimensions of $m\ V^{-1}$ (♠ see table 14.1).

Let us work with the reference field amplitude $E_{\mathrm{ref}}(t)$ and the delayed amplitude of the pulse, $E_{\mathrm{pulse}}(t - \tau)$. To obtain the pulse duration, the idea is to first obtain the amplitude of the sum frequency or the second-order autocorrelation or $(E_{A(2)}(t))$ of these fields. From the definition above, $E_{A(2)}(t)$ can be written as follows:

$$E_{A(2)}(t) \propto \int_{-\infty}^{\infty} \left| E_{\text{ref}}(t) + E_{\text{pulse}}(t - t) \right|^2 dt$$

$$\propto \int_{-\infty}^{\infty} \left[ \left| E_{\text{ref}} \right|^2 + \left| E_{\text{pulse}}(t - \tau) \right|^2 \right. \tag{18.1}$$

$$\left. + 2E_{\text{ref}}(t) \times E_{\text{pulse}}(t - \tau) \right] dt$$

In Euler's relations, the laser pulse in the reference path is represented by $E_{\text{ref}} = E(t)e^{i(\omega t + \phi(t))}$. Here, the electric field envelope $(E(t))$ and its phase $(\phi(t))$ are slowly varying functions. For a given pulse for the delayed path, i.e. $E_{\text{pulse}}(t) = E(t - \tau)e^{i(\omega(t-\tau)+\phi(t-\tau))}$, the averaged intensity $<I_{\text{AC}}>$ seen by the detector is

$$<I_{\text{AC}}> \propto \frac{1}{T} \int_{0}^{T} | [E(t)e^{i(\omega t + \phi(t))} + E(t - \tau)e^{i(\omega(t-\tau)+\phi(t-\tau))}]^2 |^2 \ dt$$

The expansion of this expression has six terms, which can be rearranged as four distinct integrals, as follows:

$$< I_{\text{AC}} > \propto \int_{-\infty}^{\infty} [E^4(t) + E^4(t - \tau)]dt + 4 \int_{-\infty}^{\infty} [E(t) \times E(t - \tau)]^2 dt$$

$$+ 4\cos\Delta\phi\cos \omega t \int_{-\infty}^{\infty} \int [E(t)E(t - \tau)] \times [E^2(t) + E^2(t - \tau)]dt \tag{18.2}$$

$$+ 2\cos (2\Delta\phi)\cos (2\omega t) \int_{-\infty}^{\infty} [E(t) \times E(t - \tau)]^2 dt$$

In equation (18.2), the first term gives the constant background, the second term is responsible for the pulse envelope, and the third and fourth terms are the interference terms at the fundamental and second-harmonic frequencies, respectively. For detection over a long time, i.e. with a slow detector, these two oscillating terms average out to zero.

If a filter is placed in front of the detector to block the remaining fundamental laser pulse, the autocorrelation of the signal (the second term) with a constant background is detected as follows:

$$<I_{\text{AC}}> \propto \int_{-\infty}^{\infty} [E^4(t) + E^4(t - \tau)]dt + 4 \int_{-\infty}^{\infty} [E^2(t)E^2(t - \tau)]dt. \tag{18.3}$$

Here, $E^2(t)$ can be denoted by $I_{\text{ref}}(t)$, while $E^2(t - \tau)$ can be denoted by $I_{\text{pulse}}(t - \tau)$, so that the actual autocorrelation intensity of the sum-frequency signal $I_{\text{SHG}}(\tau)$ can be expressed as

$$I_{\text{SHG}}(\tau) \propto \int_{-\infty}^{\infty} I_{\text{ref}}(t)I_{\text{pulse}}(t - \tau)dt. \tag{18.4}$$

The following points should be noted:

(i) For the *linear case*, a slow detector will give an output signal $E^2(t) + E^2(t - \tau)$ that is independent of the delay time ($\tau$), which therefore yields no information about $I(t)$.

(ii) As a result of introducing the nonlinear interaction in noncollinear fashion (♠ see figure 18.1), the signal $<I_{AC}>$ contains the product $E^2(t) . E^2(t - \tau)$ which depends on the pulse duration.

(iii) The signal $<I_{AC}>$ is symmetric about the zero time, which implies that the *intensity autocorrelation* does not provide any information about the asymmetry of the pulse.

**Exercise 18.1.** In the autocorrelation technique, the autocorrelated signal is measured after SFG. The amplitudes of the two incident replica pulses are $E(t)$ and $E(t - \tau)$, where $\tau$ is the varying time delay between the two pulses. What is the average detected intensity?

**Solution:** Generally, the expression for the average field amplitude resulting from two incident fields $E(t)$ and $E(t - \tau)$ with a time delay ($\tau$) is given by

$$E_{av}(\tau) = \frac{1}{T} \int_{-T/2}^{T/2} [(E(t) + E(t - \tau))]dt.$$

The intensity is given by the square of this expression. However, in autocorrelation, we measure the intensity of the signal obtained after the SFG process. Therefore, it is necessary to consider the dimensions of $\chi^{(2)}$. Hence, the amplitude of the autocorrelated signal is proportional to the square of the amplitudes of the fundamental beams, as follows. $E_{A(2)}(t) \propto \int_{-T}^{T} [(E(t) + E(t - \tau))]^2 dt$. This, in turn, gives rise to the expression given by equation (18.4):

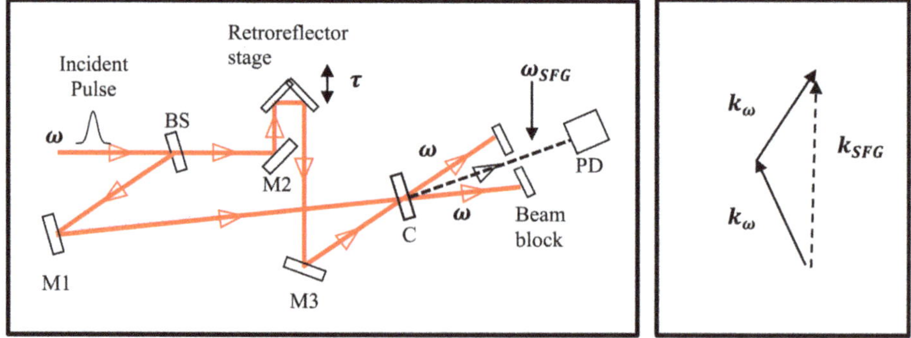


**Figure 18.1.** Background-free autocorrelation technique (left-hand panel). BS: beam splitter; M1, M2, M3: mirrors; C: nonlinear optical crystal; PD: photodiode. The sum frequency ($\omega_{SFG}$) detected by the PD is a function of $\tau$ between the two branches. The right-hand panel shows the sum-frequency wave vector ($k_{SFG}$) direction obtained by the vector sum of the incident wave vectors $k_\omega$.

$$I_{\text{autoco}}(\tau) \propto \int_{-T/2}^{T/2} I(t)I(t-\tau)dt$$

Note that even though the right-hand side of the expression has two intensity terms, the left-hand side has only one. As a result of the nonlinear contribution of $I_{\text{autoco}}$, the dimensions of the expression are balanced.

### 18.2.2 Background-free intensity autocorrelation

Figure 18.1. shows the schematic of an experimental autocorrelator setup. Here, the laser pulse to be measured is split into two replica pulses using a 50/50 beam splitter in a Michelson-type interferometer. One of the replica pulses is delayed with the help of a retroreflector mounted on a delay stage. The two pulses are focused onto a nonlinear optical crystal to generate the sum frequency (SF). The SF signal is observed with its wave vector along the bisector of the two beams. The SF signal is detected as a function of the time delay, as shown in the figure. In this autocorrelator geometry, the background signals caused by second-harmonic generation (SHG) in the individual arms of the interferometer are blocked. Therefore, this technique is also known as background-free autocorrelation.

**Exercise 18.2.**
    (i) How can one calibrate the intensity autocorrelator for the measurement of a laser pulse?
    (ii) What parameter decides the step size and the length of the translation stage used for the delay mirror?

**Solution:** The speed of light is the key parameter used to calibrate the autocorrelator. The distance moved by the translation stage of the delay mirror is converted into the time axis. This suggests that

**(i)** the intensity autocorrelator calibrates itself with the help of the step size of the translation stage of the delay mirror.

**(ii)** Light travels 10 mm in 33 ps. Both the length of the translation stage and its step size are decided by the initial estimate of the pulse duration to be measured. For example, to measure a 50 ps pulse, the delay stage can be about a few centimetres long. Note that a delay-stage length of 5 cm provides a maximum delay of 10 cm, as the beam traverses the same path twice in a single journey.

♣ As the time resolution is a function of the step size of the delay stage, the detector need not have a fast time response.

♣ If the fundamental beam is blocked with a filter, for the incident fundamental frequency $\omega$, there should be three spots corresponding to $2\omega$ on the screen placed after the crystal. Two of them are due to the generation of the second harmonics ($2\omega$) of the incident beams, and one of them is due to the sum-frequency signal $\omega_{\text{SFG}}$ (which also happens to be ($2\omega$)). Interestingly, it is easy to distinguish the three, as follows. While the first two disappear when the corresponding fundamental path is blocked, the spot due to $\omega_{\text{SFG}}$ will disappear even if any one of the beams is blocked. As illustrated by figure 18.1, the $\omega_{\mathrm{SFG}}$ spot appears in the middle of the three.

♣ Instead of a crystal, samples that have the property of two-photon excitation (2PE)-induced emission can also be used (♠ see chapter 20 for details of 2PE).

♣ The Michelson-type interferometer is extremely useful in physics. For example, it is used in gravitational-wave detectors, in which its single branch extends to several kilometers (why?).

♣ This technique of pulse-width measurement requires prior knowledge of the pulse shape.

♣ The SFG signal ($I_{\mathrm{SHG}}(\tau)$) obtained from the autocorrelator provides a temporal value ($\tau_{ac}$) corresponding to full width at half maximum (FWHM). To find the pulse duration, a functional form of $I_{\mathrm{SHG}}(\tau)$ has to be assumed. To get the actual pulse width ($\tau_p$) of the laser pulse, the obtained value of $\tau_{ac}$ has to be multiplied by a factor that depends on the assumed pulse shape. For instance, for a Gaussian pulse shape, the resultant pulse duration can be obtained by taking a simple Gaussian-shaped pulse (ignoring the constants), i.e.

$$I(t) = \exp\left(-\frac{t^2}{t_p^2}\right). \tag{i}$$

Using equation (18.4), the intensity of the autocorrelation function can be written as

$$I_{ac}(\tau) = \int_{-\infty}^{\infty} \exp\left(-\frac{t^2}{t_p^2}\right)\exp\left(-\frac{(t-\tau)^2}{t_p^2}\right)dt. \tag{ii}$$

By simple arithmetic expansion, the product inside the integral can be rearranged and the expression can be written as

$$I_{ac}(\tau) = \int_{-\infty}^{\infty} \exp\left(-\frac{2\left(t^2 - t\tau + \frac{\tau^2}{4}\right) + \tau^2 - \frac{\tau^2}{2}}{t_p^2}\right)dt, \tag{iii}$$

which can be rewritten as

$$I_{ac}(\tau) = \exp\left(-\frac{\tau^2}{2\tau_p^2}\right)\int_{-\infty}^{\infty} \exp\left(-\frac{2\left(t - \frac{\tau}{2}\right)^2}{\tau_p^2}\right)dt. \tag{iv}$$

Using the standard integral, we obtain

$$\int_{-\infty}^{\infty} \exp(-ax^2)dx = \sqrt{\frac{\pi}{a}} \equiv K. \tag{v}$$

Here, $K$ is another constant. Therefore,

$$I_{ac}(\tau) = K \exp\left(-\frac{\tau^2}{\left(\tau_p\sqrt{2}\right)^2}\right). \qquad \text{(vi)}$$

By rewriting equation (vi), we obtain

$$I_{ac}(\tau) = K \exp\left(-\frac{\tau^2}{\tau_{ac}^2}\right). \qquad \text{(vii)}$$

Comparing equations (i) and (vii), we find that $\tau_{ac} = \sqrt{2}\,\tau_p$, or

$$\frac{\tau_{ac}}{\tau_p} = 1.414. \qquad (18.5)$$

Similarly, table 18.2 gives the ratio of the FWHM of the autocorrelation signal to the actual pulse for various pulse shapes.

**Exercise 18.3.** A measured autocorrelation trace of a pulse has a temporal FWHM of 150 fs. What is the actual duration of the pulse?

   **Solution:** Depending on the assumed pulse shape, the pulse duration can be obtained from table 18.2. Thus, for a Gaussian pulse shape, according to equation (18.5), the pulse duration is 106 fs.

### 18.2.2.1 Measurement of picosecond-duration pulses

To measure laser pulse durations on the picosecond timescale, the laser pulse is split into two replica pulses using a 50/50 beam splitter, as shown in figure 18.1. The two replica pulses are focused onto a type-I nonlinear optical crystal. For instance, the phase-matching angle for SFG from the fundamental of a Nd:YAG laser beam in

**Table 18.2.** Relation between the FWHM of the measured autocorrelation signal and the actual laser pulse duration $\left(\frac{\tau_{ac}}{\tau_p}\right)$ for various functions.

| Pulse shape | Relative intensity ($I(t)$) | $\frac{\tau_{ac}}{\tau_p}$ |
|---|---|---|
| Gaussian | $\exp\left(\dfrac{-4(ln2)t^2}{\Delta t^2_{\text{FWHM}}}\right)$ | 1.414 |
| Hyperbolic sec$^2$ | $\text{sech}^2\left(\dfrac{1.76t}{\Delta t_{\text{FWHM}}}\right)$ | 1.543 |
| Lorentzian | $\dfrac{1}{\left(1+\left(\dfrac{4t^2}{\Delta t^2_{\text{FWHM}}}\right)\right)}$ | 2 |
| Exponential | $\exp\left(\dfrac{-t(ln2)}{\Delta t_{\text{FWHM}}}\right)$ | 2.421 |
| Square | $1; \text{ for } -\dfrac{\tau_P}{2} \leqslant t \leqslant \dfrac{\tau_P}{2}$ | 1 |

$\beta$–barium borate (BBO) crystal is 47.6°. The intensity of the generated sum frequency is shown as a function of time delay ($\tau$) in figure 18.2. Considering the Gaussian pulse shape, the obtained autocorrelation trace is fitted with a Gaussian function to estimate the pulse duration from table 18.2.

### 18.2.2.2 Cross-correlation technique

The principle of this technique is similar to that of autocorrelation. It is schematically shown in figure 18.3. Here, one of the laser pulses is a well-characterized reference pulse with a time duration of $\tau_{known}$. Depending on the type of phase matching, the SFG or the DFG signal of the interacting beams can be measured as a function of the time delay ($\tau$) between the interacting beams.

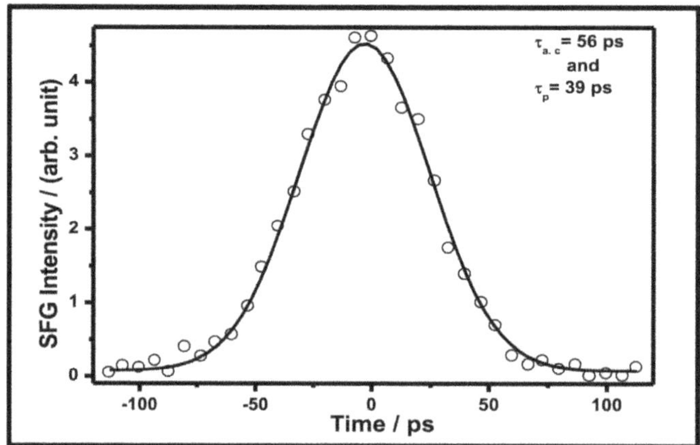


**Figure 18.2.** Autocorrelation data obtained for laser pulses (o) produced by a picosecond Nd:YAG laser. The solid line shows the fit with a Gaussian function. Note that the FWHM of the measured autocorrelator is 56 ps, which gives a pulse duration of 39 ps (♦ see table 18.2).

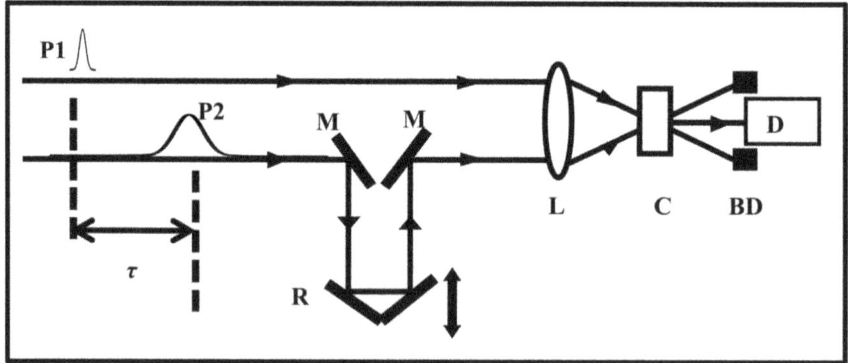


**Figure 18.3.** The cross-correlation technique used to estimate the duration ($\tau_{unknown}$) of pulse P2. P1 is a well-characterized pulse of duration $\tau_{known}$, and $\tau$ is the time delay between the two pulses. M: mirror, R: retroreflector, L: convex lens, C: nonlinear crystal, BD: beam dump, and D is the detector.

The relation between the duration of the cross-correlation trace ($\tau_{\text{cross}}$) and that of the interacting pulses is given by

$$(\tau_{\text{cross}})^2 = (\tau_{\text{known}})^2 + (\tau_{\text{unknown}})^2 \tag{18.6}$$

The pulse width of the unknown pulse ($\tau_{\text{unknown}}$) can be obtained using equation (18.6) after obtaining the value of $\tau_{\text{cross}}$ from a fit to the appropriate function.

**Exercise 18.4.** Using the available equipment in an undergraduate optics lab and a suitable SHG crystal, how can you set up an experiment to measure pulses 10 ps in duration?

**Solution:** The easiest way to measure an optical signal is to make use of an oscilloscope and a photodiode. However, the time duration of 10 ps corresponds to a frequency bandwidth of 100 GHz (♠ see table 17.1).

An undergraduate lab may only have oscilloscopes capable of 50–100 MHz. Therefore, we cannot expect to directly measure the pulse duration at the 10 ps timescale in the lab. However, by now we know that the autocorrelator works on the principle of the Michelson interferometer, which is generally available in an undergraduate optics lab. Converting 10 ps into a distance gives a value of 3 mm for the movement of the retroreflector mirror for the FWHM. The measurement of a complete pulse will need a distance of a little more than 6 mm—corresponding to a one-way movement of >3 mm, which is generally available using a Michelson interferometer. The sum-frequency signal generated as a function of the time delay of one of the mirrors through the SHG crystal is the autocorrelation signal, which gives the pulse duration. The interference pattern gives further information when its spectrum is measured, as described in the next section.

### 18.2.3 Interferometric autocorrelation

Until now, we have been using the noncollinear geometry in the background-free autocorrelator. In the interferometric method, the replica pulses are collinearly superimposed on the beam splitter (figure 18.4) and focused at the nonlinear crystal

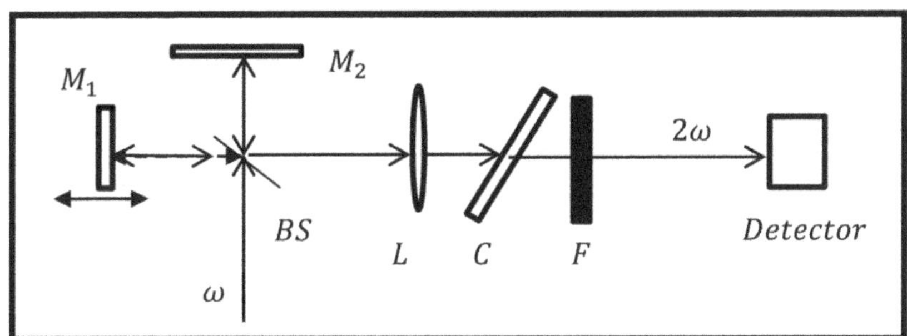


**Figure 18.4.** Schematic representation of interferometric autocorrelation. $M_1$ is the fixed mirror, $M_2$ is the delay mirror, BS is the beam splitter, $L$ is the lens, and $C$ is the nonlinear optical crystal. $F$ is the filter that blocks the fundamental ($\omega$) but allows the sum frequency ($2\omega$).

to generate the sum frequency. This geometry is known as interferometric autocorrelation.

In this geometry, the phase of the electric field has to be taken into account. Hence, all the terms in equation (18.2) contribute to the interferometric autocorrelation signal, as shown in table 18.3.

Figure 18.5 shows the typical interferometric autocorrelation trace for an 18 fs pulse at 800 nm. The collinear geometry also allows the autocorrelator signal to interfere with the second harmonic generated from each beam. The upper and lower envelopes of the obtained pulse are due to the contributions of $\omega\tau = 2\pi$ and $\pi$, respectively. The fundamental portions of the pulse envelope are blocked by a filter. The interference pattern due to the SFs is detected as a function of the delay time. The phase information is obtained from the recorded interferogram. However, the iterative algorithm used for this purpose does not actually give satisfactory phase

**Table 18.3.** Various terms and their contributions in the interferometric autocorrelation.

| Mathematical term | Contribution to the interferometric autocorrelation |
|---|---|
| Background and pulse envelope | $\left\{\int_{-\infty}^{\infty}[E^4(t) + E^4(t - \tau)]dt\right\} + \left\{4\int_{-\infty}^{\infty}[E^2(t).\ E^2(t - \tau)]dt\right\}$ |
| Fringes | $\left\{4\cos\Delta\phi\cos \omega t \int_{-\infty}^{\infty}\int E(t).\ E(t - \tau).\ [E^2(t) + E^2(t - \tau)]dt\right\}$ |
| Second-order term | $\left\{2\cos(2\Delta\phi)\cos(2\omega t) \int_{-\infty}^{\infty}[E(t).\ E(t - \tau)]^2 dt\right\}$ |

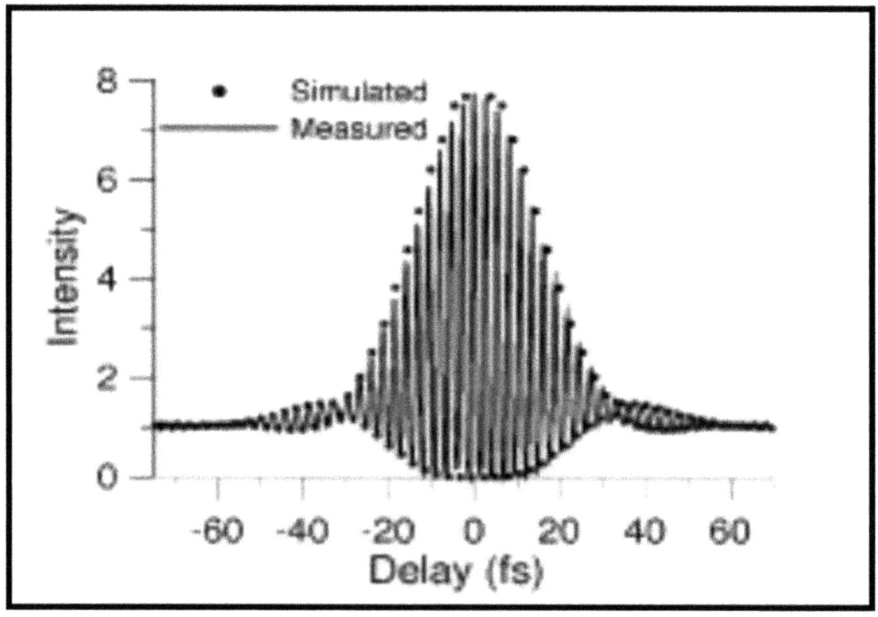


**Figure 18.5.** Interferometric autocorrelation of a 16 fs pulse. Copyright 2004 IEEE. Reprinted, with permission, from [10].

values. Therefore, in order to get the phase information, another technique known as frequency-resolved optical gating (FROG) is used, as described below.

## 18.3 Frequency-resolved optical gating

As mentioned at the beginning, autocorrelation requires prior knowledge of the shape of the laser pulse. Frequency-resolved optical gating (FROG) is a technique that can avoid this requirement. It is also known as the spectrally resolved autocorrelation technique. Using a similar approach to that of background-free autocorrelation, the main laser pulse is divided into two parts (P1 and P2). One of the replica pulses (P1) is delayed, as shown in figure 18.6. These two pulses are overlapped onto a nonlinear crystal and undergo the SFG process. The spectrum of the SFG signal is recorded by the spectrometer as a function of the delay time ($\tau$) between P1 and P2.

Mathematically, the intensity of the observed experimental trace ($I_{\mathrm{FROG}}(\omega, \tau)$) as a function of $\omega$ and $\tau$ at the sum frequency can be described by

$$I_{\mathrm{FROG}}(\omega, \tau) = \left| \int_{-\infty}^{\infty} E_{sig}(t, \tau)e^{-i\omega t}dt \right|^2 \tag{18.7}$$

where $E_{sig}(t, \tau) \propto |\, E(t)E(t - \tau)\,|^2$ is a result of the correlation of the fundamental fields.

The FROG technique uses an iterative algorithm (figure 18.7) to retrieve the electric field distribution and the spectral phase of the complex pulses. The inversion of the FROG trace can be mathematically stated as follows:

$$\sqrt{I_{\mathrm{FROG}}(\omega, \tau)}e^{i\phi(\omega, t)} = \int_{-\infty}^{\infty} E_{sig}(t, \tau)e^{-i\omega t}dt,$$

where $\theta(\omega, \tau)$ is the phase of the pulse to be determined and the field $E(\omega, \tau)$ is proportional to the square root of the intensity, $\sqrt{I(\omega, \tau)}$. Thus, the inversion of a FROG trace becomes a two-dimensional phase-retrieval problem.

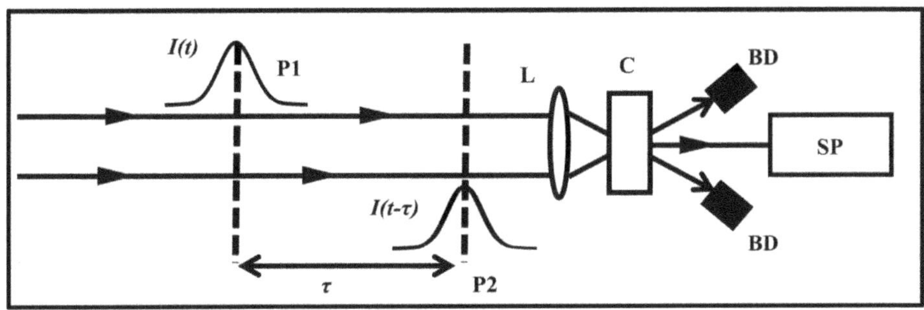


**Figure 18.6.** Schematic representation of an FROG setup. Note that except for replacing the detector with a spectrometer (SP), the other components are same as in figure 18.3.

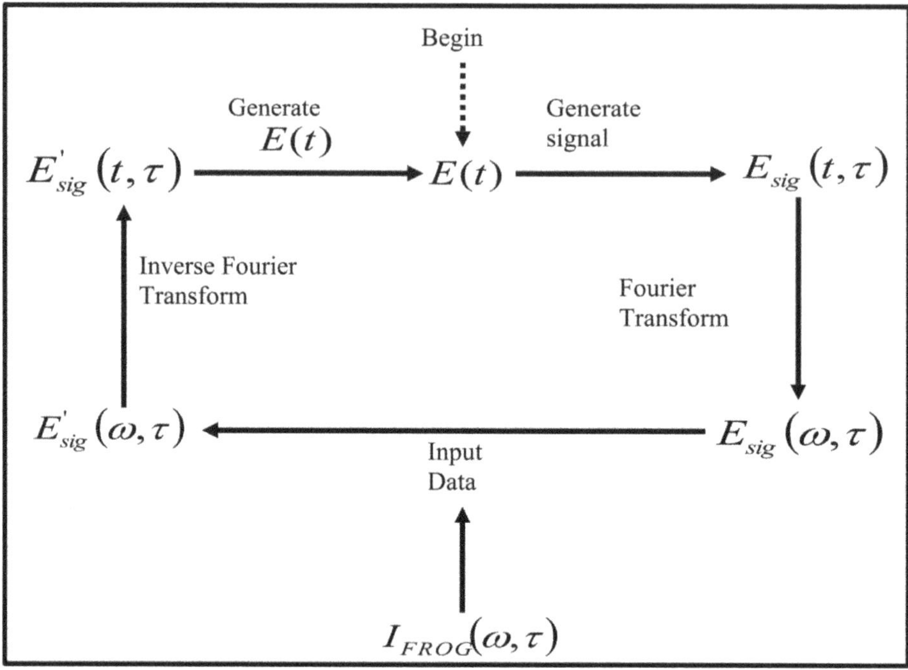


**Figure 18.7.** Schematic representation of the FROG retrieval algorithm. ♠ Reproduced with permission from [3]. Copyright The Optical Society.

The algorithm generates the theoretical FROG trace using Fourier transforms with an initial guess at the pulse shape. As information about the phase is not available, one has to proceed with the iterative algorithm to retrieve the phase information. This is carried out by iterating the parameters to obtain a new phase and field. Varying values of the frequency-dependent amplitude $E(\omega)$) and the spectral phase $(\phi(\omega))$ are used as initial inputs. The theoretical FROG trace is calculated using the equation (18.7). The measured FROG trace is then used to extract the complex electric field of the pulse by applying a Fourier transform, as indicated in figure 18.7. At each iteration of the algorithm, the theoretical and experimentally extracted pulse shapes are compared to get a better estimate of the pulse shapes. The convergence of the algorithm can be calculated using the root-mean-square error (RMS error) between the experimental measurement and theoretical FROG trace.

This technique has been successful and has been used for the measurement of ultrafast laser pulses. Figure 18.8 shows the measured FROG trace of 4.5 fs compressed pulses produced by a Ti:sapphire oscillator. Note the spectral chirping of the ultrafast pulse.

## 18.4 Spectral phase interferometry

To overcome the problem of iterative reconstruction of the pulse shape in the FROG technique, new methods have been proposed to extract the phase and the electric

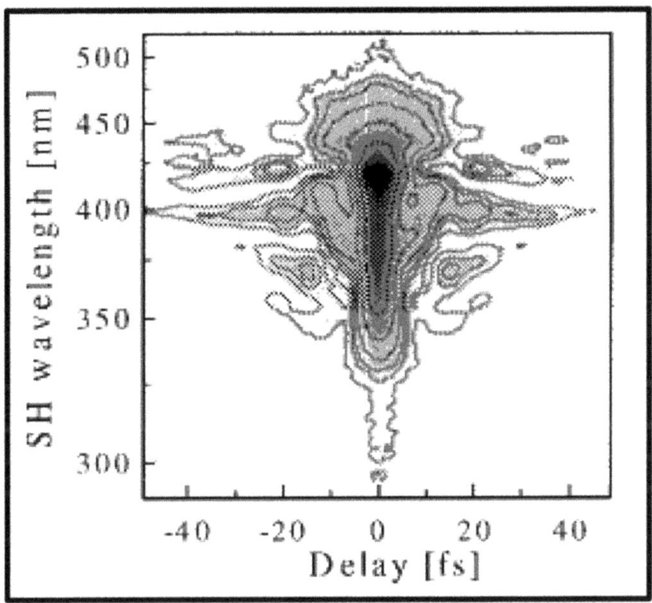

**Figure 18.8.** Experimentally measured FROG trace of 4.5 fs-duration pulse. Copyright 1999 IEEE. Reprinted, with permission, from [4].

field of the pulse. One of them is known as the 'spectral phase interferometry for direct electric field reconstruction' (SPIDER) technique. It is a self-reference technique and does not need the initial characterization of a pulse.

As shown in figure 18.9, the laser pulse split into two portions, one of which passes through a Michelson-type arrangement as described in section 18.2. The other part of the pulse is used to generate a chirped pulse; it is passed through a dispersive medium, such as a glass block.

The SPIDER procedure is further explained in figure 18.10. The two replica pulses (P1 and P2) are delayed in time. These interact with the chirped pulse (P3) in a nonlinear optical crystal. An SFG signal is observed due to the temporal and spatial matching of P1 and P2, respectively, with the leading and trailing parts of the chirped pulse.

The upconverted pulses obtained from SFG using the chirped pulse are overlapped at the detector to generate an interferogram, known here as a SPIDER pattern:

$$S(\omega) = |E(\omega)|^2 + |E(\omega + \delta\omega)|^2 + 2|E(\omega)E(\omega + \delta\omega)| \times \cos\left[\phi(\omega + \delta\omega) - \phi(\omega) + \omega\tau\right] \tag{18.8}$$

This interference pattern is similar to that discussed in section 10.4.1. Here, as usual, $E(\omega)$ denotes the electric field of the test laser pulse, $\phi(\omega)$ is its spectral phase, and $\delta\omega$

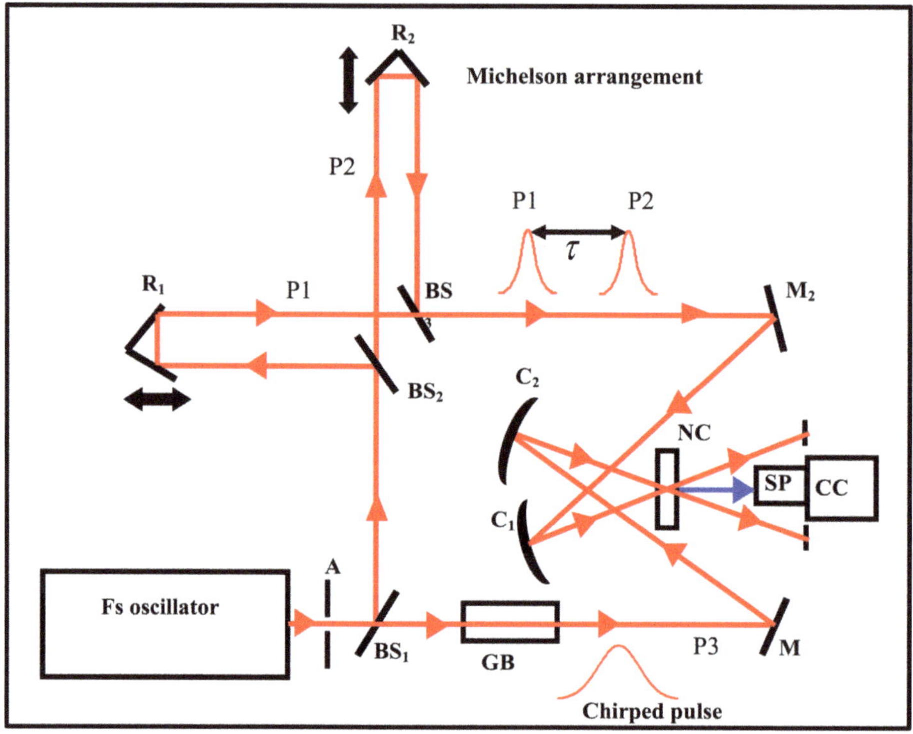



**Figure 18.9.** Schematic of the SPIDER setup. P1 and P2: replica pulses, $BS_1$, $BS_2$, and $BS_3$: beam splitters, A: aperture, GB: glass block, $M_1$, $M_2$: plane mirrors, $R_1$, $R_2$: retroreflectors, $C_1$, $C_2$: concave mirrors, NC: nonlinear crystal, SP: spectrometer, $\tau$: time delay between P1 and P2, CCD: charge-coupled device, P3: chirped pulse.

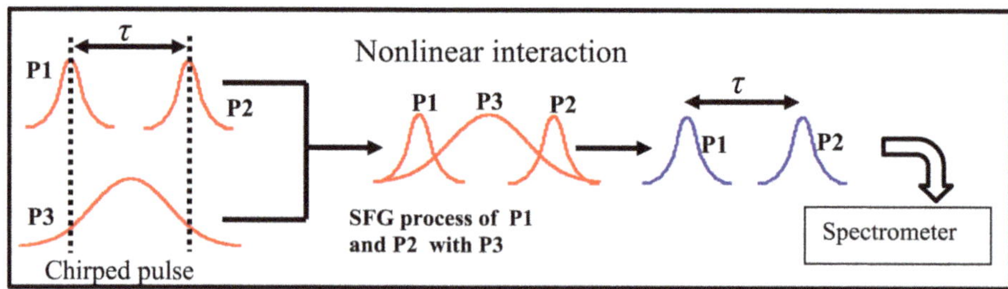



**Figure 18.10.** Detailed illustration of the SPIDER principle. The figure labels are same as those used for figure 18.9.

is the spectral shear in frequency due to chirp. The electric field and the spectral phase of the test laser pulse can be extracted from the SPIDER interferogram using a non-iterative algorithm consisting of a few Fourier transforms. This technique has been found capable of characterizing a highly structured broad supercontinuum that has a bandwidth of more than 200 THz, even at very low pulse energies of the order of less than a nanojoule.

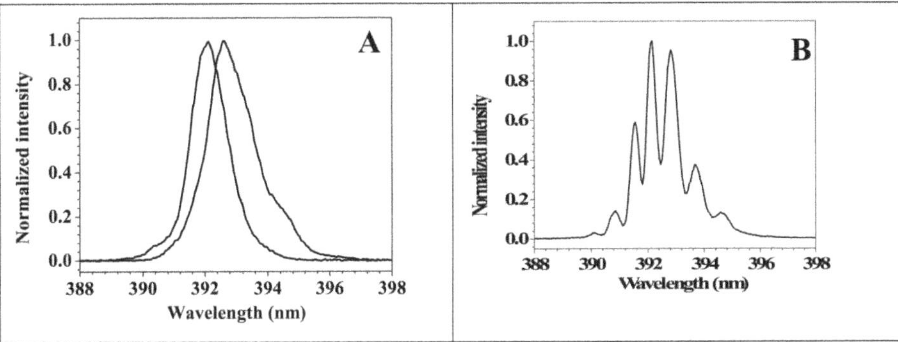

**Figure 18.11.** Spectrally sheared replica pulses as a result of the sum frequency generated using a chirped pulse (panel A); experimentally measured SPIDER interferogram (panel B) of a 100 fs-duration fundamental pulse at 784 nm. [13].

Experimentally, the spectral shear between the delayed pulses can be obtained from the sum frequency generated by the chirped pulse with the help of a nonlinear optical crystal. The obtained sum-frequency pulses are then overlapped at the entrance slit of the spectrograph to create an interference pattern. Figure 18.11 shows the recorded pulse spectra and the corresponding SPIDER interferogram of the femtosecond pulse. In order to get the spectral phase of the pulse, the spectral shear and delay between pulses are selected in such way that the SPIDER interferogram contains a well-separated maximum number of fringes. The distribution of the spectral phase is found to be flat throughout the spectrum.

## 18.5 Frequency up- and downconversion

The term *frequency upconversion* is used when a high-frequency photon is generated by a nonlinear optical process. This can happen either by SHG, SFG, or even as a result of other higher-order processes, such as multiphoton absorption-induced emission. Similarly, the process of *frequency downconversion* is a process that create photons of lower frequencies, for example, in polarization-entangled photons in quantum optics experiments (♠ see section 15.7).

As mentioned at the beginning of this chapter, several sensitive techniques can be used to measure the lifetimes of the excited states of molecular systems. However, when the decay rates of the excited states are fast (in the picosecond or subpicosecond time regimes), techniques based on traditional optics are not able to provide the required time resolution due to the limitations of their electronics. In this time range, the technique of fluorescence upconversion is used. In this method, the SF of the *fluorescence* and the excitation (*pump*) frequencies is detected with the help of a $\chi^{(2)}$ material. This is similar to the autocorrelation technique, in the sense that the intensities of the SF signals for the various portions of the *fluorescence decay* of the sample are detected as a function of the *pump* delay time.

## 18.6 Dispersion of ultrafast laser pulses

Due to the *time–bandwidth product* relation (♠ see section 8.5.2.1) the duration of a mode locked pulse is inversely proportional to the frequency bandwidth of the gain medium. Equation (17.5) reveals that a gain medium with large spontaneous spectral width has the ability to obtain shorter pulses.

♣ Ultrafast lasers with short pulses have broader spectral widths—and they do not seem to be monochromatic. Yes! Ultrashort pulsed lasers (in the visible region, say) are polychromatic.

**Exercise 18.5.** Let us assume that a Ti:sapphire laser is delivering pulses at its full gain bandwidth of 400 nm and that their peak is at 900 nm.
  (a) What is the expected pulse duration of this bandwidth-limited pulse? What is the spectral purity of the ultrafast laser pulse?
  (b) If we need a resolution of 0.01 nm resolution from a pulsed tunable dye laser (at ~600 nm, say), should the pulse duration of the dye laser be femtoseconds or nanoseconds?
  **Solution:** Recall that the time–bandwidth product $\Delta \nu \Delta t \cong$ constant.
  (a) If we convert the wavelengths of 700 nm and 1100 nm (for a spectral width of 400 nm), the bandwidth $\Delta \nu$ is $1.56 \times 10^{14}$ Hz, corresponding to a pulse duration $(\tau_p)$ of ~6 fs. The spectral bandwidth of the pulse is 400 nm.
  (b) To obtain a spectral purity of 0.01 nm at 600 nm, the calculated bandwidth of ~8.3 GHz corresponds to a pulse duration of ~12 ns.

We can see that while ultrafast femtosecond lasers are excellent tools for measuring timescales, but they are not a good choice for measuring high-resolution spectra. In fact, nanosecond lasers are used in high-resolution spectroscopy. The large spectral widths of ultrafast laser pulses experience dispersion while traveling in the medium. These effects are outlined in the following.

♣ The broad-bandwidth pulses are also useful in the chirped pulse optical parametric amplification (CPOPA) process, which is used to generate petawatt energy pulses (♠ see section 18.9).

### 18.6.1 Linear chirp

Figure 18.12 shows a schematic of the spectrum of an ultrashort laser pulse passing through a medium of refractive index $n$. According to Sellmeier's formula (♠ see section 1.4.6), the speed of light varies with the refractive index of the medium. Therefore, the spectral components of an ultrafast pulse undergo dispersion effects, as follows. The longer wavelengths (known as *red portions*) of the pulse spectrum travel faster than the shorter wavelengths (*blue portions*). This optical effect also results in the temporal broadening of the laser pulse. Such broadening effects in laser pulses fall under the category of linear chirp.

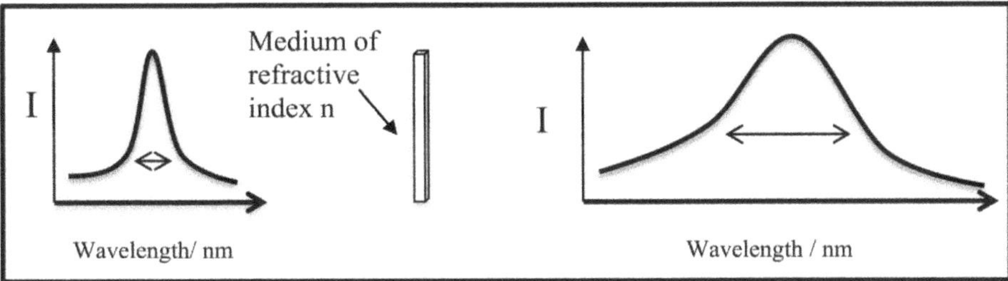


**Figure 18.12.** Schematic of the effect of linear dispersion on the propagation of an ultrafast pulse through a medium. Note the spectral width of the laser pulse (indicated by ↔) before and after it passes through the medium.

The importance of chirping in ultrafast laser pulses can be appreciated by the fact that their duration can broaden even while they propagate through air. For instance, a laser pulse that has a duration of 5 fs broadens to 7 fs after 1 m of travel through air (or 1 mm of glass). This effect, if not taken into account, can cause serious problems in experimental results.

### 18.6.2 Nonlinear chirp

The intensity-dependent refractive index (equation (16.6)) acting along the length of the medium is responsible for the effect of self-phase modulation (SPM, ♠ see chapter 16). Therefore, SPM has also been referred to as the longitudinal optical Kerr effect (LOKE), in contrast to its transverse part, which is responsible for the effect of self-focusing. The total frequency change $\Delta\omega$ is given by equation (16.7) and is responsible for the additional frequencies around the central frequency, $\omega_0$:

$$\Delta\omega = \left| \left( \frac{\omega_0}{c} \right) L \left( \frac{dn_2(I)}{dt} \right) \right|.$$

The pulse chirping due to SPM is known as nonlinear chirp. Therefore, when an ultrafast laser pulse travels through a medium, it undergoes linear as well as nonlinear chirping.

## 18.7 Dispersion compensation

We now know that ultrashort pulses have a spectral bandwidth inversely proportional to their temporal width. Further dispersion occurs as a result of linear as well as nonlinear chirp when the pulse passes through optical components. This mainly occurs because the shorter wavelengths of the pulse travel slower as compared to the longer wavelengths. Therefore, if we had a device that could reverse this effect, we could cancel the effect of dispersion. In order to achieve this, we introduce a time delay for the longer wavelengths of the laser pulse so that they catch up with the shorter wavelengths during propagation. Such a process is known as dispersion compensation. In the following, we describe two methods used for dispersion compensation.

### 18.7.1 Grating compressor

This laser pulse compressor makes use of two optical diffraction gratings. A diffraction grating consists of a large number of grooves separated by a distance $d$ placed at an angle (figure 18.13). $d$ is also known as the grating element. If a light beam is incident on such a grating at a random angle of $\theta_1$, measured from the normal, the grating equation that describes the dispersion of wavelength $\lambda$, which is diffracted at $\theta_2$, is

$$m\lambda = d(\sin \theta_1 + \sin \theta_2).$$

Here, $m$ is the order of diffraction. This equation suggests that light consisting of various wavelengths is diffracted at different values of the angle $\theta_2$. In other words, a grating can diffract white light into various colors at several orders, as shown in the figure.

♣ The colors diffracted in different orders by a compact disc (CD) are due to the presence of a large number of grooves. The same sizes of digital versatile disc (DVD) and Blu-ray disc (BD) have more orders of diffraction.

♣ Gratings are linear-dispersal devices, as can be seen from the grating equation. This is in contrast to prism dispersion, in which shorter wavelengths have higher dispersion *(why?)*.

Now let us look at an ultrafast laser pulse that has undergone the effect of dispersion and which contains a wide range of wavelengths. As can be seen in figure 18.14, after striking the first grating, the longer wavelengths (denoted by *Red*) of the pulse travel a longer path than the shorter ones (*Blue*). The second grating is geometrically placed to obtain the collimated beam parallel to the incoming beam. The distance between the grating pair can be adjusted so that the *blue portions* of the pulse 'catch up' with the *red portions*. This procedure is used for the compensation of an ultrafast pulse.

Figure 18.15 shows a geometry similar to that of figure 18.14, except that the pulse is traveling from the opposite side. As expected, this configuration of the gratings will have the opposite effect to that of a grating compressor. Therefore, we note that a grating pair can also be used to obtain a required amount of chirp.

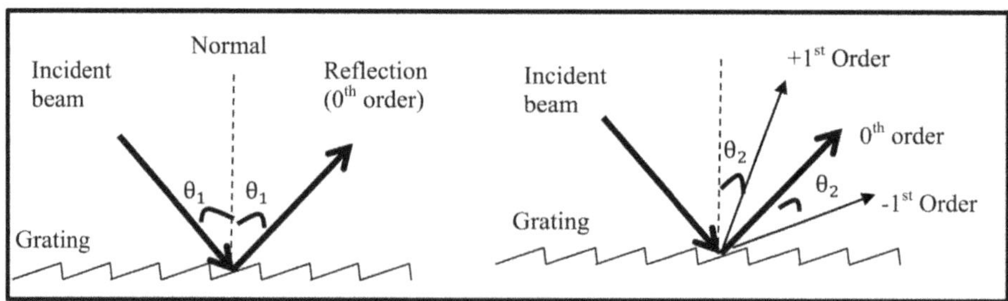


**Figure 18.13.** Diffraction of a light beam by a grating. Diffraction of the beam at an angle equal to the incident angle $\theta_1$ is known as zeroth-order diffraction (left-hand panel). Diffraction of the first order (and of the negative first order) takes place at an angle of $\theta_2$ (right-hand panel). Higher-order diffractions occur at integral multiples of $\theta_2$ (not shown in the figure).

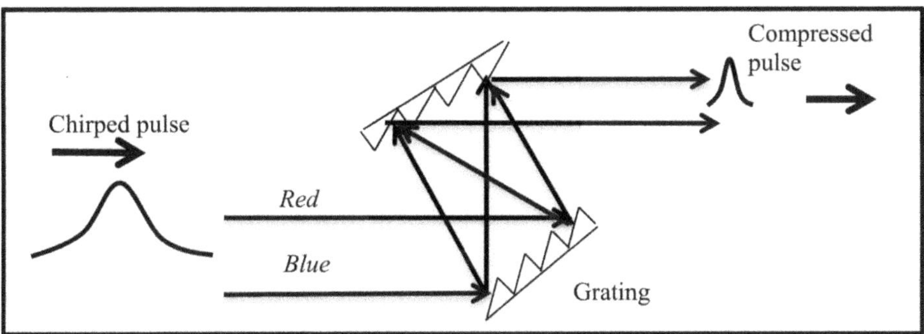

**Figure 18.14.** Grating pair used as a pulse compressor. The chirped pulse enters from the left and the compressed pulse is obtained on the right.

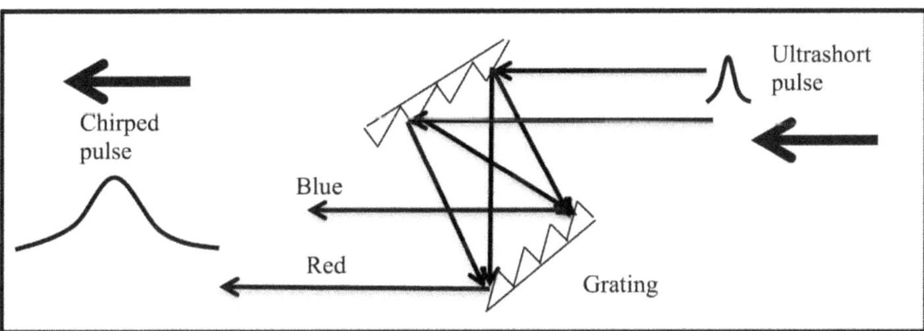

**Figure 18.15.** Grating pair used as a pulse-chirping device or a 'pulse stretcher'. (To be distinguished from figure 18.14.)

♣ In CPOPAs, the grating pair is used as a 'pulse stretcher' in order to chirp the ultrashort pulses before amplification (♠ see section 18.9). The Nobel Prize of 2018 was shared by Donna Strickland and Gerald Morou for their discovery the CPOPA principle in 1985.

**Exercise 18.6.** If a grating has 1800 grooves per mm, for incident light at 30°, what is the angle of first-order diffraction at a wavelength of 600 nm?

**Solution:** For the given incident beam, the grating equation can be written as

$$m\lambda = d(\sin \theta_1 + \sin \theta_2).$$

Here, d is the grating element, $m$ is the order of diffraction, and $\theta$ is the diffraction angle.

For 1800 grooves per mm, the grating element $d = 1/1800 \times 10^3 = 5.55 \times$ m$^{-1}$.

For $m = 1$ and sin 30°$(= 0.5)$, the value of the first-order diffraction angle $\theta_2$ is 35°45′ for a wavelength of 600 nm.

### 18.7.2 Prism compressor

We know that a single prism disperses white light into its spectrum, spanning from violet to red. In this case, the dispersion is not linear as in the case of a grating. In figure 18.16, two prisms (*Prism 1* and *Prism 2*) are placed in a manner similar to that of the classic experiment of Newton carried out using sunlight. The prisms are placed in such a way that they (i) are nearly at the Brewster angle to the beam, (ii) are at the angle of minimum deviation, and (iii) the face of *Prism 1* at which light enters is parallel to the face of *Prism 2* at which light exits and the exit-face of *Prism 1* is parallel to the entrance-face of *Prism 2*.

The red wavelengths travel faster than the blue inside the glass. Therefore, in the prism pair, the second prism is placed in such a way that the red portions of the dispersed pulse encounter larger amounts of glass than the blue portions. This facilitates the blue rays to catch up with the red rays, resulting in temporal compression of the laser pulse. This reveals that a prism pair can provide *'negative dispersion'* that helps to compensate for chirped pulses.

The distance between the prisms is a useful parameter. The chirp variations experienced by the input pulse can be compensated by adjusting this, in addition to adjusting the amount of glass in the second prism. It should be noted that the laser pulse propagates through the prism pair on its return path as well. In this configuration, a slight tilt of the mirror is required to direct the pulse out of the prism pair to the experiment.

♣ The prisms are mounted at the 'angle of minimum deviation' to obtain a collimated output free of any small variations in the direction of the incident beam.

♣ Newton verified the fact that sunlight contains colors by mixing the colors back into white light using another prism.

A simplified design of the prism compressor, consisting of four prisms, is shown in figure 18.17. The laser pulse enters the setup at *Prism 1* and exits at *Prism 4*. Here, *Prism 1* disperses the light into its components, *Prisms 2* and 3 are used to obtain the required compensation, and *Prism 4* is used to collimate the compressed pulse.

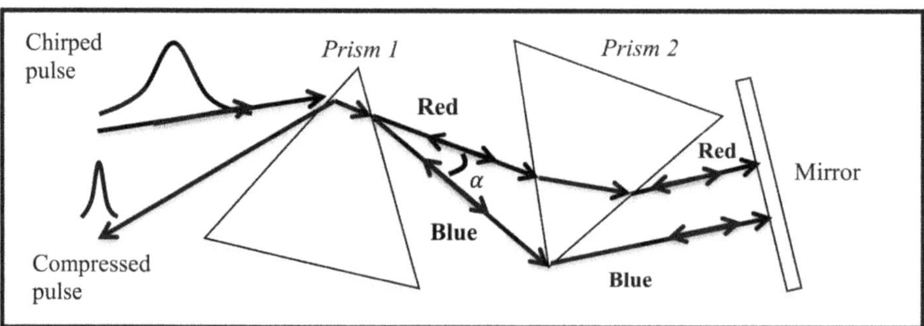


**Figure 18.16.** A prism compressor consists of two prisms, *Prism 1* and *Prism 2*. $\alpha$ is the angle between two extreme colors of a chirped pulse. The laser pulse travels back through the prism pair after reflection by the mirror, before exiting as a compressed pulse, as shown.

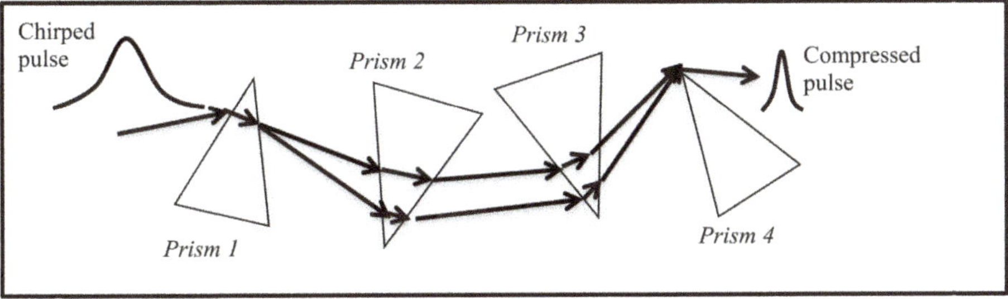



**Figure 18.17.** Prism compressor consisting of four prisms.

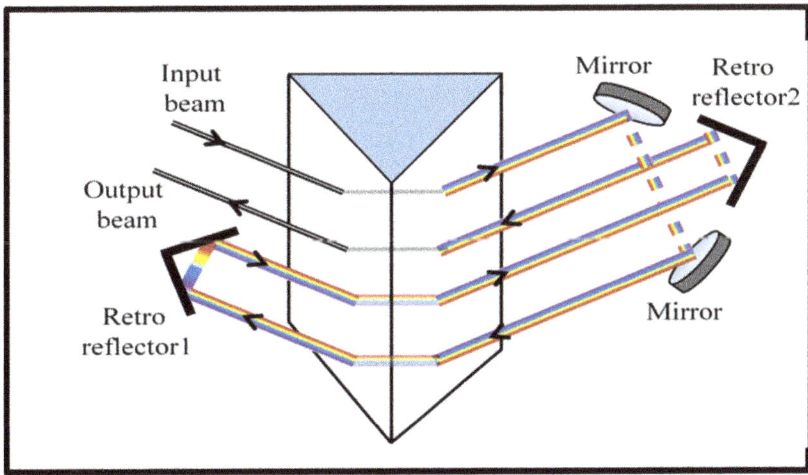



**Figure 18.18.** Mirrors and retroreflectors are needed to compress a laser pulse using a single prism.

Again, in this setup, the blue rays travel an extra distance through *Prisms 2* and 3 to compress the pulse.

Yet another variant of the prism compressor design uses a single prism. This configuration makes use of a set of mirrors along with a prism. This design uses a prism and retroreflectors and passes the beam multiple times through the same prism. Although the design is a little complex, its compactness and easier tunability make it a perfect choice for use in ultrafast optics and spectroscopy. The working principle of the pulse compressor remains the same. As mentioned earlier, in all prism compressors, the prisms are placed at the angle of minimum deviation to retain control over the direction of the output beam (figure 18.18).

## 18.8 Dispersion-free autocorrelator

As seen in the previous sections, ultrafast laser pulses experience temporal broadening while passing through optical components due to dispersion. For instance, the thickness of the optics used in the autocorrelator described in figure 18.2 viz. the beam splitter, the focusing lens, and the SHG crystal also contribute to the measured pulse duration. In order to eliminate errors due to the optics of the autocorrelator itself, reflective optics is introduced. The idea of reflective optics is that the beam splitter is replaced by a beam-splitting mirror and that the lens is replaced by a focusing mirror.

In such a setup based on reflection optics, a crystal thickness of ~50 µm is used for SFG. Based on these replacements, a compact design has been proposed which is known as the *dispersion-free configuration* of the autocorrelator. In this configuration (figure 18.19), an optical delay line is introduced by the translation (<100 µm ± 0.5 µm) of one of the half mirrors along the beam path. Using this method, ultrafast pulses of the order of ~10 fs can be measured. Figure 18.20 shows

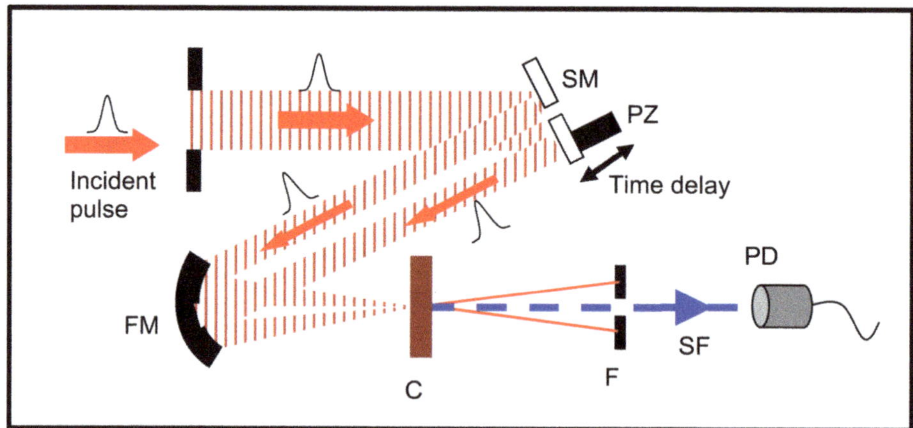


**Figure 18.19.** Schematic diagram of the dispersion-free autocorrelator. SM: split mirror, PZ: piezoelectric transducer used to generate optical delay in one of the split mirrors, FM: focusing mirror, C: nonlinear optical crystal, F: filter to block the fundamental frequency, SF: sum frequency, PD: photodiode [14].

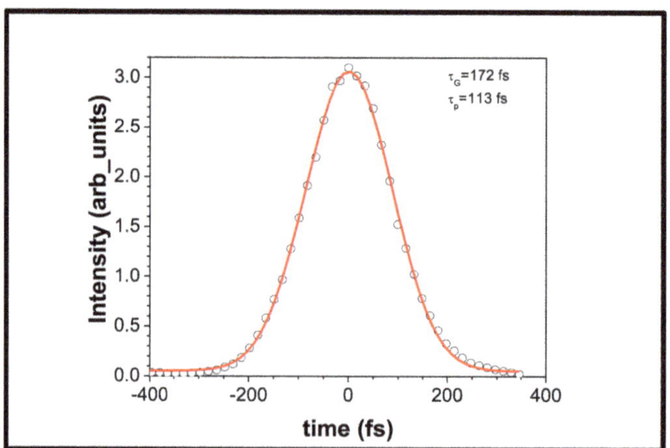


**Figure 18.20.** Autocorrelation trace for a femtosecond laser pulse obtained using a dispersion-free autocorrelator (open circles). The solid line is the fit to a Gaussian function, which gives a pulse duration of 113 fs (see table 18.2, after [14]).

an autocorrelation trace for a femtosecond laser pulse obtained using a dispersion-free autocorrelator.

♣ Splitting the Gaussian beam in the noncollinear direction can cause some concerns about preserving the shape of the Gaussian spot. For ultrafast laser pulse measurement, the required delay is very small (corresponding to ~100 μm). As long as the optics used is of the order of a few millimeters, this poses no concerns for estimates of the pulse to reasonable accuracies.

## 18.9 Chirped pulse amplification

This is a technique used to obtain high-energy ultrafast pulses from mode locked femtosecond pulses. Generally, the energy of the mode locked pulses obtained from oscillators is of the order of nanojoules to millijoules, corresponding to power levels in the range of megawatts to gigawatts. It is not possible to amplify the pulse further using an amplifier to obtain petawatt (PW) or even higher powers of laser pulses. This is due to the fact that in the later amplification stages, the peak power of the pulse being amplified can damage the amplifier itself.

Therefore, before its amplification, the pulse is first stretched in time (♠ see section 18.6.1). The resulting peak power of the laser pulse remains below the damage threshold of the amplifier. After several stages of amplification, the pulse can be recompressed. The chirped pulse can be amplified to the desired level and then recompressed to obtain high-energy pulses, as shown in figure 18.21. This is known as the chirped pulse amplification (CPA) technique. The principle of the CPA technique was introduced in 1985 by Strickland and Morou, for which they received a share of the Nobel Prize in 2018.

## Questions and problems

1. A laser pulse that has a duration of 10 ns can be accurately measured using an oscilloscope that has a bandwidth of 100 MHz. Sketch a setup that can measure the pulse duration using an oscilloscope and a detector. In this measurement, what are the limitations of the electronics?
2. Among the four terms in equation (18.2), which one is responsible for pulse duration measurement by the intensity autocorrelation technique?

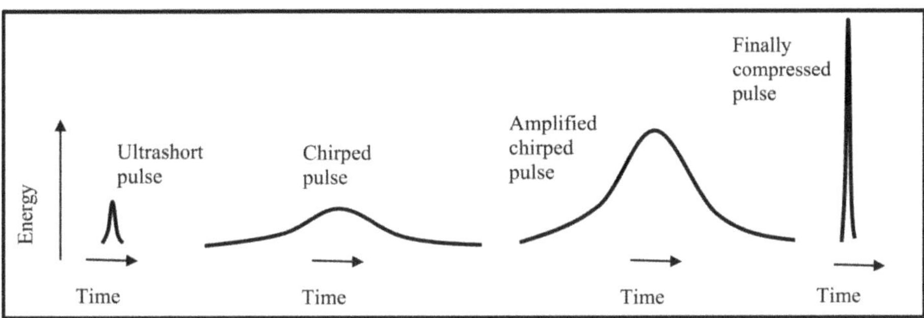


**Figure 18.21.** Principle of the CPA technique (♠ see [12]). See the text for details.

3. What is the cross-correlation technique? Under which experimental conditions does this technique become useful?

4. Very often, the pulse shapes measured by the autocorrelation technique are assumed to follow a Gaussian function. When the pulse shape is unknown, (i) how is it possible get the best estimate of the duration of an ultrafast pulse? (ii) Identify any other technique that can provide information about the pulse shape.

5. What is the advantage of interferometric autocorrelation as compared to intensity autocorrelation?

6. Write down the Fourier transform (FT) of a function $E(t)$ in the frequency domain. In which technique of ultrafast pulse measurement is the FT technique useful?

7. What is pulse chirping? With reference to the SPIDER technique, how is pulse chirping useful in the measurement of ultrafast laser pulse duration and phase?

8. How is $\chi^{(2)}$ used in the frequency upconversion technique? Differentiate frequency upconversion from photon downconversion.

9. With reference to figure 18.16, let us denote the distance between the apexes of the two prisms by $l$. The angle between the two extreme-wavelength rays emerging from *Prism 1* after the dispersion is $\alpha$. The path difference $(r)$ that contributes to the dispersion between the prism pair for a single pass is $r = l\cos\alpha$.

    (i) Obtain the expression for the second derivative of the path length with respect to the wavelength $(\lambda)$, and

    (ii) find the approximate distance that compensates for group velocity dispersion. (Hint: ♠ see [9].)

## Bibliography


[1] Andersson T and Eng S T 1983 Determination of the pulse response from intensity autocorrelation measurements of ultrashort laser pulses *Opt. Commun.* **47** 288–90

[2] Kane D J and Trebino R Feb. 1993 Characterization of arbitrary femtosecond pulses using frequency-resolved optical gating *IEEE J. Quantum Electron.* **29** 571–9

[3] DeLong K W, Trebino R, Hunter J and White W E 1994 Frequency-resolved optical gating with the use of second-harmonic generation *J. Opt. Soc. Am.* B **11** 2206–15

[4] Baltuska A, Pshenichnikov M S and Wiersma D A 1999 Second-harmonic generation frequency-resolved optical gating in the single-cycle regime *IEEE J. Quantum Electron.* **35** 459–78

[5] Trebino R 2000 *Frequency-Resolved Optical Gating: The Measurement of Ultrashort Laser Pulses* (Cambridge, MA: Academic Press)

[6] Homann C, Krebs N and Riedle E Aug. 2011 Convenient pulse length measurement of sub-20-fs pulses down to the deep UV via two-photon absorption in bulk material *Appl. Phys.* B **104** 783

[7] Mashiko H, Suda A and Midorikawa K May 2003 All-reflective interferometric autocorrelator for the measurement of ultra-short optical pulses *Appl. Phys.* B **76** 525–30

  [8] Gordon J P and Fork R L May 1984 Optical resonator with negative dispersion *Opt. Lett.* **9** 153–5

  [9] Martinez O E, Gordon J P and Fork R L Oct. 1984 Negative group-velocity dispersion using refraction *J. Opt. Soc. Am.* A **1** 1003–6

[10] Osvay K, Kovacs A P, Heiner Z, Kurdi G, Klebniczki J and Csatari M 2004 Angular dispersion and temporal change of femtosecond pulses from misaligned pulse compressors *IEEE J. Sel. Top. Quantum Electron.* **10** 213–20

[11] Diels J-C and Rudolph W 2006 *Ultrashort Laser Pulse Phenomena* 2nd edn (Cambridge, MA: Academic Press)

[12] Strickland D and Mourou G 1985 Compression of amplified chirped optical pulses *Opt. Commun.* **56** 219–21

[13] Ali A 2013 *PhD Thesis* IIT Madras

[14] Kalanoor B S 2012 *PhD Thesis* IIT Madras

# An Introduction to Photonics and Laser Physics with Applications

**Prem B Bisht**

# Chapter 19

# Optical phase conjugation

Optical phase conjugation (OPC) is one of the applications of $\chi^{(3)}$. In this *four-wave mixing* technique, three waves with wave vectors $\vec{k}_1$, $\vec{k}_2$ and $\vec{k}_3$, respectively, are incident on a sample. The fourth wave, whose wave vector is $\vec{k}_4$, is generated in the direction opposite to that of wave vector $\vec{k}_3$, as shown in the diagram. The details of this interesting phenomenon are described in this chapter. A mirror with special properties, such as aberration correction, which can be formed using the OPC process is also introduced here. In addition, the properties associated with the mixing of six or more waves are outlined.

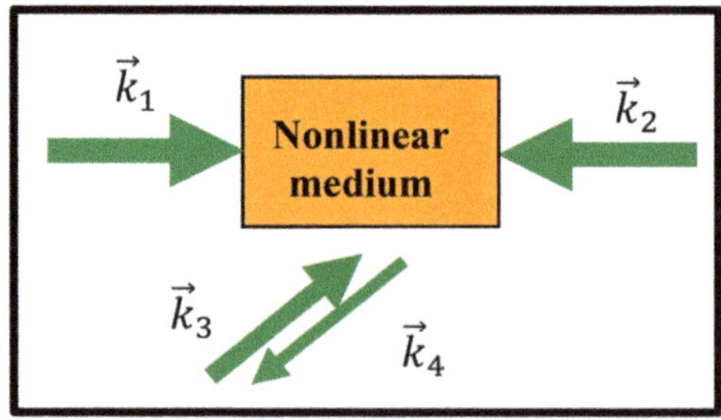


**Learning objectives**
**After reading this chapter, the learner will be able to:**
Describe the interference between two laser beams;
Identify the self-diffraction pattern that results from four-wave mixing;
Define the geometry of phase conjugation;
Demonstrate higher-order wave mixing;
Discover the phase-conjugate mirror and time reversal;
Categorize phase conjugation as a technique for aberration correction.

doi:10.1088/978-0-7503-5226-0ch19



© IOP Publishing Ltd 2022

## 19.1 Two-beam interference and Bragg diffraction

Two polarized beams A and B derived from the same laser source of wavelength $\lambda$ are said to be degenerate. These two beams, with wave vectors $\vec{k_1}$ and $\vec{k_2}$ and intensities $I_1$ and $I_2$, respectively, are shown intersecting at an angle $\theta$ in a semitransparent medium in figure 19.1. The beams create an interference pattern which is periodic, like the grooves of a diffraction grating.

This interference pattern is governed by a grating vector ($\vec{q}$):

$$\vec{q} = \vec{k_1} \pm \vec{k_2} \tag{19.1}$$

The fringe spacing or grating period ($\Lambda$) is defined as the distance between two maxima (or minima) of the interference pattern in the y-direction, as shown in figure 19.1. The value of $\Lambda$ can be expressed similarly to Bragg's law of diffraction in terms of the pump $\lambda$ and $\theta$ as

$$\Lambda = \left(\frac{\lambda}{2}\right) / \sin\left(\frac{\theta}{2}\right). \tag{19.2}$$

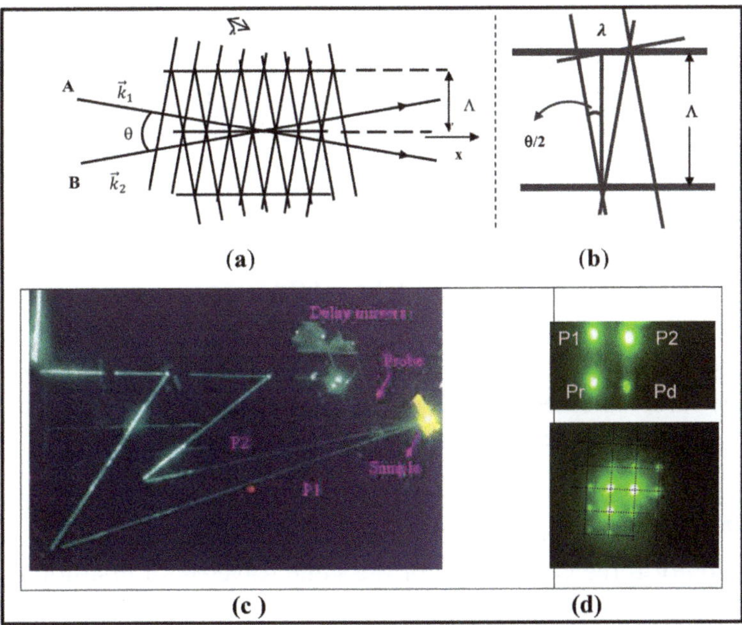


**Figure 19.1.** Grating formed by the interference of two plane polarized light beams with wave vectors $\vec{k_1}$ and $\vec{k_2}$. The grating vector ($\vec{q}$) is also shown (panel (a)). An expanded view used to calculate the fringe spacing is given in panel (b). $\Lambda$ is the grating spacing and $\lambda$ is the wavelength of the incident laser. Photo of the two pump beams P1 and P2 interfering at the sample while the 'probe' beam is on a delay stage (panel c). The top and bottom photo clips in panel (d) show the Bragg diffraction (Pd) of the probe beam (Pr) and the self-diffracted pattern, respectively [3].

When two beams are antiparallel to each other, the grating period is at its smallest, i.e. $\Lambda = \lambda/2$. For small angles, equation (19.2) becomes

$$\Lambda = \lambda/\theta. \tag{19.3}$$

Generally, the interacting beams are derived from a short-pulse-duration laser. If the incident wavelength is suitable for absorption by the medium, the sample is excited at the bright fringe positions. This results in an excited-state population grating. Such a grating decays within the lifetime of the excited state of the sample. Therefore, it is known as a laser-induced transient grating (LITG). A photograph of the experimental setup used for the LITG technique is shown in the bottom left panel of figure 19.1.

♣ In the case of acoustic gratings, the diffraction produced by thin and thick gratings is classified as being in either the Raman–Nath or Bragg scattering regime. At low frequencies and for interactions within shorter path lengths, self-diffraction (Raman–Nath diffraction) occurs. However, at high frequencies and for thick gratings, Bragg scattering takes place. For their counterparts in the LITG process, the self-diffraction pattern can be observed with thin samples, and Bragg diffraction occurs for thick samples (see the bottom right photographs in figure 19.1).

**Exercise 19.1.** For a thin grating that has path lengths of < 1 mm, self-diffraction can occur in a semitransparent sample at various angles for ultrafast pulses of the second harmonic of a Nd:yttrium aluminum garnet (YAG) laser. What is the fringe spacing for an angle of 3° between two beams derived from the same laser?

**Solution:** For self-diffraction to occur, the two beams of the second harmonic of a Nd:YAG laser (532 nm) must meet temporally and spatially at the sample. For the thin sample, the fringe spacing ($\Lambda$) is calculated using equation (19.2):

$$\Lambda = \left(\frac{\lambda}{2}\right) / \sin\left(\frac{\theta}{2}\right).$$

Here, we need to convert the value of the angle from degrees to radians ($3° = 0.052$ radians). The value obtained for $\Lambda = 10.2$ μm. As the angle is small, you can check whether we get the same result from equation (19.3).

## 19.2 Four-wave mixing: phase conjugation

The interaction of three laser beams of the same wavelength in a nonlinear optical medium, which generates a fourth beam, is known as degenerate four-wave mixing (DFWM) (♠ see section 16.4). In such a configuration, two counter-propagating pump beams of identical frequencies fall on a medium as shown in figure 19.2. These beams are called the forward pump beam, which has the wave vector $\vec{k_1}$, and the backward pump beam, which has the wave vector $\vec{k_2}$. A probe beam ($\vec{k_3}$) at the same frequency is now introduced and interacts with the nonlinear medium as shown in the figure.

The diffraction processes in this geometry of DFWM can be understood using the process of grating formation, as shown in figure 19.3. The forward pump ($\vec{k}_1$) and probe ($\vec{k}_3$) beams interfere to form an interference grating. As in the case of the $\chi^{(3)}$ process (♣ section 15.4.2), due to the conservation of momentum for *four-wave mixing*, we have

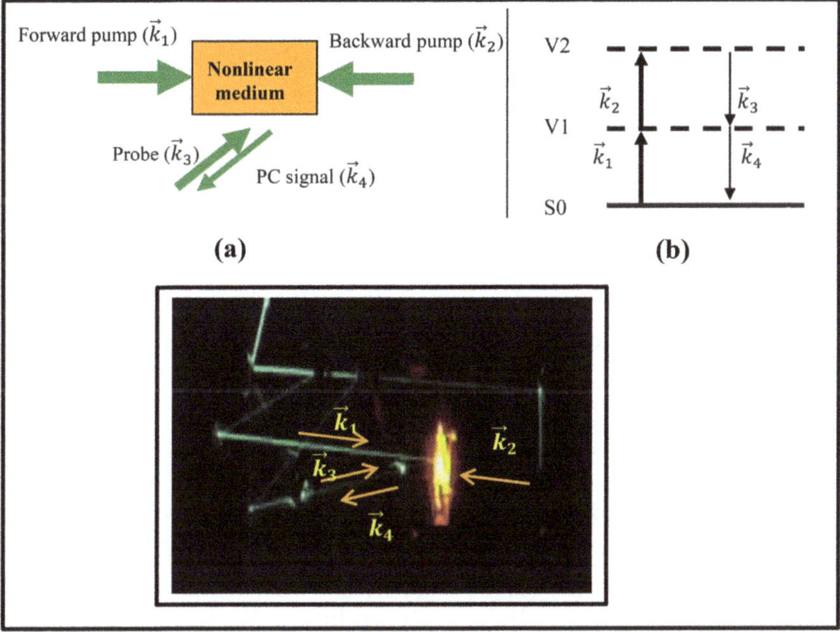


**Figure 19.2.** Interaction between forward pump, backward pump, and probe beams in a nonlinear optical medium (panel (a)). Panel (b) gives the energy-level diagram of the process generating the phase-conjugate (PC) signal ($k_4$). In the energy-level diagram (panel b), S0 is the ground state and V1 and V2 are the virtual states. The bottom panel shows all the beams in the experimental setup [5].

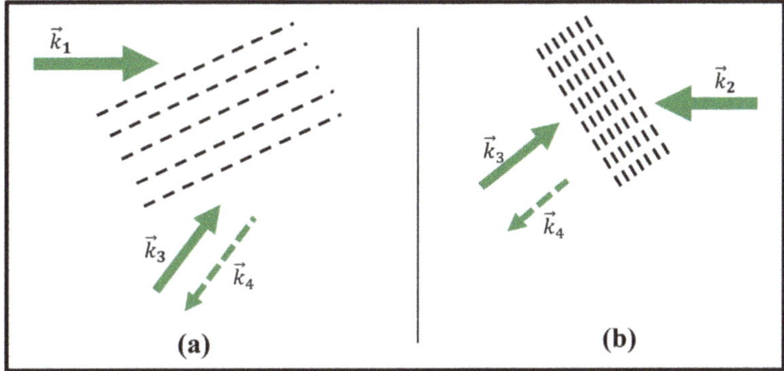


**Figure 19.3.** Grating patterns and relative spacing in phase-conjugation geometry between the forward pump ($\vec{k}_1$) and probe ($\vec{k}_3$) beams (panel (a)) and the backward pump ($\vec{k}_2$) and probe ($\vec{k}_3$) beams (panel (b)). The PC signal ($\vec{k}_4$) is also shown in both panels.

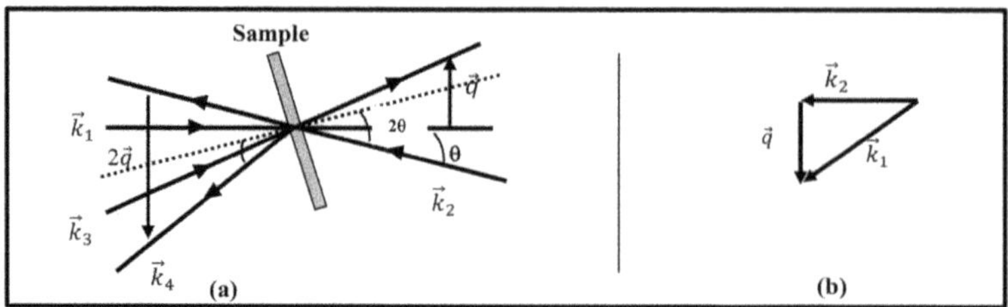

**Figure 19.4.** Phase-matching diagram for a PC signal in the six-wave mixing geometry (panel (a)). The dotted line is the normal to the sample. Panel (b) shows the grating vector $2\vec{q}$ made by $\vec{k}_2$ and $\vec{k}_4$ at an angle of $2\theta$.

$$\vec{k}_1 + \vec{k}_2 + \vec{k}_3 = \vec{k}_4; \text{ or } \Delta \vec{k} = 0.$$

If the two beams with wave vectors $\vec{k}_1$ and $\vec{k}_2$ are counter-propagating, we obtain a phase-conjugate beam (with $\vec{k}_4$). This occurs because the backward pump beam ($\vec{k}_2$) is diffracted and generates a phase-conjugate (PC) beam ($\vec{k}_4$) in the direction opposite to $\vec{k}_2$. This grating has a relatively large grating period, for which the backward pump beam ($\vec{k}_2$) satisfies the phase-matching condition for the Bragg diffraction of $\vec{k}_2$ into a PC signal beam, $\vec{k}_4$.

♣ It should be noted that the PC signal disappears if any one of the three beams is blocked.

There are other possibilities for grating formations that have a weaker contribution or no contribution. For example, $\vec{k}_2$ and $\vec{k}_3$ create a grating with a finer grating period (figure 19.3(b)). This grating diffracts $\vec{k}_1$ in the direction of $\vec{k}_4$ as the PC signal beam in the DFWM geometry. Under such conditions, the sample is said to work as a phase-conjugate mirror.

For the $\chi^{(3)}$ process, the backward pump beam makes an angle of $\theta$ with respect to the normal, as shown in figure 19.4. If we introduce $\vec{q}(= \vec{k}_3 - \vec{k}_1)$ as the grating vector for the $\chi^{(3)}$ process, the interaction of the four waves corresponding to the $\chi^{(3)}$ process is given by

$$\begin{aligned} \vec{k}_4 &= \vec{k}_2 + \vec{q} \\ &= \vec{k}_2 + \vec{k}_3 - \vec{k}_1 \end{aligned} \tag{19.4}$$

### 19.2.1 Diffraction efficiency and $\chi^{(3)}$

In the phase-conjugation geometry of DFWM, the diffraction efficiency ($\eta$) can be directly measured by taking the ratio of the intensity of the phase-conjugate beam ($I_4$) to that of the backward pump beam ($I_2$) and is given by

$$\eta = \frac{I_4}{I_2}. \tag{19.5}$$

The third-order susceptibility ($\chi^{(3)}$) is proportional to $\sqrt{\eta}$ (♠ see equation (16.3)).

♣ Theoretically, the diffraction efficiency of the thick grating is expressed as [4]:

$$\eta = \left[ \sin^2\left( \frac{\pi d \Delta n}{\lambda \cos \theta_{\mathrm{B}}} \right) + \sinh^2\left( \frac{d \Delta \alpha}{2 \cos \theta_{\mathrm{B}}} \right) \right] e^{-2\alpha d / \cos \theta_{\mathrm{B}}}, \tag{19.6}$$

where $\theta_{\mathrm{B}}$ is the Bragg angle and $\Delta \alpha$ and $\Delta n$ are the changes in absorption coefficient and refractive index, respectively. The exponential factor corresponds to the attenuation of the diffracted beam with the sample thickness $(d)$. From the measured value of $\eta$, $\chi^{(3)}$ values can be obtained as follows:

$$\chi^{(3)} = \left( \frac{\varepsilon_0 \lambda n^2}{d\pi} \right) \sqrt{\frac{\eta}{I_1 I_3}}, \tag{19.7}$$

where $\varepsilon_0$ is the permittivity of free space and $n$ is the refractive index. $I_1$ and $I_3$ are the intensities of the forward and the probe beams, respectively. For an absorbing medium, equation (19.8) can be rewritten by introducing the absorption coefficient $(\alpha)$, as follows:

$$\chi^{(3)} = \left( \frac{\varepsilon_0 \lambda n^2}{\pi} \right) \sqrt{\frac{\eta}{I_1 I_3}} \left( \frac{\alpha e^{(\alpha d / 2)}}{1 - e^{(-\alpha d)}} \right). \tag{19.8}$$

### 19.2.2 Six-wave mixing and $\chi^{(5)}$

Phase-matching conditions can also be achieved for higher-order diffraction in the DFWM geometry. For example, the case of six-wave mixing is shown in figure 19.4. Here, the angle of incidence of the backward pump beam $(\vec{k_2})$ with respect to the normal is $2\theta$. The grating vector for the phase-matching condition is shown in figure 19.4(b).

This geometry, i.e. the backward pump beam $(\vec{k_2})$ that makes an angle of $2\theta$ with respect to the normal, invokes the $\chi^{(5)}$ process. In this configuration, the direction of the diffracted beam can be written as the interaction of the six waves.

$$\begin{aligned} \vec{k_4} &= \vec{k_2} + 2\vec{q} \\ &= \vec{k_2} + 2\vec{k_3} - 2\vec{k_1} \end{aligned} \tag{19.9}$$

♣ As a general case, for multiple wave mixing, a change of angle of the backward pump beam by $m\theta$ results in the interaction of $2m + 2$ waves. Here, $m$ is a natural number. This corresponds to the $\chi^{(2m+1)}$ process. We see that four-wave mixing gives information related to $\chi^{(3)}$, six-wave mixing gives information about $\chi^{(5)}$, eight-wave mixing gives information related to $\chi^{(7)}$, and so on. It should be noted that the incident energies required to invoke the higher-order susceptibilities are higher, which imposes a constraint on experimental observations.

## 19.3 Time-reversal in phase conjugation

Let us consider a plane wave $E(\vec{r}, t)_{\mathrm{incident}}$ traveling with an angular frequency of $\omega$ and a wave vector $\vec{k}$ incident on the phase-conjugated mirror (sample) as

$$E(\vec{r}, t)_{\text{incident}} \propto [e^{i(\vec{k}.\vec{r}-\omega t)} + e^{-i(\vec{k}.\vec{r}-\omega t)}] \tag{19.10}$$

The reflected (phase-conjugate) wave $E(\vec{r}, t)_{\text{PC}}$ that retraces the path (i.e. that travels in the opposite direction to that of the incident beam) is given by

$$E(\vec{r}, t)_{\text{PC}} \propto [e^{i(-\vec{k}.\vec{r}-\omega t)} + e^{-i(-\vec{k}.\vec{r}-\omega t)}]. \tag{19.11}$$

Equations (19.10) and (19.11) show that the effect of the phase conjugation is to reverse the value of the wave-propagation direction, $\vec{k}$. The reflected wave travels with the same angular frequency $\omega$, but has a propagation vector of $-\vec{k}$. Equation (19.11) can be rewritten for $-(t)$ as follows:

$$E(\vec{r}, t)_{\text{PC}} \propto [e^{-i(\vec{k}.\vec{r}-\omega(-t))} + e^{i(\vec{k}.\vec{r}-\omega(-t))}]. \tag{19.12}$$

On comparing equation (19.12) with equation (19.10), we can see that the phase-conjugated wave is identical in form, except that the time $t$ has been replaced by $-t$, so that the reflected wave is related to the incident wave as follows:

$$E(\vec{r}, t)_{\text{PC}} \propto E(\vec{r}, -t)_{\text{incident}}. \tag{19.13}$$

This is equivalent to leaving the spatial part of $E(\vec{r}, t)_{\text{incident}}$ unchanged but reversing the sign of time $t$. In this sense, the phase-conjugation process is equivalent to the *time-reversal process* of the incident wave.

**Exercise 19.2.** A probe beam is described by $\vec{E} = E_0\hat{n}e^{i(\vec{k}.\vec{r}-\omega t)}$. Write down its phase-conjugate counterpart.

 **Solution:** The phase-conjugate beam travels in the opposite direction. This means the propagation constant has a negative sign. However, the frequency and the polarization remain the same. Provided that there is no loss of amplitude, the electric field of the phase-conjugate beam can be written as $\vec{E}_{\text{PC}} = E_0\hat{n}e^{-i(\vec{k}.\vec{r}+\omega t)}$.

## 19.4 Applications of phase conjugation

Optical phase conjugation is used to compensate for the static and dynamic distortions encountered in optical systems. The applications of phase conjugation include the use of adaptive optics to remove atmospheric aberrations, the use of holography in the real-time production of squeezed light, and the focusing of laser light for laser fusion. In the following, we discuss a few cases that illustrate the principle behind all these applications.

### 19.4.1 Reversal of the wavefront

Due to the effect of stimulated Brillouin scattering (SBS) (♠ see section 16.8.) high-pressure methane gas has been found to reflect pulsed laser beams. The observed reflection has been found unusual in the sense that it returns back to its place of origin, irrespective of the direction it was sent from.

Materials with pressure-dependent polarizability exhibit the SBS phenomenon. SBS occurs as a result of the interaction between an optical frequency ($\omega$) and a frequency ($\Omega$) corresponding to acoustic phonons. A Stokes wave similar to stimulated Raman scattering (SRS) is generated towards the longer wavelengths, as shown in figure 19.5.

The grating is created by electrostriction—a property of any medium that increases its density as a function of the incident electric field. The grating formed due to acoustic phonons in the medium is only reflective in nature. Therefore, SBS occurs in the backward direction. This property of SBS is useful in phase conjugation. The observed wavelength shift of the Stokes wave is small for SBS, of the order of 0.1 nm.

**Exercise 19.3.** An incident wavefront is made to fall sequentially on a conventional plane mirror and a phase-conjugate mirror. Keeping in mind the properties of the two mirrors, differentiate the reflected light with the help of the diagram.

**Solution:** As shown in figure 19.6, the conventional mirror follows the law of reflection. The OPC phenomenon reverses the direction of propagation of the wavefront. This happens because the phase of a reflected wavefront is reversed as compared to the case of a normal mirror. Therefore, due to the phase-conjugation property, light that is incident on a PC mirror is reflected back to the source. This property of PC mirrors can be used to make a super lens (♠ see problem 4).

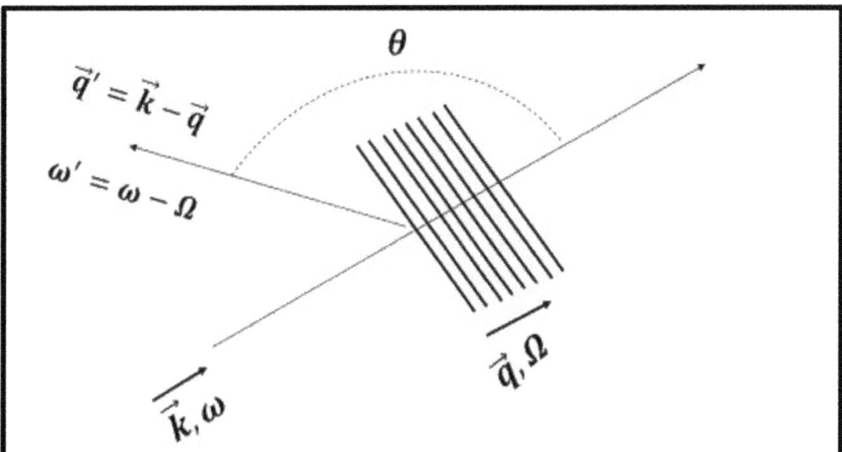


**Figure 19.5.** Phenomenon of SBS caused by back scattering of a laser frequency $\omega$ by an acoustic grating. $\Omega$ is the frequency corresponding to the grating period and $\theta$ is the angle between the grating's normal ($\vec{q}$) and the diffracted beam $\vec{q}'$.

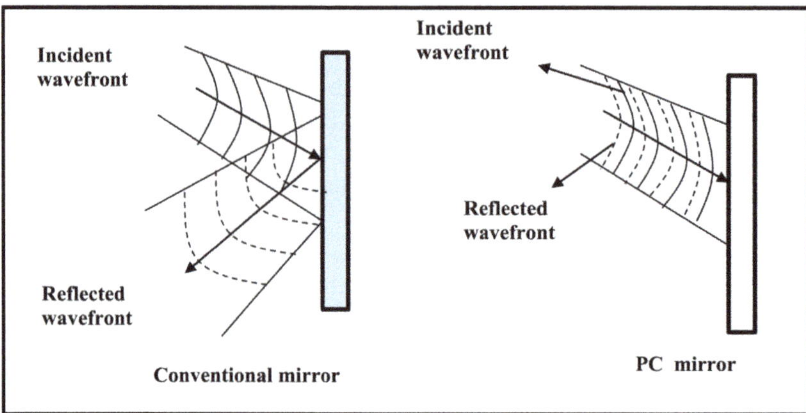


**Figure 19.6.** Incident wavefronts (solid line) and reflected wavefronts (dashed line) in the reflection of light by a conventional mirror (left) and a phase-conjugate mirror (right).

## 19.4.2 Correction of aberrations and imaging

In imaging, distortions are introduced by the defects in optical elements. Similarly, in the atmosphere, the variation between the refractive indices of the various layers leads to aberrations in images. The OPC process can be used to remove these aberrations from wavefronts, as demonstrated in figure 19.7. In the top panel, the incident beam passes through a distorting medium. After reflection by a conventional mirror in its return path, the amplitude of the distortion is twice that gained in the forward path. In contrast, in the case of the PC mirror (bottom panel), the incident wavefront is reversed *in time* on reflection, without affecting its phase. Therefore, on its return path, although the distorting medium adds equal amplitude, it is in the opposite direction to that of the forward path. In this way, an original wavefront that makes a double pass through a distorting medium is compensated for the medium's distortion.

In lightwave technology, the distortions caused by image transmission can be compensated by a combination of a PC mirror and optical fibers, as shown in figure 19.8. Here, the input beam is passing through an optical fiber (1) for a certain experimental arrangement. Before detection, it is sent to a PC mirror after passing through a beam splitter. On its return path, the beam splitter directs the beam to an optical fiber (2) with similar properties to those of fiber (1). Fiber (2) compensates for the distortions introduced by fiber (1). The same principle is also used in adaptive optics.

♣ The phase distortions inside a laser cavity can also be corrected by introducing a phase-conjugate mirror at one end. In this method, the effects of group-velocity dispersion as well as that of self-phase modulation are removed using a Kerr-like optical phase conjugator. The distortions have been compensated using a combination of doped optical fiber and a phase-conjugate mirror (♠ see [9]).

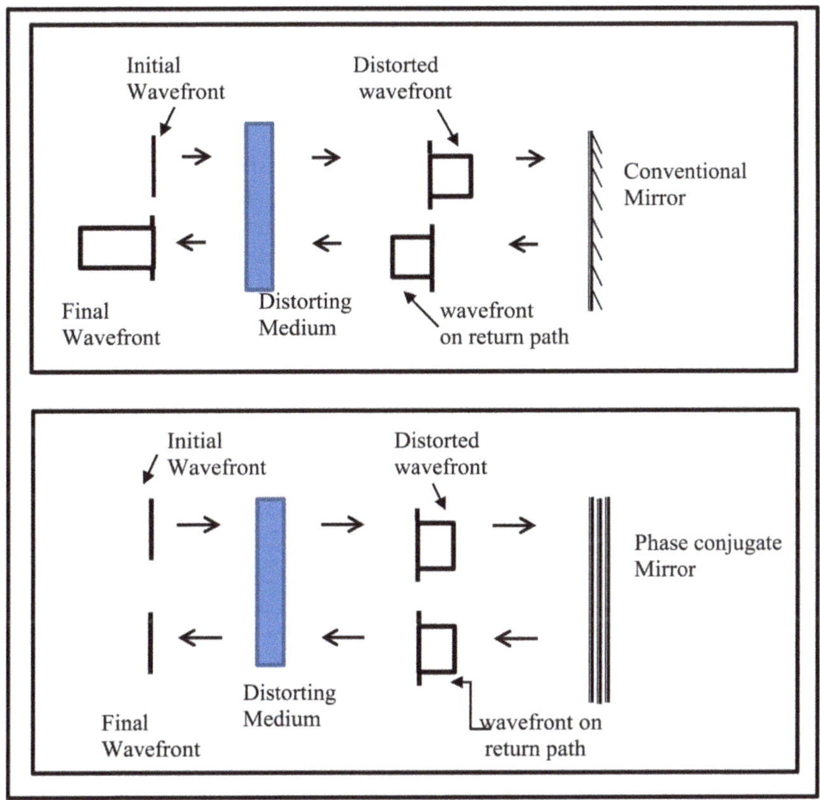


**Figure 19.7.** A light wavefront undergoing a second pass through a distorting medium after reflection by a conventional mirror (top panel) and by a phase-conjugate mirror (bottom panel). The arrows indicate the direction of wavefront propagation.

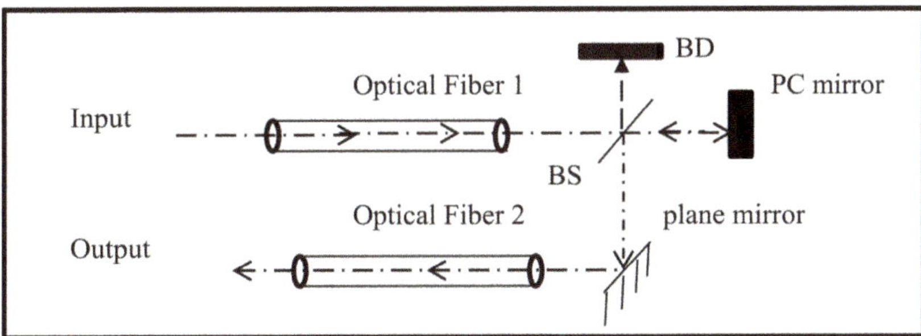


**Figure 19.8.** Schematic of the removal of distortion from an image in optical fiber communication using a phase-conjugate (PC) mirror. BS denotes the beam splitter and BD denotes the beam dump.

## Questions and problems

1. In work involving nonlinear optics, two-beam interference gives rise to self-diffraction as well as Bragg diffraction. With reference to figure 19.1, identify

the geometries that belong to the (i) Raman–Nath and (ii) Bragg scattering regimes. (♠ see [13] for details).

2. For the schematic diagram of phase-conjugation geometries shown in figures 19.2 and 19.3, create a vector diagram of the grating vectors that shows the direction of the phase-conjugate beam.

3. Using the phase-conjugation technique, it is possible to create a phase-conjugate (PC) mirror. In contrast to plane mirrors, comment on the image of your face as reflected by a PC mirror.

4. In figure 19.9, light beams are shown to originate in front of plane mirrors (a) and (c), and PC mirrors (b) and (d), respectively.

   (i) Draw the reflected beam in each case.

   (ii) If the beams in (a) and (b) are given by the equation $E_1 \propto e^{-i\vec{k}_1 \cdot \vec{r}}$ and those in (c) and (d) by $E_1 \propto \frac{1}{r} e^{-i\vec{k}_1 \cdot \vec{r}}$, write down the equations for the reflected beams in each case.

5. Two ultrashort laser pulses with electric fields $E_1 = E_0 e^{-i\vec{k}_1 \cdot \vec{r}}$ and $E_2 = E_0 e^{-i\vec{k}_2 \cdot \vec{r}}$, respectively, meet at a sample that exhibits $\chi^{(3)}$. The angle between the propagation directions of the two pulses is $180°$ (i.e. the laser beams are counter-propagating). Keeping in mind the phase matching of the wave vectors,

   (i) If another light pulse $E_3 = E_0 e^{-i\vec{k}_3 \cdot \vec{r}}$ is incident on the sample, describe the effect in terms of the generated field, $E_4$.

   (ii) If $E_3$ is given by $E_3 = E_0 e^{-i(\vec{k}_3 \cdot \vec{r} - \omega_3 t)}$, describe the time-reversal phenomenon.

6. A plane wavefront of light (W) is traveling in a medium towards a plane mirror and a phase-conjugate mirror, as shown in the left- and right-hand panels of figure 19.10, respectively. While propagating in the medium, the wavefront encounters the rectangular-shaped distortion (D) shown in each panel. What is the shape of the wavefront in each panel: (i) immediately after passing through D in the forward direction? (ii) immediately after reflection between D and the mirror? (iii) immediately after a second pass through D on its return path?

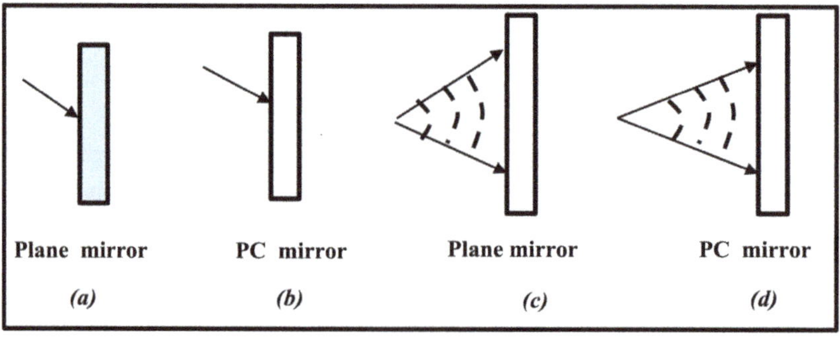


**Figure 19.9.** Diagram for problem 4.

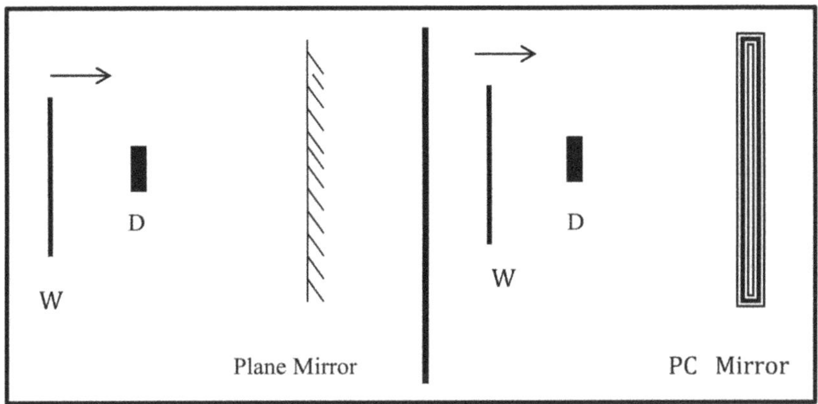



**Figure 19.10.** Diagram for problem 6.

# Bibliography


[1] Eichler H J, Günter P and Pohl D W 1986 *Laser-Induced Dynamic Gratings* (Berlin: Springer)

[2] Rajesh R J and Bisht P B 2002 Theoretical and experimental studies on photo-physical parameters of N,N′-bis(2,5,-di-tert-butylphenyl)-3,4:9-10-perylenebis(dicarboximide) DBPI by using transient grating technique *Chem. Phys. Lett.* **357** 420–5

[3] Rajesh R J 2003 *PhD Thesis* IIT Madras

[4] Kogelnik H 1969 Coupled wave theory for thick hologram gratings *Bell Syst. Tech. J.* **48** 2909–47

[5] Jena K C 2008 *PhD Thesis* IIT Madras

[6] Pepper D M 1986 Applications of optical phase conjugation *Sci. Am.* **254** 74–83

[7] He G S 2002 Optical phase conjugation: principles, techniques, and applications *Prog. Quantum Electron.* **26** 131–91

[8] Giuliano C R 1981 Applications of optical phase conjugation *Phys. Today* **34** 27

[9] Yariv A 1976 Three-dimensional pictorial transmission in optical fibers *Appl. Phys. Lett.* **28** 88–9

[10] Fisher R A, Suydam B R and Yevick D 1983 Optical phase conjugation for time-domain undoing of dispersive self-phase-modulation effects *Opt. Lett.* **8** 611

[11] Duignan M T, Feldman B J and Whitney W T 1987 Stimulated Brillouin scattering and phase conjugation of hydrogen fluoride laser radiation *Opt. Lett.* **12** 111–3

[12] Whitney W T, Duignan M T and Feldman B J 1992 Stimulated Brillouin scattering phase conjugation of an amplified hydrogen fluoride laser beam *Appl. Opt.* **31** 699–702

[13] Uchida N and Niizeki N 1973 Acoustooptic deflection materials and techniques *Proc. IEEE* **61** 1073–92

[14] Chatterjee M R and Chen S-T 1994 Multiple plane-wave scattering analysis of light diffraction by parallel Raman–Nath and Bragg ultrasonic cells with arbitrary frequency ratios *J. Opt. Soc. Am.* A **11** 637–48

**IOP** Publishing

# An Introduction to Photonics and Laser Physics with Applications

**Prem B Bisht**

# Chapter 20

# Multiphoton absorption

Following the absorption of higher-energy photons, molecules at room temperature generally emit photons of lower energies. This occurs because part of the photon energy is lost in nonradiative processes. An excited state can also be achieved by *simultaneous absorption* of two photons of a certain frequency. Under such circumstances, the emitted photon can have a higher energy than the ones absorbed by the system. The photo shows a streak of reddish-yellow emission in the visible region caused by the excitation of a dye by femtosecond pulses at near-IR wavelengths. This technique of two- or multiphoton absorption-induced emission finds application in several areas. For instance, in medical applications, healthy tissues are damaged by the excitation caused by UV–visible lasers. IR lasers are safer to tissues and can therefore be used to invoke the multiphoton absorption processes. This chapter gives details of the multiphoton process as an application of $\chi^{(3)}$. It is interesting to discover that the two processes, viz. (i) second-harmonic generation and (ii) two-photon absorption, have similarities in their energy-level diagrams, albeit these phenomena belong to different nonlinear optical susceptibilities.

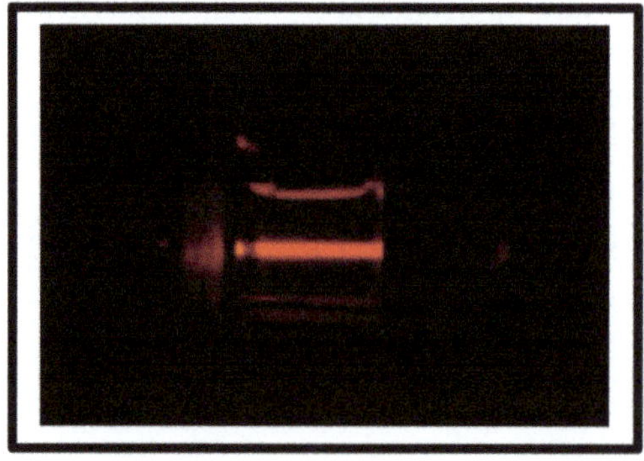

**Learning objectives**
**After reading this chapter, the learner will be able to:**
Outline the higher-order photon events;
Describe the selection rules for multiphoton absorption (MPA) and parity selection;
Understand free-carrier absorption (FCA) and excited-state absorption (ESA);
Explain the phenomenon of reverse saturable absorption (RSA);
Express MPA in terms of the RSA process;
Estimate the number of photons in an MPA process;
Distinguish between the second-harmonic generation (SHG) and 2PA processes.

## 20.1 Higher photon absorption processes

An atomic or molecular system has a short-lived state known as the excited state. An excited state can be achieved by the exposure of the molecular system to natural light followed by the absorption of a photon corresponding to an energy transition under ambient conditions. The excited state can also be achieved by simultaneous absorption of two or more photons under suitable conditions, which will be described in the following sections. Two-photon absorption (2PA) transitions occur between the two states $E_i$ and $E_f$ in the presence of intermediate virtual levels $E_k$ allowed by single-photon optical transitions. The selection rules for single-photon transitions require that the pairs of levels $E_i$ and $E_k$ or $E_k$ and $E_f$ must have opposite parity (figure 20.1). This means that the levels connected by a two-photon transition must have identical parity.

Emission induced by the absorption of two or more photons (i.e. multiphoton absorption) is a $\chi^{(3)}$ process, as such absorption takes place between the real energy levels of a molecular system and the emission is followed by relaxation in the excited state (♠ see figure 2.12).

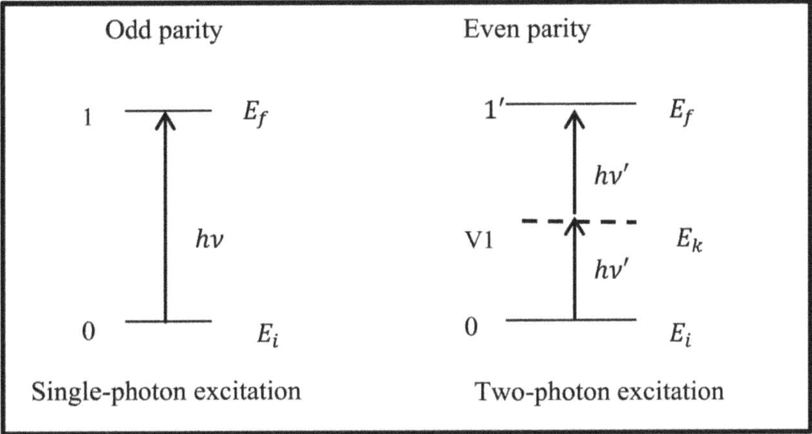


**Figure 20.1.** Different parity states (1 and 1′) can be reached with single- (left) and two-photon (right) excitations, respectively. V1 is the virtual state.

## 20.2 Units of absorption cross-sections

The Beer–Lambert law (equation (5.2)) for single-photon absorption is $\frac{\partial I}{\partial x} = -\alpha I(x)$. Here, $I$ is the light intensity falling on the sample of thickness $x$. When a laser beam passes through an optical medium and invokes multiphoton events, its attenuation is given by

$$\frac{\partial I}{\partial x} = -\alpha I - \beta I^2 - \gamma I^3 - \dots . \tag{20.1}$$

where $\alpha$, $\beta$, and $\gamma$, respectively, are the absorption coefficients for one-, two-, and three-photon absorption and so on.

By expressing the absorption coefficients in term of the corresponding cross-sections ($\sigma_a$, $\sigma_{2a}$ and $\sigma_{3a}$), we can write equation (20.1) as follows:

$$\frac{\partial I}{\partial x} = \Delta N(-\sigma_a I - \sigma_{2a} I^2 - \sigma_{3a} I^3 - \dots). \tag{20.2}$$

Here, $\Delta N$ is the difference in population between the ground state and the excited state.

Comparing this with equation (20.1), the dimensions of $\beta I$ will have the dimensions of $\sigma_{2a}$. The unit of $\sigma_a$ is $\frac{\text{cm}^2}{\text{photon}}$. Therefore, the unit of $\sigma_{2a}$ equals that of $\sigma_a/I$, i.e.:

$$\left( \left( \frac{\text{cm}^2}{\text{photon}} \middle/ \left( \frac{\text{number of photons}}{\text{area}} \right) \right) \middle/ \text{time} \right) = \text{cm}^4\,\text{s}$$

♣ The Göppert-Mayer (1 GM $= 10^{-50}$ cm$^4$ s) is used as the unit of 2PA in honor of Maria Göppert-Mayer, who pioneered the theoretical treatment of the 2PA phenomenon [1].

A rough estimate of the cross-sections of a molecule ($10^{-16}$ cm$^2$) indicates that the multiple-photon cross-sections are weaker by a factor of $10^{-32}$. This is the reason that we require higher concentrations of samples as well as large fluences to observe multiphoton events.


**Exercise 20.1.** At higher incident EM fields, the field intensity ($I$) as a function of sample thickness ($x$) is given by equation (20.1): $\frac{\partial I}{\partial x} = -\alpha I - \beta I^2 - \gamma I^3$. From this equation, obtain the coefficients for one- and two-photon absorptions.

**Solution:** The expression for the intensity decay as a function of distance for single-photon absorption is written as

$$\frac{\partial I}{\partial x} = -\alpha I.$$

Here, $\alpha$ is the single-photon absorption coefficient. To obtain the Beer–Lamberts law (equation (5.2)),

$$I(x) = I_0 e^{-\alpha x}.$$

Similarly, for two-photon absorption, the intensity variation can be written as

$$\frac{\partial I}{\partial x} = -\beta c I^2.$$

Here, $\beta$ is the 2PA coefficient and $c$ is the concentration of the sample, which includes information about the number of available molecules. We integrate the above equation to get the following useful expression:

$$I(x) = \frac{I_0}{1 + \beta c I_0 x}.$$

This expression is used to obtain the value of $\beta$ after fitting the experimental data.

## 20.3 Selection rules

Due to parity-based transitions, the selection rules for one- and two-photon absorptions are not the same. Therefore, the excited states that are not accessible under single-photon absorption can be studied by 2PA spectroscopy. Most of the popular laser dye molecules are model systems for two-photon absorption. Generally, electron donor–acceptor systems consisting of $\pi$–$\pi$ bridge electrons have larger 2PA cross-sections. Recent advancements in nanomaterial research have found carbon nanodots and semiconductor quantum dots to be stable materials.

Let us take $\epsilon(\omega)$ as the relative permittivity of a material at a frequency of $\omega$, and let $x_{mn}$ be the distance separating the charges ($q$). The transition probability ($P_{nm}$) between the initial ($m$) and the final ($n$) states is given by $P_{nm} \propto [\epsilon(\omega) \quad | qx_{mn}|^2]$ at $= \omega_{nm}$. The dipole transitions are governed by the selection rules described in chapter 2. The transitions are possible only if $qx_{mn} \neq 0$. This implies that there should be a change in the angular momentum for the transition to take place.

The change in the angular momentum quantum number ($l$) in one-photon transitions is $\Delta l = +1, -1$. Thus, an electron from the ground state ($s$ -state) can only occupy a higher $p$-state by the absorption of one photon. In contrast, when an atom simultaneously absorbs two photons, the angular momentum of the electron changes by $\Delta l = +2, 0,$ or $-2$. This is the case because each photon has an angular momentum of $+1$ or $-1$. This allows an $s$-state electron to make a transition to another $s$-state or to a $d$-state (which are the states that are unreachable by single-photon absorption). An $s$-state electron may not be able to make a transition to a $p$-state by 2PA. An important point to remember is that when two photons are simultaneously absorbed, the molecule proceeds to an intermediate state known as the virtual state. All other transitions are forbidden under the dipole approximation.

### 20.3.1 Advantage of multiphoton processes

The multiphoton absorption (MPA) processes result in multiphoton-induced emission by molecules. This aspect is exploited in biological research and medicine in the form of photoemission by dye-sensitized drugs. Tissues and cells are prone to damage by the high-energy UV–visible lasers. IR laser light, on the other hand, is not absorbed by tissues; therefore, the process of multiphoton excitation helps in the required treatment. Similarly, a multiphoton ionization (MPI) process is used to characterize molecules; this works by creating fragments of the molecules by exciting them beyond their ionization potential. The fragments can be detected according to the masses of their ions.

## 20.4 Reverse saturable absorption

Let us consider the effect of a varying photon flux from a laser beam impinging on a molecular system. At a lower incident photon flux, the process of single-photon absorption takes place and corresponding coefficient ($\alpha$) is constant, according to the Beer-Lamberts law. The variation of $\alpha$ as a function of incident intensity was shown in figure 16.10, where we saw that the effect of saturable absorption (SA) turns into reverse saturable absorption (RSA) at high intensities. This occurs because for higher values of the incident photon flux, the nonlinear absorption coefficient increases, further decreasing of the transmission of the sample. Occasionally, the process of RSA is also known as induced absorption.

To obtain an idea of the contributions of the various absorption and emission processes, see figure 20.2. Here, we consider the single-photon absorption ($\sigma_a$), spontaneous emission ($\sigma_e$), excited-state absorption ($\sigma_{es}$), and two-photon absorption (2PA) cross-sections ($\sigma_{2a}$). If we consider $\tau_2$ to be the fast relaxation lifetime and $\tau_f$ to

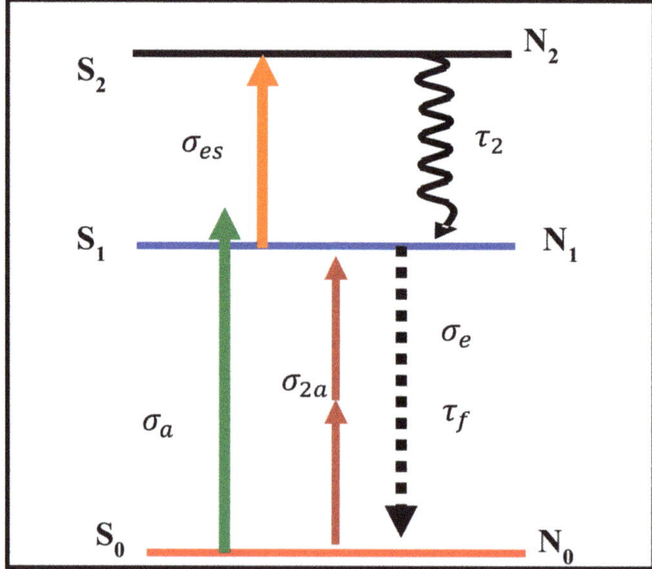



**Figure 20.2.** Energy-level diagram indicating the cross-sections of: single-photon absorption ($\sigma_a$), two-photon absorption ($\sigma_{2a}$), emission ($\sigma_e$), and excited-state absorption ($\sigma_{es}$) as indicated between the states $S_0$, $S_1$, and $S_2$. $\tau_f$ is the luminescence lifetime and $\tau_2$ is the fast relaxation from $S_2$ to $S_1$.

be the spontaneous emission lifetime, the decay rate of the first excited state ($S_1$) in terms of the indicated populations ($N_0$, $N_1$ and $N_2$) can be written as

$$\frac{dN_1}{dt} = \frac{\sigma_a N_0}{h\upsilon}I(z,\,t) + \frac{\sigma_{2a}N_0}{h\upsilon}I(z,\,t) - \frac{\sigma_{es}}{h\upsilon}N_1 I(z,\,t) - \frac{N_1}{\tau_f} + \frac{N_2}{\tau_2}. \qquad (20.3)$$

Here, the second term on the right-hand side corresponding to $\sigma_{2a}$ vanishes for small incident photon densities.

As mentioned above, one of the essential requirements for multiphoton emission is that a large number of photons must be available in the excitation (pump) beam. This must also be accompanied by a large number density of absorbing molecules in the sample. Since the MPA process takes place at high photon fluxes, several other processes can compete with it. Some of these processes are excited-state absorption in molecular systems and free-carrier absorption in semiconductors. Such competing processes give rise to an effect opposite to the SA effect and fall under a common term, namely RSA.

### 20.4.1 Excited-state absorption

Excited-state absorption (ESA) occurs when the incident photon frequencies fall within the excited-state absorption spectrum. Therefore, the presence of ESA decreases the output transmittance because the incident light is further absorbed by the molecules occupying the excited state. ESA is an irradiance-dependent nonlinear absorption process. In ESA, the transmittance change at a fixed input energy is independent of the duration of the incident laser pulse. In other words, in the case of ESA, the same fluence results in the same transmission for two different pulse durations, provided that the transient species are formed after the pulse excitation in each case.

In figure 20.2, the cross-section of ESA is represented by $\sigma_{es}$. The 'pump-probe' technique of transient absorption is used to record the ESA spectra. In this technique, the spectra of the broad wavelength probe beam are monitored in the presence of the pump pulse, which is delayed in time. Such spectra reveal the nature of the species formed as a function of time, i.e. immediately after excitation or at later times.

### 20.4.2 Free-carrier absorption

As described in section 2.9.8, in semiconductors the valence and conduction bands are separated by a bandgap of a certain energy. Here, the electrons and holes are known as charge carriers. Depending upon the type of semiconductor (n-type or p-type), the majority charge carriers can be either electrons or holes. Initially, the electrons are located in the valence band. By absorbing the required amount of energy, these are excited to the conduction band.

In the conduction band, at higher incident intensities, when an electron further absorbs a photon and is excited to a higher energy level within the same set of bands, this occurrence is known as free-carrier absorption (FCA). Such contributions to nonlinear absorption take place within a band and hence correspond to lower-energy (i.e. longer-wavelength) regions.

### 20.4.3 2PA and MPA processes

The 2PA cross-section in figure 20.2 is represented by $\sigma_{2a}$. If the laser intensity is sufficient, molecules from the ground state ($S_0$) can simultaneously absorb two photons and become excited to a suitable energy level. Similarly to ESA, the 2PA process decreases the output transmittance. Here, the transmittance change at a fixed input pulse energy is dependent on the pulse duration ($\tau_p$), hence 2PA is a fluence-dependent process. Generally, the transition strengths for multiple-photon absorption events are weak. For example, in 2PA or three-photon excitation processes, the fluorescence intensities are proportional to the second and third powers of the incident optical power, respectively. Experiments that examined the femtosecond laser pulsed-energy dependence of several organic dyes have shown the occurrence of events involving three or more photons, as described in the next section.

♣ Irradiance, $I = \frac{2W_p}{\pi\omega_0^2}$ W cm$^{-2}$. Fluence, $F = \frac{2W_p\tau_p}{\pi\omega_0^2}$ J cm$^{-2}$. Here, $W_p$ is the peak power, $\tau_p$ is the pulse duration, and $\omega_0$ is the beam waist.

## 20.5 Estimating the number of photons

The number of absorbed photons involved in an emission process can be estimated by measuring the fluorescence yield as a function of incident flux density. The relation between the intensity of fluorescence and the incident flux density is given by

$$I_{fl}^i = \eta \frac{\phi_{fl}}{i} N x \sigma_{ia} \rho_{exc}^i, \tag{20.4}$$

where $i$ is the number of photons involved in the process, $\eta$ is the detection efficiency of the experimental setup, $\phi_{fl}$ is the fluorescence quantum yield, $N$ is the molecular number density, $x$ is the thickness of the sample, $\sigma_{ia}$ is the multiphoton absorption cross-section with $i$ photons and $\rho_{exc}^i$ is the incident photon flux density.

**Exercise 20.2.** The popular laser dye rhodamine 6G can be used to observe multiphoton events. The absorption spectrum of this dye falls in the visible region. For multiphoton absorption, the excitation source selected can be the fundamental wavelength of either a pulsed Nd:yttrium aluminum garnet (YAG) laser or a Ti: sapphire laser delivering picosecond- or femtosecond-duration pulses, respectively. In figure 20.3, the observed emission intensity is plotted as a function of the fluence of the incident laser pulse. How many photons are involved in the process?

**Solution:** The dye rhodamine 6G is a model system for observing single- or multiphoton events for demonstration purposes. Due to its high quantum yield, on excitation, the lower-concentration solutions give rise to excellent fluorescence within the rhodamine 6G absorption band corresponding to single-photon absorption.

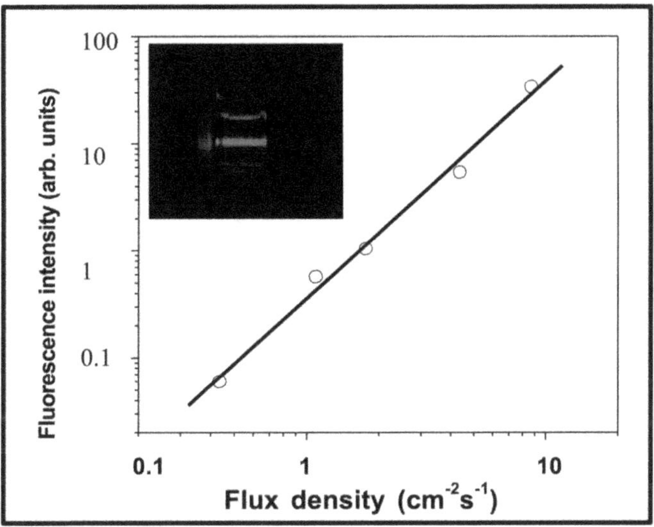

**Figure 20.3.** Log–log plot of fluorescence intensity versus incident flux density for a rhodamine 6G solution (○) in methanol at a pump wavelength of 1064 nm. The photo in the inset shows the red streak of emission from a sample in a cuvette (after [5]).

On the other hand, on excitation with ultrafast pulses in the near-IR region (a femtosecond Ti:sapphire laser at a wavelength of ~800 nm or a picosecond Nd:YAG laser at a wavelength of 1064 nm) with energies in the range of nanojoules to millijoules, respectively, excellent emission yields (inset figure 20.3) are observed for highly concentrated (>1 mM) solutions. Such emissions correspond to multiphoton events. According to equation (20.4), the number of photons can be obtained from the slope of the log–log plot of fluorescence intensity versus incident excitation power. The straight line indicates the linear fit to the data points. In figure 20.3, the slope turns out to have a value of two, indicating that it represents a two-photon event.

## 20.6 Second-harmonic or multiphoton emission?

Generally, any newly grown crystal in nonlinear optics is checked for its second-harmonic generating capability. However, if the crystal also exhibits multiphoton excited emission, the correct identification of the SHG process is not possible in a straightforward manner. This section gives a good example of such a case.

The figure below shows a new dye, *trans*-MDMASB[1], which exhibits multi-photon-induced emission under excitation by a femtosecond laser at 760 nm in solution. In its crystalline state, the dye also shows multiphoton-induced emission when pumped with the fundamental of a picosecond Nd:YAG laser, as shown in figure 20.4. The greenish color of the emission creates confusion regarding whether the radiation is due to second-harmonic generation.

---

[1] Methyl 2-(4-(dimethylamino)styryl)benzoate.

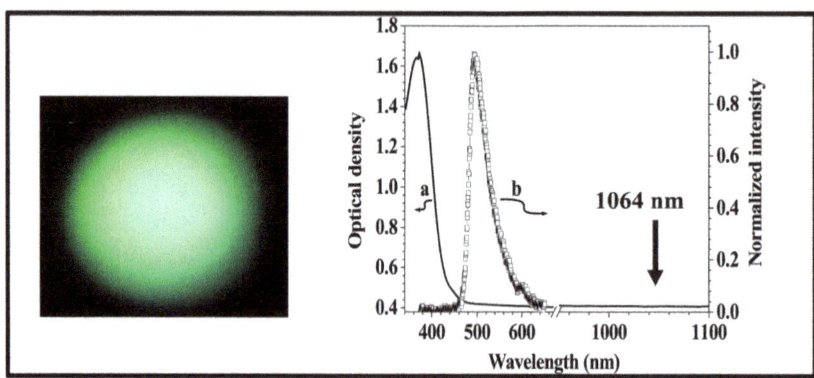

**Figure 20.4.** The left-hand image shows the emission from crystalline trans-MDMASB pumped by a picosecond Nd:YAG laser (1064 nm). The right-hand panel gives the absorption (curve a) and multiphoton-induced fluorescence (curve b) of the sample in solution on pumping at the same wavelength (after [6]).

Figure 20.4 shows that the crystal has no absorption at 1064 nm. However, a multiphoton-induced emission peak is observed at ~495 nm. This radiation is distinctly different from that corresponding to the SHG (532 nm) of the Nd:YAG laser. It is also a good match for the emission obtained under single-photon excitation.

**Exercise 20.3.** The left-hand panel of figure 20.4 shows an image of the emission observed when the sample is irradiated with the fundamental of a picosecond Nd: YAG laser. How can you identify the emission—in other words, whether it is due to SHG or a two-photon emission event?

**Solution:** The strong greenish color of the image can give the impression that the emission is due to the SHG phenomenon. However, there are two things to note: (i) the spectrum has a bandwidth broader than that generally observed for SHG and (ii) as described above, the spectrum's peak is at a different location (495 nm) from 532 nm. These points confirm that the emission is due to the MPA process. This shows the importance of measuring the spectra instead of relying on visual inspection when testing nonlinear optical crystals for their SHG efficiency. A further investigation of the slope (~2) also confirms this result (see figure 20.5).

In the above example, we see the possibility of observing the same color (with nearby emission peaks) in an image consisting of different species in different phases (say, one in the crystalline state and the other in the liquid phase). These two emissions could be due to two different mechanisms, i.e. SHG and MPA. There are cases in which these two types of radiation could peak at the same wavelength or may even fall near the SHG wavelength of the fundamental laser. These two types of radiation can be distinguished with the help of equation (20.4). An experiment carried out to estimate the number of photons involved in the emission process of the crystalline film gives a result of two for the number of photons (figure 20.5).

This observation introduces the need to distinguish between parametric and nonparametric processes in nonlinear optics, as shown in figure 20.6. The left-hand

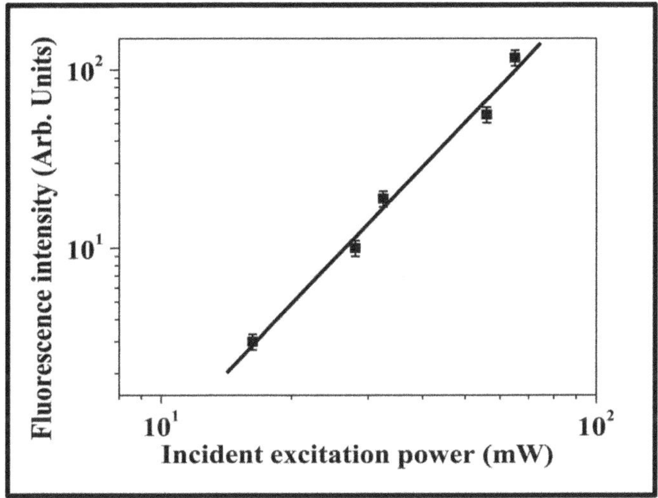



**Figure 20.5.** Observed fluorescence intensity as a function of the incident power for a crystalline pellet sample pumped by a picosecond Nd:YAG laser (1064 nm). The solid line shows the linear fit with equation (20.4), confirming 2PA events.

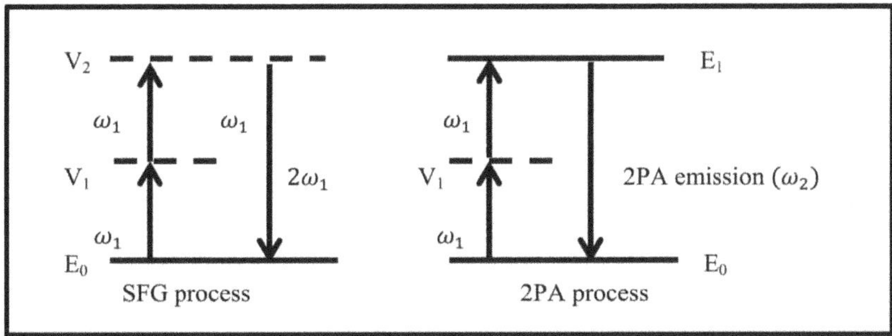



**Figure 20.6.** Energy-level diagrams for SFG (left) and 2PA (right). SFG is a $\chi^{(2)}$ process involving an initial state (E$_0$) and two virtual (V$_1$ and V$_2$) states (left-hand side). 2PA is a $\chi^{(3)}$ process that involves the real state of a molecular system (right-hand side) and an intermediate virtual state V$_1$.

side shows the SFG process, in which two photons of the same frequency, $\omega_1$, are destroyed to create the sum frequency, resulting in $2\omega_1$ through the virtual levels. This is the second-order parametric process ($\chi^{(2)}$) and is useful in SHG, SFG, and other applications as described in chapter 15.

In contrast, an excited real level is achieved by 2PA via a virtual state by simultaneous absorption of two photons of the same frequency (right panel). It is a $\chi^{(3)}$ process, i.e. a nonparametric process. The emitted photon ($\omega_2$) in this case follows the Franck–Condon principle (see chapter 2), including the excited-state relaxation processes. Such processes are important in biological systems, in which excitation by the same ultrashort pulse can allow one to differentiate between SHG- and 2PA/MPA-induced fluorescence in order to distinguish between constituents such as collagen or tissues/cells or starch in aged food ingredients. It should also be understood that 2PA emission possesses the bandwidth of the emission spectrum, while the spectral width of SFG is limited by the pump photons.

## Questions and problems

1. What are the units of 1-, 2- and 3-photon absorption cross-sections in the meter, kilogram, and second (MKS) system?
2. Differentiate between SA and RSA phenomena. How does the process of FCA in semiconductors contribute to RSA?
3. The number of absorbed photons involved in a multiphoton emission process can be estimated using equation (20.4). If the slope of the plot of emitted intensity versus incident photon flux density turns out to have a value of 3.5, how do you explain the observed process?
4. Describe how 2PA is fluence dependent, while excited-state absorption (ESA) is an irradiance-dependent process. Which one is dependent on the duration of the pump pulse?
5. In an experiment to identify the order of nonlinear optical susceptibility in a substance, picosecond pulses from the fundamental of a Nd:YAG laser are used. A beam focused onto microcrystals of the substance causes greenish light to be emitted by the microcrystals. With or without using a spectrometer, (i) identify the parameters of the experiment that can be used to distinguish the order of the optical nonlinearity ($\chi^{(2)}$ or $\chi^{(3)}$) in the obtained radiation. (ii) If the microcrystals are now dissolved in a solvent to give the same emission wavelength as in the microcrystalline form, comment on the order of the observed optical nonlinearity.
6. You know that angle-tuned phase matching for SHG is based on the selecting the angle of the $\chi^{(2)}$ material. In terms of the phase-matching condition, how do you differentiate between 2PA and SHG occurring in a crystal?
7. For the imaging of biological samples, two-photon spectroscopy (using IR lasers) is preferred as compared to single-photon spectroscopy using UV–visible lasers. Apart from the effects of scattering caused by the sample, what are the advantages of two-photon spectroscopy?
8. With reference to the previous question, the biological crystalline or amorphous components of tissues can be identified by multiphoton absorption processes. What image could be expected if a tissue (which consists of both crystalline material as well as amorphous material) were to be imaged with the help of the fundamental of a near-IR pulsed laser?

## Bibliography

[1] Göppert M 1929 Über die Wahrscheinlichkeit des Zusammenwirkens zweier Lichtquanten in einem Elementarakt *Naturwissenschaften* **17** 932
[2] Kim D Y *et al* 2005 Large two-photon absorption (2PA) cross-section of directly linked fused diporphyrins *J. Phys. Chem.* A **109** 2996–9

[3] Volkmer A, Hatrick D A and Birch D J S 1997 Time-resolved nonlinear fluorescence spectroscopy using femtosecond multiphoton excitation and single-photon timing detection *Meas. Sci. Technol.* **8** 1339

[4] Parker C A 1968 *Photoluminescence of Solutions* (New York: Elsevier Publishing Co)

[5] Sailaja R, Bisht P B, Singh C P, Bindra K S and Oak S M 2007 Influence of multiphoton events in measurement of two-photon absorption cross-sections and optical nonlinear parameters under femtosecond pumping *Opt. Commun.* **277** 433–9

[6] Ali A, Bisht P B, Senthilmurugan A and Aidhen I S 2011 A new fluorescent probe with stimulated emission and multiphoton absorption properties *Chem. Phys.* **382** 68–73

[7] Perry S W, Burke R M and Brown E B 2012 Two-photon and second harmonic microscopy in clinical and translational cancer research *Ann. Biomed. Eng.* **40** 277–91

**IOP** Publishing

# An Introduction to Photonics and Laser Physics with Applications

**Prem B Bisht**

# Chapter 21

# White-light continuum generation

Water is transparent; it neither absorbs nor emits strongly in the UV–visible region. A simple demonstration experiment can be carried out in an ultrafast laser laboratory, as follows. A near-IR laser pulse about 100 fs in duration that has energy on the microjoule scale is directed through water in a beaker. A bright white spot is observed on the screen, as shown in the figure. The spectrum of this spot is broad, ranging from the visible region to the IR region. The spot appears because of the *white-light continuum (WLC)* generation process. It is also known as the *supercontinuum* (SC), as it spans several octaves of the frequency range. A SC can be generated in several other materials, including specially engineered optical fibers known as photonic crystal fibers (PCFs). The chapter reveals the mechanism responsible for this exotic phenomenon, which finds applications in spectroscopy and laser technology, such as in the the seeding of optical parametric amplifiers, as well as in solar-cell research.

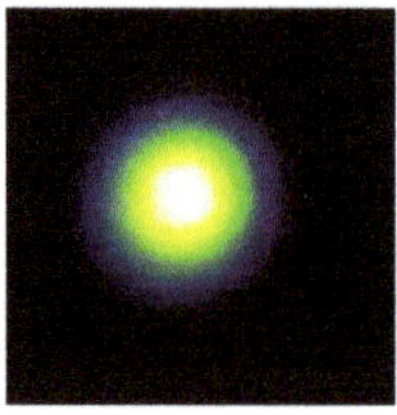

**Learning objectives**
**After reading this chapter, the learner will be able to:**
Explain the phenomenon of spatial self-phase modulation (SSPM);
Understand the WLC or SC;
Describe the mechanism of WLC generation in a water-$D_2O$ mixture;

doi:10.1088/978-0-7503-5226-0ch21



© IOP Publishing Ltd 2022

Explain the structure of PCFs;
Recognize the mechanism of WLC generation in PCFs;
Interpret conical emission.

## 21.1 Spatial self-phase modulation

In chapter 16, we introduced two aspects of self-phase modulation (SPM), the *longitudinal* and *transverse* optical Kerr effects. The longitudinal Kerr effect is associated with the broadening of ultrafast laser pulses. The transverse Kerr effect, i.e. the intensity variation of a laser-beam profile, can introduce a spatial variation of the refractive index ($\Delta n$) in the transverse direction. This effect is also known as spatial self-phase modulation (SSPM).

To get a simple picture of SSPM, let us consider a focused continuous-wave laser beam produced by an argon-ion laser. The beam has a Gaussian transverse profile with an angular frequency of $\omega$ and is propagating in the $z$-direction in a partially absorbing medium, as shown in figure 21.1. The absorbing medium is heated by the incident beam, causing changes in the density of the medium. The phase increment ($\Delta\phi\,(r, z)$) of the beam with the variation of transverse coordinate $r$ is given by $\Delta\phi(r, z) = (\omega/c)\Delta n(r)z$. For a given change in the refractive index $\Delta n(r)$, this leads to a modification of the wavefront. It can be seen that for large $z$, a diffraction pattern with multiple bright and dark rings in the far field is observed as a quasi-periodic spectrum in the $k_\perp$ space (figure 21.1). Here, $k_\perp$ is the transverse component of the wave vector of the beam.

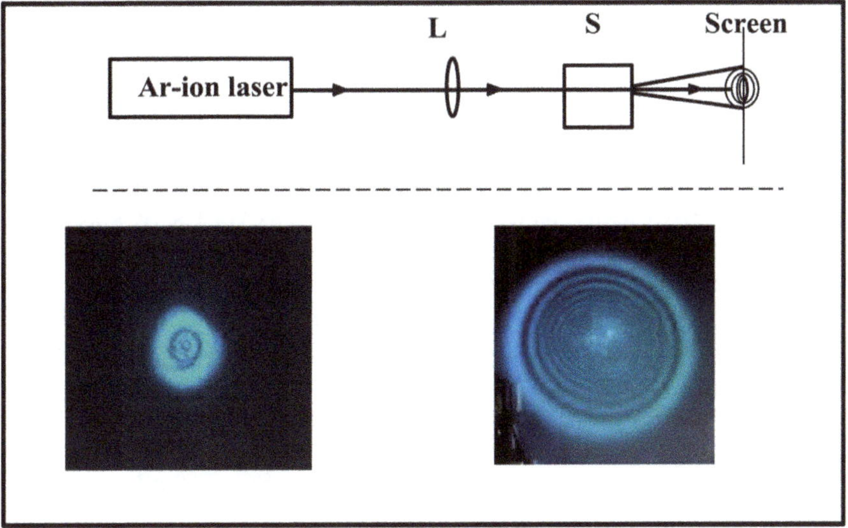


**Figure 21.1.** Top panel: experimental setup used to observe the SSPM effect. L: biconvex lens, S: dye sample. The bottom panel shows photographs of the diffraction rings observed at the screen. An incident beam wavelength of 488 nm was used with intensities of 3.7 kW cm$^{-2}$ (left) and 30.5 kW cm$^{-2}$ (right) [5].

♣ The experiment requires a sample that is partially absorbent (generally a dye) at the given pump wavelength. In figure 21.1, the sample was a dye dissolved in a solvent so that its absorption maximum suited the pump wavelength of the argon-ion laser (488 nm).

♣ For continuous-wave lasers (or even high-repetition-rate ultrashort pulse lasers), the temperature-induced changes in the refractive index ($\Delta n(r)$) dominate the nonlinear optical effects, resulting in SSPM. Understanding SSPM is extremely useful, as it generally takes place due to thermal lensing. In view of the thermal diffusion, it is also known as the Soret effect.

The bottom panel of figure 21.1 shows the typical self-diffraction patterns obtained at two different incident powers at a pump wavelength of 488 nm. With an increase in the irradiance, the number of rings increases, accompanied by a subsequent decrease in the brightness of the central ring. A large number of rings seem to be hidden in the central bright dot of the transmitted beam for larger irradiances of the pump beam. The half-cone angle $\theta_m$ of the outermost ring is given by the relation $\theta_m = \left( \left( \frac{d\Delta\phi}{dr} \right)_{\text{max}} \right) \Big/ \left( \frac{2\pi}{\lambda} \right)$.

## 21.2 White-light continuum generation

A WLC is generated when intense ultrashort laser pulses fall on transparent liquids (viz. water, $D_2O$), solids (viz. glass, sapphire, LiF, $CaF_2$, etc), gases, or specially designed optical fibers. A WLC is a quasi-continuum of light, spanning from the visible to the near-IR regions, or even further. It is also known as a supercontinuum (SC), as the frequency width of the generated continuum spans several octaves from the pump frequency.

A WLC was observed for the first time in 1970 by focusing intense picosecond-duration pulses into glass. Femtosecond SC was generated by focusing 80 fs pulses into an ethylene glycol film in 1983. A stable WLC with a broad spectrum can be obtained by pumping nonlinear optical media with high-peak-power femtosecond pulses. In the picosecond timescale, however, generation of a stable WLC requires special attention. Historically, as well as being known as WLCs, SCs were known by two other names, namely superbroadening and anomalous frequency broadening. In this chapter the acronyms WLC and SC are used interchangeably.

♣ The word *octave* in music defines notes that are multiples of lower frequencies.

♣ WLC generation is triggered by self-focusing and has a strong dependence on the bandgap of the medium used (table 21.1). For example, LiF (with a bandgap of 11.8 eV) has been reported to exhibit a supercontinuum in the region of 380–950 nm. The short-wavelength thresholds for the continuums obtained using $CaF_2$ and sapphire are 400 nm and 450 nm, respectively; their SCs extend into the near-IR region. $CaF_2$ suffers from a low damage threshold and hence its continuum is not sufficiently stable for long experiments. The continuum generated in water is stable and has a reasonably broad spectrum despite its low Kerr nonlinearity. PCFs can generate a WLC from 450 to 1700 nm or even in longer-wavelength regions, depending on the pump wavelengths.

**Table 21.1.** Some of the materials used for WLC generation and the spectral ranges obtained.

| Material | Bandgap (eV) | Wavelength range |
| --- | --- | --- |
| LiF | 11.8 | 380–950 nm |
| $CaF_2$ | 10.2 | Visible–near-IR region |
| $H_2O/D_2O$ (60/40) | 7.5 | Visible–near-IR region |
| Sapphire | 7.5 | Visible–near-IR region |
| PCF (engineered) |  | 400 nm to 4 μm |

**Exercise 21.1.** On pumping with short-duration pulses, the spectrum of an obtained supercontinuum spans from 450 nm to 850 nm. Assuming the phase to be constant over the obtained portion of the spectrum, estimate the pulse duration of the supercontinuum.

**Solution:** As the excitation is pulsed, the obtained supercontinuum is also pulsed. By taking the spectrum peak to be 650 nm, the bandwidth ($\Delta\nu$) of the given spectrum is found to be $3.14 \times 10^{14}$ Hz. Therefore, the pulse duration of the bandwidth-limited pulse can be estimated to be $\tau_p = \frac{1}{\Delta\nu}$, or ~3 fs.

## 21.3 Phenomena responsible for WLC generation

Upon pumping a medium with ultrafast powerful laser pulses, a series of events takes place. Nonlinear optics plays an important role as a result of the self-focusing of the incident beam. The large electric field associated with intense laser pulses is not only able to ionize the medium but also results in transient plasma generation. These events are known as dielectric breakdown of the medium and filamentation, respectively (♠ see section 21.6). Under pumping water or heavy water ($D_2O$) using pulse durations on the picosecond timescale, a structured WLC is obtained that has a major contribution from stimulated Raman scattering (SRS). The phenomenon of SRS also plays important role in WLC generation using photonic crystal fibers pumped by continuous lasers. In general, as described in section 16.6.2, the phenomenon of temporal self-phase modulation (SPM) is one of the key mechanisms.

In the femtosecond timescale, in addition to SRS and SPM, several other nonlinear optical parametric processes, such as cross-phase modulation (XPM, ♠ see section 16.10), four-wave mixing, cascading light up- and downconversion, four-photon parametric generation, and superfluorescence also contribute to WLC generation. The threshold power for continuum generation coincides with the critical power for self-focusing. Above the threshold power, the width of the continuum increases with the bandgap of the medium.

Whether a phenomenon mentioned above can prominently contribute to a WLC depends upon the pulse width of the pump laser and the material. For instance, while SRS contributes significantly to white-light continua on the picosecond timescale, parametric amplification of quantum noise and self-focusing do not explicitly depend on pulse duration. Multiphoton excitation is reported to be an important mechanism of free-electron generation under femtosecond pumping in condensed media. These free electrons induce a negative change in the refractive index and this contributes to anti-Stokes broadening. The phenomenon of WLC generation is still a topic of basic research due to its exotic nature and different applications. In the following sections, two examples of WLC generation, viz. (i) in water/and or D2O and (ii) in PCFs are described in detail.

## 21.4 Spectrum of the WLC in a water–$D_2O$ mixture

As mentioned in the box at the beginning of this chapter, the simplest way to obtain a WLC in the laboratory is to focus a near-IR ultrafast laser pulse of microjoule-scale energy onto a cuvette filled with water. In a picosecond timescale, the obtained WLC is structured due to contributions from stimulated Raman scattering (SRS). As outlined in chapter 16 (♠ see section 16.8), the mechanism of SRS proceeds via the vibrational excited state attained by absorption of incident photons by the medium. Let us look at the effect of incident ultrafast laser pulses in an organic solvent, such as acetone. Figure 21.2 shows the SRS spectrum of acetone pumped with the fundamental wavelength of a picosecond Nd:yttrium aluminum garnet (YAG) laser.

The observed peak positions, their full widths at half maximum (FWHMs), and their shifts from the pump laser line are given in table 21.2. The shift of the first peak (2920 cm$^{-1}$) corresponds to the CH stretch of acetone. The second and third peaks correspond to the overtones of this frequency. The peaks are sharp but their FWHM values on the order of the pump frequency ($\sim$50 $\pm$2 cm$^{-1}$) are limited by the resolution of the spectrometer.

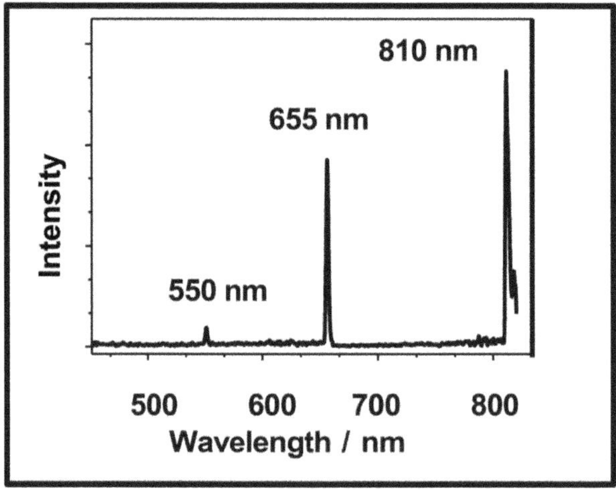


**Figure 21.2.** SRS spectrum of acetone on pumped by the fundamental wavelength of Nd:YAG laser pulses (35 ps, 20 mJ). The spectra have been corrected for the detector and grating response of the spectrometer [5].

Similar observations can be observed, for example, in water. The bandgaps of $H_2O$ and $D_2O$ are approximately the same. As $D_2O$ is heavier, its Stokes and anti-Stokes Raman frequencies are smaller than those of $H_2O$. Therefore, a mixture of these two solvents has several close-lying frequencies that make it a good choice for WLC generation under picosecond pumping. Figure 21.3 shows the experimental setup used for the generation and detection of the supercontinuum. In the image of the WLC, diffuse colored rings surround the central bright spot.

The threshold intensity for WLC generation in water can be of the order of $\sim 100$ GW cm$^{-2}$ and varies with the experimental conditions. The spectra of WLC obtained from a mixture of $D_2O$ and $H_2O$ (70:30) at various pump energies of picosecond laser pulses (1064 nm) are given in figure 21.4. As indicated above, the

**Table 21.2.** Assignments of the observed frequencies of solvents on pumping with picosecond pulses at 1064 nm[*].

| Liquid | Observed peaks (nm) | FWHM (cm$^{-1}$) | Shift from the pump (cm$^{-1}$) | Assignment |
|---|---|---|---|---|
| Acetone[#] | 810 | $70 \pm 5$ | $2920 \pm 5$ | CH str. ($\nu_1$) |
| | 655 | $50 \pm 5$ | $5840 \pm 5$ | $2\nu_1$ |
| | 550 | $50 \pm 5$ | $8760 \pm 5$ | $3\nu_1$ |
| Water | 810 | $395 \pm 30$ | $3028 \pm 5$ | OH sym. str. ($\nu_2$) |
| | 645 | $910 \pm 10$ | $6100 \pm 50$ | $2\nu_2$ |
| | 550 | $475 \pm 10$ | $8785 \pm 300$ | $3\nu_2$ |
| | 480 | $1300 \pm 50$ | $11655 \pm 400$ | $4\nu_2$ |
| $D_2O$ | 865 | $110 \pm 10$ | $2175 \pm 5$ | OD sym. str. ($\nu_3$) |
| | 725 | $300 \pm 10$ | $4365 \pm 20$ | $2\nu_3$ |
| | 625 | $855 \pm 10$ | $6580 \pm 60$ | $3\nu_3$ |
| | 555 | $780 \pm 50$ | $8670 \pm 30$ | $4\nu_3$ |
| | 500 | $325 \pm 70$ | $10600 \pm 300$ | $5\nu_3$ |

[*] The spectral broadening of the pump line in the WLC spectrum at $\sim 50$ cm$^{-1}$ is limited by spectrometer resolution. Str.: stretch; Sym. str.: symmetric stretch; like OH for water, OD is written for deuteriated water ($D_2O$).
[#] No WLC is observed in acetone.

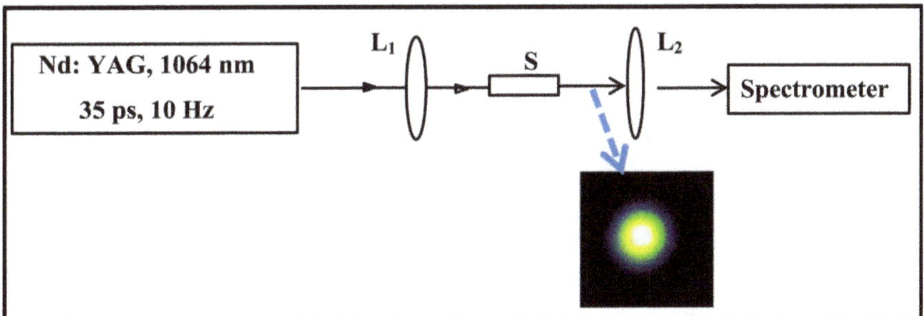


**Figure 21.3.** Experimental setup used for the generation and detection of WLC under picosecond pumping. $L_1$ and $L_2$: biconvex lens, S: glass cell containing water. The dashed arrow indicates the position near which the WLC was imaged (bottom right) [5].

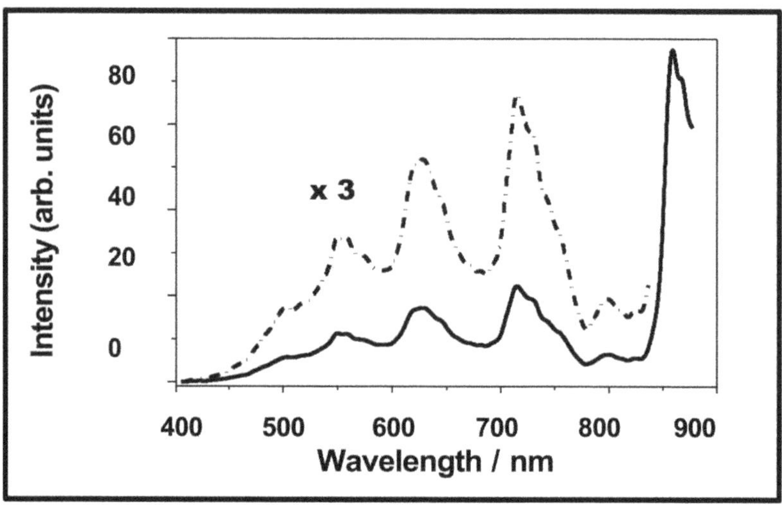


**Figure 21.4.** WLC spectrum obtained from a mixture of $D_2O$ and $H_2O$ (70:30) on pumping with 35 ps pulses produced by a Nd:YAG laser (1064 nm, 15 mJ). The spectra have been corrected for the grating efficiency and the detector response.

obtained continuum is structured due to the contributions from SRS. The spectrum of WLC in water spans the range from 450 to 970 nm.

Table 21.2 gives the assignment of the observed frequencies of a number of solvents. The fundamental wavelength of the picosecond Nd:YAG laser is 1064 nm, which corresponds to a frequency of 9398 $cm^{-1}$. As in the case of acetone, the observed SRS frequencies have been assigned for OH and OD. For instance, the wavelengths of the third overtone of the OH frequency in the spectrum correspond to 9398 $cm^{-1}$ + 4 × 2920 $cm^{-1}$ = 11680 $cm^{-1}$. This frequency falls at a wavelength of 474 nm (blue). Similarly, the fourth overtone falls in the UV region (371 nm).

♣ Generally, photodetectors used in the visible region have lower detection efficiency in the near-IR wavelengths. Similarly, gratings blazed at visible wavelengths have lower diffraction efficiency in other regions. These two effects can severely affect the observed intensities of the spectral lines scanned over a large spectral range. To obtain the actual intensities of the observed lines, the experimentally observed spectra are corrected using suitable multiplication factors.

**Exercise 21.2.** The observed peak positions of the three transitions observed in the stimulated Raman spectrum of pure acetone pumped by picosecond laser pulses at 1064 nm are shown in figure 21.2. In practice, due to the phase matching of SRS, a colored conical image is observed at low incident powers. Assuming that the higher overtones are present in the spectrum, what are the positions of the fourth and fifth transitions? How does the answer change for $D_2O$?

**Solution:** The wavelength of a Nd:YAG laser (1064 nm) corresponds to 9398 $cm^{-1}$. The colors of the first three transitions in acetone fall in the IR (810 nm), red

(655 nm), and green (550 nm) regions of the EM spectrum. Taking the CH stretch to be 2920 cm$^{-1}$, the fourth and fifth lines appear in the blue (474 nm) and near-UV (371 nm) regions, respectively.

Similarly, for $D_2O$, by taking a constant value of 2175 cm$^{-1}$ for the OD frequency, one can obtain the wavelengths of the fourth and fifth overtones, which are 493 and 445 nm, respectively.

## 21.5 Supercontinuum with photonic crystal fiber

### 21.5.1 Structure of photonic crystal fiber

The generation of SC using microstructured PCF started at the beginning of 21st century as a new area of research. PCF is an engineered glass material. We can describe its structure, which is similar to those of crystalline solids, as follows. Just as in solid-state physics, in which the electron experiences a periodic potential while traveling through crystal, in photonic crystals, the photon experiences a periodic refractive index. Figure 21.5 gives a schematic of materials that have photonic bandgaps in one dimension (1D), two dimensions (2D) and three dimensions (3D). We know that in solid-state physics, a lattice consists of an electronic bandgap with a

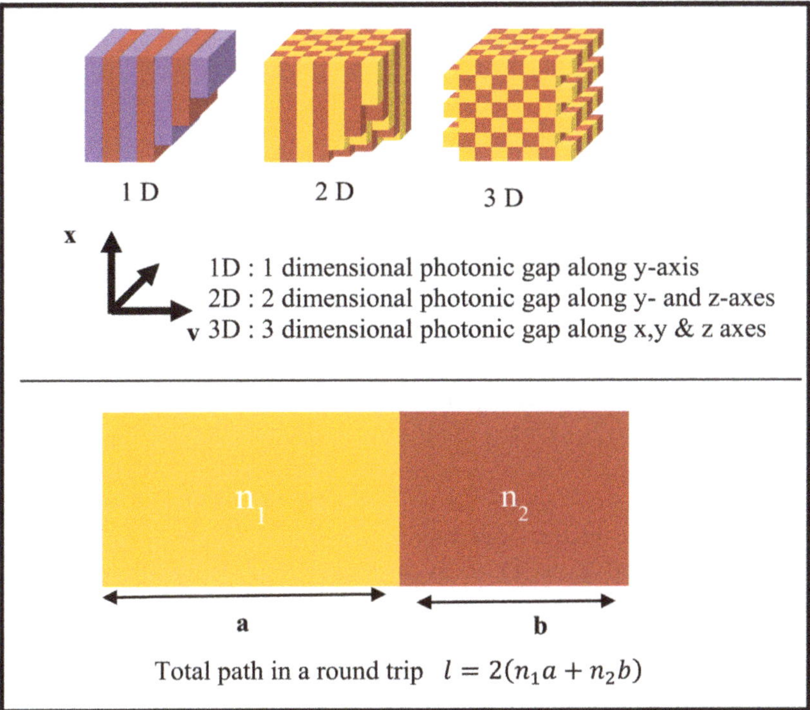


**Figure 21.5.** Structure of 1D, 2D, and 3D photonic-gap materials (top panel). The bottom panel illustrates the total path traveled by a photon in a round trip.

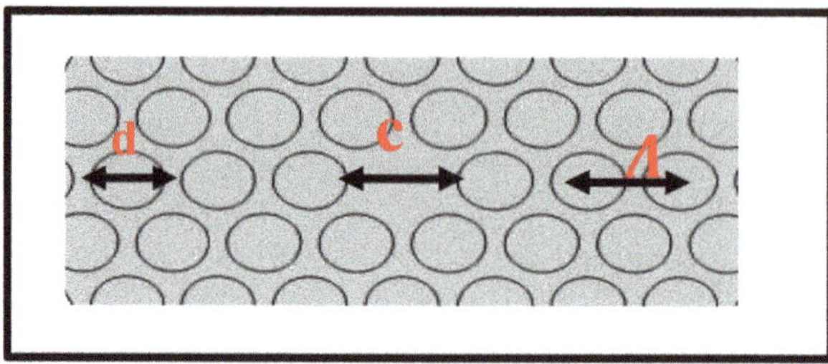


**Figure 21.6.** Schematic of the structure of a PCF showing the diameter of the air hole (d), the core diameter (c), and the pitch ($\Lambda$).

few forbidden electronic transitions. Similarly, photonic crystal has a photonic bandgap, i.e. light of a particular wavelengths does not pass through it.

Using $a$ to denote the lattice constant, the periodic potential of a lattice in the $x$ direction can be written as $V(x + a) = V(x)$. Similarly, for photonic crystal, we can write the dielectric constant ($\epsilon(x)$) as a periodic function, namely $\epsilon(x + b) = \epsilon(x)$ where $b$ is the periodicity of the dielectric constant. The methods used to obtain photonic-gap materials include the self-assembly of microspheres and wood-piled microstructures. Such methods have been shown to result in a bandgap known as the 'stop band'.

As shown in figure 21.6, photonic crystals are characterized by their core diameter ($c$, not to be confused with the speed of light), the diameter of the air hole ($d$), and the distance between the centers of two air holes, also known as the pitch ($\Lambda$). The air fraction is defined as $= d/\Lambda$.

The photonic bandgap appears as the result of multiple reflections from several dielectric surfaces. When the path difference $l$ is equal to $m\lambda$, with $m = 1, 2, 3....,$ the reflected light is superimposed on the incident light. Thus, a particular wavelength is completely reflected into the core, $c$. For complete reflection, the central wavelength of the bandgap ($\lambda_c$) is given by Bragg's law: $m\lambda_c = 2(n_1a + n_2b)$.

### 21.5.2 The mechanism responsible for the supercontinuum in PCF

Various mechanisms responsible for the broadening of the spectrum of an ultrafast pulse were outlined in section 21.3. The dominant mechanism in PCF is temporal SPM due to the periodicity of the refractive index and the concomitant optical nonlinearity along its cross-section ($\spadesuit$ see section 16.6.2). The efficiency of the SC mainly varies with the pulse duration of the pump laser. Typical SC spectra experimentally observed in a PCF at various input powers of a femtosecond oscillator are shown in figure 21.7.

From the plots of figure 21.7, it can be confirmed that the output of the PCF does not include a contribution from any substantial amount of the fundamental wavelength (760 nm). However, the spectra display a nonuniform intensity distribution throughout the observed spectral range. Table 21.3 gives a summary of the output power and the observed spectral ranges of the SC as a function of the coupled

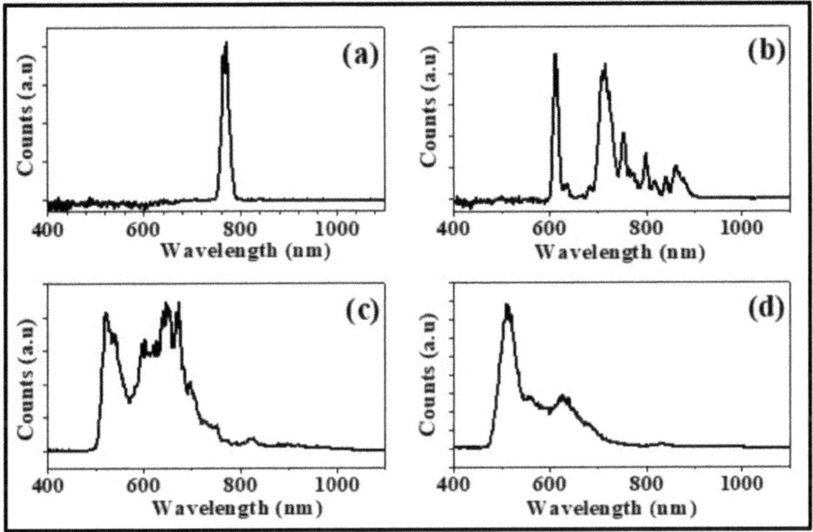

**Figure 21.7.** Spectra of the supercontinuum in PCF as a function of the incident power (a) <10 mW, (b) 30 mW (c) 170 mW, and (d) 266 mW. The input pulse duration was ~100 fs [15].

**Table 21.3.** Generated SC output power and its spectral range obtained from PCF as a function of the coupled pump power of the femtosecond laser.

| Coupled power (mW) | Generated SC power (mW) | Effective spectral range (nm) |
|---|---|---|
| < 10 | | 746–795 |
| 30 | 4 | 600–926 |
| 110 | 24 | 497–1110 |
| 230 | 48 | 454–1646 |
| 322 | 62 | 450–1673 |

input power. The measured (average) output power of the supercontinuum at a typical coupled power of 322 mW is 62 mW.

To record the full spectrum shown in figure 21.8, two spectrometers sensitive in different spectral regions were used. In contrast to the WLC produced by water, a spectral range of ~ 450–1673 nm can easily be obtained using PCF. The spectral contents of SC are demonstrated by dispersing it with a grating. An image corresponding to the obtained white-light spot is also shown in the figure. Panel B indicates the net average power of the SC with respect to the coupled input power. The linear fit indicates that the generated SC power increases linearly with the input power of the fundamental beam.

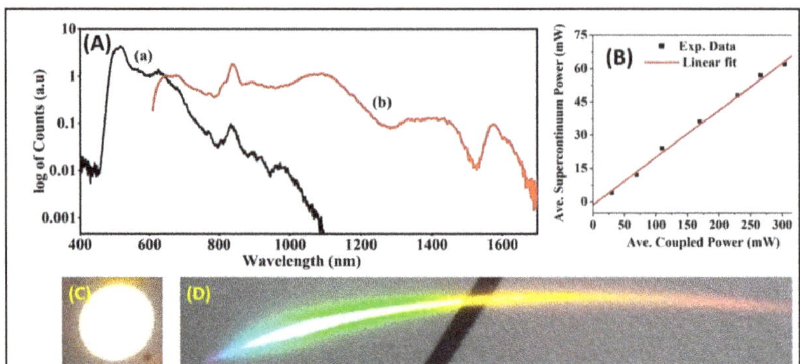

**Figure 21.8.** Top left panel (A): SC spectrum generated from PCF recorded by spectrometer1 (curve a) and spectrometer 2 (curve b). The spectra have been normalized at 663 nm. Right panel (B) gives the obtained average SC power with respect to the coupled input power at 760 nm. Panel (D) is the dispersion of the SC spot (show in *Panel C*) with the help of a diffraction grating. The slant black strip in the middle in panel D is part of the screen. [15].

## 21.6 Filamentation and conical emission

Another exotic phenomenon observed under ultrafast pumping is known as conical emission. If the incident focused power of the laser beam is greater than the critical power required for self-focusing of the material, an electron plasma is generated at the focus. This plasma acts as a defocusing medium.

In order for a material to act as a self-focusing medium, the critical power ($P_c$) is given by $P_c = \frac{3.77\lambda^2}{8\pi n_0 n_2}$, w; here, $\lambda$ is the wavelength of the laser beam, $n_0$ is the linear refractive index, and $n_2$ is the nonlinear refractive index at the working wavelength. This sequence of the events (self-focusing and defocusing) continues throughout the length of the medium. The whole process is known as filamentation. Due to the conical shape of the light emanating from this system (which has effects similar to SSPM), it is called conical emission (CE). CE manifests itself by the occurrence of emission in a well-defined direction at a small conical angle (♠ see also section 15.7) to the propagating beam, which forms one or more characteristic light cones due to the diffraction by the electron plasma.

Figure 21.9 shows the experimentally observed CE produced by a $\beta$-barium borate (BBO) crystal pumped by the fundamental of a femtosecond Ti:sapphire laser. CE is also often referred to as modulation instability (MI) due to its similarities with SSPM. CE is a result of several nonlinear optical processes that occur simultaneously in a medium, such as SPM, XPM, self-steepening, four-wave mixing (FWM), and SRS. These processes are classes as third-order nonlinear optical ($\chi^{(3)}$) processes, as described in chapter 16.

## 21.7 Dark-core beam generation

An application of the SSPM phenomenon is given here. A pump-probe experiment as shown in figure 21.10 can be used to check the effect of SSPM on a probe beam.

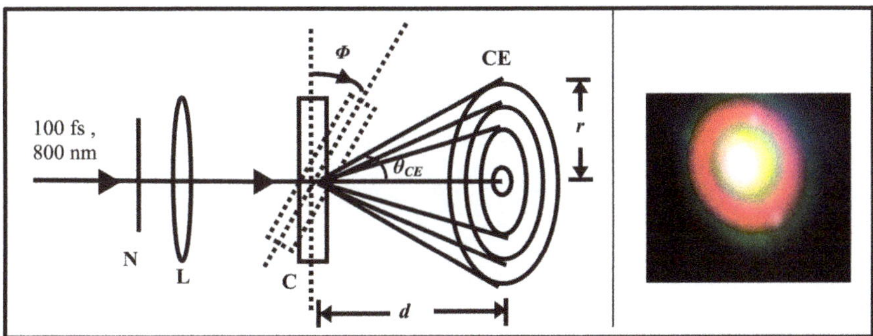

**Figure 21.9.** Schematic of an experimental setup used for CE generation (left). $N$: neutral density filter, $L$: convex lens, $C$: BBO crystal, $\Phi$: crystal tilt angle, $r$: radius of the base of the emission cone, $d$: distance between the back surface of the crystal and the detector, $\theta$: CE angle given by $\tan^{-1}(r/d)$. A photograph (right) of the CE generated in a 20 mm-thick BBO crystal (cut at 22.9 °) [16].

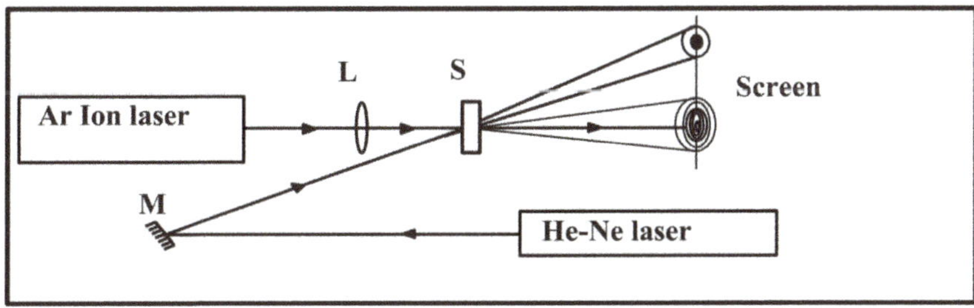

**Figure 21.10.** Pump-probe experiment used to observe the effect of SSPM (here, L: biconvex lens, S: dye sample, M: plane mirror). The beam profile of the He-Ne probe beam is modified as shown in figure 21.11 due to the thermal effects of the pump laser beam.

As indicated at the beginning of this chapter, the phenomenon of SSPM occurs when the nature of the nonlinearity induced in a medium is of the self-focusing type.

The beam profile of the He-Ne (probe) beam is modified as shown in figure 21.11 due to the thermal effects of the argon-ion laser (pump) beam. It can be seen that a dark core appears in the Gaussian beam profile of the probe beam when the pump exceeds a certain threshold power. The transmitted pump beam, however, also shows the diffraction pattern shown in figure 21.1. due to SSPM.

A photograph of the dark-core beam generated from a probe He–Ne laser beam is shown in figure 21.11. The rings are due to the effect of SSPM and are similar to those shown in figure 21.1.

The diameter of the dark core increases with the incident power. Therefore, it is suggested that at higher incident powers, even in semi-transparent media, the effect of SSPM may be present in WLC generation due to the minor heat-induced effects of high-repetition-rate lasers. It should be noted that in order to obtain a high-quality stable WLC, in addition to the effect of temporal SPM, the concurrent contribution of SSPM should be carefully balanced by controlling the size and

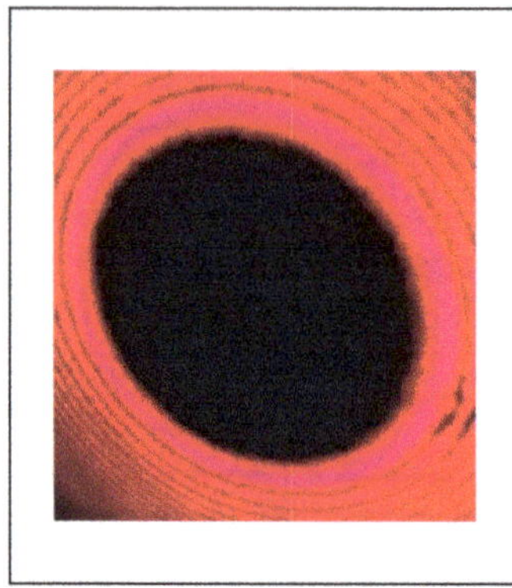

**Figure 21.11.** Photograph of a dark-core beam generated from a He–Ne laser beam (recorded at IIT Madras) (♠ see [4] and [12] for details).

power levels of the incident beam. This modified propagating beam with a dark core is different from Laguerre–Gaussian vortex (LGV) beams (♠ see ♣ under section 1.4.3); nevertheless, it can be useful in various photonic applications.

## Questions and problems

1. The images of the radiation in figure 21.1 exhibit intensity-dependent rings. Explain the observed effect and describe it in terms of a change of phase with integral multiples of $\pi$.
2. Liquid crystals are used in the display devices in our daily electronic gadgets. The molecules of liquid crystals in a medium are oriented in a particular direction. What is the mechanism responsible for the observed fringes, which are similar to those shown in figure 21.1, on pumping liquid crystals with a laser of appropriate pulse energy? (Hint: ♠ see [14]
3. Name the mechanisms responsible for WLC generation when an ultrafast laser pulse interacts with some optical materials. What is the reason for often also calling the WLC the SC?
4. The supercontinuum generated by pumping a water/$D_2O$ mixture with picosecond-duration pulses is predominantly due to SRS. (i) If the water in a cell is pumped by the picosecond-duration pulses of a Nd:YAG laser, what peak wavelengths could be present in the WLC spectrum obtained? (ii) How do these peak wavelengths change if $D_2O$ is used as the optical medium?
5. What is the similarity between photonic bandgap (PBG) materials and the periodic potential of lattices in crystals? What are the consequences of the periodic potential in crystals and in PBG materials?
6. What are the applications of photonic bandgap materials?
7. A transparent, nonlinear optical crystal is pumped by an ultrashort laser pulse. What is the mechanism of the observed conical emission as shown in figure 21.9?
8. What is the mechanism by which a dark-core beam is generated in figure 21.11?

# Bibliography

[1] Alfano R R 2016 *The Supercontinuum Laser Source* (New York: Springer)

[2] Brodeur A and Chin S L 1999 Ultrafast white-light continuum generation and self-focusing in transparent condensed media *J. Opt. Soc. Am.* B **16** 637

[3] Dey S, Bongu S R and Bisht P B 2017 Broad band nonlinear optical absorption measurements of the laser dye IR26 using white light continuum Z-scan *J. Appl. Phys.* **121** 113107

[4] Sailaja R, Sreeja V and Bisht P B 2005 Studies of self-phase modulation under CW and picosecond laser pumping: White light continuum generation in water *Indian J. Phys.* **79** 1299–304

[5] Sailaja R 2007 *PhD Thesis* (IIT Madras)

[6] Klewitz S, Leiderer P, Herminghaus S and Sogomonian S 1996 Tunable stimulated Raman scattering by pumping with Bessel beams *Opt. Lett.* **21** 248–50

[7] Russell P 2003 Photonic crystal fibers *Science* **299** 358–62

[8] Ranka J K, Windeler R S and Stentz A J 2000 Visible continuum generation in air–silica microstructure optical fibers with anomalous dispersion at 800 nm *Opt. Lett.* **25** 25

[9] Gratson G M *et al* 2006 Direct-write assembly of three-dimensional photonic crystals: conversion of polymer scaffolds to silicon hollow-woodpile structures *Adv. Mater.* **18** 461–5

[10] Ali S A *et al* 2010 Conical emission in β-barium borate under femtosecond pumping with phase matching angles away from second harmonic generation *J. Opt. Soc. Am.* B **27** 1751–6

[11] Couairon A and Mysyrowicz A 2007 Femtosecond filamentation in transparent media *Phys. Rep.* **441** 47–189

[12] Dey S, Rallabandi S, Singh S and Bisht P 2021 Study of a dark core beam generated by nonlinear thermo-optical effect *Opt. Laser Tech* **134** 106652

[13] Kudlinski A *et al* 2009 Dispersion-engineered photonic crystal fibers for CW-pumped supercontinuum sources *J. Light. Technol.* **27** 1556–64

[14] Durbin S D, Arakelian S M and Shen Y R 1981 Laser-induced diffraction rings from a nematic-liquid-crystal *Opt. Lett.* **6** 411–13

[15] Dey S 2021 *PhD Thesis* (IIT Madras)

[16] Ali A 2013 *PhD Thesis* (IIT Madras)

**IOP** Publishing

# An Introduction to Photonics and Laser Physics with Applications

**Prem B Bisht**

# Chapter 22

## Semiconductor lasers

Advancements in semiconductor technology have played a vital role in miniaturizing light sources. Modern-day photonics is flooded with light-emitting diodes (LEDs), photodetectors, amplifiers, waveguides, and so on, most of which are based on semiconductor technology. To begin with, semiconductor diode lasers were used in compact disc players and laser printers. Today, they are used as solid-state versatile pump sources in the laser industry. Following a brief introduction to the physics of semiconductors, the operational principle of semiconductor lasers is described in this chapter. The distinct spectrum of a laser is compared with that of an LED in the figure. It should be noted here that the inventors of the blue LED were awarded the Nobel Prize in 2014.

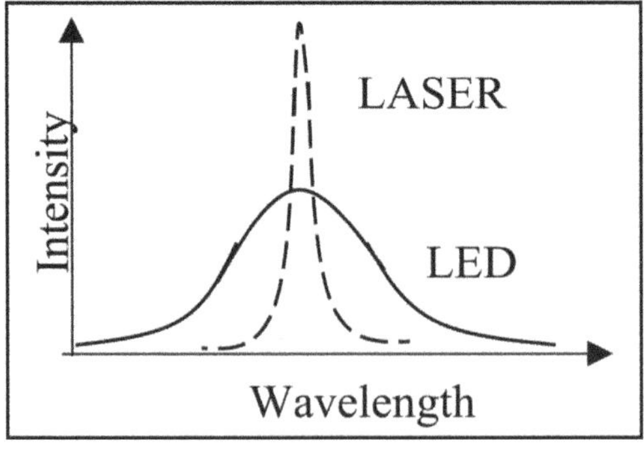


**Learning objectives**
**After reading this chapter, the learner will be able to:**
Identify direct and indirect bandgap semiconductors;

Explain $p$-type and $n$-type semiconductors, the Fermi level, and excitons;
Describe the density of states;
Understand the principles of the $p$–$n$ junction, the LED, and the diode laser;
Understand heterojunction lasers and the vertical-cavity surface-emitting laser (VCSEL);
Identify the quantum cascade laser (QCL) as an example of a unipolar device;
Differentiate between the QCL and diode lasers.

## 22.1 Semiconductors

The portion of the periodic table given in section 2.9.6.1 that includes semiconductor elements is presented in table 22.1. We previously observed that the semiconductor elements (known as intrinsic semiconductors) are placed in columns two to six. However, as will be described later, combination compounds of these elements, such as GaAs, InP, ZnTe, and CdSe, are useful as laser materials. The compound semiconductors are also known as group III–V (an example of which is GaAs) or II–VI (an example is CdSe) semiconductors.

Important applications such as various light sources or detectors require the flexibility to synthesize materials with tunable bandgaps. For this purpose, elements from different groups are required to form compounds with ternary combinations, such as AlGaAs, or quaternary combinations, such as InGaAsP. The matching of the lattice constant is an essential condition that has to be applied to two sets of binary compounds in order to obtain good combinations. As the details of this process are outside the scope of this book, the interested reader is referred to textbooks on semiconductor technology, a few of which are listed at the end of this chapter.

♣ With the introduction of nonlinear optical effects (e.g. the optical Kerr and Pockels effects) and optical waveguides, the area of semiconductor–optoelectronic devices has grown in leaps and bounds. Here, electronics and photonics are married together using the principle that the photons of a particular frequency generate mobile electrons which then control the flow of photons of another (or the same) frequency.

**Table 22.1.** Semiconductors in the periodic table.

| II | III | IV | V | VI |
|---|---|---|---|---|
|  | 13<br>Al<br>26.982 | 14<br>Si<br>28.085 | 15<br>P<br>30.974 | 16<br>S<br>32.06 |
| 30<br>Zn<br>65.38 | 31<br>Ga<br>69.723 | 32<br>Ge<br>72.630 | 33<br>As<br>74.922 | 34<br>Se<br>78.971 |
| 48<br>Cd<br>112.41 | 49<br>In<br>114.82 | 50<br>Sn<br>118.71 | 51<br>Sb<br>121.76 | 52<br>Te<br>127.60 |
| 80<br>Hg<br>200.59 | 81<br>Ti<br>204.38 | 82<br>Pb<br>207.2 | 83<br>Bi<br>208.98 | 84<br>Po<br>209 |

## 22.2 Bandgaps in semiconductors

When atoms are close to each other, for example, in solids, Pauli's exclusion principle can be used to explain their behavior qualitatively. With an increase in the interatomic spacing, the electron wave functions begin to overlap. To avoid limits imposed by Pauli's principle, the discrete energy levels of atoms come together to form new levels (i.e. energy bands) that belong to the collection of atoms. These bands have very closely spaced levels with forbidden energy gaps between them, as indicated in figure 22.1.

The highest-energy, unoccupied bands are extremely important in determining the physical properties of solids. Semiconductors have highest unfilled bands of energy as follows. The electronic structure of $_{14}$Si is $1s^2$, $2s^2$ $2p^6$, $3s^2$, $3p^2$. The outer four electrons in the crystalline matrix broaden into two distinct bands separated by an energy gap. In semiconductors, this gap, known as the forbidden energy gap ($E_g$) falls in between the set of energy levels called the valence band (VB) and the conduction band (CB). In the absence of thermal excitation ($T = 0$ K), the VB is completely filled with electrons but the CB is empty. It should be noted that in the case of metals, the VB and CB overlap.

Thermal and optical interactions can cause an electron to jump from the VB to the CB, leaving behind a hole. For semiconductors, the value of $E_g$ falls in the range of 0.1–3 eV. For example, silicon has a bandgap of 1.11 eV, while the bandgap is 1.42 eV for gallium arsenide (GaAs). If the bandgap is larger than 3 eV, the material is classified as an insulator.

Let us write the wave function of an electron in the valence band as follows:

$$\psi(\vec{r}) = U_L(\vec{r})e^{i\vec{k}.\vec{r}}.$$

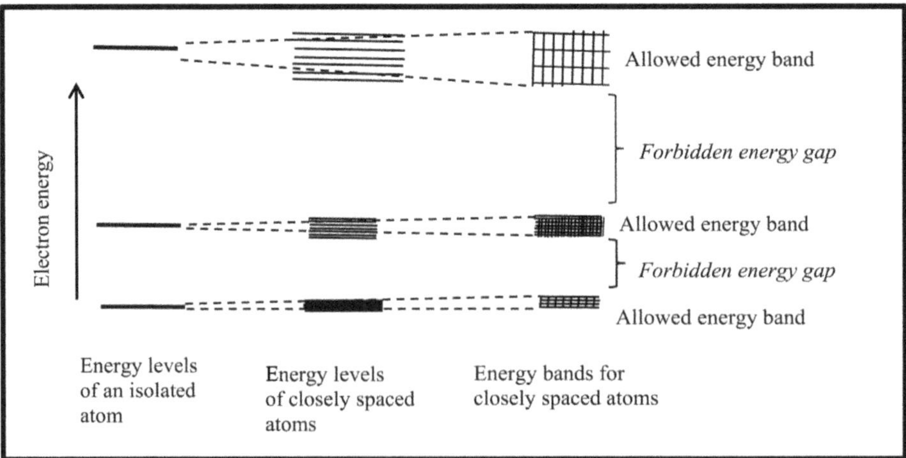


**Figure 22.1.** Energy bands and energy gaps formed in solids due to the rearrangement of electrons and the overlap of their wave functions according to Pauli's exclusion principle.

Here, the propagation constant (wave vector, $\vec{k}$) is related to the momentum of the electron $p$ by $\vec{p} = \hbar\vec{k}$, and $U_L(\vec{r})$ contains the periodicity of the crystalline lattice. If the value of $\psi(\vec{r})$ is substituted into the Schrödinger equation, it is found that the energy eigenvalues $E(\vec{k})$ versus $|\vec{k}|$ follow a parabola, as follows:

$$E(\vec{k}) = \frac{\hbar^2 |\vec{k}|^2}{2m}.$$

For an electron or a hole, the mass $m$ has to be replaced by the effective mass of an electron ($m_c$) or a hole ($m_h$), respectively. The effective mass of the electron is taken at the bottom of the conduction band ($E_c$), while that of the hole is taken at the top of the valence band ($E_v$). The corresponding values are proportional to the inverse of the second derivative of energy as a function of the wave number, namely

$$m_c = \hbar^2/\left(\frac{d^2 E_c}{dk^2}\right)_{k=0} \quad \text{and} \quad m_h = \hbar^2/\left(\frac{d^2 E_v}{dk^2}\right)_{k=0}.$$

The electrons and holes are the two carriers of current that interact with each other. As shown in figure 22.2, the absorption of a photon can create an electron–hole pair. An electron can jump from the VB to fill an existing hole by the recombination process. The recombination of the two can result in the emission of either a photon or a phonon. The phonon is a unit (quantum) of vibrational energy, generally related to lattice relaxation in solids. In contrast to photons, which have high energy, phonons have high momentum rather than energy. Phonons are useful to bridge the energy gap in indirect bandgap semiconductors.

## 22.3 Excitons

When a photon of a suitable energy collides with a semiconductor, an electron in the valence band is excited to the conduction band, leaving a hole behind it. An electron–hole pair in the elementary excited state of the semiconductor is known is an *exciton*.

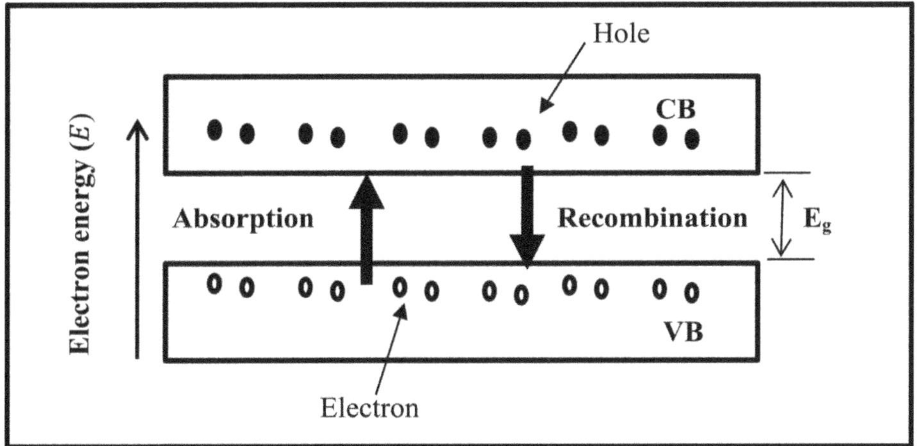


**Figure 22.2.** Electrons (open circles) and holes (filled circles) in the VB and the CB, respectively. Recombination takes place when an electron jumps from the VB to the CB and vice versa.

Here, the hole and electron are attracted to each other by a coulombic force, forming an exciton pair. Using a similar approach to the Bohr model of the atom, we can picture the exciton as a weakly bound electron and a hole orbiting around a common center of gravity. The electron mass is replaced by its reduced mass $m^*$, as follows

$$\frac{1}{m^*} = \frac{1}{m_e^*} + \frac{1}{m_h^*}.$$

The probability of recombination of an exciton depends on the pathways used to lose the excess energy. It also depends on the spatial overlap of the wave functions corresponding to electron and hole. The exciton states are situated just below the bottom of the CB. it is possible to control excitons by applying a voltage, as they are capable of moving through the lattice. Excitons are the source of luminescence in semiconductors and are also used to control optoelectronic devices.

## 22.4 Fermi level

The electronic distribution in the VB and the CB is described in terms of the 'Fermi level'. It is defined as the energy ($\epsilon_F$) at which the probability that an electron will occupy a specific level is half. Particles that obey Pauli's exclusion principle in a system follow Fermi–Dirac statistics. The probability $F(E)$, also known as the *Fermi function* of a particle with energy $E$ at a temperature $T$ is given by

$$F(E) = \frac{1}{e^{\frac{(E-\epsilon_F)}{kT}} + 1}.$$

At 0 K, due to the equal number of holes and electrons in an intrinsic semiconductor, the Fermi level lies midway between the VB and the CB (figure 22.3). In an n-type material, this is pushed up towards the CB, whereas in a $p$-type material, it is pushed down towards the VB.

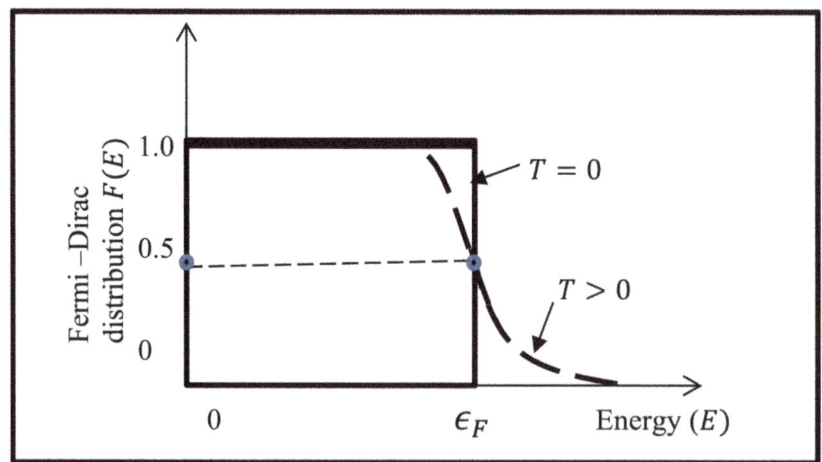


**Figure 22.3.** Plot of the *Fermi function* vs energy in the Fermi–Dirac distribution at the indicated temperatures ($T$). $\epsilon_F$ is known as the Fermi energy at $T = 0$ for an $F(E)$ value of 0.5.

## 22.5 Direct and indirect bandgaps

The variation of energy versus momentum (known as an $E$–$k$ diagram) in *direct* and *indirect bandgap* semiconductors was outlined in chapter 2 (♠ see section 2.9.8, figure 2.14). As shown by this diagram, materials with a *direct bandgap* are capable of producing luminescence. For these direct bandgap semiconductors in the II–Vth groups (such as GaAs), the momentum of an electron returning from the bottom of the CB to the top of VB changes by a very small amount. In contrast, for indirect bandgap materials, any radiative transition precedes relaxation through *phonon* emission. This process produces a significant shift in momentum when it takes place between the CB and the VB. Therefore, while direct bandgap materials have radiative transitions and can be used as laser media, intrinsic materials such as Si and Ge, which have indirect bandgaps, are not useful as lasing materials.

## 22.6 Density of states

An electron at the band edge of the CB can be approximated by a particle of mass $m$ confined to a perfectly reflective $3D$ cubic box of certain dimensions. Its quantum state is described by its energy, wave vector, and spin. To calculate the carrier concentration in the energy bands, knowledge of the distribution of the energy states as a function of energy is required. In addition, we need to know the probability that each of these states is occupied by an electron. The term *density of states* ($\rho(E)$) is used to represent the number of quantum states of energy per unit volume in an energy range.

In figure 22.4, the $E$ versus $k$ relation is shown at $T = 0$ and $T > 0$ with and without excitation. It can be seen that at 0 K, all the electrons are in the VB. On optical excitation, electrons from the VB arrive in the CB, filling all the allowed lowest excited states. In the process of excitation at higher temperatures, the thermal energy of the electrons shifts some of them into higher energy states.

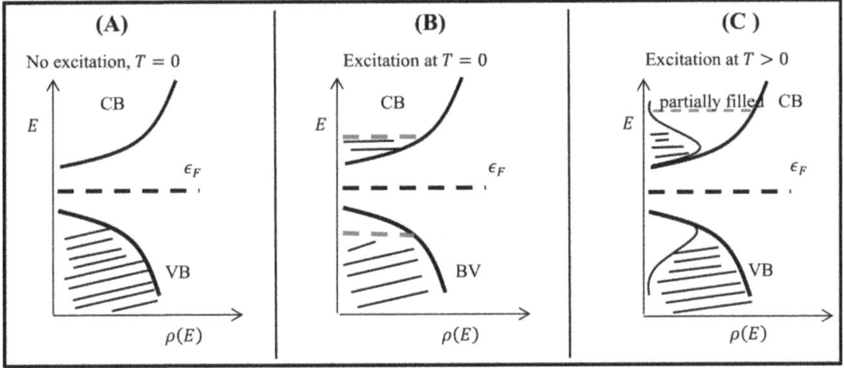


**Figure 22.4.** Concentration of electrons and holes as a function of energy at $T = 0$ without excitation (left, panel A) and on excitation (middle panel, B). At the right, panel C shows the situation on excitation at higher temperatures, i.e. $T > 0$.

## 22.7 *p*-type and *n*-type semiconductors

The electrical and optical properties of pure semiconductor elements (*intrinsic* semiconductors) can be altered by adding small amounts of impurities known as dopants. When a tiny amount of an element from a group (table 22.1) is mixed with the element of the another group, the compound becomes an *extrinsic* semiconductor.

To understand this, let us consider silicon (Si) as a model system. It has four valence electrons; each electron is shared by its neighboring atoms. When it is doped with any other element from group V, such as As, P, or Sb, that has five valence electrons, then each dopant atom provides an extra electron. Here, the dopant atom is known as the electron *donor*, while in this scenario Si is known as the *acceptor* atom. The resulting doped compound becomes an '*n*-type semiconductor', in which *electrons* are the *majority carriers*. Similarly, doping Si with any of the group III elements such as Ga, B, Al, or In with three valence electrons would provide a *hole* to the final compound. Such a compound is a '*p*-type semiconductor', in which *holes* are the majority carriers. For doped materials, additional energy levels due to impurities are created within the bands. For instance, for *n*-type semiconductors, these levels are near the conduction band, producing an excess number of electrons. Similarly, for *p*-type materials, an excess number of holes are created in the valence band.

## 22.8 The *p–n* junction and electrical excitation

A homojunction between a *p*-type semiconductor and an *n*-type semiconductor is known as a *pn junction* (figure 22.5). It has several applications in electronics and spectroscopy in light-emitting diodes (LEDs), photodetectors, and solar cells. In the narrow region of the junction, electrons and holes diffuse from areas of higher concentration (e.g. electrons in the *n*-type) towards areas of lower concentration (e.g. electrons in *p*-type) and recombine until a space charge is produced. The recombination of the charge carriers on both sides leaves a narrow region that has a shortage of mobile charge carriers. This narrow region, called the *depletion region*, has an electric field that points from *n*-type to *p*-type and therefore obstructs the diffusion of further mobile carriers through an energy barrier with a potential ($V_0$). This

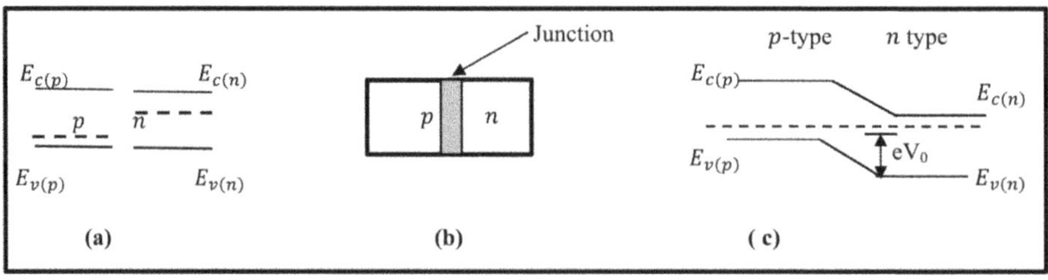


**Figure 22.5.** (a) *p*-type and *n*-type semiconductors are joined together to form a *pn* junction. $E_c$ and $E_v$ with subscripts in parentheses, respectively, denote the conduction and valence bands for the (*p*) and (*n*) types. (b) The potential energy of electrons is reduced by the barrier potential $V_0$ at the depletion region, resulting in bending of the VB and the CB at equilibrium (c). The dashed lines represent the Fermi levels for moderate doping.

potential, known as the *diffusion* or *contact* potential lowers the potential energy of an electron on the *n*-side relative to the *p*-side, resulting in bending of the energy bands. At thermal equilibrium, the Fermi level in the junction aligns over the entire structure (as shown figure 22.5(c)).

### 22.8.1 The light-emitting diode: a forward-biased p–n junction

A *pn* junction is said to be forward biased when the positive terminal of the battery is connected to its *p* side. Under forward bias, the energy barrier between the two materials is reduced as the applied voltage is dropped across the depletion region. This reduction is equal to the amount of applied voltage $V$. (figure 22.6). The electrons flow across the barrier in large numbers, resulting in radiative recombination that produces light. It should be noted that on application of the external potential, the Fermi levels are no longer aligned at the junction.

Forward-biased *pn* junctions are used as *light-emitting diodes* (LEDs). As usual, in the '*n*' region of the LED there is an excess of free electrons in the conduction band. On the '*p*' side of the junction there are excess holes. When a forward bias is applied to a *pn* junction, electrons and holes combine and emit photons. The light emitted by an *pn* junction due to the recombination process can be thought of as similar to the spontaneous emission that takes place in atomic or molecular systems. In LEDs, however, not only the excited species (as in the case of atoms, molecules or ions) but the whole crystal structure takes part in the emission process. The emitted radiation frequency depends on the bandgap of the semiconductor. The bandgap is engineered not only by choosing different semiconductors but even by changing the concentration of the constituent elements. As an example, in a heterojunction such as $GaAs_{(1-x)}P_{(x)}$, the bandgap can be tuned by varying the fraction $x$. Thus, LEDs are specially designed *pn* junction diodes which can emit in the infrared, visible, or ultraviolet region, depending on the bandgap of the materials used.

**Exercise 22.1.** What is meaning of 'biasing a semiconductor diode'? How does the character of the diode change when the bias is reversed?

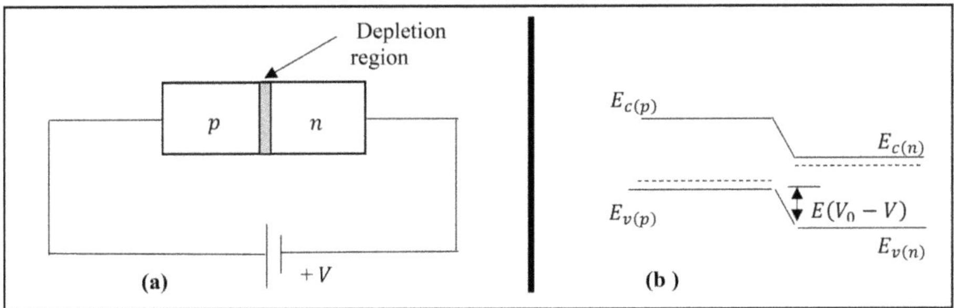


**Figure 22.6** (a) Forward biasing a p-n junction on applying the voltage $V$. (b) Forward biasing reduces the value of the potential barrier for recombination as $E(V_0 - V)$.

**Solution:** Historically, the word *diode* was used for vacuum tubes in which two terminals (the *anode and cathode*) were used for rectification of AC voltages. As the *pn* junction is also used for rectification purposes (e.g. half-wave, full-wave, or bridge rectifiers), it is also referred to as a diode. In figure 22.6, when the positive terminal of the battery is connected to the *p*-side of the *pn* junction, it is known as a *forward-biased* diode. On the other hand, reversing the polarity puts the diode in a reverse-biased circuit. While forward-biased *pn* junctions are used in the manufacture of LEDs and lasers, *reverse-biased pn* junctions, on the other hand, are used as photodetectors, an area equally important in optics and spectroscopy.

### 22.8.2 Diode lasers

A few changes in the design of LEDs paved the way to achieving lasing in special *pn* junctions. As shown in the figure 22.7, in order to achieve population inversion, heavy doping of the *n*-type and *p*-type materials is required together with forward biasing of the *pn* junction. Furthermore, higher currents (above the lasing threshold, $I_{\text{threshold}}$) are required, as indicated in the figure.

For the *pn* junction shown in figure 22.8, in heavily doped *n*-type materials, the Fermi level lies within the CB. This is in contrast to figure 22.5 (c). Similarly, for *p*-type materials, heavy doping results in the Fermi level moving into the VB.

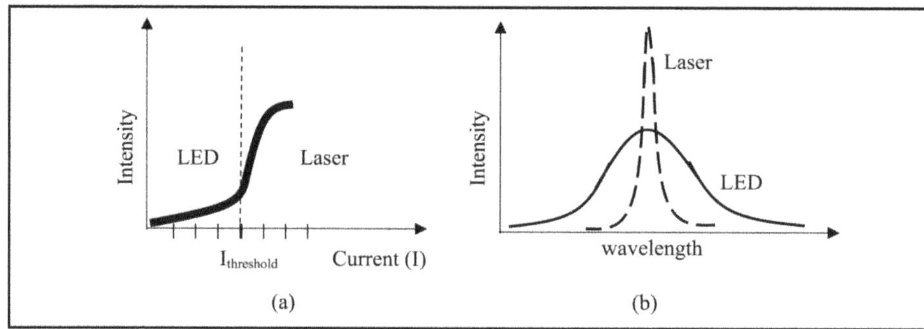


**Figure 22.7.** (a) The diode laser requires a higher current above the threshold compared to that of an LED. Spontaneous emission can be observed below the dashed line ($I_{\text{threshold}}$). The spectrum of an LED is compared with that of a laser in panel (b).

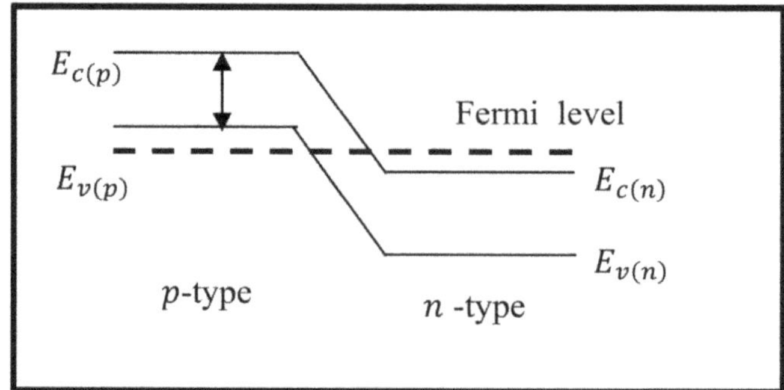


**Figure 22.8.** Heavily doped *p*- and *n*-type semiconductors and the equilibrated Fermi level in a *pn* junction.

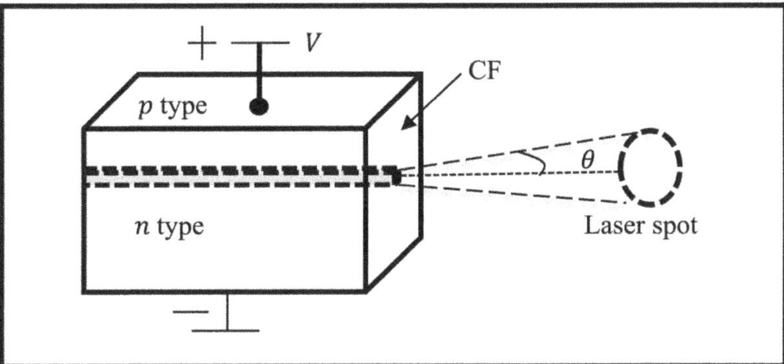


**Figure 22.9.** Schematic representation of a semiconductor laser with an applied voltage $V$ and a cleaved facet (CF) indicated on one side. The confined shaded region is the lasing medium. $\theta$ is the beam-divergence half angle.

LEDs emit light in the full solid angle, i.e. in all directions, which is similar to the behavior of lamps. In diode lasers, parallel cleaved facets of the crystal are used as the two mirrors of the Fabry–Pérot (FP) cavity. As indicated in figure 22.9, for diode lasers, these parallel surfaces produce optical feedback and directional radiation with a low divergence angle $\theta$, ($\theta \tilde{} \lambda/a$, where $a$ is size of the output aperture).

**Exercise 22.2.** The cleaved surfaces of diode lasers can be used as the mirrors of a cavity. Find the values of reflectance ($R$) for a light beam at the interface between air ($n_1 = 1.0$) and (i) glass ($n_2 = 1.5$) and (ii) GaAs ($n_2 = 3.6$).
**Solution:** From Fresnel's equations (♠ see chapter 3), for example, for transverse electromagnetic (TE) polarizations, the reflectance at normal incidence is given by $R = \left(\frac{n_2 - n_1}{n_2 + n_1}\right)^2$. For the given values, we obtain
    (a) for glass with $n_2 = 1.5$, $R = 4\ \%$;
    (b) for GaAs with ($n_2 = 3.6$), R = 31.9 %.

Therefore, to maintain the sizes of tiny diode lasers, instead of involving additional metal-coated mirrors, GaAs cleaved surfaces are used; they have reasonably high reflectance as compared to glass, for example.

## 22.9 Semiconductor heterostructures

Due to limitation of the thickness of the depletion region, it has been observed that a certain amount of heat is generated in $pn$ junctions. This adversely affects the performance of the laser. For example, this was the case for the simple GaAsP visible lasers [2] developed in the early 1960s. To eliminate the heating effects due to current flow at junctions with a thickness of a few μm, 2D heterojunctions of semiconductors (such as $n$-GaAS-$p$-GaAs-$p$-AlGaAs) or their higher versions are used for lasing at room temperature.

♣ A heterostructure of semiconductors is a structure in which the chemical composition is a function of position. The simplest heterostructure consists of an interface within a semiconductor crystal across which the chemical composition changes. $Al_xGa_{1-x}As$ is an example of such a heterostructure.

It is an interesting and obvious fact that by tuning the bandgap, the output wavelengths of semiconductor lasers can be varied over a large spectral range. The energy-level diagram of a typical heterostructure is shown in figure 22.10.

In principle, heterojunctions can be grown in such a way that the current density is confined to the active region, which is equivalent to the gain medium defined earlier. In the 2D structure mentioned above, the active region is a thin layer of the order of 10–50 nm of GaAs, thus reducing the threshold current. The active layers can even be in the form of quantum wells. A quantum well is nothing more than an active medium surrounded by material that has a larger bandgap. Heterojunctions must be made of materials with nearly similar lattice constants [3] and are widely used in the fabrication of diode lasers, quantum wells, and quantum wires. Table 22.2. summarizes a few wavelength regions of laser diodes and their semiconducting elements.

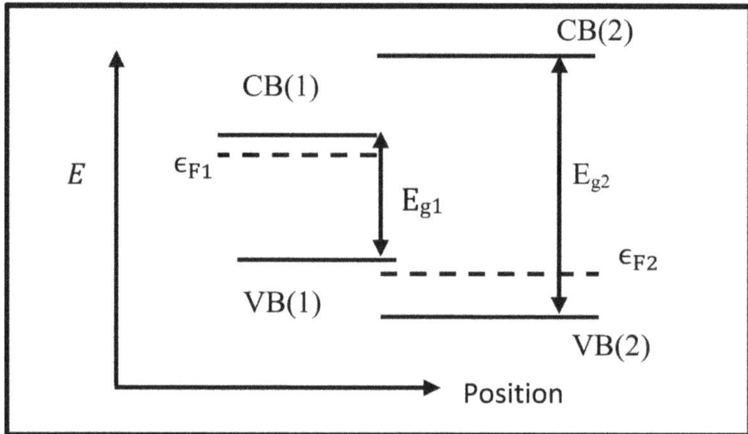


**Figure 22.10.** Energy-level diagram of a typical heterostructure, showing the locations of the CB, the VB, and the Fermi levels ($\epsilon_{F1}$ and $\epsilon_{F1)}$ of semiconductors (1) and (2), respectively.

**Table 22.2.** Typical laser wavelength ranges of semiconductor laser diodes.

| Semiconductor elements | Laser wavelengths (nm) |
| --- | --- |
| GaAs | 870 |
| GaAlAs | 750–850 |
| InGaAsP | 1100–1650 |
| InGaAsSb | 1700–4300 |
| GaN | 300–450 |
| PbSnTe | 6000–25000 |
| PbEuSeTe | 2000–4000 |

## 22.10 Vertical-cavity surface-emitting lasers

We know from Gaussian analysis of laser beams that the divergence of a laser beam is inversely proportional to the beam size (see equation (12.23)). As the diode laser is a tiny device fabricated on a wafer, its divergence is large. Nevertheless, the cavity dimensions of diode lasers are large enough (a few hundreds of $\mu m$) to support several modes; diode lasers have unstable outputs due to mode-hopping.

In order to overcome some of the problems faced by diode laser heterostructures in 2D, a configuration stacked in the vertical direction, known as a VSCEL, is used. In this configuration, a large number of monolayers of the elements that have an area equal to that of a wafer remove the constraint of the size of the laser beam, thereby reducing its divergence. The mode spacing can be controlled by the height ($h$) of the device ($c/2h$). As mentioned in a previous section ($\spadesuit$ see exercise 22.2), thin layers of cleaved semiconductor are used as mirrors. As the laser light propagates perpendicular to the surface, it is advantageous to process a large number of devices on a single wafer.

## 22.11 Quantum cascade laser: a unipolar device

### 22.11.1 Quantum size effect

In semiconductor lasers, the principle of bandgap engineering is used to achieve the tunability of laser wavelengths in the near-IR to far-IR regions [3]. This is achieved using a laser known as the quantum cascade laser (QCL).

Quantum wells are fabricated by sandwiching two different ultrathin semi-conductors. A detailed description of the quantum size effect is outside the scope of this textbook. Briefly, electrons with their wavelike properties and wavelengths comparable to the thickness of the quantum well have quantized motion perpendicular to the layers. Therefore, in place of the bands, the quantized motion gives rise to a series of discrete energy levels. The quantum size effect helps to directly control the separation of these energy levels by changing the thickness of the quantum-well layers.

In a QCL, photon emission takes place when an electron jumps from a higher to a lower energy level of a quantum well within the CB. QCL relies on only one type of charge carrier (electrons) and hence is known as a *unipolar device*. This mechanism is described in the next section and is different from that of the recombination of carriers from the CB to the VB as discussed in earlier sections.

### 22.11.2 Mechanism of lasing in the QCL

In QCL, the lasing wavelengths are not directly dependent on the bandgap. This is due to the fact that population inversion is achieved between sub-bands of a quantum well but not between a number of electrons and holes. Transition between the sub-bands is the key point of obtaining tunable emission using engineered materials. The thickness of the material device is typically of the order of a few nanometers and electrons are primarily confined to the central part of the *sandwiched well*. By varying the thickness of the quantum wells in the active regions, a QCL output in the 4 µm to 24 µm range has been obtained using a combination of Al–In–As and Ga–In–As.

There are several stages of the active region (typically ~30 to 70) in a QCL device. Each stage has (i) an injection region, (ii) an active region, and (iii) a relaxation region. The injection region injects the electrons to the higher energy state of the active region. The active region is the one in which electrons jump from the upper to the lower level, emitting stimulated photons. The relaxation region is the one in which the electrons relax to the ground state. In most QCLs, the relaxation region of one stage is the injection region for the following region. Therefore, it is also referred as an injection/relaxation region (see figure 22.11).

The initial state of each stage is populated by the resonant tunneling of electrons. The half-life of the upper state is made larger than that of the lower state by the design of the heterostructure. After the radiative transition, the electrons relax to the ground state by the emission of longitudinal optical phonons and are finally extracted from the active region by tunneling through an exit barrier. Several tens of active regions are connected by injectors, comprising a superlattice (figure 22.12). The superlattice is cascaded to enhance optical gain and thus to reduce the threshold current density at the expense of a higher operational voltage.

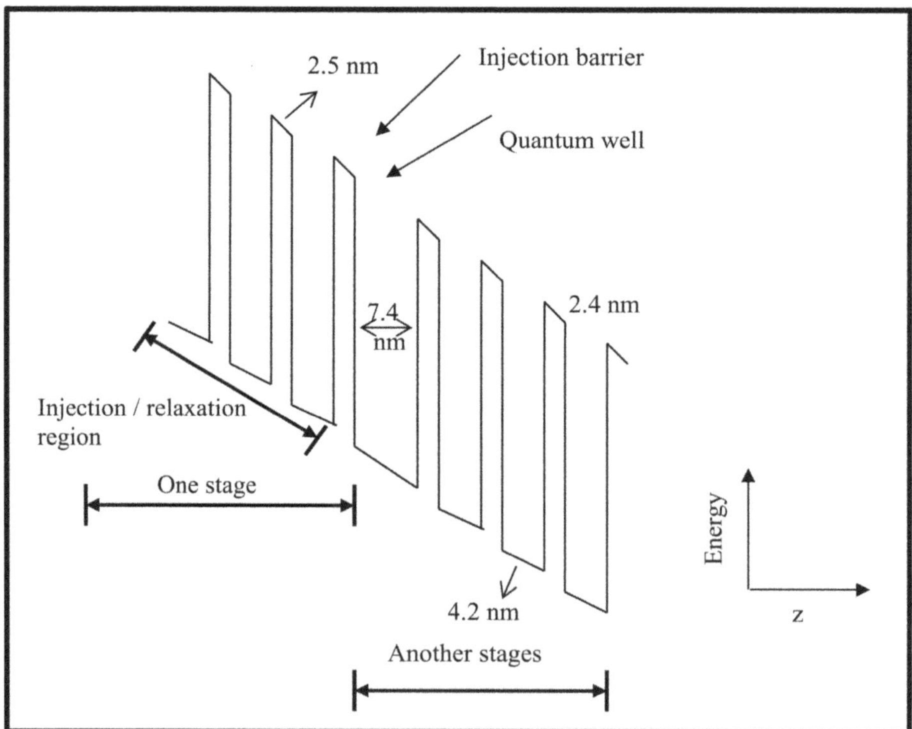


**Figure 22.11.** Schematic energy-band diagram of two periods of a QCL (after [10]). Typical thickness estimates of some of the layers are also shown. From one active region, electrons tunnel into the injector region of the identical second stage, from which they are injected into the initial state of the following active region.

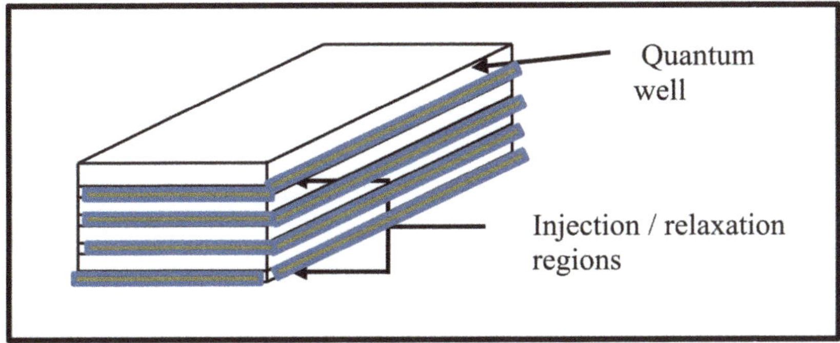


**Figure 22.12.** 3D view of a few stages of a QCL. Several identical stages are connected to make a superlattice.

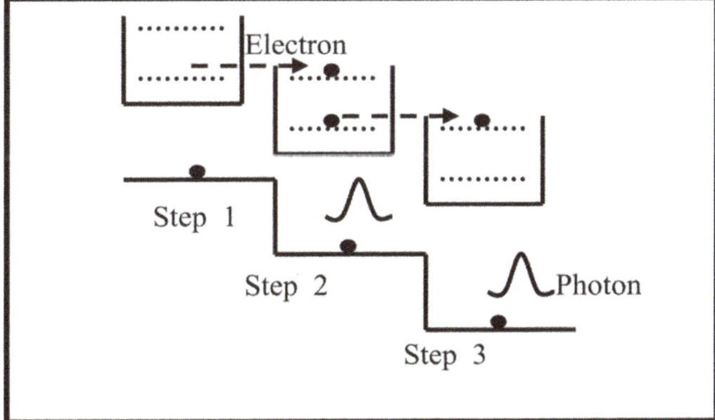


**Figure 22.13.** Electrons cascade down an energy staircase (three steps are shown here) when a voltage is applied to a QCL.

In a unipolar QCL, once a laser photon is emitted by an electron that jumps from the upper to the lower energy level of the first stage, it is recycled by injection into an adjacent **identical** stage, where a second photon is emitted, and so on. This process creates a number of photons (of same frequency) equal to the number of electronic staircases (see figure 22.13). Thus, photons are emitted by electrons cascading down an energy staircase created by the voltage applied to the device. As a number of photons are emitted as a result of a single electron cascading downwards, the efficiency of the device is $>1$.

### 22.11.3 Advantages of the QCL over laser diodes

QCLs are semiconductor lasers that emit in the mid- to far-IR region of the EM spectrum. QCLs are capable of providing a tunable output wavelength; in addition, these are reliable solid-state lasers. Therefore, QCLs offer several advantages over conventional diode lasers. The following table gives the differences between QCLs and conventional semiconductor lasers.

| QCL | Laser diode |
| --- | --- |
| A QCL is made of ultrathin semiconductor layers in a repeated stack of multiple semiconductor quantum-well heterostructures that form a superlattice; each layer of the superlattice acts like a quantum well. | Made of bulk heterostructures of semiconductors. |
| Unipolar device (uses electrons only) | Bipolar devices (use holes and electrons) |
| A photon is emitted due to the transition of an electron from a higher energy sublevel to a lower sublevel of the quantum well. | The recombination of electrons and holes between the CB and the VB gives rise to photons |
| The laser wavelength is decided by the thickness of the quantum wells, which results in different energy levels, hence it is tunable. | The laser wavelength is decided by the bandgap of the materials and is not tunable. |
| The superlattice introduces a varying electric potential across the length of the device. Hence, there is a varying probability of electrons occupying different positions over the length of the device. This is referred to as 1D multiple-quantum-well confinement. This confinement leads to a split of the band of permitted energies into a number of discrete electronic sub-bands. The engineered design of the layer thicknesses gives population inversion between the two sub-bands in the system. | Population inversion is achieved by heavy doping |
| Once an electron has undergone an inter-sub-band transition resulting in the emission of a photon, it can tunnel into the next period of the superlattice, where another photon is emitted. This *cascading* makes the quantum efficiency $>1$, which leads to higher output powers than those of semiconductor laser diodes. | The recombining electron is used in emitting just a photon. The output power depends on the doping levels and the applied current. |

## Questions and problems

1. Define the Fermi level in a semiconductor or a metal.
2. Among direct and indirect bandgap semiconductors, which is used for LEDs and which is used for laser technology?
3. For laser action to take place, why do we need heavy doping of $p$-type and $n$-type semiconductors? How is the Fermi level positioned in heavily doped semiconductors?
4. What is an exciton?

5. By measuring the spectrum of an LED and that of a diode laser, what parameter can be used to differentiate between the two?
6. In a semiconductor laser that has a wavelength of 650 nm, the half angle of the angular spread is 10 degrees. What is the slit size of the output beam?
7. In diode lasers, the mirrors of the cavity are often nothing but the surfaces of the semiconductors cleaved along a crystal plane. For a semiconductor, find the value of the refractive index if the reflectivity at normal incidence is required to be 0.18.
8. A semiconductor laser diode has a heterojunction thickness of 300 μm. The reflectivity of the cleaved surfaces (to be used as mirrors) is 40%. If the cavity losses are calculated to be 1 mm$^{-1}$, calculate the threshold value of the gain.
9. Find the expression for the longitudinal mode spacing in nanometers ($\Delta\lambda$) of a laser diode in terms of its cavity length ($L$), the refractive index of its material $n$, and its output wavelength ($\lambda$).
10. How is the efficiency of diode lasers enhanced by using heterojunctions?
11. What is a VCSEL? Identify its similarities as well as its differences from (i) quantum cascade lasers and (ii) diode lasers.
12. How is the quantum cascade laser different from heterojunction semiconductor lasers?

## Bibliography

[1] Wilson J and Hawkes J F B 1989 *Optoelectronics: An Introduction* 2nd edn (Englewood Cliffs, NJ: Prentice-Hall, Inc.)

[2] Stredman B G 2006 *Solid State Electronic Devices* (Englewood Cliffs, NJ: Prentice-Hall)

[3] Holonyak N and Bevacqua S F 1962 Coherent (visible) light emission from *Ga(As$_{1-x}$P$_x$)* junctions *Appl. Phys. Lett.* **1** 82–3

[4] Hino I, Gomyo A, Kobayashi K, Suzuki T and Nishida K 1983 Room-temperature pulsed operation of AlGaInP/GaInP/AlGaInP double heterostructure visible light laser diodes grown by metalorganic chemical vapor deposition *Appl. Phys. Lett.* **43** 987–9

[5] Holonyak N 1997 The semiconductor laser: a thirty-five-year perspective *Proc. IEEE* **85** 1678–93

[6] Sands D 2004 *Diode Lasers* (Boca Raton, FL: CRC Press)

[7] Aspnes D E, Kelso S M, Logan R A and Bhat R 1986 Optical properties of Al$_x$Ga$_{1-x}$As *J. Appl. Phys.* **60** 754–67

[8] Weber M 1998 *Handbook of Laser Wavelengths* (Boca Raton, FL: CRC Press)

[9] Koyama F, Kinoshita S and Iga K 1989 Room-temperature continuous wave lasing characteristics of a GaAs vertical cavity surface-emitting laser *Appl. Phys. Lett.* **55** 221–2

[10] Faist J *et al* 1994 Quantum cascade laser *Science (80)* **264** 553–6

[11] Faist J 2013 *Quantum Cascade Lasers* (Oxford: Oxford University Press)

[12] Gmachl C *et al* 2001 Recent progress in quantum cascade lasers and applications *Rep. Prog. Phys.* **64** 1533

**IOP** Publishing

# An Introduction to Photonics and Laser Physics with Applications

**Prem B Bisht**

# Chapter 23

## Fiber lasers

Using a beam of sunlight as a carrier, Alexander Graham Bell transmitted a telephone signal over a few hundred meters in 1880. This 'photophone' was the first demonstration of optical communication, the modern embodiment of which is today's fiber-optic communication. Free-space optical communication is not efficient due to scattering and absorption by the environment. Guiding sunlight into buildings directly or via multiple reflections is an economic way of saving energy. For instance, the photo shows sunlight being reflected into a traditional Indian house. Modern decorative devices make use of optical fibers. The first fiber laser was demonstrated as early as 1961. With subsequent developments, long-haul fiber-optic communication has revolutionized modern lifestyles. It has also helped increase the uses of optoelectronic gadgets in our daily life. The wavelength region used for fiber-optic communication is currently 1500 nm. The diagram shows a figure-of-eight laser (F8L), an example of all-fiber technology. In this chapter, we will learn about fiber lasers, which are alignment free, easy to transport, and replacing bulky lasers in science and industry.

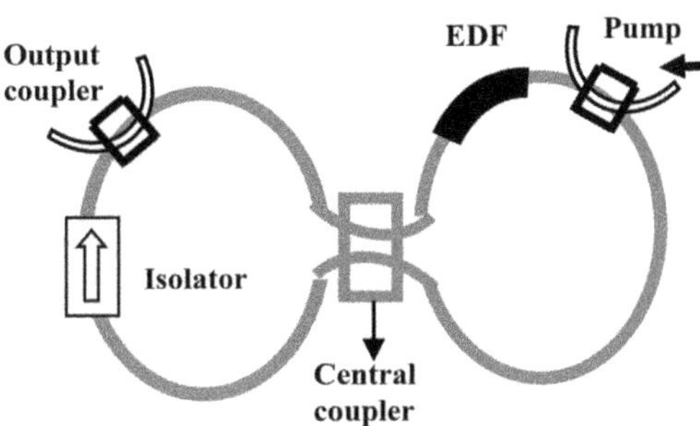


**Learning objectives**
**After reading this chapter, the learner will be able to:**
Describe the physical and optical properties of fibers;
Identify fiber gain media;
Apply the phenomena of dispersion and optical nonlinearity in optical fibers;
Understand the principle of fiber amplifiers;
Explain the working principle of a fiber laser, the figure-of-eight laser;
Describe high-power fiber lasers;
Understand the principle of the Raman laser;
Illustrate optical fiber communication;
Write down the nonlinear Schrödinger equation (NLSE).

# 23.1 Fiber laser technology

We know that the first laser (a ruby laser) was invented in 1960; it had an output wavelength of 690 nm in the visible region. Today, there is a huge range of laser media in the EM spectrum from x-rays to the microwave region. The operation of the first fiber laser was demonstrated as early as 1961 by Snitzer using neodymium-doped silicate glass fiber at 1.06 μm. It took a few decades for subsequent developments to reach present-day fiber laser technology.

It is interesting to note that for the optical fiber communication, light that has a wavelength near 1.55 μm is used. This is due to the following reasons: (i) it causes the lowest losses in the optical fiber material (silica), (ii) the negative dispersion provided by the silica fiber and (iii) the availability of the gain media (e.g. erbium) used for optical amplification. In this chapter, we will study the details of these features. As an example, a simple laser configuration (the F8L laser) is also described to aid in the understanding of an 'all-fiber laser'.

### 23.1.1 Structure of an optical fiber

A crude way to prepare an optical fiber is by simply pulling a glass or polymer material by melting it in a workshop at the required temperature. Sophisticated machines are used nowadays for this purpose. The key feature of an optical fiber is the refractive index difference between the central part (known as the core) of the fiber and its surroundings (known as the cladding). The refractive index of the core is higher than that of the cladding. Flexible, extra-strong materials in the form of a jacket are used to reinforce the walls of the optical fiber to increase its strength. An optical fiber cable thus prepared is safe from external environmental conditions, viz. while under the sea or underground. Figure 23.1. is a schematic diagram of an optical fiber.

♣ The Central Glass and Ceramic Research Institute (CGCRI) in Kolkata has excellent facilities for optical fiber fabrication.

♣ Optical fibers are also laid on the sea bed for intercontinental high-speed optical communication.

### 23.1.2 Acceptance angle and numerical aperture

Light propagates in an optical fiber due to the effect of total internal reflection (TIR) (see chapter 3). There are two kinds of fiber: (a) graded-refractive-index fiber, (b) step-index fiber. As their names suggest, in graded-index fiber, the refractive index profile of the cladding changes gradually, while in step-index fiber, this change takes the form of a step. Typical refractive index values for the cladding and core of a silica fiber are 1.1 and 1.45, respectively (figure 23.2).

As shown in the figure, the critical angle ($\theta_c$) for TIR depends on the acceptance angle ($\theta_{in}$) of the fiber.

Using the parameters shown in figure 23.3, it is easy to obtain the expression for the numerical aperture (NA) of the optical fiber (♠ see question 2). It is left as an

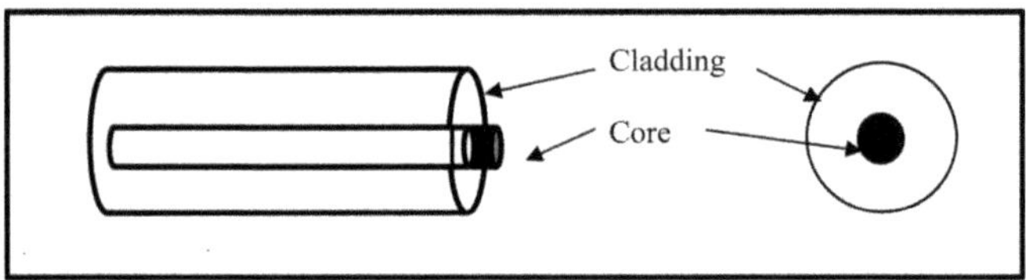


**Figure 23.1.** Side view (left) and a cross-sectional view of an optical fiber (right). The fiber core is surrounded by cladding that has a smaller refractive index.

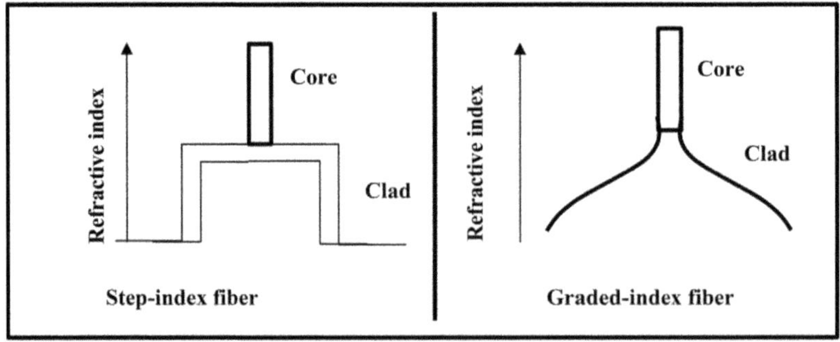


**Figure 23.2.** Refractive index profiles along the cross-section for the core and cladding of step-index fiber (left) and graded-index fiber (right).

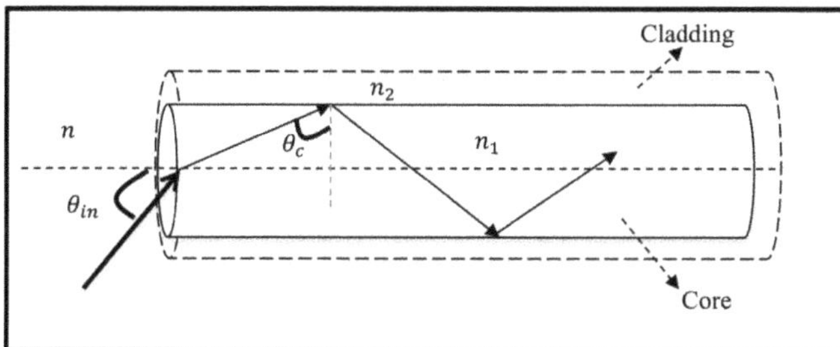


**Figure 23.3.** The angle of acceptance ($2\theta_{in}$) for an optical fiber. Here, $n_1$ and $n_2$ ($<n_1$) are the refractive indices of the core and the cladding, respectively. The incident beam is considered to be propagating in a refractive index of $n$ ($= 1$ for air) before entering the fiber. The angle $\theta_{in}$ is made bigger in this schematic for clarity (see exercise 23.2 for basic estimates).

exercise for the reader to show that the refractive index ($n$) of the medium and the incident angle ($\theta_{in}$) are related to the NA as follows:

$$\mathrm{NA} = n\ \sin\ \theta_{in} \equiv \sqrt{(n_1^2 - n_2^2)}.$$

**Exercise 23.1** . A fish tank is completely filled with water ($n = 4/3$). (a) Find the angle of total internal reflection for the fish at a depth of $h$ inside the tank to be able to see through an area of radius of $r$. (b). If the tank is covered with glass that has a refractive index of 1.5 and the water touches the surface of the glass, what is the critical angle?

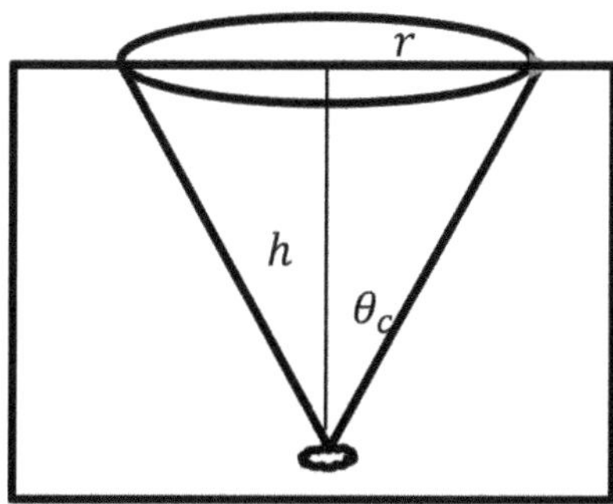


**Solution:** With reference to the diagram, we need to estimate the critical angle ($\theta_c$):
(a) Generally, the angles for the refraction of a light beam from water ($\theta_w$) to air ($\theta_a$) are given by

$$\frac{\sin \theta_w}{\sin \theta_a} = \frac{1}{1.333}.$$

Here, $\sin \theta_a$ has to be one for TIR and $\theta_w$ is $\theta_c$. Therefore,

$\sin \theta_c = \dfrac{1}{n} = \dfrac{1}{1.333}$. This implies that $\theta_c = 49°$.

For a certain height $h$, for the same $\theta_c$, the required radius ($r$) is given by $\tan \theta_c = \dfrac{r}{h}$:

$$r = h \times \tan \theta_c = h\frac{\sin \theta_c}{\cos \theta_c} = h\frac{\sin \theta_c}{\sqrt{(1 - \sin^2\theta_c)}} = \frac{h}{\sqrt{n^2 - 1}}.$$

(b) For a water–glass interface, the respective refractive indices are $n_2 = 1.333$ and $n_1 = 1.5$. Hence

$$\theta_c = \sin^{-1}\left(\left(\frac{1.333}{1.5}\right)\right) = 62.7°$$

Therefore, the radius $r = h \times \tan \theta_c = 1.94 \times h$.

**Exercise 23.2.** The NA of an optical system gives us an idea of its acceptance angle. Generally, microscope objective lenses are characterized by their NA values as well as by their resolving power. Find the acceptance angles for the following two cases: (i) a microscope objective that has an NA of 0.15; (ii) an optical fiber that has a core refractive index of 1.47 and a cladding refractive index of 1.465.

**Solution:** (i) For the microscope objective, the NA $= n \sin \alpha$. For air, the semi-acceptance angle is given by $\sin^{-1}(\text{NA})$. Therefore, the acceptance angle $= \sin^{-1} 0.15 \equiv 9°$.

(ii) For the optical fiber, the NA $= \sqrt{(n_1^2 - n_2^2)}$. Here, $n_1 = 1.47$ and $n_2 = 1.465$. Therefore, the NA $= 0.121$ and the acceptance angle is $\sim 14°$.

## 23.2 Gain media for fiber lasers

Depending on the required laser wavelength, the core of the fiber is doped with ions of rare-earth elements, as shown in table 23.1. Trivalent erbium ions ($Er^{+3}$) can be doped into silica fiber as well as into crystals. Trivalent erbium ions are the most common active media and have a large gain near 1550 nm. The gain medium is not only used for lasing in fiber lasers but also for amplifying optical signals in long-haul optical communication.

**Table 23.1.** Some of the gain media used in fiber lasers.

| Gain medium | Amplifier wavelength |
| --- | --- |
| Erbium (Er) | 1550 |
| Neodymium (Nd) | 1064 |
| Ytterbium (Yb) | 1050 |

♣ A related concept in optical fiber technology is the erbium-doped fiber amplifier (EDFA), in which doped fibers are used for the amplification of signals over several tens of kilometers.

Figure 23.4 schematically presents the spectral range of an $Er^{+3}$-doped gain medium and its energy-level diagram. $Er^{+3}$ absorbs over a large spectral range (from the UV–visible region to ~1480 nm). Therefore, any wavelength below this range can be used for optical pumping. Commonly available pump wavelengths in this case are those of either diode lasers (980 nm) or argon-ion lasers.

## 23.3 Chromatic dispersion and nonlinear effects

As also discussed in chapter 18 (♠ see section 18.6), the effect of dispersion affects optical pulse propagation in a medium. The same is true of optical fiber communication, in which the signal is distorted in the time domain as well as in the frequency domain. Figure 23.5 gives a schematic of this effect, which is relevant in the propagation of optical pulses at a certain repetition rate. The main reason for this dispersion (due to linear or nonlinear optical effects) is the fact that the light speeds are different for different colors in a given medium. While linear dispersion (chromatic dispersion) follows Sellmeier's formula (equation (1.16)) based on the refractive index, the nonlinear

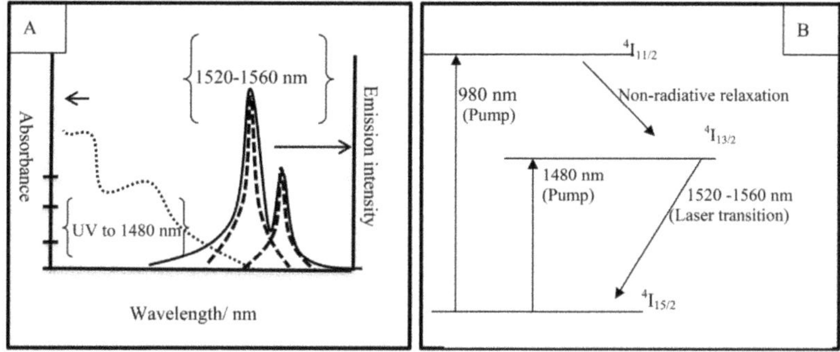



**Figure 23.4.** (A). Schematic of the absorption (dots) and spontaneous emission (continuous line) spectra of $Er^{+3}$-doped fiber. The gain spectra of $Er^{+3}$ are shown by dashed lines. The corresponding wavelength scales are given in brackets. Panel (B) shows the energy-level diagram of $Er^{+3}$. Pumping can be performed within the absorption band using wavelengths shorter than 1480 nm, as indicated.

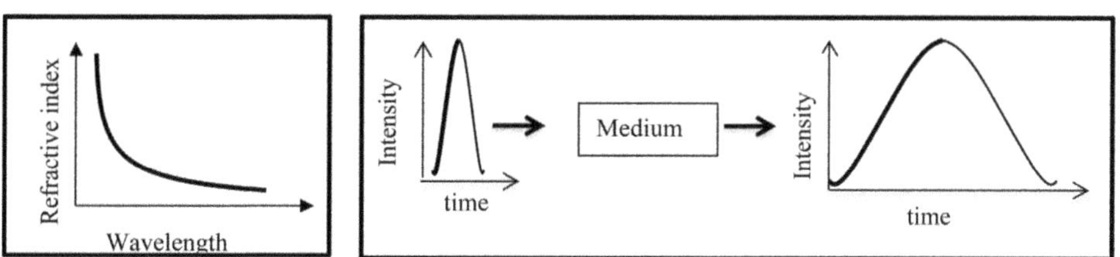



**Figure 23.5.** Variation of the refractive index for silica glass (left-hand panel). Effect of chromatic dispersion on a light pulse due to a medium (right-hand panel).

effects are due to the optical Kerr effect ($\spadesuit$ see chapter 16). It should be noted that the cross-section of the fiber core is very small (of the order of 10 $\mu m^2$). Therefore, even a miniscule laser power coupled with the core can invoke optical nonlinearity.

$\clubsuit$ Optical fiber propagation takes advantage of the interplay between the distortions of the optical pulse due to dispersion and its compensation due to nonlinear optical effects. This results in a soliton-like pulse that does not dissipate at longer propagation lengths. However, signal amplifiers may be needed every several tens of kilometers.

Apart from chromatic dispersion, there are two more types of dispersion: (i) material dispersion and (ii) polarization dispersion, as described in the following sections.

### 23.3.1 Material dispersion

If a light pulse containing a group of wavelengths with different speeds travels in a fiber, the different wavelengths reach the detector at different times. This results in temporal broadening of the light pulse. This kind of chromatic dispersion is known as material dispersion, as it depends on the refractive index of the medium. The dispersion coefficient ($D$) of the material used in a single-mode fiber is defined as $D = \frac{\Delta\tau}{L\Delta\lambda}$, where $\Delta\tau$ is the broadening of the optical pulse duration while passing through material of length $L$. $\Delta\lambda$ is the spectral width of the optical pulse. The unit of $D$ is generally ps (nm km)$^{-1}$.

### 23.3.2 Mode dispersion

Chromatic dispersion is negligible when light travels smaller distances at low-frequency bandwidths (also known as transmission speeds). For multimode fibers, the various modes give rise to small dispersion effects. The introduction of the single-mode fiber does not cause mode dispersion over large distances, but at higher transmission speeds, the chromatic dispersion introduces distortions into the signal.

Ideal laser light does not undergo any *material dispersion* effects. However, *mode dispersion* definitely takes place, as shown in figure 23.6. Light modes traveling at different incident angles ($\theta_{in}$) reach the detector at different times ($\delta\tau_i$). This broadening of the light pulse is said to arise from the *transit-time* effect. For an optical pulse that has an initial time duration $\Delta\tau$, if the broadening due to

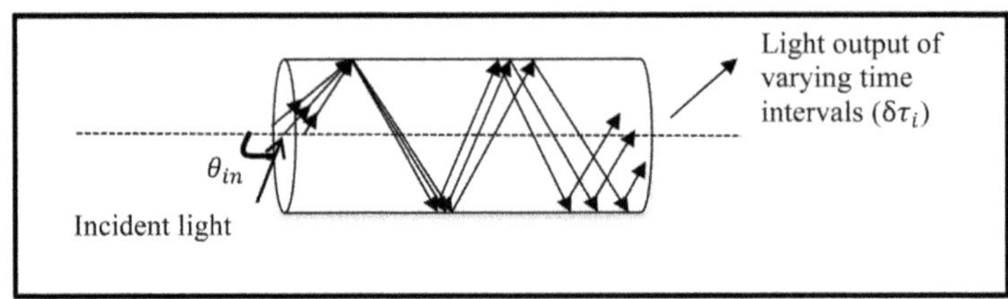


**Figure 23.6.** Schematic of incident light beams at different incident angles ($\theta_{in}$) to the fiber input arriving at the detector with corresponding varying time intervals ($\delta\tau_i$).

intermodal dispersion is given by $\Delta\tau_{id}$, the net broadening of the pulse duration $(\Delta\tau_{net})$ is given by $\Delta\tau_{net} = \sqrt{(\Delta\tau + \Delta\tau_{id})^2}$.

### 23.3.3 Polarization dispersion

Polarization dispersion takes place due to a change of thickness or the shape of the core of a fiber along its length. This happens either during manufacture or transport. This may result in changes to the optic axis at this point, which could give rise to local birefringence. Birefringence causes two components (or modes) to travel at different velocities, resulting in dispersion similar to mode dispersion.

## 23.4 Optical nonlinearity

The propagation constant $(\beta)$, is commonly known as the wave vector $(k)$. It is a function of frequency $(\omega)$ and in a medium that has a refractive index of $n(\omega)$, it is given by

$$\beta(\omega) = \frac{2\pi}{\lambda} \equiv \frac{n(\omega)\omega}{c}. \tag{23.1}$$

For small changes in $n(\omega)$ about the central frequency $(\omega_0)$ of the laser, we can express the propagation constant using a Taylor series expression as follows:

$$\beta(\omega) = \beta_0 + \frac{\beta_1(\omega - \omega_0)}{1!} + \frac{\beta_2(\omega - \omega_0)^2}{2!} + \frac{\beta_3(\omega - \omega_0)^3}{3!} + \dots . \tag{23.2}$$

Here, $\beta_1 = \left(\frac{\partial\beta}{\partial\omega}\right)_{\omega_0}$ and $\beta_2 = \left(\frac{\partial^2\beta}{\partial\omega^2}\right)_{\omega_0}$ are expressed in units of ps km$^{-1}$ and ps$^2$ km$^{-1}$.

♣ The Taylor series: for small changes in $h$, the function $f(x)$ can be written as
$f(x + h) = f(h) + \frac{f'(h)}{1!}h + \frac{f''(h)}{2!}h^2 + \frac{f'''(h)}{3!}h^3 + \dots\dots$ Here, primes denote the differentiation w.r.t. parameter $h$.

Using equation (23.1), the expression for $\beta_1$ can be written as follows:

$$\beta_1 = \frac{1}{c}\left(n(\omega) + \omega\frac{\partial n}{\partial\omega}\right). \tag{23.3}$$

The quantity in brackets is known as the *group index* $(n_g)$. Therefore, $\beta_1$ can also be written as

$$\beta_1 = \frac{n_g}{c} \equiv \frac{1}{v_g}.$$

Here, $v_g$ is the *group velocity*. Similarly, we can express the higher term $\beta_2$ as the *group velocity dispersion* (GVD) parameter, $\beta_2 = \frac{\partial\beta_1}{\partial\omega}$. Using equation (23.3), we get

$$\beta_2 = \frac{1}{c}\left(2\frac{\partial n}{\partial\omega} + \omega\frac{\partial^2 n}{\partial\omega^2}\right).$$

This can also be expressed in terms of $v_g$ as

$$\beta_2 = \frac{\partial\left(1/v_g\right)}{\partial\omega} \equiv \frac{1}{v_g^2}\frac{\partial v_g}{\partial\omega}. \tag{23.4}$$

Plots of $n$, $n_g$, $D$, and $\beta_2$ as a function of wavelength are shown in figure 23.7. It can be seen that while $n$ falls with the wavelength, $n_g$ proceeds to an asymptotic value at 1.6 μm.

One of the important observations obtained from figure 23.7 (right panel) is that the value of $D$ increases while that of $\beta_2$ decreases with wavelength. It should be noted that for silica, the values of both of these parameters become zero at 1300 nm. On further increasing the wavelength, while the value of the former continues to increase on the positive scale, $\beta_2$ becomes negative. For wavelengths above 1300 nm, negative group velocity dispersion takes effect. The behavior of these two parameters is useful in optical fiber communication in various bands, as shown in table 23.2. Just for interest, figure 23.7 also shows a plot of a walk of parameter $(d_{12})$ as the difference between the peak wavelengths of two pulses as a function of the wavelength.

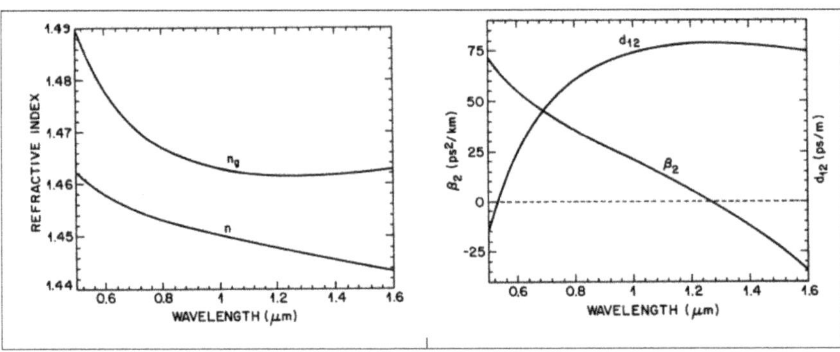

**Figure 23.7.** Variation of the refractive index ($n$) and group index ($n_g$) as a function of wavelength (left panel). Variation of the GVD parameter ($\beta_2$) and dispersion coefficient ($D$) as a function of wavelength (right panel). Note that $\beta_2 = 0$ at 1300 nm. (Reprinted from [15], copyright (1989), with permission from Elsevier.)

**Table 23.2.** Popular nomenclature for wavelength ranges in fiber-optic communication.

| Band | Range | Description |
| --- | --- | --- |
| O-band | 1260–1360nm | Original |
| E-band | 13600–1460 nm | Extended |
| S-band | 1460–1530 nm | Short wavelength |
| C-band | 1530–1565 nm | Conventional |
| L-band | 1565–1625 nm | Long wavelength |
| U-band | 1625–1675 nm | Ultra-long wavelength |

♣ Incidentally, negative dispersion can also be achieved by a prism or a grating compressor (♠ see section 18.7).

**Exercise 23.3.** From the modeling of passive FP resonator cavities, we know that the free spectral range (FSR) is given by the ratio of the speed of light to the twice the distance between the mirrors. In a similar optical system, if $n_g$ is the frequency-dependent refractive index, what is the frequency spacing ($\Delta\nu$) in terms of the group velocity ($v_g$)?

**Solution:** The frequency ($\nu$) is expressed in terms of wavelength ($\lambda$) by $\nu = \frac{c}{\lambda}$. For small changes in the parameters, this implies that

$$d\nu = \left| \frac{-c}{\lambda^2} \right| d\lambda \tag{i}$$

also

$$d\nu = \frac{v}{2L} = \frac{c}{n_2 L} \tag{ii}$$

Therefore, from equations (i) and (ii), $\left| \frac{c}{\lambda^2} \right| d\lambda = \frac{c}{n2L}$, so that $d\lambda = \frac{\lambda^2}{2nL}$. For the frequency-dependent refractive index ($n_g(\nu)$), we can write $d\lambda = \frac{\lambda^2}{2n_g L}$. The propagation constant $\beta$ is given by $\beta = \frac{2\pi}{\lambda} = \frac{n(\nu)\nu}{c}$. For ($n_g(\nu)$), it is written as $\beta_1 = \frac{n_g(\nu)\nu}{c}$.

Therefore, based on equation (23.3), $\beta_1 = \frac{1}{c}\left(n(\nu) + \nu\frac{\partial n}{\partial \nu}\right)$. In this expression, the quantity in the brackets is known as the *group index* ($n_g$). $\beta_1$ can also be written as $\beta_1 = \frac{n_g}{c} = \frac{1}{v_g}$. Here, $v_g$ is known as the *group velocity*. Therefore, from (ii), the frequency spacing in term of $n_g$ is given by $d\nu_g = \frac{c}{n_g L} = \frac{v_g}{2L}$.

## 23.5 Fiber amplifiers and lasers

Consider the optical fiber to be a flexible long rod that has the thickness of a human hair. These flexible optical fibers can be closely coupled to light sources and detectors in a compact fashion. This is a developed area of science and technology due to the industrial applications of optical communication in the modern era. It is also due to the fact that bulky lasers may suffer from misalignment in transit and require vibration isolation for their performance. On the other hand, due to the spliced connections between the pump, gain fiber, and output coupler, fiber-based optical setups can easily be transported without any misalignment. Given the advantage of the on-site performance of fiber optics along with the capability of harmonic generation in nonlinear optical crystals, this technology has immense industrial potential. Because of its high level of application in optical communication, the hot topics also include fiber amplifiers. Fiber lasers are also used as seed lasers in modern femtosecond lasers due to their long life and the stability of their output power.

Fiber amplifiers are used in long-haul optical communication repeaters to amplify the signal. A schematic of an optical fiber network is shown in figure 23.8. A portion of the glass fiber is doped with an amplifier material such as $Er^{+3}$ (see table 23.1). The pump light can be coupled through one side of the long fiber, a design known as *end-pumped configuration*. It can be coupled through a *pump coupler* without disturbing the input signal beam. The amplified signal can be suitably coupled on the other side of the glass fiber.

As mentioned before, due to the refractive index distribution and TIR, the light propagates through the core region with no sideways losses, even when the fiber is bent. One of the typical geometries of fiber laser is the F8L, which is an all-fiber laser and a good example with which to discuss the various aspects of fiber lasers in the following sections.

## 23.6 Figure-of-eight laser

The basic structure of an F8L is given in figure 23.9. It looks similar to the Arabic numeral eight, hence the name. An F8L consists of a *nonlinear amplifying loop mirror* (NALM), an *optical isolator* for unidirectional propagation of the laser beam, and an output coupler for the laser output. The NALM, which contains an $Er^+$-doped fiber (EDF), acts as a gain medium as well as a passive mode locker in F8Ls. The functions of these elements are discussed in the following sections.

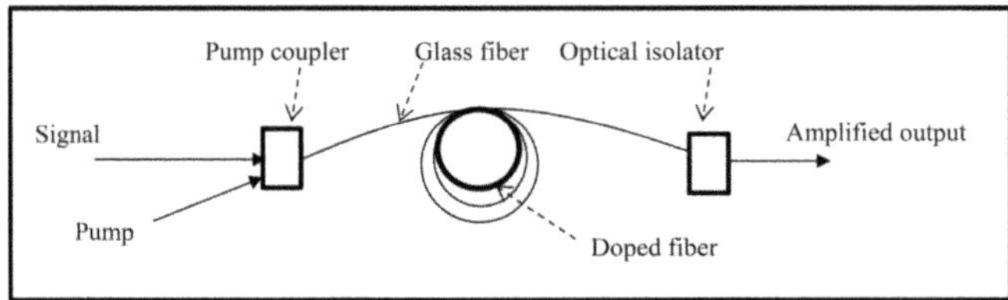


**Figure 23.8.** Schematic diagram of a fiber amplifier.

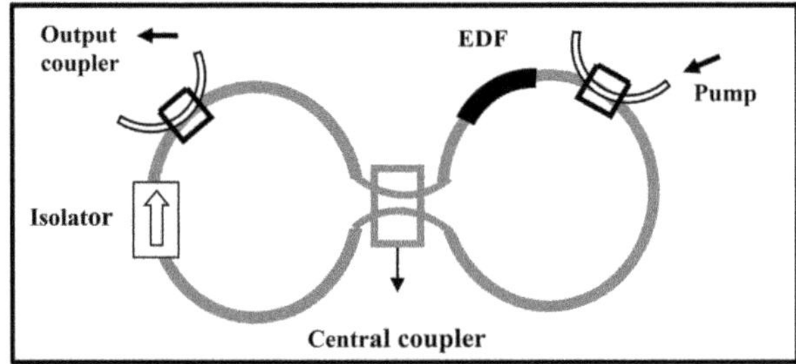


**Figure 23.9.** The F8L. 'EDF' denotes an $Er^+$-doped fiber.

### 23.6.1 Optical isolator

Figure 23.10 shows the optical isolator. It consists of two polarizers (P1 and P2) and a Faraday rotator (FR) (♠ see section 3.7.2) placed in between them, as shown in the figure. The first polarizer is set with its polarization axis vertical ($\theta = 0°$) and the other has its polarization axis set at an angle of 45° to the first. When light is incident on the isolator from left to right, as shown in the figure, P1 makes it vertically polarized and the FR rotates it by 45°, which is parallel to the direction of P2. Hence, the beam passes through the isolator. On the other hand, when the beam travels from right to left, with P2 and the FR in its path, its polarization is rotated by 90° before it reaches P1. As a result, it is blocked by P1.

### 23.6.2 Nonlinear amplifying loop mirror

The NALM is a switching device which operates on a nonlinear phase induced by self-phase modulation (SPM). Figure 23.11 shows a schematic diagram of an NALM. It consists of an *X-fiber coupler* with a power coupling ratio of $C_r: 1 - C_r$. The coupler has two input ports (1 and 2) and two output ports (3 and 4). The two output ports are joined together. A single input is sent to port 1 and split into two counter-propagating fields. These fields are recombined at the coupler. The optical path lengths of the two beams are precisely the same, as they travel the

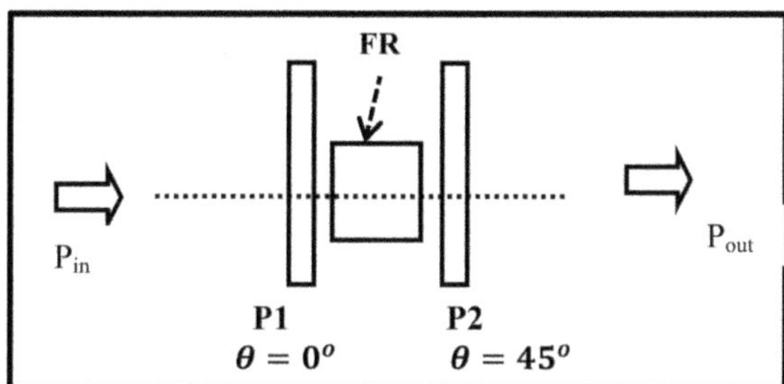


**Figure 23.10.** Schematic diagram of the optical isolator: P1 and P2 are polarizers and FR is a Faraday rotator.

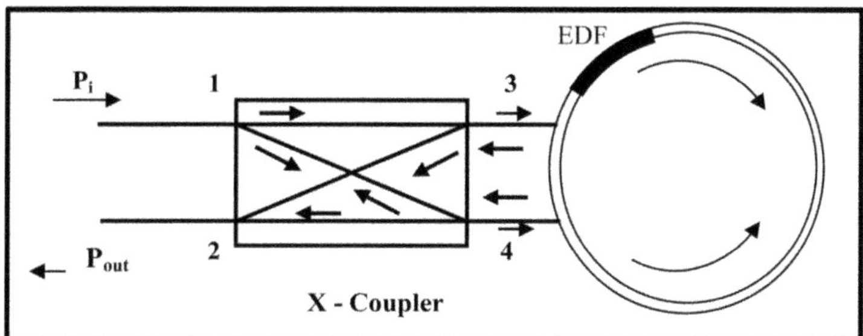


**Figure 23.11.** The NALM: 1 and 2 are input ports and 3 and 4 are output ports. 'EDF' denotes a section of $Er^+$-doped fiber (shown as a dark patch).

same path but in opposite directions. Depending upon the power level, the counter-propagating pulses enter the coupler and leave the coupler via port 1 or port 2.

In the NALM, an EDF is placed asymmetrically in the loop with respect to the X-coupler. The phase shifts experienced by the clockwise ($\delta\phi_1$) and counterclockwise components ($\delta\phi_2$) around the loop are given by

$$\delta\phi_1 = \frac{\pi}{\lambda} n_2 g \, I_{s0} L \tag{23.5}$$

$$\delta\phi_2 = \frac{\pi}{\lambda} n_2 \, I_{s0} L, \tag{23.6}$$

where $I_{s0}$ is the signal intensity at port 3 of the central coupler, $L$ is the length of the fiber loop and $g$ is the gain of the amplifier. Pulse switching takes place when the relative phase difference between the counter-propagating components is $\pi$.

### 23.6.3 Mode-locked operation in the F8L

With the benefit of the previous two sections, the working principle of the F8L is easy to understand, as follows. It is interesting to note that the phenomenon of mode locking is inherent in the F8L, without the use of any external modulator. Port 1 of the NALM is coupled to the output coupler and port 2 is connected to port 1 through an optical isolator (figure 23.10). As explained above, the optical isolator only transmits light in one direction, i.e. from port 2 to port 1. The amplified spontaneous emission (ASE) of the gain medium (the EDF), acts as a seed light in the laser. The seed is amplified by making use of the gain in the EDF itself as an amplifier until the nonlinear phase difference between the two counter-propagating components satisfies the transmission condition of the NALM. In this way, the NALM operates in pulsed mode and produces mode-locked output pulses. The negative GVD parameter of glass at 1550 nm (♠ see section 23.4) helps to compensate for the distortions caused by optical nonlinearity. Due to this interplay between the nonlinearity and dispersion, F8L output is mode locked and has soliton-like pulses (♠ see section 23.9.4).

## 23.7 High-power fiber lasers

High-power fiber lasers are designed with double cladding as shown in figure 23.12. The core of the fiber is surrounded by the first cladding, which has a lower refractive index as usual. The second nonsymmetrical cladding, also known as the *pump cladding*, surrounds the core and the first cladding (panels (b) and (c)). The second cladding is made of a multimode waveguide of high NA and has a larger area as shown in panel (c). The pump light propagates along the fiber in the second cladding and couples to the active medium longitudinally.

Efficient coupling of the pump light takes place due to the asymmetric design of the pump cladding, which breaks the symmetry of the structure. The refractive index of the core is chosen to guide a single, fundamental mode. The design of the cladding also includes polarization control. Polarization control is obtained using control

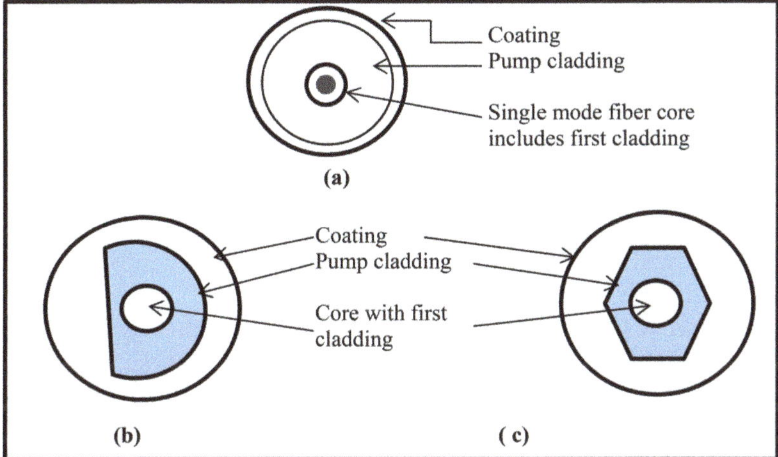



**Figure 23.12.** (a). Double cladding for high-power lasers and symmetric *pump cladding* for the maximum pump power couplings (b) and (c).

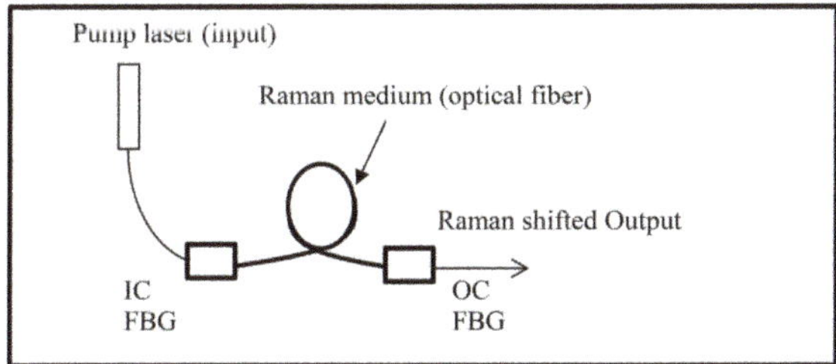



**Figure 23.13.** Schematic of a Raman fiber laser. The input and output FBG couplers used to select the required wavelengths are indicated by IC and OC, respectively.

over birefringence of the order of $10^3$ between the components of the polarizer. It should be noted that photonic crystal fibers (PCFs) are designed in the same way for both power scaling and supercontinuum generation (♣ see section 21.5.1).

## 23.8 Raman fiber laser

As a result of the Raman effect, Stokes and anti-Stokes frequencies are also generated in fiber lasers. For example, in figure 23.13 the pump beam is coupled to a fiber with the help of a grating element attached to the fiber known as fiber Bragg grating (FBG). Another FBG selects the output. Depending on the shift of the Stokes' line and the Raman efficiency of the medium, the laser output can be controlled. Multistage, cascaded Raman lasers can produce high output powers.

♣When a piece of fiber that has a core containing a photosensitive material is illuminated by the interference pattern of an intense laser, this results in the inscription of a permanent grating in the core, which is known as an FBG. FBG structures that have a desired grating spacing can also be made by directly stepwise micromachining the fibers using lasers. FBGs have immense industrial applications as Bragg reflectors. FBGs work on the principle of Bragg's law. They allow a particular frequency in the forward direction but block other frequencies depending on the grating spacing.

## 23.9 Optical fiber communication

The principle of optical fiber communication (OFC) is shown in Figure 23.14. The microphone of a home telephone system converts the sound waves from the human voice into an electrical signal. The electronic signal is converted into the corresponding optical signal with the help of an electrical-to-optical (EO) converter; the resulting signal can now can be transported by the optical fiber. At the receiver end, the optical signal is converted back to voice via an optical-to-electrical (OE) signal converter.

Distortions of the light pulse occur while it propagates in the fiber. These distortions can occur because of light propagation losses or linear or nonlinear dispersion. In the following, some of the important parameters necessary for OFC are outlined.

### 23.9.1 Losses due to fiber absorption

Consider the glass of a window that has a thickness of a few mm. Depending on the quality of the glass, part of the incident light will be either reflected, scattered, or transmitted. Glass fibers in OFC run to several kilometers before an amplification stage (known as a repeater) is installed. Let us assume that 10% of the power of the incident light is transmitted for a 10 m journey, as shown in figure 23.15).

Following the Beer–Lamberts law, in propagating over this length, the output power ($P_{out}$) is written in terms of the input power ($P_{in}$) using a net loss parameter $\alpha$, as follows:

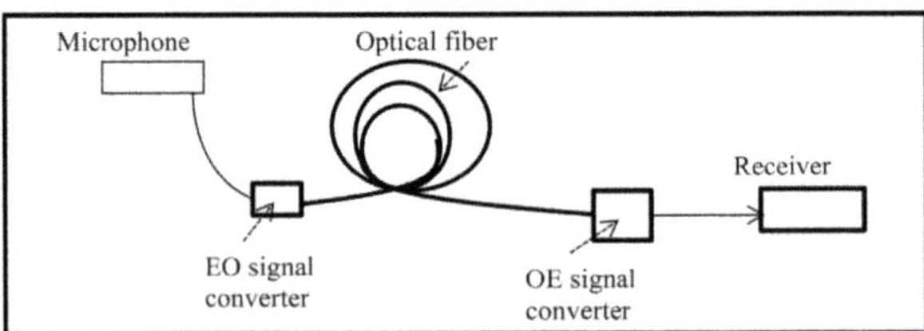


**Figure 23.14.** Principle of optical fiber communication in the house-to-house telephone system. EO: electrical-to-optical, OE: optical-to-electrical.

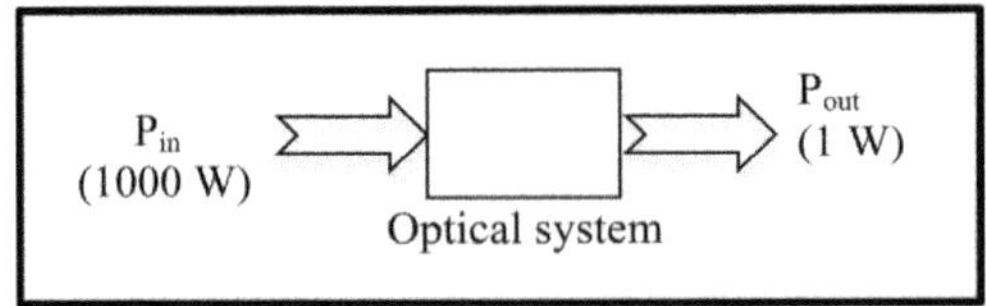


**Figure 23.15.** Schematic of a loss of input power ($P_{in}$) in an optical component that gives 10% output power ($P_{out}$).

**Table 23.3.** Loss per unit length ($\alpha$ in dB) for silica at 1300 nm for typical values of $P_{out}/P_{in}$.

| $P_{out}$ (mW) | $P_{out}/P_{in}$ | $\alpha$ ( dB) | $P_{out}/P_{in}$ | $\alpha$ ( dB) |
| --- | --- | --- | --- | --- |
| 1 | 0.001 | 30.0 | 0.8 | 1.0 |
| 100 | 0.1 | 10.0 | 0.9 | 0.5 |
| 500 | 0.5 | 3.0 | 0.99 | 0.04 |

$$P_{out} = P_{in}e^{-\alpha L} \text{ or } \frac{P_{out}}{P_{in}} = e^{-\alpha L}$$

We define the loss in dB (per unit length) as

$$\alpha = -10\log\left(\frac{P_{out}}{P_{in}}\right). \tag{23.8}$$

For $\frac{P_{out}}{P_{in}} = 10^{-3}$, we have a loss of 30 dB. Following technological developments by the glass industry, modern glass at 1300 nm has a loss of about 0.4 dB or even a factor of 10 less. Therefore, it is usual to obtain an output of the order of 99% of the input signal, as indicated in the following exercise.

**Exercise 23.4** . Find the loss per unit length ($\alpha$) of silica at 1300 nm for an input power of 1 W when the output power is 1 mW, 100 mW, and 500 mW.

**Solution:** Using equation (23.8), the calculated values of $\alpha$ for various output powers are given in table 23.3.

### 23.9.2 Fiber scattering losses

The loss profile of silica is indicated in figure 25.16. It is known that due to fluctuations in local parameters during the manufacturing process, the fiber produced can have variations in scattering losses. In the lower (UV–visible) wavelength region, the losses are dominated by Rayleigh scattering, which is inversely proportional to the fourth power of the wavelength ($\propto 1/\lambda^4$). In addition, due to water remaining in the silica, the -OH vibrations absorb in certain regions of the EM spectrum as indicated in the figure. From the point of view of optical communication, there are two dips in the loss profile of silica at ~ 1300 nm and ~1500 nm. At longer wavelengths (indicated in the figure as IR absorption), there are losses due to phonons or lattice vibrations.

### 23.9.3 Gain medium

We know that the gain medium is the key to obtaining laser lines as well as for amplifying the signal for long-haul communication. As discussed in section 23.2,

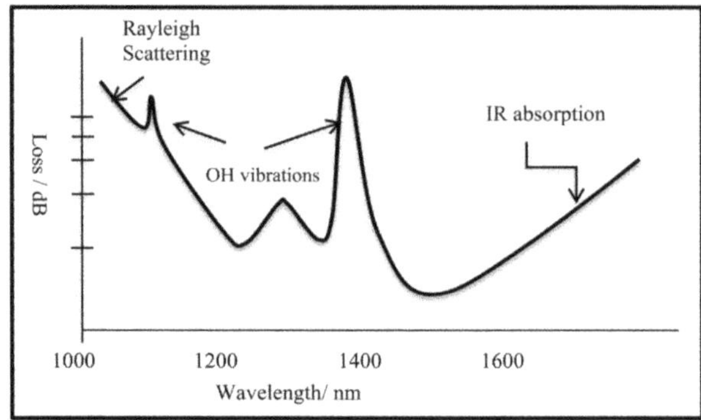



**Figure 23.16.** Schematic of the loss profile of silica fiber showing various contributions (after [15]).

rare-earth ions have a large gain near 1550 nm. It should be noted that losses are minimized at these wavelengths, as shown in figure 23.16.

### 23.9.4 Soliton formation

As outlined in the working principle of the F8L discussed above, the interplay between nonlinearity (equation (23.2) and dispersion (equation (23.4)) in the fiber results in dispersion-free pulse propagation. This happens mainly due to the negative dispersion of the fiber at 1550 nm (♠ see figure 23.7) and the inherent property of optical nonlinearity, even for weak light propagation, in the tiny core of the fiber. By choosing a suitable length of the fiber for compensation of the negative dispersion, soliton-like pulses can be obtained that do not dissipate for a considerable length. If $L_D$ and $L_{NL}$ respectively, are the dispersion and the nonlinear lengths, the parameter $N_s$ is given by

$$N_s^2 = L_D/L_{NL} \tag{23.9}$$

The integer value of the parameter $N_s$ is related to the soliton order. This parameter governs the relative importance of the SPM and GVD effects on pulse evolution along the fiber. For extreme values of $N_s$, either dispersion or the SPM dominates light-pulse propagation. However, for values of $N_s \sim 1$, both SPM and GVD play an equally important role during pulse evolution.

Usually, SPM adds new frequency components which are red shifted near the leading edge and blue shifted near the trailing edge of the pulse (♠ see chapter 16). Since the red components travel faster than the blue components of the optical pulse in the normal dispersion regime, SPM leads to an enhanced rate of pulse broadening compared to that expected from GVD alone. This in turn affects spectral broadening, as the SPM-induced phase shift ($\phi_{NL}$) becomes smaller than the shift that would occur if the pulse shape were to remain unchanged.

In contrast, for a pulse propagating in the anomalous dispersion region of the fiber, the situation is different. Due to SPM, the spectrum broadens and the induced chirp is positive. However, the dispersion-induced chirp is negative in the anomalous dispersion region. These two chirp contributions nearly cancel each other out along the center portion of the Gaussian pulse for $L_D = L_{NL}$. Therefore, during propagation, the pulse shape adjusts itself to make such cancellation as complete as possible. Thus, GVD and SPM cooperate with each other to maintain a chirp-free soliton-like pulse.

### 23.9.5 Higher-order dispersion and other effects

In chapter 12, we encountered beam-like solutions of the transverse Helmholtz equation (equation (12.15)). The propagation of the electric field ($E$) in the z-direction of an optical fiber is again given by a similar equation, as follows. By including the loss $\alpha(>0)$ and GVD parameter $\beta_2(>0)$ and subsequent replacement of the transverse Laplacian ($\nabla_T^2$) with the second derivative of $E$ with respect to time, the equation takes the following form

$$i\frac{\partial E(z,\,t)}{\partial z} = \frac{\beta_2}{2}\ \frac{\partial^2 E(z,\,t)}{\partial t_1^2} - \left(\gamma\,|\,E(z,\,t)\,|^2 - i\frac{\alpha}{2}\right)E(z,\,t). \qquad (23.10)$$

The first term on the right describes the dispersion ($\beta_2$), while the second term introduces the intensity-dependent nonlinearity parameter $\gamma$, and the third term represents the loss due to absorption processes ($\alpha$). This equation is similar in appearance to the time-dependent Schrödinger equation (2.4b) and is known as the nonlinear Schrödinger equation (NLSE).

Here, $\gamma = \left(\frac{2\pi}{\lambda}\right)n_2\,r_f$, where $n_2$ is the nonlinear refractive index and $r_f$ ($\cong 1/2$) is a reduction factor due to the variation of light intensity in the cross-section of the fiber. Although this equation has been successful in explaining several nonlinear effects, it needs to be modified, depending upon experimental conditions, to account for the presence of stimulated Raman scattering (SRS) and stimulated Brillouin scattering (SBS) by further adding the contributions of these terms.

## Questions and problems

1. What is the NA of an optical device? From figure 23.3, find the value of the NA of the fiber using $\sqrt{(n_1^2 - n_2^2)}$.
2. The refractive index of the core ($n_1$) of an optical fiber is larger than that of the cladding ($n_2$). For this fiber, a parameter ($\Delta$) is defined as $\Delta = \frac{n_1^2 - n_2^2}{2n_1^2}$. For close values of the two refractive indices (as in case of the fiber core and cladding), this can be approximated by $\Delta = \frac{n_1 - n_2}{n_1}$. (i) Verify that the NA is given by $\text{NA} = n_1\sqrt{(2\Delta)}$ . (ii) For the light beam incident on a silica fiber from air, for typical values of $\Delta = 0.01$, find the NA.
3. Explain the transition denoted by $4I_{11/2}$ in terms of spin, orbital, and total angular momentum (♣ hint: see chapter 2).
4. The output of a fiber is 0.9 W when the input is 1 W. Calculate the attenuation (loss) in dB.
5. The first solid-state laser was invented in 1960. Why is it never used in modern optical communication technology?

6. Show that the group index $n_g$ can be given in terms of wavelength $\lambda$ by $n_g = \frac{c}{v_g} = n(\lambda_0) - \lambda_0\frac{dn}{d\lambda}$. Here, $v_g$ is the group velocity $(\Delta\omega/\Delta k)$. Find the time taken by a pulse to travel a medium of length $L$.

7. Pulse broadening is given by $\Delta\tau = \frac{d\tau}{d\lambda_0}\Delta\lambda_0$. Show that the dispersion coefficient $(D_m)$ is given by $D_m = \frac{1}{c\lambda_0}\lambda_0^2\left(\frac{d^2n}{d\lambda_0^2}\right)$.

8. The chromatic dispersive power $(p)$ of a lens is defined in terms of its focal length $f_1$ and its change of the refractive index from red to blue wavelengths as follows: $p = \frac{\Delta n}{f}$. The chromatic dispersion can be removed by a combination of two thin convex lenses whose focal lengths are $f_1$ and $f_2$ separated by a distance $d$. If $p_1$ and $p_2$ are the dispersive powers of the lenses, prove that in order to obtain $p_1 = p_2$, we must have $d = \frac{f_1 + f_2}{2}$. It should be noted that in this case, the focal length $F$ of the lens combination is given by $\frac{1}{F} = \frac{1}{f_1} + \frac{1}{f_2} - \frac{d}{f_1 f_2}$.

9. Name some of the gain media used as fiber amplifiers in the wavelength region of 1550 nm. In this context, what are (i) full form and (ii) the working principle of the EDFA?

10. Optical fibers can be suitable for invoking optical nonlinearity, even at low coupled laser powers. Can you justify this statement? In an experiment, the time broadening due to material dispersion for an optical pulse is 20 ps km$^{-1}$. If, after traveling a distance of 1 km, the pulse broadens to 30 ps, what is the initial pulse duration?

11. What is the reason for the presence of asymmetric pump cladding around the core in high-power fiber lasers?

12. What is the use of a Bragg grating in Raman amplifiers?

13. The structures embedded in sensors, such as fiber Bragg gratings (FBGs) are known as 'smart structures'. How does a FBG help in identifying a new developing crack in an oil pipeline or a long bridge?

14. What is Rayleigh scattering and how does it affect the propagation of light in optical fibers?

15. Water can be present as an impurity in silica, causing losses. Which wavelength region is affected in optical fibers due to absorption by -OH ions?

16. By which mechanism do SPM and GVD compensate for each other in the F8L?

## Bibliography

[1] Snitzer E 1961 Optical maser action of Nd$^{+3}$ in a barium crown glass *Phys. Rev. Lett.* **7** 444–6

[2] Stone J and Burrus C A 1973 Neodymium-doped silica lasers in end-pumped fiber geometry *Appl. Phys. Lett.* **23** 388–9

[3] Townsend J E S *et al* 1987 Solution-doping technique for fabrication of rare-earth-doped optical fibers *Electron. Lett.* **23** 329–31

[4] Krupke W F 2000 Ytterbium solid-state lasers. The first decade *IEEE J. Sel. Top. Quantum Electron.* **6** 1287–96

[5] Mears R J, Reekie L, Poole S B and Payne D N 1986 Low-threshold tunable CW and Q-switched fiber laser operating at 1.55 μm *Electron. Lett.* **22** 159–60

[6] Okayasu M *et al* 1989 High-power 0.98μm GaInAs strained quantum well lasers for $Er^{3+}$-doped fiber amplifier *Electron. Lett.* **25** 1563–5

[7] Hasegawa A and Tappert F 1973 Transmission of stationary nonlinear optical pulses in dispersive dielectric fibers. I. Anomalous dispersion *Appl. Phys. Lett.* **23** 142–4

[8] Mollenauer L F, Stolen R H and Gordon J P 1980 Experimental observation of picosecond pulse narrowing and solitons in optical fibers *Phys. Rev. Lett.* **45** 1095–8

[9] Sindhu T G, Bisht P B, Rajesh R J and Satyanarayana M V 2000 Effect of higher order nonlinear dispersion on ultrashort pulse evolution in a fiber laser *Microw. Opt. Technol. Lett.* **28** 196–8

[10] Duling I N 1991 Subpicosecond all-fiber erbium laser *Electron. Lett.* **27** 544–5

[11] Richardson D J, Nilsson J and Clarkson W A 2010 High power fiber lasers: current status and future perspectives [invited] *J. Opt. Soc. Am.* B **27** B63–92

[12] Russell P 2003 Photonic crystal fibers *Science (80–).* **299** 358–62

[13] Hill K O and Meltz G 1997 Fiber Bragg grating technology fundamentals and overview *J. Light. Technol.* **15** 1263–76

[14] Grubb S G *et al* 1995 High-power 1.48 μM Cascaded Raman laser in germanosilicate fibers *Optical Amplifiers and Their Applications 18, SaA4* (Washington, D.C.: Optical Society of America)

[15] Agrawal G P 1989 *Nonlinear Fiber Optics* (New York: Academic)

**IOP** Publishing

# An Introduction to Photonics and Laser Physics with Applications

**Prem B Bisht**

# Chapter 24

# Coherent radiation obtained using special geometries

**Summary:** In this book we have discussed the principles and properties of some of the lasers and other coherent light sources given in the following table (table 24.1). Details of various lasers covering the broad region of the UV–visible–IR spectrum are summarized in the table. Coherent radiation produced using nonlinear optics techniques such as second-harmonic generation/third-harmonic generation (SHG/THG) and optical parametric amplifiers/optical parametric oscillators (OPAs/OPOs) contribute to this; hence, it is by no means an exhaustive list. In this chapter, the principles of a few special types of laser and an up-to-date laser technology, viz. the generation of laser pulses on the attosecond timescale ($10^{-18}$s) as well as laser power levels of petawatts ($10^{15}$W) are discussed. Among the techniques described here are (i) the mirrorless distributed feedback (DFB) laser, (ii) the free-electron laser (FEL), and (iii) high harmonic generation (HHG) and soft x-ray lasers.

**Learning objectives**
**After reading this chapter, the learner will be able to:**
List the gas, solid, and liquid lasers;
Identify the lasers used to obtain light in various regions of the electromagnetic (EM) spectrum;
Understand the principle of mirrorless lasers;
Identify light sources based on the acceleration of electrons;
Explain the principle of the FEL;
Describe the mechanism of HHG;
Confirm that attosecond pulses can be generated in the deep-UV/x-ray region;
Summarize the future outlook of lasers.

**Table 24.1.** The gain media, pumping mechanisms, and output wavelengths of some of the lasers and coherent radiations referred to in this book.

| S. No. | Source of coherent radiation | Gain-medium type and (pumping[*]) | Gain medium | Major wavelengths | Chapter no. |
|---|---|---|---|---|---|
| 1 | He–Ne laser | Gas (Elect) | He:Ne (8:1) | 626.8, 543.5 nm | 6 |
| 2 | Argon laser | Gas (Elect) | Argon gas | 514.5, 488 nm | 6 |
| 3 | Nitrogen laser | Gas (Elect) | Air/nitrogen gas | 336 nm | 7 |
| 4 | Excimer laser | Gas (Elect) | XeCl | 308 nm | |
| 5 | $CO_2$ laser | Gas (Elect) | $CO_2$ | 9.4, 10.6 $\mu m$ | 6 |
| 6 | Dye laser | Liquid (Opt) | Dyes | Near UV–VIS–near IR | 2 |
| 7 | Nd:YAG laser | Solid (Opt) | $Nd^{+3}$ in YAG | 1064 nm | 6 |
| 8 | Nd: Glass laser | Solid (Opt) | $Nd^{+3}$ in glass | 1050 nm | 8 |
| 9 | Ti:sapphire laser | Solid (Opt) | $Ti^{+3}$: $Al_2O_3$ | 750–1100 nm | 20 |
| 10 | Ruby laser | Solid (Elect) | $Cr^{+3}$: $Al_2O_3$ | 694.3 nm | |
| 11 | Diode laser | Solid (Elect) | Semiconductors | Various | 22 |
| 12 | Quantum cascade laser | Solid (Elect) | Quantum wells, electrons | Tunable in IR | 22 |
| 13 | OPO/OPA | Solid (Opt) | NLO crystals | Tunable in VIS/IR | 16 |
| 14 | Fiber laser | Solid (Opt) | $SiO_2$ (Glass) | 1550 nm region | 23 |
| 15 | EDFA | Solid (Opt) | $Er^{+3}$ in fibers | 1550 nm | 23 |
| 16 | DFB laser | Solid, liquid (Opt, elect) | Semiconductors/dyes | Tunable in VIS | 22 |
| 17 | FEL | Gas/(Elect) | Accelerated electrons | Tunable | 24 |
| 18 | Raman laser | Solid/liquid/gas (Opt) | Several materials | Several | 24 |
| 19 | EUV and x-ray lasers | Gas/solid (Opt) | Argon, other gases | Soft x-ray | 24 |
| 20 | VCSEL | Solid (Elect) | Semiconductors | Various | 22 |
| 21 | F8L | Solid | $Er^{+3}$ doped fiber | 1550 nm | 23 |

[*] Elect: electrical pumping, Opt: optical pumping, EDFA: erbium-doped fiber amplifier, EUV: extreme UV, VCSEL: vertical-cavity surface-emitting laser, F8L: figure-of-eight laser, YAG: yttrium aluminum garnet, NLO: nonlinear optics, VIS: visible

## 24.1 Mirrorless laser cavities

In conventional lasers, the two cavity mirrors provide the feedback necessary for laser oscillation. In distributed feedback (DFB) lasers, there are no mirrors. Here, the spatially periodic structure in the gain medium not only gives the necessary feedback for the oscillation but also provides the mechanism that restricts the oscillation to a narrow frequency range. Their compact design, mechanical stability, fine frequency tunability, and ultrafast pulse generation capability make DFB lasers versatile and indispensable. The principle of DFB is used in several applications in photonics.

### 24.1.1 Principle of DFB lasers

Laser oscillation can be generated by incorporating a periodic, grating-like structure in the gain medium. The grating-like structure is obtained either by direct interference methods or by inscribing a permanent grating at the required spacing using lithographic techniques. In direct interference methods, two replica beams of a pump laser (focused by a cylindrical lens to prepare light-sheets) can interfere in an absorbing medium as shown in figure 24.1. Due to interference and the absorption of light, a grating-like structure is formed in the medium (figure 24.2).

♣ In the lithographic technique, micro-fabricated patterns are transferred from a prefabricated photo-sensitive mask onto the substrate. Excimer lasers with pulses in the far UV region are used to create the minimum features.

The periodicity of the spatial modulation ($\Lambda$) or the fringe spacing is determined by the half angle ($\theta/2$) between the two interfering beams as follows:

$$\Lambda = \lambda_{\text{pump}}/2 \sin (\theta/2) \tag{24.1}$$

where $\lambda_{\text{pump}}$ is the wavelength of the pump laser. A light wave undergoes Bragg reflection at this periodical structure, which provides the feedback necessary for laser action (figure 24.2).

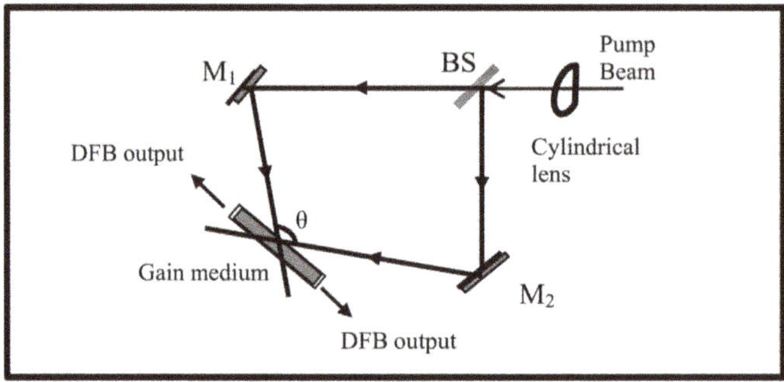


**Figure 24.1.** Typical setup for a distributed feedback laser. M1, M2: mirrors, B.S: beam splitter. The interference region at the gain medium is shown on an expanded scale in figure 24.2.

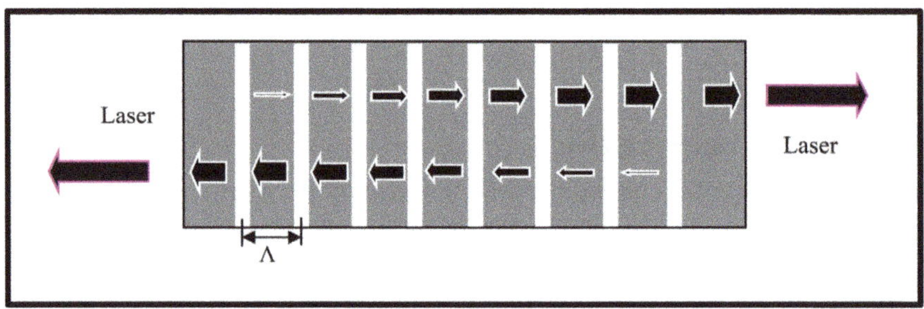


**Figure 24.2.** Increase of laser photons in a DFB laser. The interference fringes (i.e. the grating) are shown as white stripes. The thicknesses of the arrows indicate the amplification of light along the length of the gain medium.

In DFB lasers the feedback is provided by backward Bragg scattering from the spatially periodic modulation of the laser medium. Due to Bragg-angle dependence, the feedback is strongly frequency selective and the resonant mode for laser oscillation corresponds to the wavelength ($\lambda_{\mathrm{DFB}}$) that satisfies Bragg's law for various orders ($m$).

$$\lambda_{\mathrm{DFB}} = \frac{n\lambda_{\mathrm{pump}}}{m \sin{(\theta/2)}}$$

Here, $n$ is the refractive index of the gain medium. Finally, only photons with wavelengths that can propagate along the length of the gain medium are responsible for amplification (figure 24.2) . Hence, the equation for the *first order* can be rewritten as

$$\lambda_{\mathrm{DFB}} = 2n\Lambda \tag{24.2}$$

Equation (24.2) shows that the laser can be tuned by varying the refractive index or the intersection angle that decides the fringe spacing.

The stimulated emission in the distributed gain medium gives rise to two counter-running waves. The total field ($E$) in the medium is the sum of these waves (figure 24.3) with complex amplitudes $R(z)$ and $S(z)$ respectively, as follows:

$$E = R(z)\mathrm{e}^{-ikz/2} + S(z)\mathrm{e}^{ikz/2}. \tag{24.3}$$

where $k = 2\pi/\Lambda$. These two waves experience gains as they travel in the population-inverted region through the gain medium. The inversion builds up due to pump photons until the field increases sufficiently for gain saturation to become dominant. At this moment, the Bragg reflectivity decreases rapidly, allowing the field to escape as a beam with a narrow linewidth restricted according to equation (24.2).

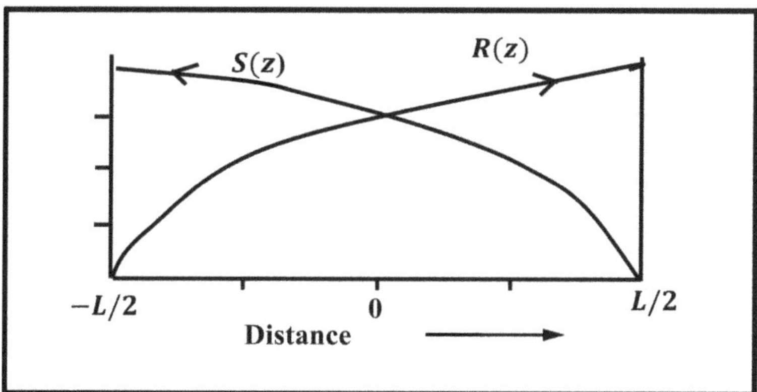


**Figure 24.3.** Spatial dependence of the amplitudes of the counter-propagating waves ($R(z)$ and $S(z)$) in a gain medium of length ($L$).

### 24.1.2 Ultrashort pulses and the tunability of DFB lasers

DFB lasers are used to generate short-duration pulses from longer ones. DFB lasers are also capable of fine-tuning laser frequencies of the order of 1 KHz or even less due to Bragg selectivity. DFB techniques have been applied to dye lasers and semiconductor lasers. If dye is used as a gain medium, the resulting laser is known as the distributed feedback dye laser (DFDL). Short pulses can be obtained on pumping with pulsed lasers due to the self-Q-switching effect. Self-Q-switching derives its name from the effect of saturable absorption as described in chapter 17. When the energy stored in the gain medium overcomes the losses and can then be transmitted in the form of an optical pulse, this is known as self-Q-switching. If this occurs in gain-coupled pulsed DFB lasers, the self-Q switching mechanism leads to transform-limited pulses that are shorter than the pump pulse by a factor of 50–100.

## 24.2 Coherent radiation based on acceleration of charge

An oscillating charge behaves like a dipole that radiates in the manner described in section 1.4.4. Here, we will describe two kinds of laser based on this principle, albeit in entirely different contexts. The first is the FEL, while the second is based on the phenomenon of HHG.

### 24.2.1 Free-electron lasers

As outlined in section 1.4.4, an accelerating charge near the speed of light gives rise to EM radiation. The power of this radiation is proportional to the square of the acceleration of the charge. As shown schematically in figure 24.4, in FEL, an array of alternating magnetic fields in the form of a grating structure is built to oscillate the charge. The strengths of the magnetic fields are of the order of 0.01–1 T. The electrons are sent into this magnetic field at relativistic speeds to oscillate in vacuum. In this configuration, electron beam energies of the order of gigaelectron volts are used.

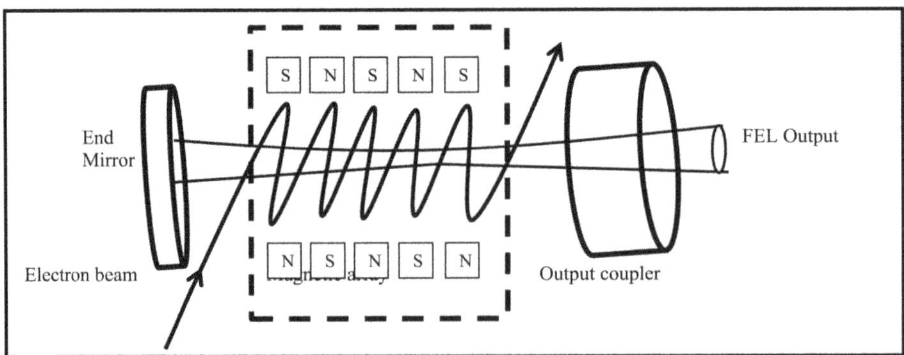


**Figure 24.4.** Schematic diagram of a free-electron laser (FEL). N and S are the magnetic poles. The electron beam enters the magnetic array and exits from the cavity as indicated by arrows.

The radiation predominantly propagates in the direction perpendicular to the dipole vector. As we know, in conventional lasers such as the He–Ne laser, the laser's output wavelength is decided by the discrete transition frequencies of the gain media. In contrast, in FELs, the lasing wavelength is decided by the kinetic energy of the oscillating electrons passing through an array of magnetic structures. Hence, the laser frequencies result from transitions between continuous states. As a result, the FEL is a coherent light source with wide wavelength tunability, especially in the far-IR region.

Depending on the length as well as the strength of the magnetic array, a large spectral range of wavelengths from nanometers (the x-ray region) to a few tens of millimeters (in the microwave range) can be obtained. The nature of the electron beam used in an FEL selects either pulsed or continuous-wave operation. The output powers of various FELs can be in the gigawatt range for pulsed operation. The source of electrons can be a storage ring or a linear accelerator similar to those used in synchrotron sources.

♣ In synchrotron sources the radiation obtained can be monochromatic but not coherent. Using FELs, due to fast interactions, the micro bunching of electrons, and Compton scattering (♠ see section 2.8), coherent monochromatic radiation is obtained.

Figure 24.4 shows a schematic of an FEL, in which the cavity mirrors are placed in the perpendicular direction to that of the oscillating electrons. The high-energy electrons are prevented from hitting the mirrors by magnetic reflectors. The radiation traveling between the mirrors is amplified at the desired frequency, which is set by the speed and the grating period of the magnetic array.

### 24.2.2 Extreme UV and soft x-ray lasers

Figure 24.5 gives an expanded view of figure 2.1 in the vacuum UV (10 eV) to hard x-ray (10 keV) region. It should be noted that the photons of vacuum ultraviolet (VUV) can be absorbed by less than 1 mm of air and are able to ionize several substances. All materials are ionized by extreme UV (EUV) radiation. Soft x-rays interact with the core electrons of materials. The wavelengths of hard x-rays fall

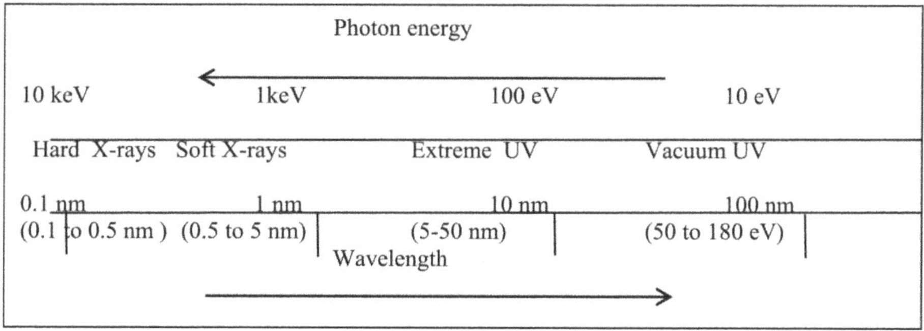


**Figure 24.5.** Expanded version of figure 2.1 in the deep-UV and x-ray regions.

within the size range of the lattice spacings of crystals, which is the reason why x-ray sources are used for the determination of crystal structure.

EUV is an emerging research field in the scientific and technological applications of photonics. Therefore, we will now learn about two methods used to obtain radiation in the EUV and soft x-ray region using conventional lasers. These methods fall under the category of HHG and use the principles of (i) the tunnel ionization and recombination process, and (ii) the cascading Raman scattering process.

### 24.2.2.1 The Keldysh parameter and the HHG process

As described in chapter 16, most of the perturbative-type nonlinear optical phenomena are described by equation 16.6. However, if the electric field strength is greater than the Coulomb field of the atom, the effects of perturbative-type nonlinearity are no longer manifested, due to a new type of atom–field interaction. In such cases, incident field-induced tunneling of electrons takes place through the atomic potential. Following the Lorentz-Drude model of the atom, the atomic field strength is given by $E_{\text{atomic}} \sim \text{m}^{-1}$ (♠ see section 14.2.1). For larger incident fields, a new dimensionless adiabaticity parameter known as the Keldysh parameter ($\gamma$) needs to be introduced. This parameter helps us to identify the two regions of nonlinear optics. It is defined as the ratio between the laser-field frequency ($\omega$) and the tunneling frequency ($\omega_t$). Therefore, $\gamma$ relates the strength of the laser field and the ionization potential of the atom.

For free electrons generated by high field interactions, the values of the electric fields generated by lasers fall in the intensity range of the order of $>10^{16}$ W cm$^{-2}$. Lasers can produce fields several orders higher than such values. As indicated in figure 24.6, the effects at these high energies do not fall under the perturbative approach of quantum mechanics. A detailed description of such effects is outside the scope of this textbook. Briefly, a new approach to nonlinear optics has to be considered for a free electron in an intense laser field, as follows.

The time-averaged kinetic energy associated with the motion of a free electron (of charge $q$ and mass $m$) in an electric field $E$ with frequency $\omega$ is given by

$$K = \frac{q^2 E^2}{m\omega^2}.$$

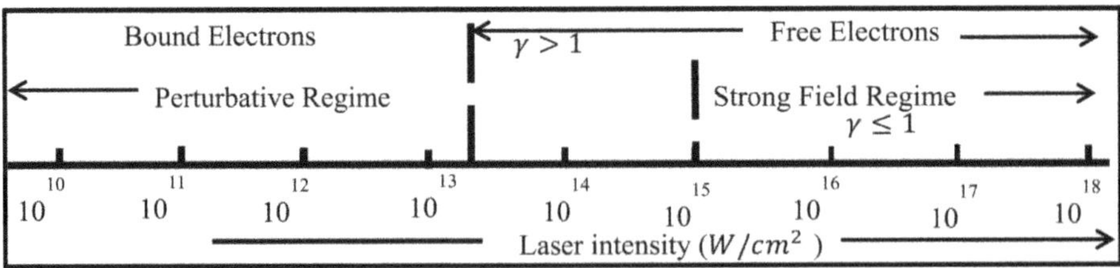


**Figure 24.6.** Regions of nonlinear optics in terms of the Keldysh parameter ($\gamma$) as a function of the incident laser intensity.

This energy, known as ponderomotive energy, is associated with electron oscillation about an equilibrium position. The energy available due to the recombination of photons is the difference between the kinetic energy of the electron and the ionization energy of the atom.

Beyond the perturbative region, a plateau region and a cutoff region for the HHG have been experimentally observed, as described by the following equation:

$$\hbar\omega = I_p + 3.17K. \tag{24.4}$$

Here, $I_p$ is the ionization potential of the atom. The ponderomotive energy or the potential $(K)$ is given by $K \propto I\lambda^2$. Here, $I$ and $\lambda$ are the laser intensity and the wavelength, respectively. The Keldysh parameter in terms of $K$ is also given by $\frac{1}{\gamma} = \sqrt{\frac{K}{2I_p}}$.

**Exercise 24.1.** A Ti:sapphire amplifier laser at a wavelength of 800 nm, a pulse duration of 10 fs, and a power output of 10 W at 1 kHz is used in an experiment. The ponderomotive energy for Ar gas is given as 45 eV. For a laser pulse beam spot of 1 mm, how many harmonics are generated? The given value of the Ar ionization potential is 15.8 eV.

**Solution.** We can find the peak power (W cm$^{-2}$). However, for the given data and using equation (24.4), we can obtain the cutoff energy of the harmonic, which is 158.45 eV. Dividing this value by the corresponding energy of the pump wavelength (800 nm) we obtain the harmonic number, which is 92. (For details, see [16].)

The experimental scheme for HHG using tunnel ionization is shown in figure 24.7. Here, a high-intensity laser beam is focused on a jet stream of a gas, generating high harmonics as indicated in figure 24.8. The intensity of the laser beam can be as large as $(10^{18} \sim 10^{20}$ cm$^{-2})$, which allows a large number (i.e. a few hundred) of odd harmonics to be obtained.

There are three regimes in the observed HHG spectrum: (i) the perturbative region, (ii) the plateau region and (iii) the cutoff region. It should be noted that the intensities of the various orders of higher harmonics are of the same order of magnitude in the plateau region (figure 24.8). There is a cutoff value below which HHG is not obtained. For example, in figure 24.8, the cutoff is approximately near the 28th harmonic.

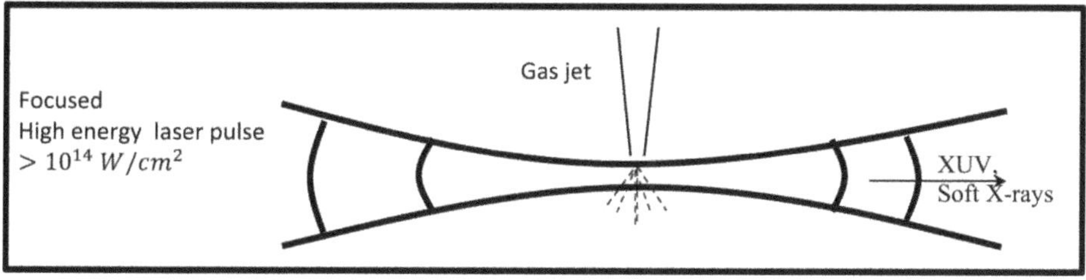


**Figure 24.7.** An experimental scheme for high harmonic/soft x-ray generation.

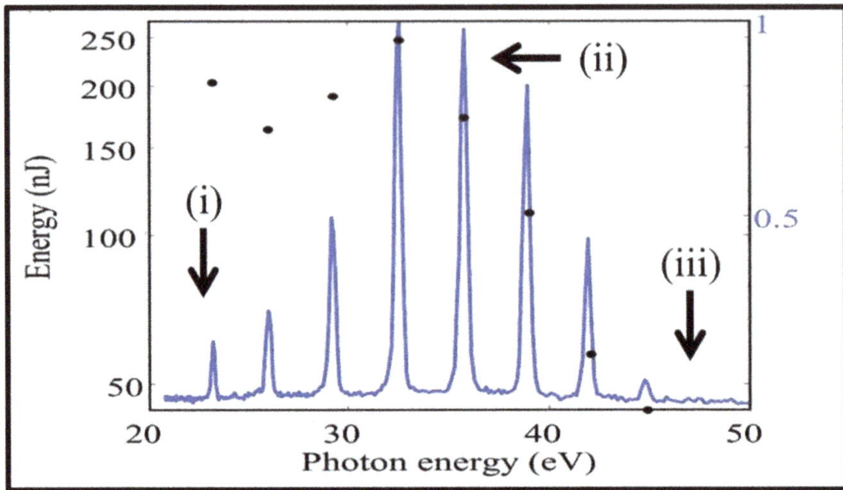


**Figure 24.8.** Typical HHG spectrum obtained from argon gas pumped by a femtosecond laser. The pulse energy per individual harmonic is indicated by dots (reprinted from [10] with the permission of AIP Publishing). The perturbative region (i), the plateau region (ii), and the cutoff region (iii) are also indicated.

### 24.2.2.2 Odd harmonics in HHG

The symmetry properties of centrosymmetric materials, such as liquid and gaseous media (♠ see chapter 16), are especially important to our understanding of the observed behavior. It should be noted that for centrosymmetric systems, only odd harmonics can be obtained, as $\chi^{(2)}$ is zero. However, for non-centrosymmetric systems, both odd and even harmonics are obtained.

### 24.2.2.3 Tunnel ionization model of HHG

Ultrafast femtosecond laser pulses carry the electric field required for the generation of high harmonics. Due to the inverse square law dependence of the coulomb potential, the electron feels an appreciable force in the field of the incident laser pulse. Since the nucleus is heavy compared to electrons, the electrons are accelerated very close to the nucleus. Due to the oscillatory nature of the incident optical field, the electrons approach the nucleus during each period, as shown in figure 24.9(A). This process modifies the potential well, facilitating electron tunneling as shown in figure 24.9(B). The frequency of the radiation emitted during the recombination process is proportional to the incident field acceleration and its optical period. This model, known as the *tunnel ionization model*, is used to explain the generation of high harmonics. It is summarized as follows:

1. An atom irradiated by the strong laser field of a laser pulse can be ionized into a parent ion and a free electron with no initial kinetic energy.
2. The electric field of the laser accelerates the electron.
3. When the electron collides with the parent ion, the ion and electron can recombine. The total extra energy (i.e. the kinetic energy of the electron plus the binding energy of the atom, *Ip*) is released by the emission of a photon of a higher harmonic.

### 24.2.2.4 Sub-femtosecond-duration pulse generation

In the HHG process, a sequence of pulses is produced within a short-duration femtosecond pump pulse. A train of higher harmonic pulses is obtained; these pulses are of even shorter (i.e. sub-femtosecond or attosecond ($10^{-18}$ s)) duration, as shown in figure 24.10. In a collection of atoms, the maximum number of electrons is ejected at the positive or the negative half cycles of the optical pulse. Therefore, the sequence of pulses is generated at double the repetition rate of the fundamental pulse. In the cascaded Raman process, however, the Fourier transform of the

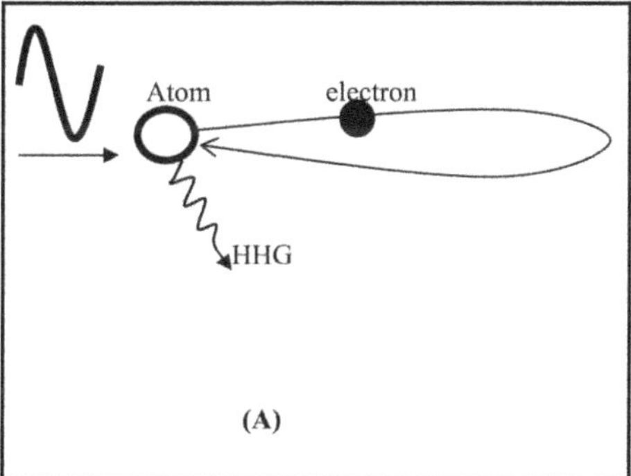


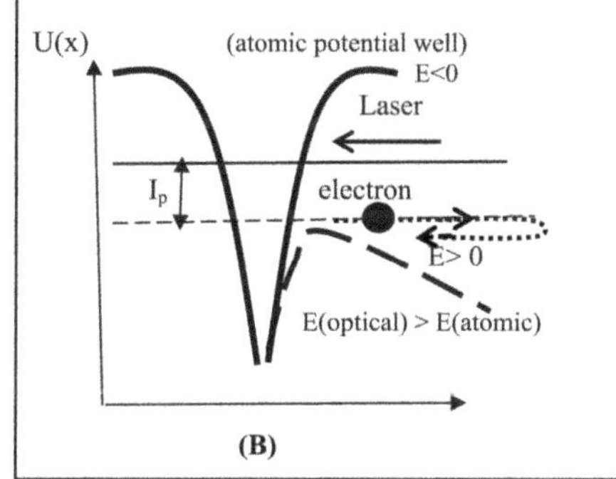


**Figure 24.9.** Mechanism of HHG: in an optical cycle of a laser pulse, an electron from an ionized atom is accelerated and recombined (A). The modification of the potential well that occurs when the intense laser beam strikes the material is shown in (B).

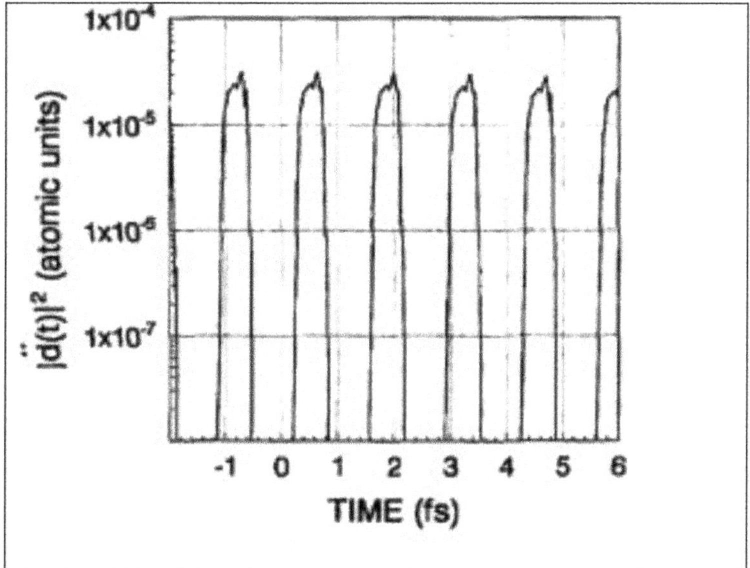


**Figure 24.10.** Attosecond-duration pulses produced by HHG in neon, seen in the temporal domain. (Reproduced with permission from [9], copyright The Optical Society.)

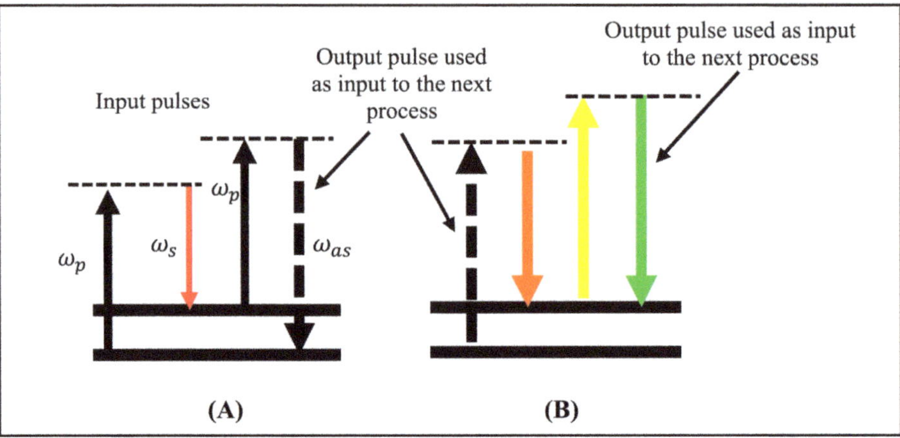


**Figure 24.11.** High harmonics generated by the stimulated Raman scattering process: the anti-Stokes frequency of the first Raman process (A) acts as the pump pulse of the second process (B), and so on. $\omega_p$, $\omega_s$, and $\omega_{as}$ are the pump, Stokes, and anti-Stokes frequencies, respectively.

sidebands gives rise to pulses at the same repetition rate, as described in the following section.

### 24.2.2.5 Cascading Raman effect

The concept of the Raman effect was introduced in chapter 16 (section 16.8). In the sequential or cascaded Raman process, two lasers are used, one of which is tunable in frequency. For example, a Nd:YAG laser (with frequency of $\omega_p$) along with a tunable Ti:sapphire laser can be used in such a way that the tunable frequency difference is equal to the fundamental vibrational frequency ($\Delta\omega_v$) of the molecular system.

In the first state of the process, due to stimulated Raman scattering (SRS), the input pulses give rise to an output anti-Stokes pulse. In the subsequent Raman process, this output pulse is used as the input pulse, as shown in figure 24.11. At high intensity, for two input frequencies that are nearly resonant with a Raman resonance, the process cascades several times, yielding a series of equally spaced modes according to $\Delta\omega = \omega_p \pm n\Delta\omega_v$, as shown in the image in figure 24.12.

## 24.3 Present and future outlook

High peak power laser pulses can be generated by the principle of chirped-pulsed optical parametric amplification (CPOPA). When this is combined with the techniques of HHG in the soft x-ray and EUV regions, pulse lengths on timescale of the hundreds of attoseconds can be obtained. These pulse durations correspond to the order of the revolution time of an electron in the first Bohr orbit of the hydrogen atom.

♣ Part of the Noble prize was awarded to Strickland and Mourou in 2018 for inventing the CPOPA technique. The other part was awarded to Arthur Ashkin for the optical trapping of particles.

♣ In hydrogen, electron completes a revolution in its orbit in ~124 as.

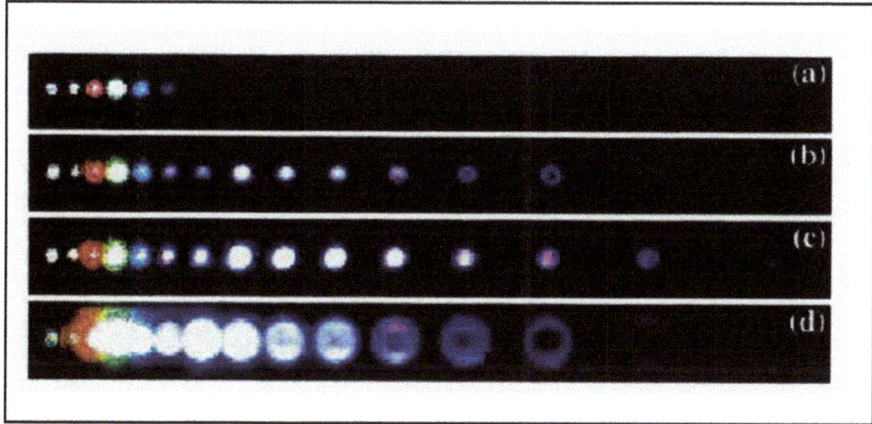


**Figure 24.12.** Raman frequency spots produced by the cascaded Raman process: (a) to (c) at 70 Torr, (d) is at 350 Torr of pressure. The first two spots are the driving frequencies of the Nd:YAG and Ti:sapphire lasers attenuated by several orders. Reprinted (Fig. 2) with permission from [11], copyright (2000) by the American Physical Society.

Gas-filled capillary waveguides are phase matched to obtain high harmonics. It should be mentioned that the durations of attosecond pulses can also be measured using the same principle as those of the pump-probe autocorrelation techniques described in chapter 18. In these regions, the autocorrelation techniques are known as the *reconstruction of attosecond beating by interference of two-photon transitions* (RABBIT) and *frequency-resolved optical gating for complete reconstruction of attosecond bursts* (FROG-CRAB)[1].

The future of laser physics and photonics continues to be interesting and highly promising, as can be seen from the application areas of high-power laser research summarized in figure 24.13 by Mourou and Tajima. The figure shows the inverse linear dependence between the pulse duration of coherent light emission and the obtained laser intensity for more than 18 orders of magnitude .

## Questions and problems

1. Name a few laser gain media consisting of transition metals or rare-earth ions as constituents in a host system.
2. A laser cavity typically has two mirrors. What is the principle of the 'mirrorless laser'?
3. The cyclotron is a particle accelerator. What are the major differences between the synchrotron, the cyclotron, and the free-electron laser?
4. You must have done experiments using a Michelson interferometer in undergraduate laboratory classes. The Michelson interferometer is also used to detect gravitational waves at the Laser Interferometric Gravitational-Wave Observatory (LIGO). In the LIGO, why are the arm lengths of the Michelson interferometer several kilometers long?

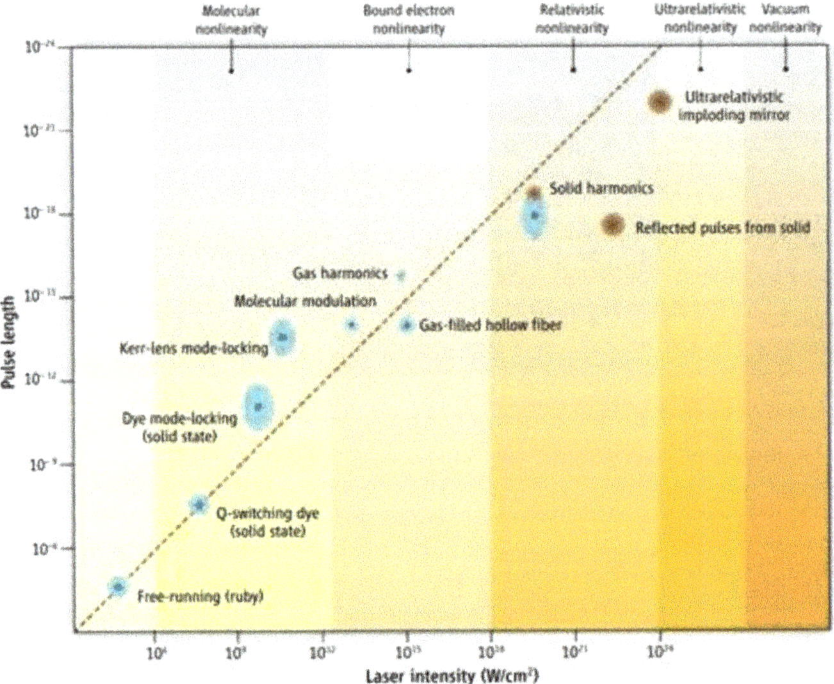



**Figure 24.13.** Phenomena that are accessible with the development of short-duration intense pulses (from [15], reprinted with permission from AAAS).

5. In the FEL, electrons are accelerated to relativistic speeds. Describe the mechanism that allows FELs to obtain a large degree of frequency tunability from the UV to millimeter regions.

6. Obtain the expression for the time-averaged kinetic energy (i.e. the ponderomotive energy) associated with the motion of a free electron (of charge $e$ and mass $m$) in an electric field $E$ whose frequency $\omega$ given by $E = E_0 e^{-i\omega t}$.

7. The cutoff frequency for HHG is given by equation (24.4). For a particular system, the ponderomotive energy is given in eV by $K = 9.33 \times 10^{-14} I_L$ (W cm$^{-2}$)$\lambda_L^2$($\mu$m$^2$). A typical experiment uses a Ti:sapphire laser at a wavelength of 800 nm, a pulse duration of 33 fs, and an output power of 6 W at 1 kHz. If the laser pulse has a focused spot size of 2 mm in He gas inside a capillary, estimate the number of harmonics generated. Take the ionization potential of He to be 24.6 eV. (♠ Also see [16].)

8. What is the three-step model of high harmonic generation? Under what conditions are only the odd harmonics generated?

9. How does the *stimulated* Raman scattering from a material contribute to the generation of new frequencies?

10. What is the mechanism by which high harmonics are generated using the cascaded Raman effect?

# Bibliography


[1] Kogelnik H and Shank C V 1972 Coupled-wave theory of distributed feedback lasers *J. Appl. Phys.* **43** 2327–35

[2] Scifres D R, Burnham R D and Streifer W 1974 Distributed-feedback single heterojunction GaAs diode laser *Appl. Phys. Lett.* **25** 203–6

[3] Sailaja R and Bisht P B 2007 Tunable multiline distributed feedback dye laser based on the phenomenon of excitation energy transfer *Org. Electron. physics, Mater. Appl.* **8** 175–83

[4] Huang S *et al* 2017 Dual-cavity feedback assisted DFB narrow linewidth laser *Sci. Rep.* **7** 1185

[5] Attwood D S 1999 *X-Rays and Extreme Ultraviolet Radiation: Principles and Applications* (Cambridge: Cambridge University Press)

[6] Popov V S 2004 Tunnel and multiphoton ionization of atoms and ions in a strong laser field (Keldysh theory) *Phys. Usp.* **47** 855

[7] Augst S, Strickland D, Meyerhofer D D, Chin S L and Eberly J H 1989 Tunneling ionization of noble gases in a high-intensity laser field *Phys. Rev. Lett.* **63** 2212–5

[8] Corkum P B 1993 Plasma perspective on strong field multiphoton ionization *Phys. Rev. Lett.* **71** 1994–7

[9] Corkum P B, Burnett N H and Ivanov M Y 1994 Subfemtosecond pulses *Opt. Lett.* **19** 1870–2

[10] Rudawski P *et al* 2013 A high-flux high-order harmonic source *Rev. Sci. Instrum.* **84** 73103

[11] Sokolov A V, Walker D R, Yavuz D D, Yin G Y and Harris S E 2000 Raman generation by phased and antiphased molecular states *Phys. Rev. Lett.* **85** 562–5

[12] Strickland D and Mourou G 1985 Compression of amplified chirped optical pulses *Opt. Commun.* **56** 219–21

[13] Durfee C G *et al* 1999 Phase matching of high-order harmonics in hollow waveguides *Phys. Rev. Lett.* **83** 2187–90

[14] Trebino R 2000 *Frequency-Resolved Optical Gating: The Measurement of Ultrashort Laser Pulses* (New York: Academic)

[15] Mourou G and Tajima T M 2011 Intense, shorter pulses *Science (80–)* **331** 41 LP–2

[16] Gibson E A *et al* 2004 High-order harmonic generation up to 250 eV from highly ionized argon *Phys. Rev. Lett.* **92** 33001

**IOP** Publishing

An Introduction to Photonics and Laser Physics
with Applications

**Prem B Bisht**

# Appendix A

## Suggested further reading

### *Laser and photonics*

| | |
|---|---|
| Agrawal G P | *Nonlinear Fiber Optics*, 4th edn (New York: Academic, 2007) |
| Boyd R W | *Nonlinear Optics*, 2nd edn (New York: Academic, 2003) |
| Diels J-C and Rudolph W | *Ultrashort Laser Pulse Phenomena: Fundamentals, Techniques, and Applications on a Femtosecond Time Scale* (New York: Academic, 1996) |
| Duarte F J | *Tunable Laser Applications*, 2nd edn (Boca Raton, FL: CRC Press, 2009) |
| Eichler H J, Günter P, Pohl D W | *Laser-Induced Dynamic Gratings* (Berlin: Springer, 1986) |
| Hasegawa A | *Optical Solitons in Fibers* (Berlin: Springer, 1989) |
| Hetch J | *The Laser Guidebook* (New York: McGraw-Hill, 1986) |
| Hitz B, Ewing J J, Hetch J | *Introduction to Laser Technology* 3rd edn (Piscataway, NJ: IEEE Press, 1991) |
| Kroechner W | *Solid-State Laser Engineering* 5th edn (Berlin: Springer, 1999) |
| Menzel R | *Photonics: Linear and Nonlinear Interactions of Laser Light and Matter* (Berlin: Springer, 2001) |
| Miloni P W and Eberley J H | *Laser Physics* (New York: Wiley, 2010) |
| O'Shea D C | *Introduction to Lasers and their Applications* (Englewood Cliffs, NJ: Prentice-Hall, 1977) |
| Rullière C | *Femtosecond Laser Pulses: Principles and Experiments* (Berlin: Springer, 2003) |
| Saleh B E A and Teich M C | *Fundamentals of Photonics* (New York: Willey, 1991) |
| Sands D | *Diode Lasers* (Boca Raton, FL: CRC Press, 2004) |
| Schäfer F P | *Dye Lasers* (Berlin: Springer, 1990) |
| Shen Y R | *The Principles of Nonlinear Optics* (New York: Wiley-Interscience, 2003) |

(*Continued*)

| | |
|---|---|
| Siegman A E | *Lasers* (Sausalito, CA: University Science Books, 1986) |
| Silfvast W T | *Laser Fundamentals* (Cambridge: Cambridge University Press, 1999) |
| Sirohi R S | *A Course of Experiments with He-Ne Laser* (Hoboken, NJ: Wiley Eastern Ltd, 1985) |
| Sutherland R L | *Handbook of Nonlinear Optics* (New York: Marcel Dekker, 2003) |
| Svelto O | *Principles of Lasers* translated from Italian and edited by Hana D C (Berlin: Springer, 1998) |
| Thyagarajan, K and Ghatak A | *Lasers Fundamentals and Applications* (Berlin: Springer, 1981) |
| Trebino R | *Frequency-Resolved Optical Gating: The Measurement of Ultrashort Laser Pulses* (Berlin: Springer, 2000) |
| Verdeyen, J T | *Laser Electronics* (New Delhi: PHI, 1981) |
| Weber M | *Handbook of Laser Wavelengths* (Boca Raton, FL: CRC Press, 1998) |
| Wilson J and Hawkes J F B | *Lasers Principles and Applications* (Englewood Cliffs, NJ: Prentice-Hall, 1987) |
| Wilson J and Hawkes J F B | *Optoelectronics: An Introduction* (Englewood Cliffs, NJ: Prentice-Hall, 1983) |
| Yariv A | *Quantum Electronics* 3rd edn (New York: Wiley, 1989) |

## *Optics and spectroscopy*

| | |
|---|---|
| Abramczyk H | *Introduction to Laser Spectroscopy* (Amsterdam: Elsevier, 2005) |
| Banwell C N and McCash E M | *Fundamentals of Molecular Spectroscopy* (New Delhi: Tata McGraw-Hill, 1997) |
| Demtröder W | *Laser Spectroscopy* (Berlin: Springer, 2003) |
| Fleming G R | *Chemical Applications of Ultrafast Spectroscopy* (Oxford: Oxford University Press, 1986) |
| Fowles G R | *Introduction to Modern Optics* 2nd edn (New York: Dover Publications, 1975) |
| Ghatak A | *Optics* 4th edn (New Delhi: Tata McGraw Hill Education, 2009) |
| Kroechner W | *Solid-State Laser Engineering* 5th edn (Berlin: Springer, 1999) |
| Parkar CA | *Photoluminescence of Solutions* (Amsterdam: Elsevier, 1968) |
| Streetman B G and Banerjee S | *Solid State Electronic Devices* (New Delhi: PHI, 2006) |
| Struve W S | *Fundamentals of Molecular Spectroscopy* (New York: Wiley, 1989) |
| Thakur S N and Rai D K | *Atom, Laser and Spectroscopy* (New Delhi: PHI, 2010) |
| White H E | *Introduction to Atomic Spectra* (New York: McGraw-Hill, 1934) |

**IOP** Publishing

# An Introduction to Photonics and Laser Physics with Applications

**Prem B Bisht**

# Appendix B

## Luminescence

Luminescence is the inherent property of spontaneous emission by an object (known as a *fluorophore* or a *chromophore*) as a result of excitation by an external agency. This is in contrast with *incandescence*, in which light is emitted from a hot object such as a filament of a bulb. Materials exhibiting the property of luminescence are called *phosphors* or luminescent materials. The word 'luminescence' was introduced by Eilhard Wiedemann in the 1880s. In Latin, 'lumen' is a synonym for light. Depending on the lifetime of the excited fluorophore, luminescence is further divided into fluorescence and phosphorescence (see chapter 2). Luminescence is also categorized according to the external source of excitation. Some of these categories are as follows:

| | |
|---|---|
| Bioluminescence | A result of biochemical reactions in living organisms, such as fireflies |
| Cathodoluminescence | Observed when high-energy electrons hit a luminescent material |
| Chemiluminescence | Certain materials emit light as a result of chemical reactions |
| Crystalloluminescence | A type of luminescence observed during the crystallization of some materials |
| Electroluminescence | Observed when an electric field is applied across a material; this is in contrast to the phenomenon of *incandescence* |
| Lyoluminescence | A kind of *chemiluminescence* observed when a solid is dissolved in a solvent |
| Photoluminescence | Observed when a material is excited by light (EM radiation) |
| Piezoluminescence | Emission occurs when certain piezoelectric crystals are subjected to pressure; also known as *mechanoluminescence* |
| Radioluminescence | Occurs due to the bombardment of a material by ionizing radiation |
| Sonoluminescence | A kind of *mechanoluminescence* generated when gas bubbles in a liquid are excited by sound |
| Thermoluminescence | Crystals pre-irradiated by ionizing radiation (such as gamma rays that produce defect levels) emit light upon heating |
| Triboluminescence | Occurs in some materials when subjected to stress or strain; can take place due to friction between two stones |

doi:10.1088/978-0-7503-5226-0ch26  © IOP Publishing Ltd 2022

**IOP** Publishing

# An Introduction to Photonics and Laser Physics with Applications

**Prem B Bisht**

# Appendix C

## Physical constants

| Quantity | Symbol | Value in MKS units |
|---|---|---|
| Avogadro's number | $N_A$ | $6.02214 \times 10^{23}\,\text{mole}^{-1}$ |
| Boltzmann constant | $K_B$ | $1.3806 \times 10^{-23}\,\text{J K}^{-1}$ |
| Elementary charge | $e$ | $1.60218 \times 10^{-19}\,\text{C}$ |
| Electron mass | $m_e$ | $9.10939 \times 10^{-31}\,\text{kg}$ |
| Proton mass | $m_p$ | $1.672621 \times 10^{-27}\,\text{kg}$ |
| Permeability of free space | $\epsilon_0$ | $8.85 \times 10^{-12}\,\text{C}^2\text{N}^{-1}\text{m}^{-2}$ |
| Permittivity of free space | $\mu_0$ | $4\pi \times 10^{-7}\,\text{NA}^{-2}$ |
| Planck's constant | $h$ | $6.2607 \times 10^{-34}\,\text{J s}$ |
| Rydberg constant | $R_\infty$ | $1.097373 \times 10^5\,\text{cm}^{-1}$ |
| Speed of light in vacuum | $c$ | $2.997925 \times 10^8\,\text{ms}^{-1}$ |
| Stefan–Boltzmann constant | $\sigma$ | $5.68 \times 10^{-8}\,\text{Wm}^{-2}\text{K}^{-4}$ |

Abbreviations: J: joules, K: kelvin, C: coulombs, N: newtons, A: ampere, m: meters, kg: kilograms, s: seconds, W: watts.

www.ingramcontent.com/pod-product-compliance
Ingram Content Group UK Ltd.
Pitfield, Milton Keynes, MK11 3LW, UK
UKHW051939150726
7214IPUK00005B/48

9 780750 352253